Advances in Intelligent Systems and Computing

Volume 1273

The series "Advances in Intelligent Systems and Computing" contains publications on theory, applications, and design methods of Intelligent Systems and Intelligent Computing. Virtually all disciplines such as engineering, natural sciences, computer and information science, ICT, economics, business, e-commerce, environment, healthcare, life science are covered. The list of topics spans all the areas of modern intelligent systems and computing such as: computational intelligence, soft computing including neural networks, fuzzy systems, evolutionary computing and the fusion of these paradigms, social intelligence, ambient intelligence, computational neuroscience, artificial life, virtual worlds and society, cognitive science and systems, Perception and Vision, DNA and immune based systems, self-organizing and adaptive systems, e-Learning and teaching, human-centered and human-centric computing, recommender systems, intelligent control, robotics and mechatronics including human-machine teaming, knowledge-based paradigms, learning paradigms, machine ethics, intelligent data analysis, knowledge management, intelligent agents, intelligent decision making and support, intelligent network security, trust management, interactive entertainment, Web intelligence and multimedia.

The publications within "Advances in Intelligent Systems and Computing" are primarily proceedings of important conferences, symposia and congresses. They cover significant recent developments in the field, both of a foundational and applicable character. An important characteristic feature of the series is the short publication time and world-wide distribution. This permits a rapid and broad dissemination of research results.

**** Indexing: The books of this series are submitted to ISI Proceedings, EI-Compendex, DBLP, SCOPUS, Google Scholar and Springerlink ****

More information about this series at http://www.springer.com/series/11156

Miguel Botto-Tobar · Willian Zamora ·
Johnny Larrea Plúa · José Bazurto Roldan ·
Alex Santamaría Philco
Editors

Systems and Information Sciences

Proceedings of ICCIS 2020

Springer

Editors
Miguel Botto-Tobar
Eindhoven University of Technology
Eindhoven, Noord-Brabant, The Netherlands

Willian Zamora
Universidad Laica Eloy Alfaro de Manabí
Manta - Manabí, Ecuador

Johnny Larrea Plúa
Universidad Laica Eloy Alfaro de Manabí
Manta - Manabí, Ecuador

José Bazurto Roldan
Universidad Laica Eloy Alfaro de Manabí
Manta - Manabí, Ecuador

Alex Santamaría Philco
Universidad Laica Eloy Alfaro de Manabí
Manta - Manabí, Ecuador

ISSN 2194-5357 ISSN 2194-5365 (electronic)
Advances in Intelligent Systems and Computing
ISBN 978-3-030-59193-9 ISBN 978-3-030-59194-6 (eBook)
https://doi.org/10.1007/978-3-030-59194-6

This Springer imprint is published by the registered company Springer Nature Switzerland AG
The registered company address is: Gewerbestrasse 11, 6330 Cham, Switzerland

Preface

The 1st International Conference on Systems and Information Sciences (ICCIS) was held on the main campus of the Universidad Laica Eloy Alfaro de Manabí, in Manta, Ecuador, during July 27th until 29th, 2020, and it was organized jointly by Universidad Laica Eloy Alfaro de Manabí, in collaboration with GDEON. The purpose of ICCIS is to bring together systems and information sciences researchers and developers from academia and industry around the world to discuss cutting-edge research. The content of this volume is related to the following subjects:

- AI, Expert Systems and Big Data Analytics
- Cloud, IoT and Distributed Computing
- Communications
- Database System and Application
- Financial Technologies (FinTech), Economics and Business Engineering
- m-Learning and e-Learning
- Security
- Software Engineering
- Web Information Systems and Applications
- General Track

ICCIS 2020 received 99 submissions written in English by 342 authors coming from 12 different countries. All these papers were peer-reviewed by the ICCIS 2020 Program Committee consisting of 193 high-quality researchers. To assure a high-quality and thoughtful review process, we assigned each paper at least three reviewers. Based on the peer reviews, 37 full papers were accepted, resulting in an 37% acceptance rate, which was within our goal of less than 40%.

Steering Committee

Miguel Botto-Tobar	Eindhoven University of Technology, The Netherlands
Ángela Díaz Cadena	Universitat de Valencia, Spain

Publication Chair

Miguel Botto-Tobar	Eindhoven University of Technology, The Netherlands

Program Committee

A. Bonci	Marche Polytechnic University, Italy
Ahmed Lateef Khalaf	Al-Mamoun University College, Iraq
Aiko Yamashita	Oslo Metropolitan University, Norway
Alejandro Donaire	Queensland University of Technology, Australia
Alejandro Ramos Nolazco	Instituto Tecnólogico y de Estudios Superiores Monterrey, Mexico
Alex Cazañas	The University of Queensland, Australia
Alex Santamaria Philco	Universitat Politècnica de València, Spain
Alfonso Guijarro Rodriguez	University of Guayaquil, Ecuador
Allan Avendaño Sudario	Escuela Superior Politécnica del Litoral (ESPOL), Ecuador
Alexandra González Eras	Universidad Politécnica de Madrid, Spain
Ana Núñez Ávila	Universitat Politècnica de València, Spain
Ana Zambrano	Escuela Politécnica Nacional (EPN), Ecuador
Andres Carrera Rivera	The University of Melbourne, Australia
Andres Cueva Costales	The University of Melbourne, Australia
Andrés Robles Durazno	Edinburg Napier University, UK
Andrés Vargas Gonzalez	Syracuse University, USA
Angel Cuenca Ortega	Universitat Politècnica de València, Spain
Ángela Díaz Cadena	Universitat de València, Spain
Angelo Trotta	University of Bologna, Italy
Antonio Gómez Exposito	University of Sevilla, Spain
Aras Can Onal	TOBB University Economics and Technology, Turkey
Arian Bahrami	University of Tehran, Iran
Benoît Macq	Université Catholique de Louvain, Belgium
Bernhard Hitpass	Universidad Federico Santa María, Chile
Bin Lin	Università della Svizzera italiana (USI), Switzerland
Carlos Saavedra	Escuela Superior Politécnica del Litoral (ESPOL), Ecuador
Catriona Kennedy	University of Manchester, UK

César Ayabaca Sarria	Escuela Politécnica Nacional (EPN), Ecuador
Cesar Azurdia Meza	University of Chile, Chile
Christian León Paliz	Université de Neuchâtel, Switzerland
Chrysovalantou Ziogou	Chemical Process and Energy Resources Institute, Greece
Cristian Zambrano Vega	Universidad de Málaga, Spain/Universidad Técnica Estatal de Quevedo, Ecuador
Cristiano Premebida	Loughborough University, ISR-UC, UK
Daniel Magües Martinez	Universidad Autónoma de Madrid, Spain
Danilo Jaramillo Hurtado	Universidad Politécnica de Madrid, Spain
Darío Piccirilli	Universidad Nacional de La Plata, Argentina
Darsana Josyula	Bowie State University, USA
David Benavides Cuevas	Universidad de Sevilla, Spain
David Blanes	Universitat Politècnica de València, Spain
David Ojeda	Universidad Técnica del Norte, Ecuador
David Rivera Espín	The University of Melbourne, Australia
Denis Efimov	Inria, France
Diego Barragán Guerrero	Universidad Técnica Particular de Loja (UTPL), Ecuador
Diego Peluffo-Ordoñez	Yachay Tech, Ecuador
Dimitris Chrysostomou	Aalborg University, Denmark
Domingo Biel	Universitat Politècnica de Catalunya, Spain
Doris Macías Mendoza	Universitat Politècnica de València, Spain
Edison Espinoza	Universidad de las Fuerzas Armadas (ESPE), Ecuador
Edwin Quel	Universidad de las Américas, Ecuador
Edwin Rivas	Universidad Distrital de Colombia, Colombia
Ehsan Arabi	University of Michigan, USA
Emanuele Frontoni	Università Politecnica delle Marche, Italy
Emil Pricop	Petroleum-Gas University of Ploiesti, Romania
Erick Cuenca	Université Catholique de Louvain, Belgium
Fabian Calero	University of Waterloo, Canada
Fan Yang	Tsinghua University, China
Fariza Nasaruddin	University of Malaya, Malaysia
Felipe Ebert	Universidade Federal de Pernambuco (UFPE), Brazil
Felipe Grijalva	Escuela Politécnica Nacional (EPN), Ecuador
Fernanda Molina Miranda	Universidad Politécnica de Madrid, Spain
Fernando Almeida	University of Campinas, Brazil
Fernando Flores Pulgar	Université de Lyon, France
Firas Raheem	University of Technology, Iraq
Francisco Calvente	Universitat Rovira i Virgili, Spain
Francisco Obando	Universidad del Cauca, Colombia
Franklin Parrales	University of Guayaquil, Ecuador
Freddy Flores Bahamonde	Universidad Técnica Federico Santa María, Chile

Gabriel Barros Gavilanes	INP Toulouse, France
Gabriel López Fonseca	Sheffield Hallam University, UK
Gema Rodriguez-Perez	LibreSoft/Universidad Rey Juan Carlos, Spain
Ginger Saltos Bernal	Escuela Superior Politécnica del Litoral (ESPOL), Ecuador
Giovanni Pau	Kore University of Enna, Italy
Guilherme Avelino	Universidade Federal do Piauí (UFP), Brazil
Guilherme Pereira	Universidade Federal de Minas Gerais (UFMG), Brazil
Guillermo Pizarro Vásquez	Universidad Politécnica de Madrid, Spain
Gustavo Andrade Miranda	Universidad Politécnica de Madrid, Spain
Hernán Montes León	Universidad Rey Juan Carlos, Spain
Ibraheem Kasim	University of Baghdad, Iraq
Ilya Afanasyev	Innopolis University, Russia
Israel Pineda Arias	Chonbuk National University, South Korea
Ivan Izonin	Lviv Polytechnic National University, Ukraine
Jaime Meza	Universiteit van Fribourg, Switzerland
Janneth Chicaiza Espinosa	Universidad Técnica Particular de Loja (UTPL), Ecuador
Javier Gonzalez-Huerta	Blekinge Institute of Technology, Sweden
Javier Monroy	University of Malaga, Spain
Javier Sebastian	University of Oviedo, Spain
Jawad K. Ali	University of Technology, Iraq
Jefferson Ribadeneira Ramírez	Escuela Superior Politécnica de Chimborazo, Ecuador
Jerwin Prabu	BRS, India
Jong Hyuk Park	Korea Institute of Science and Technology, Korea
Jorge Charco Aguirre	Universitat Politècnica de València, Spain
Jorge Eterovic	Universidad Nacional de La Matanza, Argentina
Jorge Gómez Gómez	Universidad de Córdoba, Colombia
Juan Corrales	Institut Universitaire de France et SIGMA Clermont, France
Juan Romero Arguello	University of Manchester, UK
Julián Andrés Galindo	University Grenoble Alpes, France
Julian Galindo	Inria, France
Julio Albuja Sánchez	James Cook University, Australia
Kelly Garces	Universidad de Los Andes, Colombia
Kester Quist-Aphetsi	Center for Research, Information, Technology and Advanced Computing, Ghana
Korkut Bekiroglu	SUNY Polytechnic Institute, USA
Kunde Yang	Northwestern Polytechnic University, China
Leonardo Chancay García	Universidad Técnica de Manabí, Ecuador
Lina Ochoa	CWI, The Netherlands

Lohana Lema Moreira	Universidad de Especialidades Espíritu Santo (UEES), Ecuador
Lorena Guachi Guachi	Yachay Tech, Ecuador
Lorena Montoya Freire	Aalto University, Finland
Lorenzo Cevallos Torres	Universidad de Guayaquil, Ecuador
Luis Galárraga	Inria, France
Luis Martinez	Universitat Rovira i Virgili, Spain
Luis Urquiza-Aguiar	Escuela Politécnica Nacional (EPN), Ecuador
Maikel Leyva Vazquez	Universidad de Guayaquil, Ecuador
Manuel Sucunuta	Universidad Técnica Particular de Loja (UTPL), Ecuador
Marcela Ruiz	Utrecht University, The Netherlands
Marcelo Zambrano Vizuete	Universidad Técnica del Norte, Ecuador
María José Escalante Guevara	University of Michigan, USA
María Reátegui Rojas	University of Quebec, Canada
Mariela Tapia-Leon	University of Guayaquil, Ecuador
Marija Seder	University of Zagreb, Croatia
Mario Gonzalez Rodríguez	Universidad de las Américas, Ecuador
Marisa Daniela Panizzi	Universidad Tecnológica Nacional – Regional Buenos Aires, Argentina
Marius Giergiel	KRiM AGH, Poland
Markus Schuckert	Hong Kong Polytechnic University, Hong Kong
Matus Pleva	Technical University of Kosice, Slovakia
Mauricio Verano Merino	Technische Universiteit Eindhoven, The Netherlands
Mayken Espinoza-Andaluz	Escuela Superior Politécnica del Litoral (ESPOL), Ecuador
Miguel Botto Tobar	Eindhoven University of Technology, The Netherlands
Miguel Fornell	Escuela Superior Politécnica del Litoral (ESPOL), Ecuador
Miguel Gonzalez Cagigal	Universidad de Sevilla, Spain
Miguel Murillo	Universidad Autónoma de Baja California, Mexico
Miguel Zuñiga Prieto	Universidad de Cuenca, Ecuador
Milton Román-Cañizares	Universidad de las Américas, Ecuador
Mohamed Kamel	Military Technical College, Egypt
Mohammad Al-Mashhadani	Al-Maarif University College, Iraq
Mohammad Amin	Illinois Institute of Technology, USA
Monica Baquerizo Anastacio	Universidad de Guayaquil, Ecuador
Muneeb Ul Hassan	Swinburne University of Technology, Australia
Nam Yang	Technische Universiteit Eindhoven, The Netherlands
Nathalie Mitton	Inria, France
Nathaly Orozco	Universidad de las Américas, Ecuador

Samiha Fadloun	Université de Montpellier, France
Sergio Montes León	Universidad de las Fuerzas Armadas (ESPE), Ecuador
Stefanos Gritzalis	University of the Aegean, Greece
Syed Manzoor Qasim	King Abdulaziz City for Science and Technology, Saudi Arabia
Tatiana Mayorga	Universidad de las Fuerzas Armadas, ESPE), Ecuador
Tenreiro Machado	Polytechnic of Porto, Portugal
Thomas Sjögren	Swedish Defence Research Agency (FOI), Sweden
Tiago Curi	Federal University of Santa Catarina, Brazil
Tony T. Luo	A*STAR, Singapore
Trung Duong	Queen's University Belfast, UK
Vanessa Jurado Vite	Universidad Politécnica Salesiana, Ecuador
Waldo Orellana	Universitat de València, Spain
Washington Velasquez Vargas	Universidad Politécnica de Madrid, Spain
Wayne Staats	Sandia National Labs, USA
Willian Zamora	Universidad Laíca Eloy Alfaro de Manabí, Ecuador
Yessenia Cabrera Maldonado	University of Cuenca, Ecuador
Yerferson Torres Berru	Universidad de Salamanca, Spain/Instituto Tecnológico Loja, Ecuador
Zhanyu Ma	Beijing University of Posts and Telecommunications, China

Organizing Institutions

Uleam
UNIVERSIDAD LAICA
ELOY ALFARO DE MANABÍ

Facultad de Ciencias
Informáticas

Impulsando la investigación

Collaborators

Contents

AI, Expert Systems and Big Data Analytics

Bringing Machine Learning Predictive Models Based on Machine Learning Closer to Non-technical Users

Pablo Pico-Valencia[1,2(✉)], Oscar Vinueza-Celi[1], and Juan A. Holgado-Terriza[2]

[1] Pontificia Universidad Católica del Ecuador, Esmeraldas, Ecuador
pablo.pico@pucese.edu.ec
[2] Universidad de Granada, Granada, Spain

Abstract. Today, data science has positioned as an area of interest for decision makers in many organizations. Advances in Machine Learning (ML) allow training predictive models based on the analysis of datasets in multiple domains such as: business, medicine, marketing, among others. These models are able to learn and predict future behaviors which helps in the decision-making process. However, many of the ML tools such as Python, Matlab, R Suite, and even their libraries, require that every action must be performed as a sequence of commands by means of scripts. These software packages require extensive technical knowledge of statistics, artificial intelligence, algorithms and computer programming that generally only computer engineers are skilled at. In this research we propose the development of a process complemented with the assistance of a set of user graphic interfaces (GUIs) to help non-sophisticated users to train and test ML models without writing scripts. A tool compatible with Python and Matlab was developed with a set of GUIs adapted to professionals of the business area that generally require to apply ML models in their jobs, but they do not have time to learn programming.

Keywords: Machine learning · Supervised learning · Unsupervised learning · GUI · Python · Matlab

1 Introduction

Today, data science is an area of interest for decision-makers in both private and public institutions. Multiple are the applications that data scientists have given to Artificial Intelligence (AI), through the technique of Machine Learning (ML), to solve problems in disciplines [14] such as: medicine, marketing, security and business. In these disciplines, solutions have been oriented to solve problems such as: product and service recommendations [16], sentiment analysis [2], disease diagnosis [5], fraud detection [9], financial crisis prediction [12], among others.

In general terms, the ML technique is mainly focused on creating models based on supervised and unsupervised algorithms that learn from large historical

M. Botto-Tobar et al. (Eds.): ICCIS 2020, AISC 1273, pp. 3–15, 2021.
https://doi.org/10.1007/978-3-030-59194-6_1

datasets. These datasets are the input to apply the existing learning algorithms. These algorithms allow the creation of predictive models based on three type of ML such as regression, classification and clustering. For the application of these types of ML, statistical and mathematical models have been proposed and implemented by means of supervised and unsupervised learning algorithms provided by APIs compatible with most popular programming languages.

The implementation of the supervised (e.g., linear regression; decision trees; support vector machine, SVM; K-Nearest Neighbor, KNN; Naïve Bayes; Support Vector Machine, SVM) and unsupervised (e.g., k-means) learning algorithms [4,10,11] have been implemented through specialized libraries integrated into programming languages such as Python, Matlab, R Studio and Java. Scikit-learn for Python, the ML and data science module for Matlab, MLR for R Studio and Weka for Java. However, almost all these algorithms must be applied using scripts.

One of the software tools that provides a Graphical User Interface (GUI) instead of command line interface to apply the ML algorithms is Weka. Nonetheless, currently the most popular and employed tools in the field of ML are Python for AI scientists, Matlab for engineers and R Studio for statisticians. But these tools are oriented to sophisticated users with solid knowledge in Statistics, Theory of Computation, AI and Computer Programming keeping professionals from other areas (e.g., Economy, Business, Medicine, Education, Psychology) out of their scope.

The aim of this study is focused on designing and implementing an intuitive learning process with easy-to-use GUIs that can be employed by non-sophisticated users (e.g., economists, doctors, businessmen, teachers) in order to load different datasets from which they generate their own ML models and thus make predictions based on historical data. This is beneficial because currently any professional requires the use of this type of technology to predict behavioral patterns based on data that can be used in decision-making at organizational, educational or scientific level.

This paper is organized into six sections. In the Sect. 2, the works related to the implementation of tools with GUIs oriented to the creation of ML models are presented. In Sect. 3, tools used to create the GUIs in Python and Matlab are described as well as the main algorithms of ML. Section 4 describes a systematic process to create ML models that non-sophisticated users must follow. This process is implemented by means of a GUI-based tool developed for Python and Matlab. In Sect. 5, the results are presented in relation to the training and testing of ML models using the developed tool. These results as well the experience experimented are also briefly discussed in Sect. 6. Finally, in Sect. 7 the conclusions and future works are outlined.

2 Related Works

Data scientists have wide expertise in Mathematics, Statistics and Computer Science. Consequently, they can develop algorithms and create ML models using the most popular programming languages for data science such as: Python, Matlab, R Studio and Java. These languages are compatible with specialized ML libraries. However, their use requires solid levels of programming and scripts writing. This is the case of the library Pandas and Scikit-learn for Python, the ML and data science module for Matlab and MLR for R Studio. All these libraries are oriented to the programming of scripts instead of providing graphical-assisted tools that facilitate the tasks involved in ML [3]. Just, Weka [8] has an user-friendly GUI-tool to perform data analysis using ML algorithms.

Many personalized systems are proposed to apply ML for non-sophisticated users such as [18–20]. However, such systems are limited and only exploit a trained ML model in production systems for making predictions in specific problems. Then, they are not designed to create new models.

On the other hand, there are proposals oriented to provide tools for the creation of ML models using user-friendly GUIs. There are several proposals for the most popular ML languages such as Matlab, R Suite and Python. In the case of Matlab, Murphy et al. [13] propose the development of a specialized toolbox with an intuitive interface for the use of ML algorithms. However, not all the ML tools available for Matlab have an interface that does not require the writing of scripts. This is the case of the work proposed by Pelckmans et al. [15] in which the use of the Support Vector Machine (SVM) algorithm is enabled in Matlab but with an interface that requires some programming tasks.

In the case of R Studio, several graphic tools have also been developed to make easier for users the creation of ML models. The work of Graham [7] proposes Rattle (the R Analytical Tool To Learn Easily) as a GUI tool to exploit the ML algorithms through the Data Mining technique. This tool is oriented for statisticians, professors and students. That is why, it has been widely employed at universities to teach how to do sophisticated data analyses without needing users to write R scripts by applying the ML algorithms. Therefore, Rattle has provided a steppingstone toward using R as a programming language for ML. This is the same case of the Weka tool but coded in Java programming language. This tool allows users to apply ML algorithms just by clicking over friendly GUIs.

ML techniques are widely used nowadays. Then, there are implementations of ML techniques in other popular programming language such as C or C++. This is the case of the Dlib-ML library for C++ which enables data scientists to work in an ML environment like Python, R Studio o Matlab. However, it also lacks a GUI to facilitate its use with just a few simple clicks. In the same way, the Scikit-learn library provides a base for the development of ML in Python, but it does not have any graphical tool that facilitates its use to non-sophisticated users. Consequently, for Python we did not find a tool with a GUI such as in Matlab and R Studio.

Finally, it is important to point out that the use of graphic tools in ML makes the tasks easier for non-sophisticated users. However, GUIs also provides a flexible and intuitive way for modelling workflows involving learning algorithms. In this sense, some software tools have been developed. Some of these tools are: Darwin [6] for Matlab, KnowledgeFlow in Weka [8], Rattle [7] in R Studio and

TensorFlow [1] in Python. These technological tools help data scientists to model complex analysis processes which are widely required in today's information society.

3 Methodology

3.1 Learning Algorithms Used

In ML it is possible to differentiate between two types of algorithms, supervised and unsupervised [14]. Both types of algorithms base the learning process on two phases. The first phase corresponds to the training of the model based on a dataset and a learning algorithm. Following the training phase, it is possible to test the model by the data scientist. To perform this phase, usually different data than those used to perform the training are used. The results obtained depend on the level of accuracy provided by the algorithm in the training phase. It is also important to note that the model generated will be employed by end-user in production systems [5,14].

Supervised learning algorithms are widely used to perform regression and classification tasks. To carry out the training phase, these algorithms use large historical datasets in which the outputs are known. A set of rules is defined on these data as a learning result, on which the model is based, to later carry out predictions using non-training data. Next, the most extended algorithms among the developers of ML applications are briefly described [10].

- **Linear regression.** Mathematical method that allows to approximate the dependency relationship between a dependent variable (also referred to as "output", "target" "class" or "label") and a set of independent variables (also referred to as "inputs", "features" or "attributes") by calculating a line equation in the hyperplane.
- **Decision tree.** The decision tree identifies the most significant variable and the value that provides the best homogeneous sets of population. This algorithm makes a set of tree decisions, so that the intermediate nodes represent solutions and the final nodes determine the prediction of the model.
- **k Nearest Neighbor (kNN).** The kNN algorithm assumes that similar things exist in the vicinity. This algorithm allows classifying values by looking for the most similar data points (by proximity) and specifying a value of k that corresponds to the number of neighbours.
- **Suport vector machine (SVM).** Discriminatory classifier formally defined by an optimal hyperplane of separation in dimensional spaces. This hyperplane is a line that divides a plane into two parts where in each class it is on one side.
- **Naïvy Bayes.** Method based on Bayes' theorem used to discriminate different objects according to certain characteristics. It focuses on finding the probability of occurrence of A, given that B has occurred. In this case, B is the evidence and A is the hypothesis.

On the other hand, unsupervised learning algorithms are generally used in clustering operations. They are also trained from large historical datasets; but in this case the output labels are unknown. Their goal is focused on extracting meaningful information by exploring the structure of such unlabeled data. Next, one of the most widely used unsupervised algorithms among data scientists is briefly described.

- **k-Means.** Grouping method, which aims at partitioning a set of n observations into k groups in which each observation belongs to the group whose mean value is closest.

3.2 Technologies Used

Most popular software tools for creating predictive ML models are based on Python. For this specific programming language, two very popular, powerful, and accessible libraries widely employed by data scientists, called Pandas and Scikit-learn, have been developed [17]. Pandas is an open source library providing high-performance, easy-to-use data structures, manipulation and analysis of quantitative data from Python. In contrast, Scikit-learn provides an open source ML library that includes classical learning algorithms and a model evaluation into the same programming language [3].

On the other hand, unlike Python, a Matlab tool has also been selected. Matlab is a desktop environment oriented to carry out iterative analysis and design processes with a programming language that expresses matrix and array mathematics. However, in terms of data science, Matlab also has certain limitations to allow non-sophisticated users to create their own ML models without programming. The Matlab complement called ML and data science module allows the following tasks: modelling and simulating processes integrating a wide variety of ML and deep learning algorithms, automated feature selection and hyper parameter adjustment, and data analysis with Matlab graphics. In addition, data science module automatically transforms ML models into C/C++ code, making it easy to deploy such models in production systems.

4 Proposal

In this work, a new process assisted by graphical user interfaces is proposed to facilitate the creation of predictive models of ML in Python and Matlab so that no scripts need to be written. In order to validate this proposal a new graphical tool was designed and developed. The creation of interfaces was done by using the *tkinter* library in Python and the visual editor tools (GUIDE) in Matlab. In the case of Python, *tkinter* has been combined with the *pandas* and $Scikit-learn$ libraries for encapsulating the functions and procedures that execute the data analysis within graphical elements. On the contrary, in the case of Matlab, no additional libraries to the standards were used.

Before designing the graphical tool, each ML algorithms was analyzed theoretically (using online documentation) and practically (using scripts). Five supervised learning algorithms (linear regression, decision tree, kNN, SVM and Naïve Bayes) and one unsupervised learning algorithm (k-Means) were analyzed in depth.

Figure 1 illustrates the process that end-users must follow to create an ML model using any of the learning algorithms considered in this study. These steps must be executed sequentially by means of the assistance of the new designed GUIs. In general terms, the process to be followed was implemented in three GUIs, the first one to load dataset form which the created ML model will learn, the second one to train the model, and the third one to test the trained model. By means of the three proposed GUIs it is possible to create an ML model in Python or Matlab without writing a single line of code.

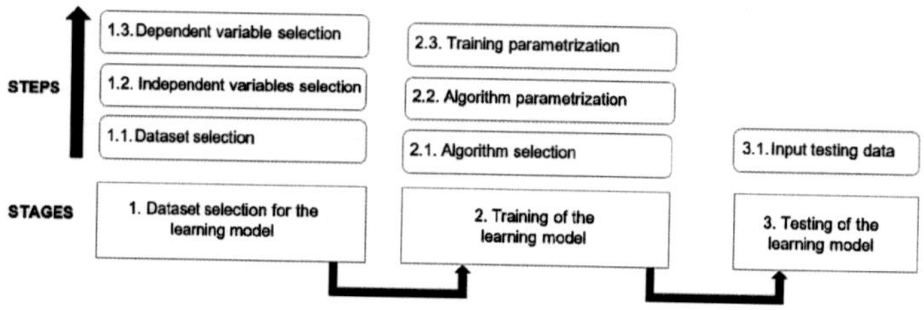

Fig. 1. Systematic process for non-technical users to create a ML model

In order to illustrate the process and the guidance provided by the GUIs, a first example is described along with the followed process. This example consists of creating a linear discriminant to distinguish between species of iris flowers such as setosa, virginica and versicolor. The dataset used, called *iris_dataset*, is a dataset widely used in the demonstration of the use of existing ML algorithms. This dataset contains 50 samples of each of three species of iris flowers (setosa, virginica and versicolor) which are described by four attributes measured in centimetres such as: sepal length and width, and petal length and width.

Before applying the process, we assume that the non-expert user has downloaded the described dataset (*iris_dataset*). From this dataset, the non-technical user should analyze all features (variables) in detail in order to determine which variables and algorithm will use. In this example, the decision tree learning algorithm was selected to create the ML model. Then, the user followed the proposed three stages illustrated in Fig. 1: a) dataset selection for the learning model, b) training the learning model, and c) testing the trained learning model. It is impotant to note that this process can be applied to use the decision tree as well any other studied algorithm.

4.1 STAGE 1: Dataset Selection for the Learning Model

Stage description: This phase is performed following the process by the user using the GUI illustrated in Fig. 2a. The user must start selecting and loading the dataset that will be used to train the predictive model. At this point the user must also select the dependent variable and the set of independent variables. The selection of these variables is determined according to the data that is selected for creating the learn model. These variables are displayed in the GUI according to the dataset loaded (Fig. 2a).

Stage application: Following the guidelines of the example, the computer's file system was explored through the GUI in Fig. 2a and the file *iris_dataset.csv* previously downloaded and stored was selected. Then, the features of the data were analyzed. The features *sepal_length*, *sepal_width*, *petal_length*, *petal_width* and *flower* were identified. Of these variables, the feature *flower* was selected as a dependent variable because the model proposed tries to determine the type of flower based on the remaining features. Then, the four remaining features (*sepal_length*, *sepal_width*, *petal_length*, *petal_width*) were selected one by one as independent variables.

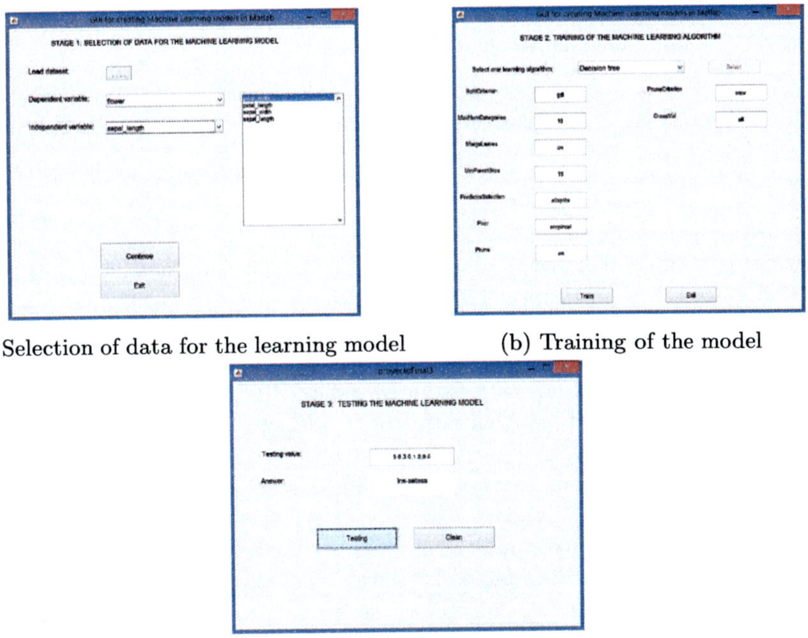

(a) Selection of data for the learning model

(b) Training of the model

(c) Testing of the model

Fig. 2. Proposed Matlab's GUI for the creation of a ML model base on the decision tree algorithm

4.2 STAGE 2: Training of the Learning Model

Stage description: At this stage the user must select the learning algorithm to be used as the baseline for the creation of the predictive model. Accordingly, it is necessary to input the values of the arguments that the selected algorithm supports if it is required. If not, these arguments assume the values that the algorithm gives them by default as is shown in Fig. 2b.

Stage Application: It was decided to apply the decision tree algorithm in the proposed example. Particularly for the *decisiontree* algorithm, as illustrated in Fig. 2b, the user can modify the default values loaded in the GUI. Also, for this algorithm it was possible to update the value of its arguments. The supported arguments for the decision tree algorithm were the following: Split-Sreation, MaxNumCategories, MergeLeaves, MinParentSize, PredictorSelection, Prior, Prune, PruneCtriterion and CrossVal. Descriptions of these arguments as well as the arguments of the other studied algorithms are described in detail in the official documentation of Matlab (available at https://es.mathworks.com/help/stats/index.html) and Scikit-Learn library (available at https://scikit-learn.org/stable/_downloads/scikit-learn-docs.pdf). To finish this stage the user must click on the option *createmodel*. Depending on the accuracy of the model the user could decide to retrain the model with other values of its properties or to accept it and continue to the next stage.

4.3 STAGE 3: Testing of the Learning Model

Stage description: The last stage for the creation of the ML model that the user must carry out before using the model created in production environments is the testing stage. At this level the user can validate the ML model by inserting testing data to the trained model as shows Fig. 2c. It can be used also to make predictions with the ML model created after executing the previous stages.

To evaluate the accuracy and quality of the model, the results provided by the predictive model use other data than those used for training. If the model complies with expectations, it can be considered validated and the ML model can be used in production environments to support decision in real scenarios. Otherwise the model must be retrained with different values in the arguments of the algorithm used until acceptable results are achieved.

Stage Application: The graphical interfaces illustrated in Fig. 2 correspond to Matlab compatible interfaces. However, the study also proposes Python equivalent GUIs. Figures 3a–3c show the execution of the three phases already described; but to create an ML model using the linear regression algorithm in Python. In this case a dataset called *salary.csv* dataset was employed. Linear regression in Python included arguments such as: copy_x, fit_intercept, n_jobs and normalize. In addition, we defined the feature *salary* as dependent variable and *years_of_experience* as independent variable. Also, he training data was split into 80% for training and 20% for testing.

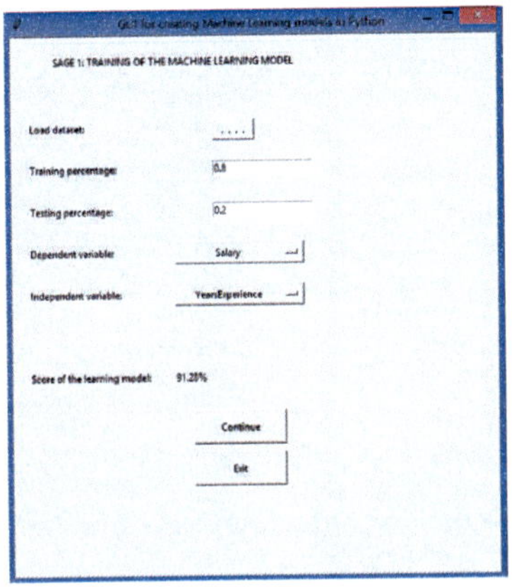

(a) Selection of dataset for the model

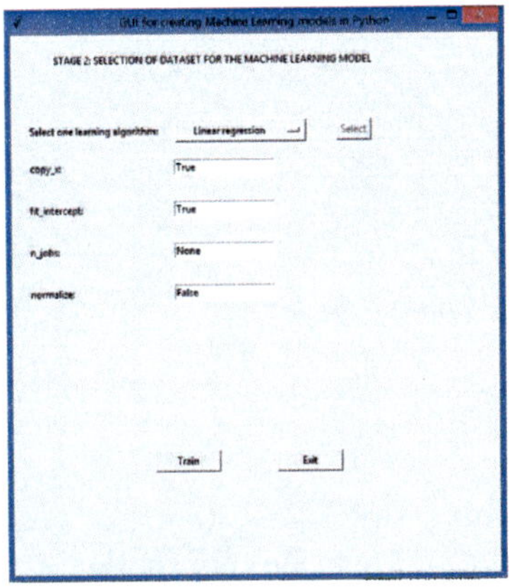

(b) Training of the model

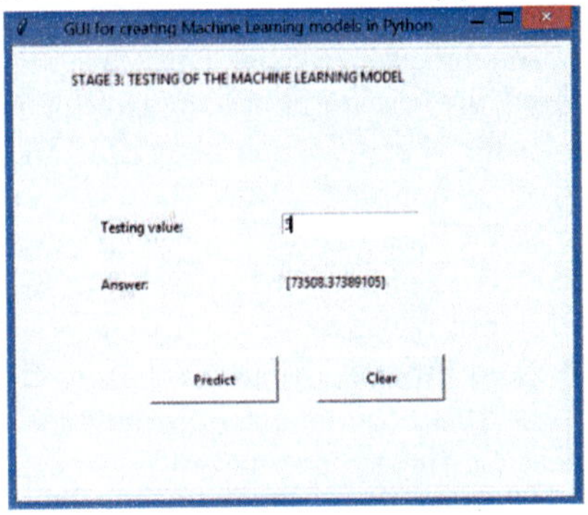

(c) Testing of the model

Fig. 3. Proposed GUI for the creation of a machine learning model base on the linear regression

It is important to highlight that in terms of usability there were no marked differences between the GUIs for Python and Matlab programming languages. The complexity of each tool used has been abstracted in our proposal tool. In short, both environments allow non-technical users to apply the six ML algorithms considered in this study.

5 Results

The results obtained demonstrated it is possible to create models of ML using the learning algorithms under study without programming scripts in Python and Matlab. In addition, the proposed tools-based GUI do not restrict the opportunity to handle certain datasets and default values for arguments supported by the algorithms. Tools proposed in this paper were used to create predictive models using five supervised learning algorithms (items 1–5) and one unsupervised

Table 1. Machine learning algorithms applied using GUIs for Python and Matlab.

#	Estudied algorithms	# Arguments Python	# Arguments Matlab	Datasets used	Applied arguments	Training accuracy
1	Linear regression	4	0	salary.csv	By default	95%
2	Decision tree	13	9	iris_dataset.csv	max_depth = 3	100%
3	kNN	8	3	iris_dataset.csv	n_neighbors = 3	95%
4	SVM	14	8	iris_dataset.csv	kernel = 'linear'	99%
5	Gaussian Naïve Bayes	2	2	iris_dataset.csv	By default	95%
6	K-Means	11	8	iris_dataset.csv	n_clusters = 3; max_iter=1000	90%

learning algorithm (item 6). These algorithms are shown in the first column (*Studied algortims*) in Table 1.

In the same table, columns *#Arguments Python* and *#Arguments Matlab* show the total of the arguments supported by each studied learning algorithm which can be personalized by users GUIs. For instance, important arguments that are often customized are: n_neighbors for the kNN algorithm in the case of Python and k_clusters for the K-Means algorithm in the case of Matlab. In addition, the proposed guided process with the assistance of GUIs also allow both loading default values of arguments and modifying them if required. This makes it possible to create customized ML models according to data loaded. In this sense, the proposal is not oriented to support a specific dataset, but it also allows loading datasets according to the user's needs. Datasets loaded for each learning algorithm are shown in the column *Dataset*. We have used public and popular datasets for testing the proposal tool such as *iris_dataset*. However, we have also stored these datasets in a GitHub repository (https://github.com/picopabant1984/GUI4ML).

Finally, last column of Table 1 shows the metric of the supervised and unsupervised tested algorithms. This metrics is the *accuracy* of training for regression (item 1), classifiers (items 2–5) and clustering (item 6) algorithms. These results were computed according the specific arguments of each algorithm as is shown on column *Applied arguments* of Table 1. We have not made efforts to compare the algorithms in terms of metrics as stated in [4,9]. For the proposed study, we have only focused on determining the flexibility that the new assistive process can provide to non-technical users.

For testing the proposed tool, seven users tested our proposal. They were students of engineering who knew the theoretical basis of each studied algorithm. Therefore, they were able to create models and test them without reporting inconveniences. The applied tests are preliminary experiments that must be systematized with a greater number of volunteers.

6 Discussion

Results demonstrated the technical feasibility of the proposed new tool to plan, train and test ML models in Python and Matlab without the need to write complex specific scripts in these languages, just by making a few clicks. On the other hand, as pointed out by the users who participated in the preliminary tests, with this tool it was easier to directly select the ML algorithm (according to the types of data), to customize the selected algorithms to create user-defined models (changing the value of the arguments of each algorithm to create more accurate models), to know the metrics of each algorithm (on which to base the trial-and-error tests) and to make predictions from users models (making predictions with new data). However, we think that this kind of tools is oriented to professionals from areas such as Accounting Sciences, Economics, International Trade and Business Administration, who have the need to create predictive models –to base their decisions on historical data– and because of their limited programming knowledge it is very complex or impossible for them to do it through Python or Matlab scripts.

The evaluation of the proposed tool shows that the use of the tool requires some technical knowledge to create predictive models based on data in a simple way. Therefore, the tool must be complementary to specific courses in which non-technical users are taught the possibilities provided by machine learning techniques for obtaining data models from which predictions can be made using a working methodology. In this way professionals for example in the field of management will be able to create predictive models based on ML and thus make decisions based on data analysis and past experiences.

7 Conclusions and Future Works

Python had enough online documentation to create the proposed interfaces. This documentation allowed us to give the information of each of the arguments that support the studied supervised and unsupervised ML algorithms, such as: simple regression, decision tree, k-nearest neighbour (kNN), Naïve Bayes, support vector machine (SVM) and K-means. With the available information, it was possible to implement the GUIs so that non-sophisticated users can create ML models in Python without having to have solid programming basis.

To carry out the creation of an ML model using the studied algorithms, it is enough that the user only has knowledge of the theoretical bases of each algorithm. The proposed guided process with the assistance of a GUI allow training and testing ML models in an intuitive way with few clicks. To do so, it is necessary to perform basic tasks among which are: selection of data, training of the model and testing the trained model.

As future work, it is proposed to extend this study in testing with more people including students and professionals from other areas of knowledge different to Computer Science. Likewise, it is proposed to enrich the graphic elements that provide the developed GUIs so that they are even more friendly and usable.

References

1. Abadi, M., Barham, P., Chen, J., Chen, Z., Davis, A., Dean, J., Devin, M., Ghemawat, S., Irving, G., Isard, M., Kudlur, M., Levenberg, J., Monga, R., Moore, S., Murray, D.G., Steiner, B., Tucker, P., Vasudevan, V., Warden, P., Wicke, M., Yu, Y., Zheng, X.: Tensorflow: a system for large-scale machine learning. In: 12th USENIX Symposium on Operating Systems Design and Implementation (OSDI 16), pp. 265–283. USENIX Association, Savannah, GA, November 2016. https://www.usenix.org/conference/osdi16/technical-sessions/presentation/abadi
2. Agarwal, B., Mittal, N.: Machine learning approach for sentiment analysis. In: Prominent Feature Extraction for Sentiment Analysis, pp. 21–45. Springer (2016)
3. Buitinck, L., Louppe, G., Blondel, M., Pedregosa, F., Mueller, A., Grisel, O., Niculae, V., Prettenhofer, P., Gramfort, A., Grobler, J., et al.: Api design for machine learning software: experiences from the scikit-learn project. arXiv preprint arXiv:1309.0238 (2013)
4. Caruana, R., Niculescu-Mizil, A.: An empirical comparison of supervised learning algorithms. In: Proceedings of the 23rd International Conference on Machine Learning, pp. 161–168 (2006)
5. Fatima, M., Pasha, M., et al.: Survey of machine learning algorithms for disease diagnostic. J. Intell. Learn. Syst. Appl. **9**(01), 1 (2017)
6. Gould, S.: Darwin: a framework for machine learning and computer vision research and development. J. Mach. Learn. Res. **13**, 3533–3537 (2012)
7. Graham, J.W.: Rattle: a data mining GUI for R. R J. **1**, 45 (2009). https://doi.org/10.32614/rj-2009-016
8. Hall, M., Frank, E., Holmes, G., Pfahringer, B., Reutemann, P., Witten, I.H.: The weka data mining software: an update. ACM SIGKDD Explor. Newslett. **11**(1), 10–18 (2009)
9. Itoo, F., Singh, S., et al.: Comparison and analysis of logistic regression, Naïve Bayes and KNN machine learning algorithms for credit card fraud detection. Int. J. Inf. Technol. 1–9 (2020)
10. Kotsiantis, S.B., Zaharakis, I., Pintelas, P.: Supervised machine learning: a review of classification techniques. Emerg. Artif. Intell. Appl. Comput. Eng. **160**, 3–24 (2007)
11. Kotsiantis, S.B., Zaharakis, I.D., Pintelas, P.E.: Machine learning: a review of classification and combining techniques. Artif. Intell. Rev. **26**(3), 159–190 (2006)
12. Lin, W.Y., Hu, Y.H., Tsai, C.F.: Machine learning in financial crisis prediction: a survey. IEEE Trans. Syst. Man Cybern. Part C (Appl. Rev.) **42**(4), 421–436 (2011)
13. Murphy, K., et al.: The Bayes net toolbox for matlab. Comput. Sci. Stat. **33**(2), 1024–1034 (2001)
14. Paluszek, M., Thomas, S.: MATLAB Machine Learning. Apress, New York (2016)
15. Pelckmans, K., Suykens, J., Van Gestel, T., De Brabanter, J., Lukas, L., Hamers, B., De Moor, B.: LS-SVMlab: a MATLAB/C toolbox for least squares support vector machines. Internal Report ESAT-SISTA (2002)
16. Piletskiy, P., Chumachenko, D., Meniailov, I.: Development and analysis of intelligent recommendation system using machine learning approach. In: Integrated Computer Technologies in Mechanical Engineering, pp. 186–197. Springer (2020)
17. Raschka, S., Mirjalili, V.: Python Machine Learning: Machine Learning and Deep Learning with Python, scikit-learn, and TensorFlow 2. Packt Publishing Ltd., Birmingham (2019)

18. Guo, T., Li, G.-Y.: Neural data mining for credit card fraud detection. In: 2008 International Conference on Machine Learning and Cybernetics, vol. 7, pp. 3630–3634 (2008)
19. Tara, K., Sarkar, A.K., Khan, M.A.G., Mou, J.R.: Detection of cardiac disorder using matlab based graphical user interface (GUI), pp. 440–443. IEEE (2017)
20. Vishwakarma, H.O., Sajan, K.S., Maheshwari, B., Dhiman, Y.D.: Intelligent bearing fault monitoring system using support vector machine and wavelet packet decomposition for induction motors, pp. 339–343. IEEE (2015)

Aphids Detection on Lemons Leaf Image Using Convolutional Neural Networks

Jorge Parraga-Alava[1,2,6](✉), Roberth Alcivar-Cevallos[2], Jaime A. Riascos[3,4,6], and Miguel A. Becerra[5,6]

[1] Facultad de Ciencias Informáticas, Universidad Técnica de Manabí, Portoviejo, Ecuador
jorge.parraga@usach.cl
[2] Universidad de Santiago de Chile, Santiago, Chile
roberth.alcivar@usach.cl
[3] Corporación Universitaria Autónoma de Nariño, Pasto, Colombia
jarsalas@inf.ufrgs.br
[4] Universidad Mariana, Pasto, Colombia
[5] Institución Universitaria Pascual Bravo, Medellín, Colombia
miguel.becerra@pascualbravo.edu.co
[6] SDAS Research Group, Ibarra, Ecuador
http://www.sdas-group.com

Abstract. Ecuador has been recognized for the export of high-quality plant products for food. Plant leaves disease detection is an important task for increasing the quality of the agricultural products and it should be automated to avoid inconsistent and slow detection typical of human inspection. In this study, we propose an automated approach for the detection of aphids on lemon leaves by using convolutional neural networks (CNNs). We boarded it as a binary classification problem and we solved it by using the VGG-16 network architecture. The performance of the neural network was analyzed by carrying out a fine-tuned process where pre-trained weights are updated by unfreezing them in certain layers. We evaluated the fine-tuning process and compared our approach with other machine learning methods using performance metrics for classification problems and receiver operating characteristic (ROC) analysis, respectively and we evidenced the superiority of our approach using statistical tests. Computational results are encouraging since, according to performance metrics, our approach is able to reach average rates between 81% and 97% of correct aphids detection on a real lemons leaf image dataset .

Keywords: Aphids · Lemon plants · Convolutional Neural Networks · Image classification · Supervised learning

1 Introduction

Citrus, especially orange, lemon and tangerine, are among the essential fruit plants worldwide. Its cultivation and consumption are carried out equally in the

M. Botto-Tobar et al. (Eds.): ICCIS 2020, AISC 1273, pp. 16–27, 2021.
https://doi.org/10.1007/978-3-030-59194-6_2

five continents, being exploited commercially in all countries where the climate conditions are optimal for its development [10]. Lemon plants are affected by a large number of pests, which produce considerable losses in the quantity and quality of their productions. Aphids are among the most destructive pests of lemon plants. They are small insects that suck plant sap [17] and affect crop growth. They can reproduce in days, with the nymphs hiding on the lower surface of leaves [5]. For example, in Ecuador in 2016, lemon production decreased mainly due to the presence of pests such as aphids that prevented the flowering of the plantations. According to *Ministerio de Agricultura y Ganadería* (MAG), during that year, only 55094 tons of lemon were harvested. It was necessary to buy this product from Colombia to supply the national demand, representing an increase of 321% in the importation.

Early detection and identification of lemon plant diseases is necessary to minimize the loss of plants, for better treatment and preventing the spreading of the disease, and so improve the crop production. Currently, the traditional method of aphids detection is to identify the small insects manually. However, human inspection is a tedious task that can be affected by decisions such as expertise in disease symptoms or some degree of subjectivity regarding "ground truth". Moreover, applying manual detection is expensive in large-scale crops. Therefore, more advanced and sophisticated methods should be used to support the aphids detection on lemons leaf. In this sense, machine learning (ML) emerges as an alternative that plays a dominant role in plant disease detection and identification [4,15,21].

Image classification task refers to a process that can classify an image according to its visual content [11]. Convolutional Neural Networks (CNNs) are the current state-of-art architecture for image classification tasks. In citrus plant aphids detection works are scarce, but in the few cases that there are, CNNs have been employed successfully. In [5], Chen et al. exposed a method for segmentation and counting of aphid nymphs on leaves. They achieve a Precision of 95.6% for automatic counting of aphids. Xuesong et al. [24] used a contour detection method to real-time enumeration of aphids through smartphones. They achieve a counting Accuracy of 95% and 92.5% for greenhouse and outside scenarios, respectively. Xing and Lee [20] adopted Inception-ResNet-V3 architecture to classify 10 citrus pests (but no aphids detection). Results suggest that they reach Accuracy rates between 94% and 98%.

Although the works mentioned above have demonstrated to be adequate to carry out citrus plant disease detection, they present some issues that should be tackled. No studies have focused on the automatic aphids detection from lemon leaf images in real conditions such as background with multiple lighting brightness. Also, it is important the proper selection of parameters in CNN architecture since a poor choice can lead to a performance degradation of the neural model. In this paper, we propose a CNNs-based approach to classify lemon leaves regarding whether or not aphids are present. Here, we tackled the detection of aphids on lemon leaves as a binary classification problem. We selected the CNNs-based approach due to compared with the manual method, CNNs can detect the presence of aphids in an objective and Accuracy way by analyzing the leaf images in real conditions. In our approach, we adopted the VGG-16 [22] architecture and modified the pre-trained weights to fit our model. The results showed that our CNNs-based approach outperforms traditional classification methods.

The main contributions of this work are: (1) a small lemon leaf images dataset of images with visible aphids and images without such disease (healthy); and (2), an automated CNN-based approach to detect aphids on lemon leaves, which can achieve significant performance compared to other machine learning methods for classification tasks. Our CNN-based approach can be considered as a start point towards big scale pest detection on lemons crops. The paper is organized as follows. Section 2 shows a brief description of the image dataset and CNN architecture as well as the theoretical backgrounds of the classification problems. Section 3 presents computational experiments and results. Finally, Sect. 4 contains the conclusions of this work.

2 Materials and Methods

2.1 Classification Problems

Classification problems have been considered a key component of machine learning, with a huge amount of applications in diverse areas. A classification problem can be formally defined as the task of estimating the label y of a K-dimensional input vector x, where $x \in \chi \subseteq \mathbb{R}^K$ and $y \in \Upsilon = \{C_1, C_2, \ldots, C_Q\}$. This task is accomplished by using a classification rule or function $g : \chi \longrightarrow \Upsilon$ ables to predict the label of new patterns [18]. Note that Q corresponds to the number of possible labels or classes to estimate. When the output variable that represents the label to estimate has two possible values, the problem is called *binary classification*. When it has more than two possibles values, the problem is called *multi-class classification*.

2.2 Dataset

Images of leaves were taken from a lemon crop located in Junín, Ecuador, for two weeks to obtain the plants' real conditions, such as multiple lighting brightness (cloudy, sunny, and windy days) and temperature levels (high and low humidity levels). The images were captured by using a smartphone camera with a resolution of 800×600 pixels at a working distance of 20–30 cm without zoom. Thus, the dataset contains 150 imagery of the upper and back sides of lemon leaf images (70 healthy and 80 with aphids presence). Besides, we carried out an annotation data process to obtain a fully-labeled of the lemon leaf images using a data labeling web-tool called Labelbox[1]. Annotations correspond to a state (healthy and aphids presence). The dataset can be download https://data.mendeley.com/datasets/4b6vr2zkbm/1.

[1] http://www.labelbox.com.

2.3 Proposed CNN Approach

There are various architectures of CNNs available on literature that have demonstrated excellent results for classification tasks. In this work, we use a CNN based on the pre-trained CNNs called VGG-16 [22] to tackle the detection of aphids on lemon leaves as a binary classification problem analyzing the whole image without applying a segmentation to the same. In Fig. 1, we show a representation of the proposed CNN-based architecture for aphids detection.

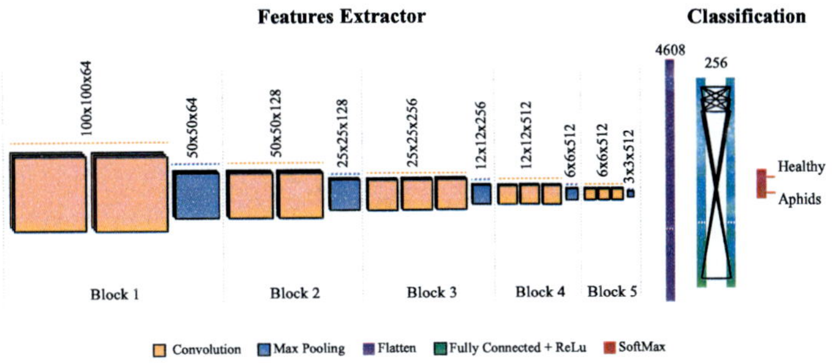

Fig. 1. Architecture of the proposed Convolutional Neural Network (CNN) approach to aphids detection on lemons leaves.

The network in Fig. 1 is based on a model proposed by K. Simonyan and A. Zisserman in [22]. In our implementation of CNN for aphids detection on lemon leaves, we considered five blocks for convolutional and pooling layers, which are used as *feature extractor*. Moreover, to perform the *classification* process, we append to the output of feature extractor diverse layers organized as a stack, which includes a layer to flatten results into a vector and to use it as input in a fully connected layer corresponding to an MLP. Finally, the model includes a layer representing an activation function of type softmax to classify the outputs of MLP as healthy leaf or leaf with aphids. Note that the fully connected layer includes as activation function of the Rectified Linear Unit (ReLU), which take all negative values yield by the MLP and convert them to zero.

2.4 Performance Evaluation

We consider the well-known measures *Accuracy*, *Precision*, *Recall* and *F-measure*. These last are also known as *sensitivity* and *F-score*, respectively. Here, the aphid presence on lemon leaves was called positive (P) and the healthy leaf negative (N). Moreover, we include receiving operating characteristic (ROC) curves to compare the performance of our approach with other classification methods. It is a two-dimensional description of classifier performance, which allows us to compare classifiers at different sensitivity/specificity trade-offs [13]. From the ROC curve, we use the area under the curve (AUC), a single scalar value used to measure the overall performance of classifiers.

3 Results and Discussion

The goal of the experiments is to carry out the aphids detection on lemons leaves from leaf images using a convolutional neural network (CNN). Also, experiments are intended to identify the best parameters of the CNN to maximize the number of leaves with aphids correctly identified.

3.1 Implementation Details

Our CNN was implemented by using R version 3.6.0, R Studio version 1.2.1335. To use the VGG-16 architecture, the tensorflow library and Keras framework were used through "tensorflow" [2] and "keras" [1] R libraries, respectively. The SVM (Support Vector Machine) and Random Forests algorithms were implemented using "svm" [14] and "scikit-learn" [16] R and Python libraries, respectively. ROC curves were plotted using "pROC" [19] R library. To generate predictions of new images, we used Python version 3.7.3 with Spyder version 3.3.3. We performed our CNN approach with parameters: number of iterations = 30, number of epochs = 50, batch size = 32, optimizer = "adam" [12], learning rate = 0.0001 and loss function = "cross entropy". Parameters were set up considering default configurations in the Keras R package and recommendations for binary classification problems in image recognition [6].

3.2 Preprocessing

We carried out a preprocessing stage to adapt the images of input properly to the experiment. We resized the images in our *LeLePhid* dataset from 800×600 pixels to 100×100 pixels. Furthermore, since *LeLePhid* contains colorful image, we applied a rescaling factor of 1/255 to transform every pixel value to [0,1] range. We split *LeLePhid* into three subsets: training, test, and validation. Each one has 60%, 20%, and 20% of the images, respectively. As *LeLePhid* dataset has a small number of images, over-fitting is an issue that can arise during the model training. To tackled it, we considered a *data augmentation* process. It is a powerful technique to increase the size of a training subset by creating more images by altering the rotations, brightness, zoom, and other operations over the images in the dataset.

3.3 Results

Effect of Unfreeze Early Layers: Our CNN approach is based on VGG-16 architecture, which was trained in the ImageNet dataset [8]. It is possible to classify images belonging to 1000 classes using this architecture with its pre-trained weights. However, since these weights are adjusted using images taken in typical activities of our day-to-day lives but very opposed to scenarios such as crops, it is necessary to re-adjust these weights. The idea is that our CNN can detect aphids in lemon leaves images captured with a typical smartphone camera and in diverse backgrounds. The weights of VGG-16 are pre-trained, i.e., frozen in each block of our CNN model. To evaluate this neural architecture's ability to detect aphids on lemons leaf image, we performed a fine-tuning process by unfreezing the convolution layers available in the blocks shown in Fig. 1. Thereby, when the network is trained, it updates the unfrozen layer's weights and keeps the pre-trained weights in the other layers.

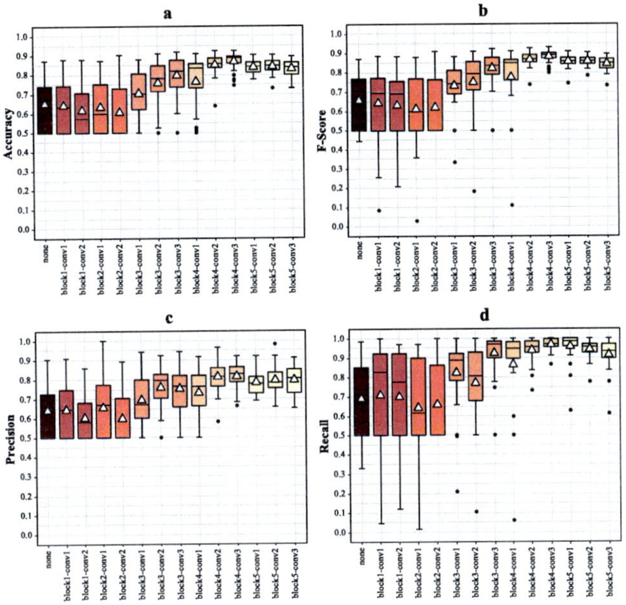

Fig. 2. Performance regarding to (**a**) Accuracy, (**b**) F-Score, (**c**) Precision and (**d**) RRecall of our CNN approach when each layer is unfrozen. The horizontal axis indicates an unfrozen layer, and triangles represent the average values of each metric.

In Table 1, we show the mean values of four metrics over 30 runs of different cases of unfreeze early layers for *LeLePhi* dataset. Here, values highlighted in gray represent maximum values for each metric, revealing that on average, our CNN approach achieves better levels of aphids detection when the network weights are re-adjusted in the block four and convolution layer three.

Finally, in Fig. 3, several images with their classification are shown. Such classification was obtained considering the best model of our CNN approach when it was trained with layers unfrozen from convolution three in block four. We implemented a simple code of our CNN approach, and we load the best model of the training phase. After that, we selected the images, and we carried out the prediction of them.

Table 1. Mean values of classification measures over 30 runs of our CNN model for different cases of unfreeze first layers. In gray, we highlight the highest values.

Unfreezed Layer	Accuracy	F-Score	Precision	Recall
Nothing	0.66	0.66	0.64	0.69
Block 1 Convolution 1	0.64	0.64	0.64	0.71
Block 1 Convolution 2	0.62	0.63	0.60	0.70
Block 2 Convolution 1	0.63	0.61	0.66	0.64
Block 2 Convolution 2	0.61	0.62	0.60	0.66
Block 3 Convolution 1	0.70	0.73	0.69	0.82
Block 3 Convolution 2	0.76	0.75	0.76	0.77
Block 3 Convolution 3	0.80	0.82	0.75	0.92
Block 4 Convolution 1	0.76	0.77	0.73	0.86
Block 4 Convolution 2	0.85	0.87	0.81	0.94
Block 4 Convolution 3	0.87	0.88	0.81	0.97
Block 5 Convolution 1	0.84	0.86	0.78	0.96
Block 5 Convolution 2	0.84	0.86	0.79	0.94
Block 5 Convolution 3	0.83	0.84	0.79	0.91

Comparison with Other Classification Techniques: To demonstrate the advantage of our CNN approach of aphids detection on lemons leaf image, we compared it with SVM (Support Vector Machine) [7] and Random Forest [3], which are techniques commonly used in image classification. In SVM, we evaluated the linear, polynomial, and sigmoid kernels. We selected the radial kernel since it achieved the best performance. Furthermore, in comparison, we used the CNN with pre-trained weights, i.e., original VGG-16 architecture. In Table 2 the main values of classification metrics used in comparison are shown.

Table 2. Performance of classification methods (mean values) for aphids detection on lemons leaf image over 30 runs. In gray, we highlight the highest values for each metric.

Classification method	Accuracy	Sensitivity	Specificity	AUC
Random Forest	0.80	0.75	0.70	0.81
SVM	0.76	0.97	0.55	0.76
VGG-16[a]	0.66	0.69	0.64	0.64
Our CNN approach	0.87	0.97	0.74	0.85

[a]Using pre-trained weights.

To compare the performance of classifiers, we measured the area under the ROC curves (AUC). In Fig. 4, we show the ROC curves considering the best

Fig. 3. Examples of lemon leaf images classified using our best CNN model regarding the *Accuracy* metric. Correct classification labels are green, and the incorrect ones are red.

values for each classification method used in comparison to the AUC metric. The ROC curves were obtained by showing reciprocal relationships between sensitivity and 1-specificity at various threshold settings.

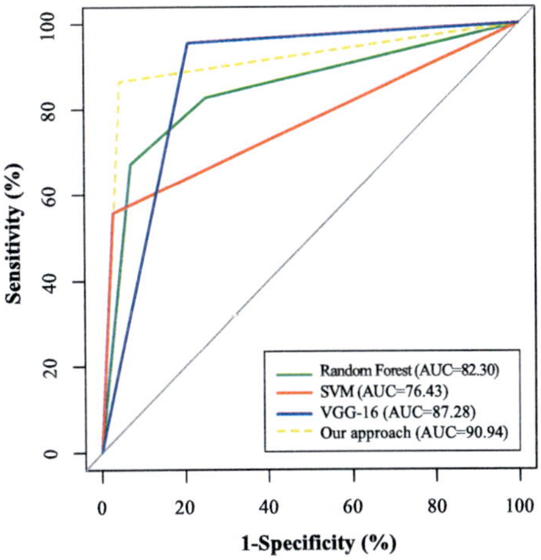

Fig. 4. Receiver Operating Characteristic (ROC) curve of classification methods for aphids detection on lemon leaf images used in the comparison.

3.4 Discussion

The experimental results aim to demonstrate that our method can better distinguish between healthy and unhealthy (aphids presence) lemon leaves images with a high probability. In Fig. 2, we show boxplot of obtained values by four metrics over 30 runs of different cases of unfreeze first layers of our CNN approach. Based on it can be observed that using pre-trained weights of VGG-16 architecture, i.e., when none layer is unfrozen, allowing CNN to achieves values between 0.50 and 0.85 in four classification metrics considered. When the convolution layers of two first blocks are unfrozen, the performance of CNN is similar to the case of pre-trained weights (none). However, from block three to block five can be observed that unfreeze convolution layers to re-adjust the weights of CNN seems to improve its performance. Results showed in Fig. 2 and Table 1 seem to demonstrate that, at least visually, the performance of the network improves when the layers are unfrozen from block three. To detect whether such situations allow the CNN model to operate similarly or not from the statistical point of view, we carry out the ANOVA test. The test verifies whether the classification metrics mean values achieved by the CNN with at least one unfrozen layers are significantly different from those achieved by the CNN with original pre-trained weights, i.e., with none unfrozen layer.

Once we verified the normality, homoscedasticity, and independence of data, we conducted the ANOVA test and found *p-values* of 2e−16, 2e−16, 2e−15, and 2e−13 for Accuracy, F-score, Precision and Recall, respectively. As they are lower than our threshold of 0.05, we can say that there is a statistically significant difference in the CNN performance when at least one layer is unfrozen and none unfrozen. Then, we carried out the Tukey posthoc test [23] with Holm correction [9] to find out which of the unfrozen layers are responsible for such differences. We have the following cases:

1. **Accuracy**. When the layers block3-conv2, block3-conv3, and all block four and block five are unfrozen, the performance of the CNN differs significantly from original pre-trained weights, i.e., when none layer is unfrozen with *p-values* of 0.0074, 0.0001 and <2e-7 for other cases, respectively. While among them, the unfrozen layer whose configuration makes the CNN's performance differs is <u>block4-conv3</u> with *p-values* <40.05 for all cases except with the layers block4-conv2, and all layers included in block five.
2. **F-Score**. When the layers block3-conv3 and all layers in block four and block five are unfrozen, the performance of the CNN differs significantly to original pre-trained weights, i.e., when none layer is unfrozen with *p-values* of 0.0005, 0.0256 and <2e-6 for other cases, respectively. While among them, the unfrozen layer whose configuration makes the CNN's performance differs is <u>block4-conv3</u> with *p-values* <0.05 for all cases except with the layers block3-conv3, block4-conv1, block4-conv2 and all layers included in block 5.
3. **Precision**. When the layers block3-conv2, block3-conv3, block4-conv2, block4-conv3, and all layers in block five are unfrozen, the performance of the CNN differs significantly from original pre-trained weights, i.e., when none

layer is unfrozen with *p-values* <2e-6. While among them, the unfrozen layer whose configuration makes the CNN's performance differs is block4-conv3 with *p-values* <0.05 for all cases except with the layers block3-conv2, block3-conv3, block4-conv1, block4-conv2 and all layers included in block five.

4. **Recall**. When the layers block3-conv3 and all layers in block five and block five are unfrozen, the performance of the CNN differs significantly to original pre-trained weights, i.e., when none layer is unfrozen with *p-values* <0.05. While among them, the unfrozen layer whose configuration makes the CNN's performance differs is block4-conv3 with *p-values* <0.05 for all cases except with layers block3-conv1, block3-conv3, block4-conv1, block4-conv2 and all layers included in block five.

The proved statistical significance of the CNN approach when the layer convolution 3 in block 4 (block4-conv3) is unfrozen demonstrates the superiority to update the weights from this layer. It case has the best average value in the four metrics considered (Table 1). These situations indicate that to unfreeze layers from convolution three in block four has a positive effect in CNN approach performance to settle aphids detection on lemons leaf image.

In Fig. 3, using our best CNN approach, it is possible correctly to classify 9 out of 10 images of lemon leaves evaluated. In this sense, the *Precision* and *Accuracy* metrics that answering the questions: *What proportion of positive (healthy) identifications was correct?* and *What percentage of predictions (healthy and aphids) was correct?* achieve a 90% in both cases. These values indicate that our approach can detect aphids on lemons leaf image with a high performance.

In Table 2, our approach obtained the best mean values showing a Sensitivity in recognition of positive images (aphids) equal to 97%, while concerning the ability to identify the negatives (healthy), it showed a specificity of 74%. Note that SVM also has an excellent performance for aphids detection (Sensitivity), but it is degraded when healthy leaves should be detected (Specificity). From the comparison of our CNN approach with the other classification methods using the same dataset, the potentialities of the method for the detection of healthy leaves and leaves with the presence of aphids can be deduced. An 87% of Accuracy implies that globally our CNN-based classifier is doing a good job identifying aphids on lemons leaf images.

In Fig. 4 larger, the area under the curve (AUC) corresponds to the better the classification effect. Here, we can observe that AUC values of the CNN approach are higher than those of other classification methods; visually, it seems to have the best performance for image classification. To detect whether classification methods operate similarly or not from a statistical point of view, we carried out the ANOVA test. It verifies whether the average performance regarding AUC of the methods are significantly different between them. For the SVM case, we used only its best version, i.e., with a radial kernel. We conducted the test and found *p-value* of *0.00019*. As they are lower than our threshold of 0.05, we can say that there is a statistically significant difference in the performance of classification methods regarding AUC value. As the ANOVA test is significant, we can compute Tukey HSD (Tukey Honest Significant Differences) to perform multiple pairwise comparisons between the means of groups. We conducted the test and found *p-values* lower than our threshold of 0.05 when compared to our approach with SVM and VGG-16, so we can say that there is a statistically significant difference between these methods. However, when our CNN approach was compared with Random Forest, we found no differences.

4 Conclusions

In this paper, we designed a CNN approach for aphids detection on lemons leaf image. It is based on the VGG-16 pre-trained network architecture. By computational experiments, we demonstrate that unfreeze a convolutional layer in the block 4 to fine-tune the process of weights update contributed to better predictive capacity of our CNN compared to original VGG-16 architecture. Results show that the CNN approach achieved average between 81% and 97% of Accuracy, F-score, Precision and Recall for the binary classification problem of detection healthy leaves or presence of aphids in images of lemon leaves. Using validation data, the best model of our CNN approach was able to identify healthy, and with aphids presence leaves in 96.87% (specificity) and 100% (sensitivity) of the cases, respectively. Statistical tests demonstrated that the proposed CNN approach could detect aphids on lemons leaf images with higher levels than other machine learning methods for the detection of plant leaf diseases used in the comparison. As a future work, we would like to continue improving the structure and configuration of our CNN approach to increase classification rates. Furthermore, we would like to apply our work to multi-class problems related to detect levels of severity of other plant leaf diseases such as coffee leaf rust (CLR) and red mites. Our significant objective is to use a tuned CNN-based model in a device such a drone to carry out large-scale pest detection on coast Ecuadorian crops.

Acknowledgment. This work is supported by the Smart Data Analysis Systems Group - SDAS Research Group (http://sdas-group.com)

References

1. Allaire, J., Chollet, F.: keras: R Interface to Keras, R package version 2.2.4.1.9001 (2019)
2. Allaire, J., Tang, Y.: TensorFlow: R Interface to TensorFlow, R package version 1.13.1 (2019)
3. Breiman, L.: Random forests. Mach. Learn. **45**(1), 5–32 (2001)
4. Carrillo, J.M., Parraga-Alava, J.: How predicting the academic success of students of the ESPAM MFL?: a preliminary decision trees based study. In: 2018 IEEE Third Ecuador Technical Chapters Meeting (ETCM), pp. 1–6, October 2018
5. Chen, J., Fan, Y., Wang, T., Zhang, C., Qiu, Z., He, Y.: Automatic segmentation and counting of aphid nymphs on leaves using convolutional neural networks. Agronomy **8**(8), 129 (2018)
6. Chollet, F., Allaire, J.: TensorFlow for R: image classification on small datasets with keras (2017)

7. Cortes, C., Vapnik, V.: Support-vector networks. Mach. Learn. **20**(3), 273–297 (1995)
8. Deng, J., Dong, W., Socher, R., Li, L., Li, K., Fei-Fei, L.: Imagenet: a large-scale hierarchical image database. In: 2009 IEEE Conference on Computer Vision and Pattern Recognition, pp. 248–255, June 2009
9. Holm, S.: A simple sequentially rejective multiple test procedure. Scand. J. Stat. **6**(2), 65–70 (1979)
10. Instituto Colombiano Agropecuario (ICA): Manejo fitosanitario del cultivo de cítricos Medidas para la temporada invernal (2016)
11. Kaeli, D., Mistry, P., Schaa, D., Zhang, D.P.: Chapter 9 - case study: image clustering. In: Kaeli, D., Mistry, P., Schaa, D., Zhang, D.P. (eds.) Heterogeneous Computing with OpenCL 2.0, pp. 213–228. Morgan Kaufmann, Boston (2015)
12. Kingma, D.P., Ba, J.: Adam: a method for stochastic optimization. arXiv preprint arXiv:1412.6980 (2014)
13. Melo, F.: Receiver Operating Characteristic (ROC) Curve, pp. 1818–1823. Springer, New York (2013)
14. Meyer, D., Dimitriadou, E., Hornik, K., Weingessel, A., Leisch, F.: e1071: Misc Functions of the Department of Statistics, Probability Theory Group (Formerly: E1071), TU Wien, R package version 1.7-2 (2019)
15. Parraga-Alava, J., Dorn, M., Inostroza-Ponta, M.: A multi-objective gene clustering algorithm guided by apriori biological knowledge with intensification and diversification strategies. BioData Min. **11**(1), 16 (2018)
16. Pedregosa, F., Varoquaux, G., Gramfort, A., Michel, V., Thirion, B., Grisel, O., Blondel, M., Prettenhofer, P., Weiss, R., Dubourg, V., Vanderplas, J., Passos, A., Cournapeau, D., Brucher, M., Perrot, M., Duchesnay, E.: Scikit-learn: machine learning in python. J. Mach. Learn. Res. **12**, 2825–2830 (2011)
17. Pickett, J.A., Allemann, R.K., Birkett, M.A.: The semiochemistry of aphids. Nat. Prod. Rep. **30**(10), 1277–1283 (2013)
18. Pérez-Ortiz, M., Jiménez-Fernández, S., Gutiérrez, P.A., Alexandre, E., Hervás-Martínez, C., Salcedo-Sanz, S.: A review of classification problems and algorithms in renewable energy applications. Energies **9**(8) (2016)
19. Robin, X., Turck, N., Hainard, A., Tiberti, N., Lisacek, F., Sanchez, J.C., Müller, M.: pROC: an open-source package for R and S+ to analyze and compare ROC curves. BMC Bioinform. **12**, 77 (2011)
20. Shuli, X., Marely, L.: A study of tangerine pest recognition using advanced deep learning methods. Preprints (2018)
21. Silva, C.F., Siebra, C.A.: An investigation on the use of convolutional neural network for image classification in embedded systems. In: 2017 IEEE Latin American Conference on Computational Intelligence (LA-CCI), pp. 1–6, November 2017
22. Simonyan, K., Zisserman, A.: Very deep convolutional networks for large-scale image recognition. In: 3rd International Conference on Learning Representations, pp. 1–14 (2015)
23. Tukey, J.W., et al.: Comparing individual means in the analysis of variance. Biometrics **5**(2), 99–114 (1949)
24. Xuesong, S., Zi, L., Lei, S., Jiao, W., Yang, Z.: Aphid identification and counting based on smartphone and machine vision. J. Sens. **2017**, 1–7 (2017)

Check for updates

Prediction of Monthly Electricity Consumption by Cantons in Ecuador Through Neural Networks: A Case Study

Jorge I. Guachimboza-Davalos(✉), Edilberto A. Llanes-Cedeño, Rodolfo J. Rubio-Aguiar, Diana B. Peralta-Zurita, and Oscar F. Núñez-Barrionuevo

Universidad Internacional SEK, Quito, Ecuador
jiguachimboza.mee@uisek.edu.ec

Abstract. The present work shows a multivariate regression through deep learning related to energy consumption in the province of Pichincha-Ecuador in relation to its cantons. This is done to establish consumption habits in relation to the existing population. For this, a data preprocessing stage was implemented and later an artificial neural network of a hidden layer and 15 neurons. As a result, a mathematical model with an absolute error of 2.2% is obtained.

Keywords: Power consumption · Multivariate regression · Neural network

1 Introduction

Currently, one of the biggest government concerns is global warming. Since this causes an abrupt change in the environmental conditions of the planet, altering the living conditions of people, animals and plants [1]. This problem is mainly caused by the excessive emission of polluting gases (carbon dioxide CO_2, carbon dioxide CO, hydrocarbons, oxides of nitrogen, sulfur, among others) of companies that use fossil fuels for the development of a product or service [2,3]. Related to this, is the production of emissions due to the electricity demand of the population. The same, which is a key element for the socio-economic development of each country. That said, 41% of the world's CO_2 emissions are attributed to electricity generation. The same ones that contribute the majority (around 99.5%) of CO_2.

In Latin America, there is an average of 0.65 tons of CO_2 for each MWh generated. [4,5]. In Ecuador, this result is 397.5 g of CO_2 per KWh, where the highest value corresponds to 2010 since the generation of energy from non-renewable sources was (52.2%) [6]. Currently, with the change in the production matrix, 60.85% of electricity production is based on renewable energy. Where 58.53% is related to Hydroelectric sources [7]. This production is oriented 30.93% to

M. Botto-Tobar et al. (Eds.): ICCIS 2020, AISC 1273, pp. 28–37, 2021.
https://doi.org/10.1007/978-3-030-59194-6_3

residential consumption and 26.01% to the Industrial sector [6]. With this, these forms of use have an important factor for the projection of electrical consumption and the planning of new sources of electricity generation [8]. However, the use of this resource in homes and industries show a high variation that depends on the lifestyle of the users and the type of good or service that is generated [9]. For this reason, it is necessary to implement prediction models that allow searching for people's consumption patterns in relation to a variety of attributes (consumer characteristics). However, these studies have been focused on consumption by independent households with the growth of the Internet of Things. Therefore, there are no adequate metrics for consumption trends organized by types of home users (rural and rural) or from different provinces of a country or region. In addition, The main problem to obtain a reliable forecast of Electric power consumption is choosing a forecast model that accurately presents initial data. The demand pattern is very complex due to the deregulation of the energy markets. Therefore, finding an appropriate forecasting model for a specific power grid is not an easy task. Although many forecasting methods were developed, none can be generalized for all demand patterns.

To do this, machine learning algorithms allow mathematical-computational calculations to be made to extract knowledge from the data. This criterion is widely used since it seeks to emulate certain functioning of the human brain for decision-making based on experience. However, data analysis has its limitations due to the poor structure of information storage by government entities. This is generally caused by the use of categorical variables (they do not contain numbers). That is why it is necessary to code these variables to numerical form. However, this results in increasing the size of the database as it uses ones and zeros to convey the necessary information and can confuse the learning model by providing non-existent information to the model. However, with an adequate methodology, sub-processes of data cleaning, normalization, selection of characteristics, among others, can be carried out. As a result, the mathematical model can learn from true information and dismiss the variables that are of lesser contribution [10]. With this, consumption parameters can be obtained and the future demand for electricity needed to satisfy the increase in this service can be forecast in the future.

However, there are classic approaches to machine learning that can represent the study of a phenomenon but that depends on the type of data to be used (categorical and numerical). This learning model requires feature extraction and a monitoring algorithm to generate a model. On the other hand, deep learning is based on the use of neural networks, since it allows the aforementioned stages to be carried out in a computational model. This new approach is based on neural networks [11]. This criterion allows you to organize what you have learned (self-organization). Therefore, they have the ability to be fault-tolerant, since the more times it comes into operation (iterate) it can solve its error (update of weights). For this reason, you cannot experience sudden system crashes [12].

Some works such as [13,14] present systems for measuring electricity consumption in homes to store and analyze them in the cloud. On the other hand, they do not present a sector-specific solution and are based on personal consumer trends. In this case, the application is focused on the end-user and not on a state-level analysis. It is for this reason that the present work carries out an analysis of the different consumption by cantons of the Pichincha province in Ecuador with the objective of determining parameters and consumption trends to present proposals for energy demand towards the future. For this, the information from the Electricity Regulation and Control agency has been compiled to implement deep learning through the use of neural networks. For this, a preprocessing is carried out with processes for the selection of characteristics and normalization of the database. Subsequently, an analysis of the activation functions in the different layers of a neural network is performed to determine the appropriate model that can provide the best prediction of desired values of electrical consumption. For this, different metrics are performed to calculate the forecast error. As relevant results, the model had a forecast error of 2.2%.

The rest of the document is structured as follows: Sect. 2 shows the methodological scheme for data analysis. Section 3 presents the analysis of the neural network to provide the best forecast for electricity consumption. Section 4 shows the results obtained with the different error analysis. Finally, Sect. 5 presents the results and future work.

2 Materials and Methods

This section shows, on the one hand, the acquisition of data (Sect. 2.1), preprocessing of the information using the selection of characteristics (Sect. 2.2) and the normalization of the database (Sect. 2.3). In addition, it presents the data analysis scheme (Sect. 2.4).

2.1 Data Acquisition

Data collection is through the information presented by the Ministry of Electricity and Renewable Energy in its monthly accounts during 2019. In these reports, the company that distributes electricity to the provinces, cantons, and parishes is explained in detail. In this specific case, the consumption information for the years 2019 and 2018 is taken. All data is stored in the matrix $\mathbf{Y} \in \mathbb{R}^{m \times n}$ where m represents the number of samples and n the attributes of the database. In this case, $m = 2{,}550$ and $n = 11$. Attributes refer to: Year of consumption, month, a company that bills, province, canton, parish, type of equipment with a voltage of 220, number of customers, energy billed, consumption increase, residential consumption, subsidy energy generated, billing of the service and value of the subsidy.

2.2 Feature Selection

The choice of a training set for the prediction algorithms is a crucial aspect since the response time must be adequate [15]. Typically, for a supervised data analysis task, a large number of attributes can be candidates for characterization purposes. However, many of these attributes may be irrelevant or redundant and consequently, the algorithm may be over-fitting and some important features may be affected. Therefore, the performance or acceptance of the model may not meet what was expected [16]. In fact, in many applications involving large data sets, predictions do not work properly until the unwanted features are necessarily removed [15].

In this sense, the feature selection task reduces the number of attributes that represent a data set that best fits a model or machine learning task. There is a wide range of criteria and algorithms that can be used for this task. However, depending on the nature of the data and the objective of the data analysis, some feature selection criteria may be more appropriate. The most important criteria are taken into account in this work. On the one hand, the correlation of variables is taken into account in a statistical model under the criteria of p_{value} and R^2 with the use of the `backward elimination` criterion. This combination indicates that changes in the predictors are related to changes in the response variable and that the model largely explains the variability of the response. With this, an estimate can be obtained of how the model represents the phenomenon studied [16].

2.3 Database Normalization

Index normalization means adjusting the measured values on different scales with respect to one in common. This process is done prior to the process of making mathematical models. This avoids the existence of variables that, according to their nature of the data, contribute a large amount to the model as they have very high values in relation to the rest. In addition, it prevents the use of categorical variables from being diminished due to its low scale (0 to 1). That is why there are different methods and forms of standardization.

2.4 Data Analysis Scheme

The present work is aimed at finding the best model that fits the study presented where the power consumption of the Ecuadorian population can be prognostic with the most representative variables. The proposed scheme is shown in Fig. 1.

2.5 Neural Network Training

The information obtained by the neural network cannot transmit everything learned, it needs an activation function representing an activation potential (a binary form of event probability). In this sense, the activation function allows the update of weights (learning of the neuron). The Eq. 1 shows the operation of a neural network for prediction. Furthermore, it is shown in Fig. 2.

$$Y = f(\sum_{i=0}^{i=m} \sum_{j=0}^{j=n} x_i w_{ij}) \tag{1}$$

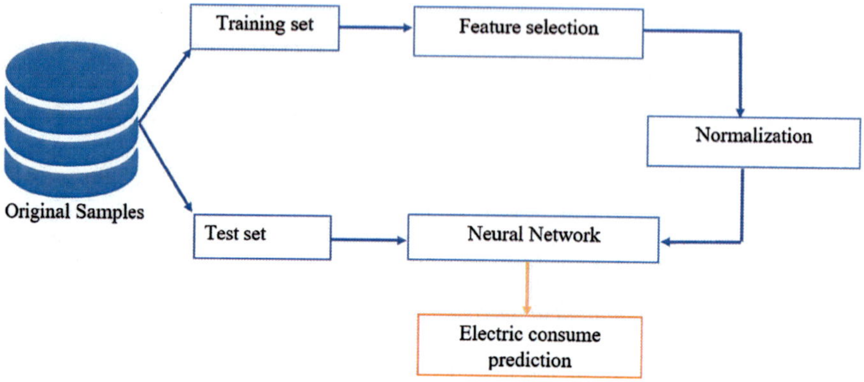

Fig. 1. Data analysis scheme

3 Results

The results of the present work are divided into: (Sect. 3.1) selection of characteristics and (Sect. 3.2) result of the neural network.

3.1 Feature Selection

The codification of the categorical variables in relation to the canton, since by parish the model would have a very high codification and would provide more information (33 rural and 32 urban parishes) than the numerical variables. Consequently, the coding of cations is shown in Table 1.

In this sense, we proceed to eliminate these variables in order to improve the model. Fig. 4 shows the result of the analysis of covariance where it indicates that the model fit has not changed (R squared) when eliminating the variables previously mentioned.

3.2 Neural Network

Once the variables have been selected to implement them in the neural network. A minimum of 10 neurons is established to train the model and test its operation. As a result, 15 neurons are established to represent a lower forecast dispersion error (MSE) and the absolute mean deviation error (MAD). This value is monetary. Table 2 shows this analysis (Fig. 3).

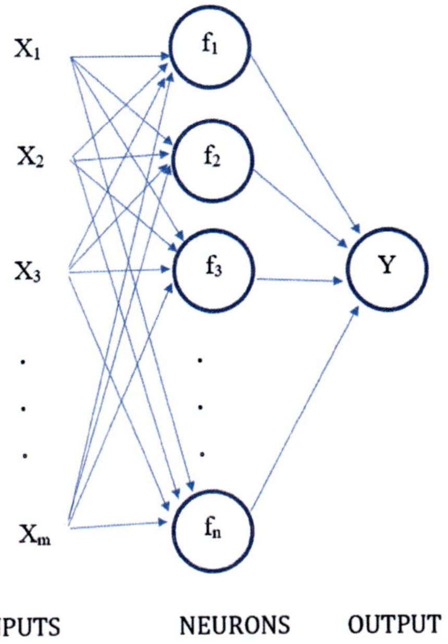

Fig. 2. Neural network designed

Table 1. Pichincha cantons categorical codification

Cantons	Cod1	Cod2	Cod3	Cod4	Cod5	Cod6	Cod7
Cayambe	1	0	0	0	0	0	0
Quito	0	1	0	0	0	0	0
Pedro Vicente Maldonado	0	0	1	0	0	0	0
Puerto Quito	0	0	0	1	0	0	0
Rumiñahui	0	0	0	0	0	1	0
San Miguel de los Bancos	0	0	0	0	0	0	1

Table 2. Error analysis for numbers of neurons

Neurons	MSE	MAD
10	4.5%	125
12	3.8%	75
15	2.2%	45
17	2.2%	45
20	2.2%	45

```
Coefficients:
                            Estimate Std. Error t value Pr(>|t|)
(Intercept)                4.968e+01  2.446e+01   2.031   0.0426 *
ï..Canton2                -2.136e+01  4.352e+01  -0.491   0.6238
ï..Canton3                 8.331e+00  3.431e+01   0.243   0.8082
ï..Canton4                -6.127e+01  4.065e+01  -1.507   0.1322
ï..Canton5                 5.440e+01  6.688e+01   0.813   0.4162
ï..Canton6                 1.893e+02  8.257e+01   2.293   0.0221 *
ï..Canton7                -1.960e+01  9.091e+01  -0.216   0.8294
Equipamiento              -1.672e-01  3.208e-01  -0.521   0.6023
Numero.Clientes           -1.391e+00  8.446e-02 -16.471   <2e-16 ***
Energia.Facturada.k.Wh     1.071e-01  1.294e-03  82.750   <2e-16 ***
Consumo.Incremental.kWh.  -1.232e-02  8.861e-04 -13.902   <2e-16 ***
Consumo.Residencial.kWh.  -7.286e-03  8.133e-04  -8.958   <2e-16 ***
Energia.Subsidio.PEC.kWh.  1.575e+02  1.341e+02   1.175   0.2404
ValorSubsidio.PEC.USD.    -1.750e+03  1.490e+03  -1.175   0.2404
---
Signif. codes:  0 '***' 0.001 '**' 0.01 '*' 0.05 '.' 0.1 ' ' 1

Residual standard error: 269.3 on 818 degrees of freedom
Multiple R-squared:  0.9997,	Adjusted R-squared:  0.9997
F-statistic: 2.186e+05 on 13 and 818 DF,  p-value: < 2.2e-16
```

Fig. 3. Covariance analysis with attributes of matrix **Y**

```
Coefficients:
                           Estimate Std. Error t value Pr(>|t|)
(Intercept)               3.251e+01  1.372e+01   2.369   0.0181 *
ï..Canton2               -1.582e+01  4.325e+01  -0.366   0.7147
ï..Canton3                1.156e+01  3.415e+01   0.339   0.7350
ï..Canton4               -5.909e+01  4.064e+01  -1.454   0.1463
ï..Canton5                5.637e+01  6.677e+01   0.844   0.3987
ï..Canton6                1.905e+02  8.253e+01   2.308   0.0213 *
ï..Canton7               -1.220e+01  9.088e+01  -0.134   0.8932
Numero.Clientes          -1.344e+00  7.744e-02 -17.362   <2e-16 ***
Energia.Facturada.k.Wh    1.084e-01  1.048e-03 103.433   <2e-16 ***
Consumo.Incremental.kWh. -1.197e-02  8.430e-04 -14.196   <2e-16 ***
Consumo.Residencial.kWh. -7.690e-03  7.775e-04  -9.890   <2e-16 ***
---
Signif. codes:  0 '***' 0.001 '**' 0.01 '*' 0.05 '.' 0.1 ' ' 1

Residual standard error: 269.6 on 821 degrees of freedom
Multiple R-squared:  0.9997,	Adjusted R-squared:  0.9997
F-statistic: 2.837e+05 on 10 and 821 DF,  p-value: < 2.2e-16
```

Fig. 4. Covariance analysis with attributes of matrix **X**

With this, a prediction of electricity consumption can be established in relation to the payment of the return. Fig. 5 shows the prognosis generated by the neural network.

Finally, a summary of all the parishes is made to have a final summary of the consumption by canton in relation to the year 2019. As can be seen in Fig. 6, the forecast is very accurate for true consumption.

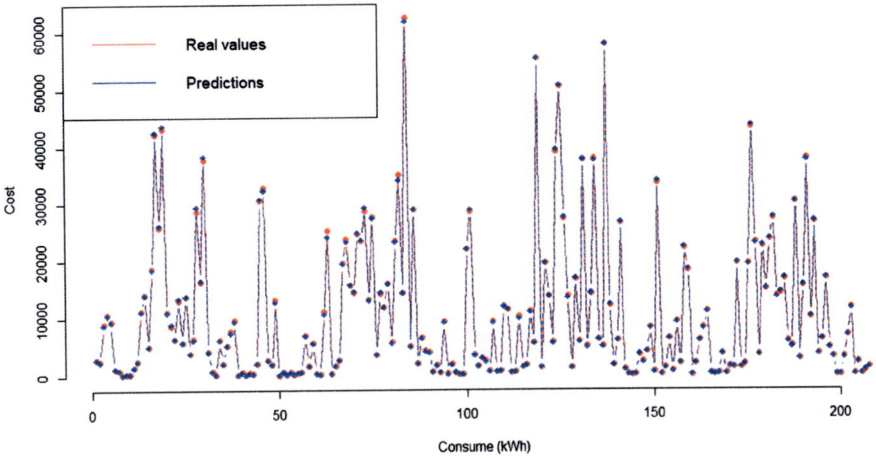

Fig. 5. Cost predictions since neural network analysis

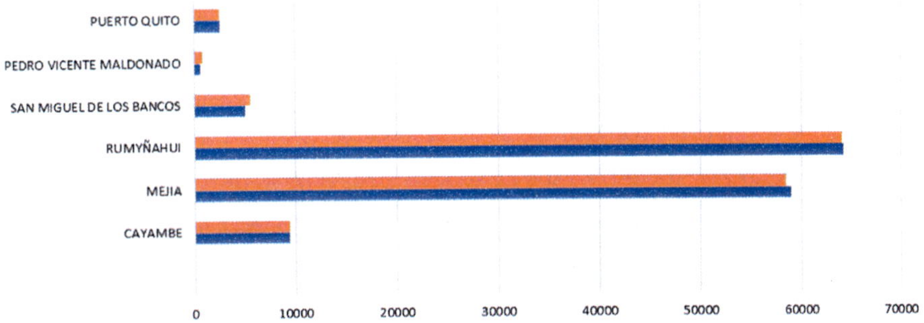

Fig. 6. Final regression model

4 Conclusions and Future Work

The present work was able to verify the functioning of learning from neural networks. With this, the prediction of the target variable was very close to the real value. In this sense, it can be seen that by using these criteria, strategies for the future can be defined with the aim of improving electricity service and establishing clear policies for better use of the country's energy resource.

As future works, it is proposed to carry out a deeper analysis of the information in relation to the energy consumption of the country related by parishes.

References

1. Ye, X., Wu, X., Guo, Y.: Real-time quality prediction of casting billet based on random forest algorithm. In: Proceedings of the 2018 IEEE International Conference on Progress in Informatics and Computing, PIC 2018, pp. 140–143. Institute of Electrical and Electronics Engineers Inc., May 2019
2. Liu, Y., Wang, W., Ghadimi, N.: Electricity load forecasting by an improved forecast engine for building level consumers. Energy **139**, 18–30 (2017)
3. UEM en Funcionamiento - Ministerio de Educación. https://educacion.gob.ec/uem-en-funcionamiento/
4. Zhang, X.M., Grolinger, K., Capretz, M.A., Seewald, L.: Forecasting residential energy consumption: single household perspective. In: Proceedings - 17th IEEE International Conference on Machine Learning and Applications, ICMLA 2018, pp. 110–117. Institute of Electrical and Electronics Engineers Inc., January 2019
5. Diao, L., Sun, Y., Chen, Z., Chen, J.: Modeling energy consumption in residential buildings: a bottom-up analysis based on occupant behavior pattern clustering and stochastic simulation. Energy Build. **147**, 47–66 (2017)
6. Ministerio de Electricidad y Energía Renovable - Ente rector del Sector Eléctrico Ecuatoriano. http://historico.energia.gob.ec/
7. Parra Narváez, R.: Factor de emisión de CO¿sub¡2¿/sub¡ debido a la generación de electricidad en el Ecuador durante el periodo 2001–2014. In: Avances en Ciencias e Ingeniería, vol. 7, no. 2, December 2015
8. Baquero, M., Quesada, F.: Eficiencia energética en el sector residencial de la Ciudad de Cuenca, Ecuador. Maskana **7**(2), 147–165 (2016)
9. Zhao, H.X., Magoulès, F.: A review on the prediction of building energy consumption. Renew. Sustain. Energy Rev. **16**(6), 3586–3592 (2012)
10. Toapanta-Lema, A., Gallegos, W., Rubio-Aguilar, J., Llanes-Cedeño, E., Carrascal-García, J., García-López, L., Rosero-Montalvo, P.D.: Regression models comparison for efficiency in electricity consumption in ecuadorian schools: a case of study. In: Botto-Tobar, M., Zambrano Vizuete, M., Torres-Carrión, P., Montes León, S., Pizarro Vásquez, G., Durakovic, B. (eds.) Applied Technologies, pp. 363–371. Springer, Cham (2020)
11. Wang, X., Lai, C., Yu, D., Xu, Y., He, Y.: Application of energy performance contracting in distribution transformer based on neural network prediction. In: 2019 IEEE 3rd Information Technology, Networking, Electronic and Automation Control Conference (ITNEC), pp. 42–46, March 2019
12. Petkovic, I., Balaban, N.: Detecting defaulters for consumed electric energy with neural network clustering. In: 2008 6th International Symposium on Intelligent Systems and Informatics, pp. 1–4, September 2008
13. Rezaei, R., Sharghi, A., Motalebi, G.: A framework for analysis affecting behavioral factors of residential buildings' occupant in energy consumption. J. Sustain. Archit. Urban Des. **5**(2), 39–58 (2018). http://jsaud.sru.ac.ir/article_895.html%0A, http://jsaud.sru.ac.ir/pdf_895_9f6e3441f7399851f2185e37696ed98e.html
14. Johnson, B.J., Starke, M.R., Abdelaziz, O.A., Jackson, R.K., Tolbert, L.M.: A method for modeling household occupant behavior to simulate residential energy consumption. In: IEEE PES Innovative Smart Grid Technologies Conference, ISGT 2014 (2014)

15. Zhai, Y., Song, W., Liu, X., Liu, L., Zhao, X.: A chi-square statistics based feature selection method in text classification. In: 2018 IEEE 9th International Conference on Software Engineering and Service Science (ICSESS), pp. 160–163, November 2018
16. Peker, M., Arslan, A., Şen, B., Çelebi, F.V., But, A.: A novel hybrid method for determining the depth of anesthesia level: combining relieff feature selection and random forest algorithm (relieff+rf). In: 2015 International Symposium on Innovations in Intelligent SysTems and Applications (INISTA), pp. 1–8, September 2015

Using Multivariate Time Series Data via Long-Short Term Memory Network for Temperature Forecasting

Jorge L. Charco(✉), Telmo Roque-Colt, Kevin Egas-Arizala, Charles M. Pérez-Espinoza, and Angélica Cruz-Chóez

Universidad de Guayaquil, Delta and Kennedy Av., P.B. EC090514 Guayaquil, Ecuador
{jorge.charcoa,telmo.roquec,kevin.egasa,charles.pereze, angelica.cruzc}@ug.edu.ec

Abstract. This paper presents a Long-Short Term Memory, as a special kind of Recurrent Neural Network, capable of learning long-term dependencies to estimate the temperature. In order to improve the performance of the proposed model, a multivariate time series is considered. A Multivariate time series has more than one time-dependent variable. Each variable depends not only on its past values but also has some dependency on other variables (i.e., temperature, humidity, atmosphere pressure, wind speed, cloud cover and among others). A pre-processing is applied to dataset provided by the National Institute of Meteorology and Hydrology to obtain daily weather variables. The proposed model is compared with a previous approach using standard metrics, obtaining better results on the average temperature in Guayaquil city.

Keywords: Recurrent neural networks · Temperature · Prediction · Long short-term memory · Multivariate time series

1 Introduction

Climate change is currently one of the main problems of concern to humanity due to the environmental impacts by the increase in greenhouse gases, which are the results of human activities. Temperature is one of the main meteorological variables, which is related to climatic events and it is used to explain the state of the atmosphere. The increase in global temperature caused by the greenhouse effect is the main responsible for increasing in sea level, the decreasing in snow and ice cover, as well as the change in the trend of precipitation of water [7,11,13]. The change in climate variables have led to the development of mathematical physical models to get patterns that allow predicting these variables and helping government and institutions to be prepared to avoid economic losses and humans. In Ecuador, rising temperatures have affected activities such

M. Botto-Tobar et al. (Eds.): ICCIS 2020, AISC 1273, pp. 38–47, 2021.
https://doi.org/10.1007/978-3-030-59194-6_4

as agriculture, health and food security. There are two seasons: the rainy season called winter from January to April and the dry season known as summer from May to December. However, the climate in Ecuador changes according to the regions and it not depend of the season i.e., the coast region can be raining but the climate is hot. The climate varies all year round due it is influenced by two types of Pacific currents: a cold current known as the Humboldt current and a hot current known as El Niño. One of these cities of the coast region is Guayaquil, where different persons depend of weather state to plan their activities. It is one of the most populous cities in Ecuador, with different commercial activities such as: tourism, agriculture, livestock farming, forestry, fisheries and among other.

Different techniques for climate prediction have been applied, specifically about temperature, such as: mathematical models, support vector machine, bayesian network, decision tree algorithms and among other; obtaining each time better results. However, the accuracy of results depend of amount of data available and the level of complexity involved in the methods. Many years of information are necessary to obtain better results. Furthermore, the temperature changes significantly due to the increase in greenhouse gases, making its prediction more complicated.

The technological advances and deep learning techniques have allowed development new algorithms to predict the temperature, shifting from classical approaches to new models based on artificial intelligence. Recurrent Neural Network (RNN) is a class of artificial neural networks for time series data analysis that considers the dimension of time to get better results. It has been used in some sequence analysis approaches such as text, sound or video processing [2,5,14]. An extension of the RNN is Long Short-Term Memory (LSTM), which has memory to store relevant information previous to improve their results [6,12].

In the current work, LSTM architecture is used for temperature prediction using dataset provided by the National Institute of Meteorology and Hydrology (INAMHI). This architecture is trained using dataset, which contains information of the temperature (e.g., minimum, maximum, atmospheric pressure and humidity) and it is taken each five minutes during a year. The remainder of the paper is organized as follows. In Sect. 2 previous works are summarized; then, in Sect. 3 the proposed approach is detailed together with a description of the used dataset. Experimental results are reported in Sect. 4 and comparisons with a previous approach are also presented. Finally, conclusions and future works are given in Sect. 5.

2 Related Work

During the last years, artificial intelligence algorithms have been used in text, sound or video processing tasks, due to their capability for features extraction and time series data analysis. In that direction, some works have been proposed. In [10], the authors propose a Support Vector Machine (SVM) algorithm for weather prediction by using time series data of the daily maximum temperature and it is used to obtain the maximum temperature of the next day, considering n

previous days (between 2 to 10 days) referred to as order of the input. The SVM is trained by using Non-linear regression method. In [3], the authors introduce Bayesian Networks to model the spatial and temporal dependencies among the different stations using a directed acyclic graph, allowing to obtain a probabilistic model consistent with all the available information and whose obtained model is combined with numerical atmospheric predictions. Other approach has been in [4], where the authors present a hybrid approach to combine discriminatively trained predictive models with a deep neural network for space-time inferences about climate. Similarity function is used as kernel based to learn automatically from the data to ensure that the inferences respect natural laws. Their aim is jointly predicted temperature, pressure, dew point and winds across space and time. Integrated Global Radiosonde Archive dataset is used to evaluate proposed approach. This dataset contains historical weather data spanning decades compiled from eleven source datasets. In [8], the authors propose models based on Deep Belief Network and Restricted Boltzmann Machine for the weather parameters prediction such as temperature, pressure and humidity. An unsupervised manner is applied for each proposed model to learn non-linear hierarchical features from the dataset, and then, the models are retrained in supervised manner with an output layer to predict the appropriate output. The mean squared error is used as loss function for training and testing of the proposed models.

In [1], the authors present an effective method based on Decision Tree (DT) to predict climate phenomena such as total cold, total heat and snowfall. The DT data mining techniques and algorithms are used to classify meteorological parameters such as maximum temperature and minimum temperature that are mainly responsible for the weather prediction. The dataset has been collected from meteorologist's centre. The case data cover the period of 2012 to 2015 from different cities. In [15], the authors propose to use Recurrent Neural Network, specifically, Long Short-Term Memory (LSTM) for Sea Surface Temperature (SST) prediction. The SST problem is formulated as a time series regression problem. The LSTM layer is used to model the time series relationship; and the fully-connected layer is used to obtain the final prediction. NOAA High resolution dataset is used to train the proposed model. This dataset contains daily values from September 1981 to November 2016. The root mean squared error is used as evaluation metric to measure the obtained prediction. Other approach is presented in [9], where the authors propose a comparison between MultiLayer Perceptron (MLP) and LSTM models. The dataset is provided by INAMHI. It contains 8.397 values corresponding to the minimum and maximum temperature from February 1992 to November 2015. These values have been taken once a day. The MLP model is trained with different parameters until an acceptable estimation error, while the LSTM model is trained with different number of nodes (e.g., between 50 and 70 hidden nodes) to obtain better results. The Euclidean estimation standard is used to evaluate the proposed models.

3 Proposed Approach

This section details the main contribution of current work. LSTM architecture proposed to predict the minimum and maximum temperature based on Multivariate Time Series data analysis, is presented. The network takes as input temperature, atmospheric pressure and humidity; including previous days (e.g., 10 days) to predict the temperature of the next day (see Fig. 1).

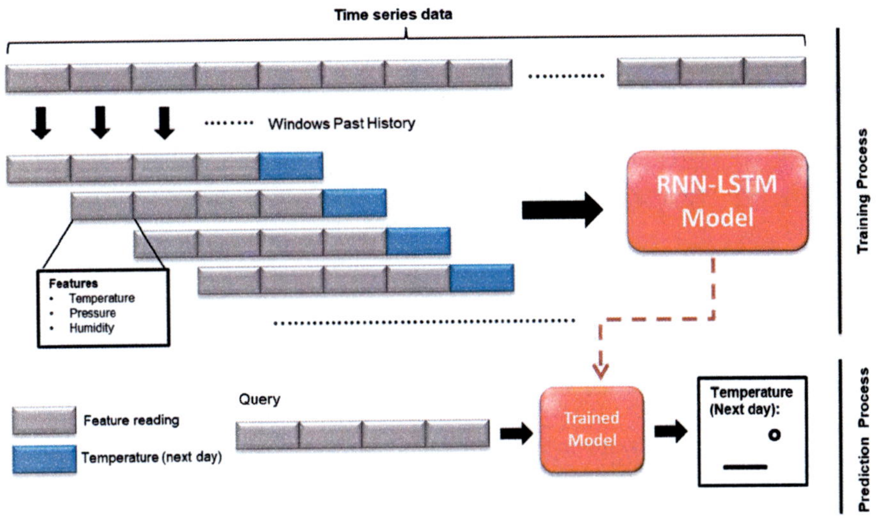

Fig. 1. Training and prediction process

3.1 Dataset

In order to train the proposed model a large dataset is necessary. In the current work temperature time series data provided by INAMHI, has been used. Temperature and humidity values are obtained by using *Pt100 RTD 1/3 Class B IEC 751* sensor. This device allows a measurement range from 0 a 100% RH and from -40 to +60 °C. For atmospheric pressure values, class B sensor is used, with pressure range from 500 to 1100 hPa, precision of ±0.1 hPa and temperature range from -20 to +60 °C.

The weather station M1271 captures information mentioned above. It is located in the Faculty of Natural Sciences within the University of Guayaquil (see Fig. 2). The climate variables are captured by the weather station (e.g., temperature, humidity, atmospheric pressure and speed winds). Only temperature, humidity and atmospheric pressure values are used by the proposed model. The dataset is composed with 103948 values. These values were taken each 5 min throughout the year 2019 (i.e., from January to December) using the weather station mentioned previously. In Table 1, twelve records are presented.

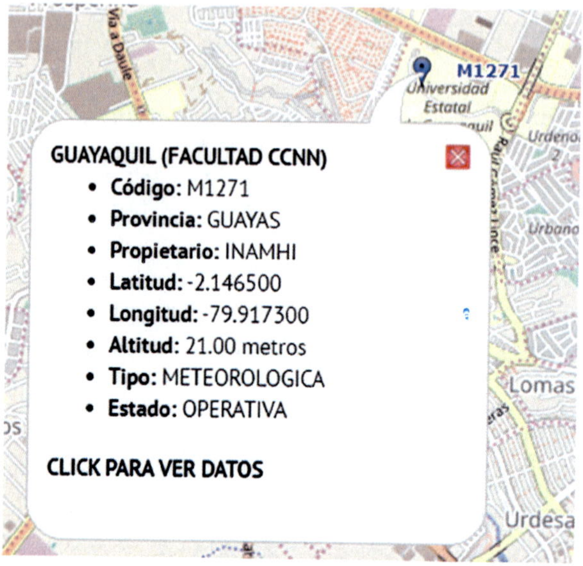

Fig. 2. Location of weather station M1271 in Guayaquil City.

Table 1. Data captured each 5 min throughout a hour from the weather station M1271

Date	Temperature (°C)		Humidity (%)		Atmospheric pressure (hPa)	
	Min	Max	Min	Max	Min	Max
2019-12-01 01:00:00	24.20	24.32	83.78	83.99	1009.11	1009.18
2019-12-01 01:05:00	24.19	24.31	83.61	84.12	1009.13	1009.26
2019-12-01 01:10:00	24.07	24.25	83.81	84.31	1009.22	1009.34
2019-12-01 01:15:00	24.06	24.25	84.20	84.51	1009.27	1009.39
2019-12-01 01:20:00	24.19	24.29	83.97	84.44	1009.33	1009.45
2019-12-01 01:25:00	24.06	24.22	83.97	84.53	1009.40	1009.53
2019-12-01 01:30:00	24.10	24.21	84.04	84.51	1009.48	1009.61
2019-12-01 01:35:00	24.12	24.21	84.19	84.48	1009.58	1009.68
2019-12-01 01:40:00	24.01	24.21	84.23	84.57	1009.62	1009.71
2019-12-01 01:45:00	24.02	24.15	84.30	84.74	1009.62	1009.80
2019-12-01 01:50:00	23.99	24.15	84.59	84.84	1009.74	1009.82
2019-12-01 01:55:00	24.06	24.19	84.30	84.67	1009.77	1009.90

Considering the table mentioned above, one day has 288 observations with maximum and minimum values of temperature, humidity and atmospheric pressure. In order to obtain a new dataset with daily maximum and minimum values of weather variables (i.e., temperature, humidity and atmospheric pressure) is necessary to realize a pre-processing of the dataset. This process is to choose the maximum and minimum value of temperature among 288 available observations. With process previous, humidity and atmospheric pressure values are chosen based on temperature, since there is a relationship between them. In Table 2, ten daily observations of new dataset are showed.

Table 2. Daily weather variables values obtained as result of pre-processing

Date	Temperature (°C)		Humidity (%)		Atmospheric pressure (hPa)	
	Min	Max	Min	Max	Min	Max
2019-01-16	23.97	32.70	67.88	96.35	1007.81	1009.61
2019-01-17	22.79	31.59	71.48	96.37	1010.37	1012.19
2019-01-18	24.35	31.89	64.52	87.61	1010.08	1010.44
2019-01-19	22.79	31.42	63.72	97.18	1009.14	1009.32
2019-01-20	23.65	30.40	68.35	97.06	1007.91	1008.83
2019-01-21	24.07	34.76	55.62	89.58	1006.03	1009.28
2019-01-22	24.92	33.08	62.62	90.50	1005.85	1008.70
2019-01-23	25.08	31.49	65.73	82.70	1004.82	1008.15
2019-01-24	25.43	31.97	67.37	93.36	1007.15	1007.41
2019-01-25	24.43	32.82	62.86	89.72	1006.55	1009.41

As result of pre-processing, the new dataset contains 365 observations of weather variables (i.e., temperature, humidity and atmospheric pressure values).

3.2 Network Architecture

The proposed approach is a RNN-LSTM, which contains two LSTM layers and a Dense layer. The first one, with 32 internal units, hyperbolic tangent as activation function, including return sequences to stack multiple hidden LSTM layers. The second one, contains 16 internal units and ReLU as activation function. Finally, the output of the last LSTM layer is used to feed a Dense Layer to predict the temperature for the next day (see Fig. 3).

Mean Absolute Error (MAE) is used as loss function to measure the effectiveness of proposed method, since it is more robust when the training data is corrupted with outliers. It is defined as follows:

$$MAE = \frac{\sum_{i=1}^{n} |y_i - \hat{y}_i|}{n}. \quad (1)$$

where y_i is temperature value, $\hat{y}_i$ is predicted temperature value of proposed model and n is batch size for training process.

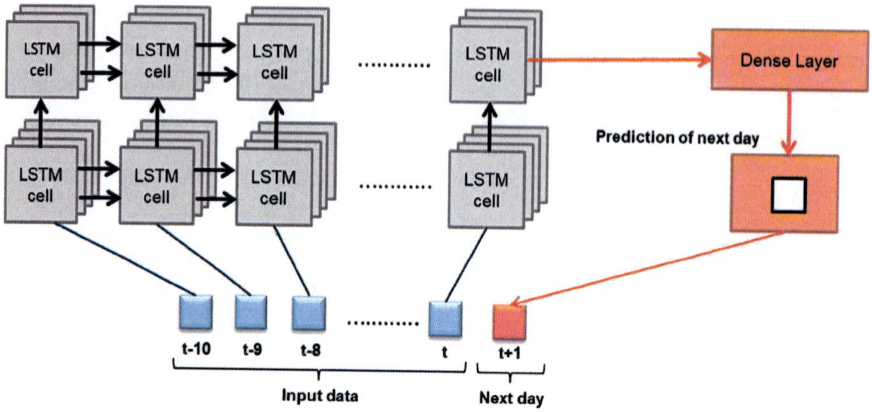

Fig. 3. RNN-LSTM is fed with time series data of last ten days, which correspond to temperature atmospheric pressure and humidity values to predict the temperature of next day.

4 Experimental Results

As mentioned above, a RNN-LSTM based architecture is proposed for temperature estimation. A dataset of daily weather variables (i.e., temperature, humidity and atmospheric pressure values) is used as input to the model. This input has a shape of WxF, where W is Windows Past History (WPH) and F corresponds to the weather variables. The WPH is the number of days, which are used as input to the proposed model. Each day of WPH contains three weather variables, which are embedding to train the architecture.

The proposed approach was implemented with the TensorFlow and trained with Intel Core I5 2.5 GHz CPU. RMSProp optimizer is used to train the network with a learning rate of 10^{-3} during 100 epochs. The training process took approximately about 3 hours using a batch of 282.

The network architecture was trained on daily weather variables dataset (see Table 2), obtained after of pre-processing of original dataset. The training set is composed of 282 observations, embedding WPH and weather features (see Fig. 4), which were used to feed to the network; and the validation set is composed of 83 observations to validate the learning process.

4.1 Results

In this section, experimental results obtained with the proposed model are presented. Additionally, the obtained results are compared with a state-of-the-art RNN-based method [9] on weather variables dataset provided by INAMHI. Different quantitative metrics have been used to evaluate the proposed approach. One of the metrics is percentage error. In each month, this metric is applied to measure the difference between the estimated temperature and ground truth.

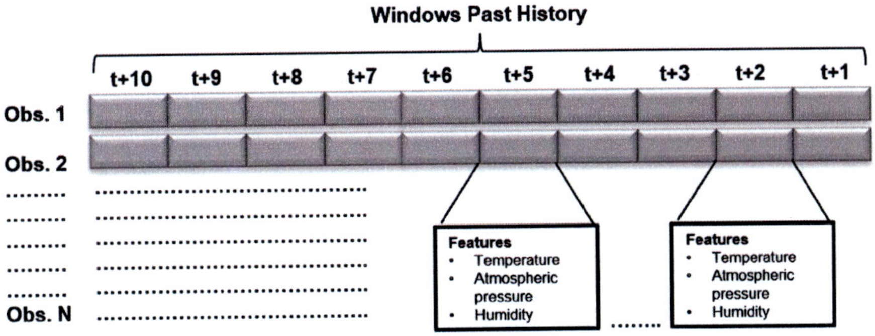

Fig. 4. Data structure for training and validation process

Table 3. Comparison of average temperature error of each month between our model (TempNet) and proposed model by [9].

Month	Real Temperature (°C)			Our model TempNet (°C)			Proposed model by [9]	Error TempNet	Error [9]
	Min	Max	Avg	Min	Max	Avg	Avg	Avg	Avg
January	24.09	31.54	27.81	24.23	32.20	28.22	27.51	1.46%	**1.10%**
February	24.26	32.23	28.24	24.11	33.51	28.81	27.14	**2.00%**	3.91%
March	24.27	32.71	28.49	23.93	33.37	28.65	26.89	**0.56%**	5.62%
April	23.91	33.52	28.71	23.59	33.31	28.45	27.53	**0.92%**	4.13%
May	24.01	32.72	28.36	23.44	32.39	27.92	29.68	**1.57%**	4.64%
June	22.81	30.76	26.78	22.29	31.00	26.65	28.52	**0.50%**	6.48%
July	21.53	31.16	26.34	20.93	32.03	26.48	29.20	**0.51%**	10.84%
August	20.24	31.41	25.82	19.82	31.28	25.55	29.23	**1.06%**	13.18%
September	20.20	30.51	25.35	21.19	31.18	26.19	29.69	**3.29%**	17.10%
October	21.45	30.59	26.02	20.94	30.86	25.90	29.07	**0.46%**	11.72%
November	22.79	31.83	27.31	22.98	32.30	27.64	26.80	**1.21%**	1.87%
December	23.48	33.65	28.56	23.56	35.61	29.59	28.15	**3.59%**	1.45%
Mean error estimation								**1.43%**	6.84%

Additionally, mean error is used to evaluate the performance of the proposed approach over all months of the year, and they are depicted on Table 3.

Minimum and maximum temperature of each month of the year were obtained by our trained network, and for a fair comparison with proposed model by [9], the average was also calculated. The proposed approach obtains more accurate results. The average temperature error obtained by our model (TempNet) improves the results of proposed model by [9] in about 378%. The results with large errors are generated by other weather variables such as wind speed, UV rays and among others, which are not considered to train the proposed model. Fig. 5 show comparisons of the results obtained by our proposed model

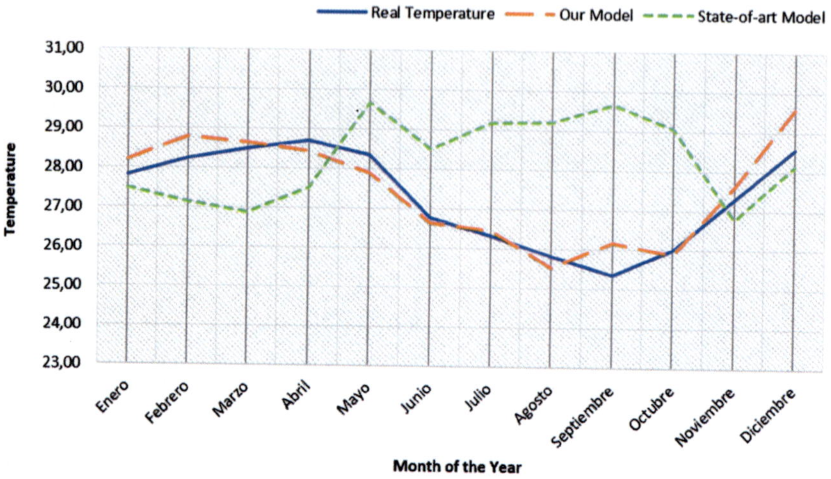

Fig. 5. Comparison of temperature average error between proposed approach and approach proposed by [9] on dataset provided by INAMHI.

with respect to the average temperature for each month of the year and the proposed approach by [9].

5 Conclusion

This paper addresses the challenging problem of estimating the temperature in Guayaquil city. A RNN based architecture is proposed to accurately estimate the temperature. The manuscript shows how features extracted from daily weather variables using LSTM and multivariate time series can help the estimation of temperature. Experimental results show that the proposed model helps to reduce error in the obtained results (average temperature). Future work will be focused on extending the usage of weather dataset, by increasing the dataset with other weather variables such as wind speed, UV rays and among others. Additionally, different models based on RNNs will be considered.

Acknowledgements. This work has been partially supported by the National Institute of Meteorology and Hydrology - Ecuador, whose weather variables dataset has been provided by them.

References

1. Bhatkande, S.S., Hubballi, R.G.: Weather prediction based on decision tree algorithm using data mining techniques. Int. J. Adv. Res. Comput. Commun. Eng. **5**(5), 483–487 (2016)
2. Cakır, E., Virtanen, T.: Convolutional recurrent neural networks for rare sound event detection. Deep Neural Networks for Sound Event Detection **12** (2019)

3. Cofino, A.S., Cano Trueba, R., Sordo, C.M., Gutiérrez Llorente, J.M.: Bayesian networks for probabilistic weather prediction (2002)
4. Grover, A., Kapoor, A., Horvitz, E.: A deep hybrid model for weather forecasting. In: Proceedings of the 21th ACM SIGKDD International Conference on Knowledge Discovery and Data Mining, pp. 379–386 (2015)
5. Hang, R., Liu, Q., Hong, D., Ghamisi, P.: Cascaded recurrent neural networks for hyperspectral image classification. IEEE Trans. Geosci. Remote Sens. **57**(8), 5384–5394 (2019)
6. Karim, F., Majumdar, S., Darabi, H., Harford, S.: Multivariate lstm-fcns for time series classification. Neural Netw. **116**, 237–245 (2019)
7. Liu, D., Guo, X., Xiao, B.: What causes growth of global greenhouse gas emissions? Evidence from 40 countries. Sci. Total Environ. **661**, 750–766 (2019)
8. Narejo, S., Pasero, E.: Meteonowcasting using deep learning architecture. Int. J. Adv. Comput. Sci. Appl. 8 (2017)
9. Pérez-Espinoza, C.M., Sanchez-Guerrero, J., Samaniego-Cobos, T., Beltran-Robayo, N.: Comparison between two deep learning models for temperature prediction at guayaquil. In: International Conference on Technologies and Innovation, pp. 17–29. Springer (2019)
10. Radhika, Y., Shashi, M.: Atmospheric temperature prediction using support vector machines. Int. J. Comput. Theory Eng. **1**(1), 55 (2009)
11. Seneviratne, S.I., Donat, M.G., Mueller, B., Alexander, L.V.: No pause in the increase of hot temperature extremes. Nat. Clim. Change **4**(3), 161 (2014)
12. Sherstinsky, A.: Fundamentals of recurrent neural network (rnn) and long short-term memory (lstm) network. Physica D: Nonlinear Phenomena **404**, 132306 (2020)
13. Thacker, I., Sinatra, G.M.: Visualizing the greenhouse effect: restructuring mental models of climate change through a guided online simulation. Educ. Sci. **9**(1), 14 (2019)
14. Wang, R., Li, Z., Cao, J., Chen, T., Wang, L.: Convolutional recurrent neural networks for text classification. In: 2019 International Joint Conference on Neural Networks (IJCNN), pp. 1–6. IEEE (2019)
15. Zhang, Q., Wang, H., Dong, J., Zhong, G., Sun, X.: Prediction of sea surface temperature using long short-term memory. IEEE Geosci. Remote Sens. Lett. **14**(10), 1745–1749 (2017)

Failure Detection in Induction Motors Using Non-supervised Machine Learning Algorithms

David S. Toscano and Enrique V. Carrera(✉)

Departamento de Eléctrica, Electrónica y Telecomunicaciones Universidad de las Fuerzas Armadas ESPE, Sangolquí 171103, Ecuador
{dstoscano,evcarrera}@espe.edu.ec

Abstract. Induction motors are widely used around the world in many industrial and commercial applications. Early detection of faults in these devices is important to avoid service disruption and increase their useful life. Thus, many non-invasive schemes have been proposed to detect failures in induction motors using machine learning techniques mainly. Many of these machine learning algorithms are based on supervised training, requiring examples of motors with and without failures. Collecting training data with positive and negative examples is certainly a time consuming and costly task at the field. Hence, this work proposes to use non-supervised training algorithms to detect failures in induction motors using exclusively measurements in healthy motors. Specifically, two non-supervised learning algorithms, namely k-means and self-organizing maps, are employed to detect broken bars and faulty bearings. Final results are compared to supervised training algorithms, showing that non-supervised learning algorithms are also able to detect motor failures with high accuracy, and probing that robust models can be built using only healthy-motor data.

Keywords: Failure detection · Induction motors · Machine learning algorithms · Non-supervised learning.

1 Introduction

Induction motors are widely used in many industrial and commercial applications since they are self-starting, reliable and economical. These motors consume around 60% of the electrical energy produced in the world [2], given that they can be found in industrial compressors, medical pumps, and even household appliances such as fans. Like any electromechanical device, a induction motor can present failures during your normally long life time. In general, a failure produces service disruption and reduces the efficiency of the motor, shorting its useful lifetime [12].

M. Botto-Tobar et al. (Eds.): ICCIS 2020, AISC 1273, pp. 48–59, 2021.
https://doi.org/10.1007/978-3-030-59194-6_5

In many cases, a failure is only detected when the motor stops working. Thus, in order to allow us an early detection of failures in induction motors, many non-invasive schemes have been proposed and studied [8]. Most of these schemes rely on the measurement of electrical current, voltage, power, mechanical vibration and/or temperature. Using these measurements, mathematical models can be built to detect specific motor failures. In particular, machine learning techniques have been used to classify these failures [8]. However, the most popular machine learning algorithms are based on a supervised training, requiring positive and negative examples.

Nonetheless, the measurement of some electrical or mechanical variables in operative motors at the field is a difficult task, since most of these motors do not present failures or their failures have not been identified yet. In other words, the process of failure detection in the industry cannot depend on having measurements for identified faulty motors [6]. In fact, practical motor failure detection can be seen as the analysis of historical data in search of behavioral changes that can be detected using the expertise of the maintenance people, without knowing exactly what the behavior of a faulty motor is.

Based on that, this work proposes to create a mathematical model that relies exclusively on data from motors without failures. For that, non-supervised machine learning techniques [10] will be used to discriminate between healthy and faulty motors. The goal is to build models for failure detection using data from the same motors that are being monitored. In particular, this paper employs two non-supervised learning algorithms, namely k-means and self-organizing maps (SOM) [10], to detect broken bars and faulty bearings. The results of these non-supervised training algorithms are also compared with a supervised algorithm, namely artificial neural networks, that is our current baseline.

Experimental results show that non-supervised learning machine algorithms based on SOM are successful detecting motor failures, but there are some design criteria that needs to be replaced for faulty motors in order to propose a model fully based on healthy motors, and then to be ready for a field implementation.

The rest of this paper is organized as follows. Section 2 introduces the main failures in induction motors and some common detection techniques. Section 3 presents the testbed utilized to obtaining our training data. The main results and their implications are analyzed in Sect. 4. Finally, Sect. 5 concludes this paper and discuss some future works.

2 Materials and Methods

2.1 Failures in Induction Motors

There are several types of failures in induction motors. In general terms, motor failures can be classified in [14]:

- Failures related to the stator (38%)
- Failures related to the rotor (10%)
- Failures related to the bearings (40%)
- Failures related to other components (12%)

Stator related failures involve short circuits in its windings. It has been shown that these faults can be detected successfully from voltage and current signals [13]. On the other hand, rotor failures are normally associated to the breaking of one or more bars. The lack of electrical continuity in a bar produces a variation in the magnetic flux, which in turn produces changes in the electrical current of the stator [11].

A broken bar causes inefficiencies in the motor, increasing the starting current and decreasing its power factor [5]. In fact, if the number of broken bars increases, the motor could not start due to lack of torque [11]. This type of failure also produces the appearance of certain frequencies at the stator current. The new frequencies are defined according to the equation [1]:

$$f_r = f_s \left[\frac{k}{p}(1-s) \pm s \right]$$

where f_s is the electrical supply frequency, $k/p = 1, 5, 7, 11, 13$, etc. is the normal winding configuration, and s is the per-unit slip [15].

In addition, bearing failures are related to the breakage of one or more of their sub-components, which can be the outer race, inner race, cage or balls. When one of these sub-components fails, mechanical vibrations are produced generating an increment in the number of frequency components at the stator current. The mechanical vibrations associated to failures in the bearings produce the frequencies [16]:

$$f_i = \frac{n f_e}{2} \left[1 + \frac{d}{D} \cos \alpha \right]$$

$$f_o = \frac{n f_e}{2} \left[1 - \frac{d}{D} \cos \alpha \right]$$

$$f_b = \frac{D f_e}{d} \left[1 - \frac{d^2}{D^2} \cos^2 \alpha \right]$$

where f_i is the frequency related to a failure in the bearing inner race, f_o is the frequency related to a failure in the bearing outer race, f_b is the frequency related to a failure in the bearing balls, D is the diameter of the cage, f_e is the speed of the axis in Hz, d is the diameter of the balls, α is the contact angle between the balls and the races, and n is the number of balls in the cage.

As we can see, these equations need specific characteristics of the bearing geometry [16]. Furthermore, the mechanical vibrations caused by these faults influence the stator current at frequencies $fv = |f_s \pm k f_j|$, where f_j is any of the previous frequencies f_i, f_o or f_b, with $k = 1, 2, 3$, etc.

All of the physical variables mentioned in this section do not involve any intrusive measurement, although the measurement of mechanical vibrations implies an extra effort and cost when compared to the measurement of electrical voltages and currents, since these sensors must be next to the motor and must be installed with the corresponding wiring. On the other hand, electrical currents and voltages are variables that can be obtained from the industrial boards. Based on this, the present study detects motor failures using electrical currents basically.

2.2 Fault Detection in Induction Motors

There are different failure detection methods for induction motors. Most of them can be classified into 3 categories: model based, signal based, and knowledge based [8], although we could also consider hybrid techniques such as the one described in [7].

The model-based methods compare a value measured in the field with a value obtained through a mathematical model. Their main difficulty is to model the system correctly. On the other hand, the signal-based methods use frequency and/or time analysis techniques to estimate abnormal behaviors based on historical data. Finally, the knowledge-based methods use machine learning or symbolic intelligence to carry out automatic classification/detection.

2.3 Non-supervised Training Algorithms

Non-supervised machine learning algorithms basically perform clustering [10]. They produce as result of their training a set of centroids corresponding to the behavior of currents in healthy motors. It is possible to identify a failure comparing the position where appears a new motor with respect to the previous centroids. If this distance is larger than a predefined safe radius, the motor has a failure. The selected radius has direct influence in the values of sensitivity and specificity, and its value depends on the utilized non-supervised training algorithm.

- The k-means algorithms has been applied to abnormality detection before [9]. In this case, the predefined radius is defined as the average distance of the training samples associated to each centroid and multiplied by a parameter M.
- SOM has also been applied to abnormality detection before [3]. In this case, the predefined radius is the maximum distance of the training samples associated to each centroid and increased by a given *margin*. However, this maximum distance is limited to a maximum value that is also input of the algorithm and called *threshold*.

We can see that the performance of the k-means model only depends on the parameter M, while the SOM model depends on the *margin* and *threshold*. These values are tuned using faulty motor data, since the sensitivity and specificity of the algorithms is conditioned by M, *margin*, and *threshold*. If we do not have faulty-motor data, we can analyze specificity, but not sensitivity.

3 Testbed and Configurations

3.1 Data Acquisition

For data acquisition, the testbed includes a new ABB motor, model 3GQA 082302-AXA. The three-phase motor is mechanically coupled to a DC generator

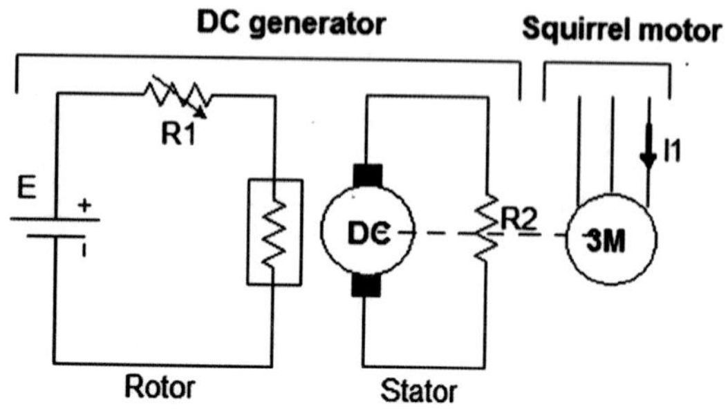

Fig. 1. Testbed used for data acquisition, where R_2 is the resistive load and R_1 is a variable resistance to control the voltage connected to the load.

connected to a resistive load where the rotor current I_1 is changed from 0 to a maximum value in order to obtain data for a wide per-unit slip range. The connection scheme is shown in Fig. 1.

In order to generate failures, an extra rotor, identical to the original but with two consecutive bars drilled, obtained from other 3GQA 082302-AXA motor, is used to emulate broken bars. An extra bearing, 6204-2RSH, whose outer race has been longitudinally cut in one side is also used to emulate faulty bearings.

In all the cases, the electrical current of a single phase (I_1) is measured with a sampling frequency of 50 kHz during 2 seconds. This experiment is repeated until having a total of 600 samples for the healthy motor. In the same way, other 600 samples are obtained for the motor with a defective rotor, and 600 samples with the faulty bearing motor.

3.2 Feature Extraction

In order to train the machine learning algorithms, the electrical current measurements are transformed into some useful features. For that, previous knowledge about the nature of the failures and their influence on the spectrum components is applied.

Broken Bars. First, the frequencies of interest must be defined according to the value of f_r which depends on the value of k/p. In fact, working with all the possible values is not a good alternative, since that deteriorates the performance of the algorithm and increases its computational cost. Quick tests were performed to select the most applicable values for k/p.

The parameters for these quick tests are summarized in Table 1, and the corresponding performance results (*i.e.*, sensitivity) are shown in Fig. 2. From Table 1 and Fig. 2, cases B, C, D can be highlighted and $k/p = 5$ will be used along this paper as the band of interest. Furthermore, lower results can be seen in case A because the frequency band is too close to the fundamental frequency f_s. This is the main reason why values greater than 1 should be chosen for k/p.

Table 1. Frequency bands of interest when broken bars are characterized.

Study case	Used values	Frequency band
A	$f_s = 60$ Hz, $s = 0.08$, $k/p = 1$	50.4–60 Hz
B	$f_s = 60$ Hz, $s = 0.08$, $k/p = 5$	271.2–300 Hz
C	$f_s = 60$ Hz, $s = 0.08$, $k/p = 7$	381.6–420 Hz
D	$f_s = 60$ Hz, $s = 0.08$, $k/p = 11$	602.4–660 Hz

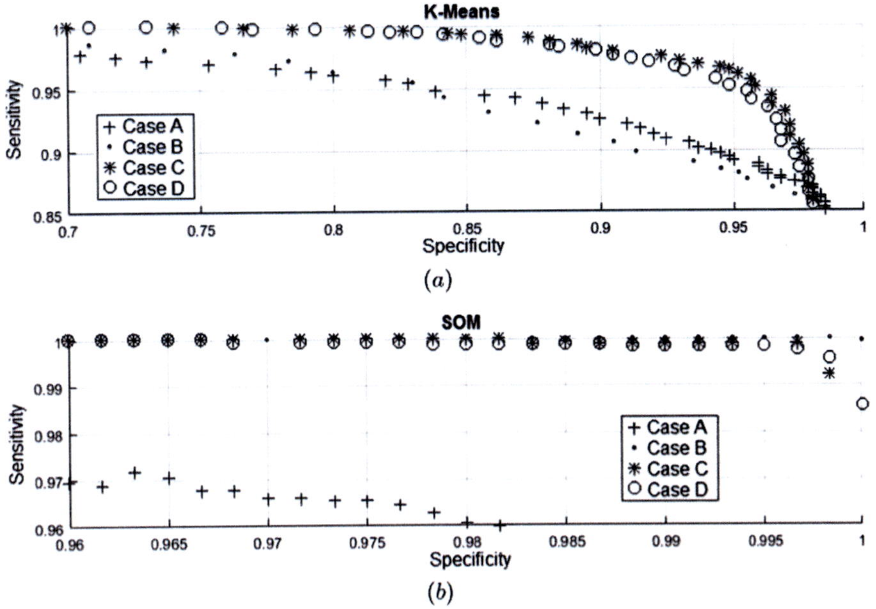

Fig. 2. Performance of basic versions of the failure detection algorithm based on (*a*) *k*-means, and (*b*) SOM for the cases specified in Table 1.

Finally, from the Fourier transform, only the coefficients that are within the frequency band of interest in Table 1 are important. With these specific values, an array is created to compute its average and their four highest values in it. This operation allows us to reduce the number of inputs to the machine learning algorithm.

Faulty Bearings. The Fourier transform has proved to be inefficient to detect faulty bearings when using current signals, but the energy of the nodes in a discrete wavelet decomposition has shown to be successful. In previous studies, a decomposition that reaches 3.9 Hz [16] or 7.5 Hz [4] has been successful. Thus, a decomposition that satisfies nodes with 3.05 Hz of resolution is proposed for this study.

Table 2. Specification of the motor bearings.

Variable	Value	Description
f_s	60 Hz	Supply frequency
n	8	Number of bearing balls
D	33.7 mm	Diameter of the cage
d	7.94 mm	Diameter of the balls
p	4	Number of poles
α	0	Contact angle of the failure
s	0–0.06	Per-unit slip
k	± 1	Variable associated to armonics

Table 3. Nodes of interest from the wavelet decomposition.

Number	Nodes	k	Frequencies
1	(13,8)	–1	24.41–27.47 Hz
2	(13,9)	–1	27.47–30.52 Hz
3	(13,10)	–1	30.52–33.57 Hz
4	(13,47)	+1	143.43–146.48 Hz
5	(13,48)	+1	146.48–149.54 Hz
6	(13,49)	+1	149.54–152.59 Hz

Consequently, it is proposed to use 6 nodes according to the bearing data presented in Table 2. Hence, the resulting bands of interest are from 26.2 to 31.7 Hz and from 146.2 to 151.7 Hz, whereby the nodes of interest are defined considering a wavelet decomposition of 6 levels, which are detailed in Table 3.

The choice of the mother wavelet is also made by testing different wavelets through basic versions of the algorithms. The results shown in Table 4 define that the reverse bi-orthogonal 3.5 mother wavelet is the best option in this application.

Final Features. The 14 features selected for being used by the machine learning algorithms are detailed in Table 5. From them, 3 features correspond to the time domain: RMS, kurtosis and standard deviation. These are parameters normally used to describe an electrical signal. The other parameters are selected because of the specific spectral conditions explained above.

4 Main Results

There are some parameters that need to be tuned in order to get optimal performance in the motor failure detection. These parameters depend on the type

Table 4. Specificity, in percent, for different mother wavelets (considering a sensitivity of 100%).

Wavelet	SOM	*k*-means	Wavelet	SOM	*k*-means	Wavelet	SOM	*k*-means
Db1	86.8	9.2	Fk8	38.2	7.0	Rbio1.1	86.5	8.2
Db2	69.7	11.0	Bior1.1	87.5	8.7	Rbio1.3	95.8	24.5
Db3	60.8	17.3	Bior1.3	30.0	4.7	Rbio1.5	82.8	18.8
Db4	59.7	9.8	Bior1.5	19.5	3.5	Rbio2.2	77.0	11.3
Db5	23.3	2.7	Bior2.2	66.0	8.5	Rbio2.4	98.8	25.2
Coif1	63.7	13.0	Bior2.4	52.5	7.0	Rbio2.8	83.8	14.3
Coif2	19.0	1.3	Bior2.8	39.2	1.8	Rbio3.1	13.2	0
Coif3	11.3	2.8	Bior3.1	22.3	5.2	Rbio3.3	82.0	23.2
Coif4	19.5	2.3	Bior3.3	3.5	1.3	Rbio3.5	99.5	40.3
Sym1	86.3	9.7	Bior3.5	7.7	0	Rbio3.7	98.2	29.0
Sym2	69.8	9.2	Bior3.7	12.7	0.3	Rbio3.9	91.0	20.2
Sym3	68.2	14.2	Bior3.9	17.3	0.7	Rbio4.4	79.2	4.7
Dmey	1.5	0	Bior4.4	68.7	5.2	Rbio5.5	31.3	11.2
Fk4	66.8	8.7	Bior5.5	35.8	3.8	Rbio6.8	78.8	9.3
Fk6	64.7	12.8	Bior6.8	77.3	7.5			

Table 5. Final selected features.

Features	Number of values
RMS of the original signal	1
Kurtosis of the original signal	1
Standard deviation of the original signal	1
Average of selected coefficients in the Fourier transform	1
Highest selected coefficients in the Fourier transform	4
Energy of nodes in the wavelet decomposition	6

of non-supervised algorithm to use. For instance, in the case of *k*-means, the number of centroids and the value of M should be defined. In the case of SOM, the number and geometry of the nodes, and the *threshold* and *margin* values, must be specified.

The selection of these parameters is performed using data from a faulty motor, although the final goal is to omit the requirement of these data in the training phase.

4.1 Neural Network

As our baseline for motor failure detection, a supervised learning algorithm has also been implemented. In this paper, the implemented algorithm is an artificial neural network with 8 processing elements in the hidden layer, a learning rate of 0.0005, and a maximum number of epochs of 100000.

The evaluation of the neural network uses a k-fold cross-validation with $k = 10$, repeating this process at least 5 times to get average and standard deviation values. These results are summarized in Table 6. Note that the neural network only distinguishes between healthy an faulty motors, it does not discriminate between failures.

Table 6. Summary of the performance reached by the neural network.

Parameter	Sensitivity	Specificity
Average	99.75%	99.97%
Standard deviation	0.08%	0.07%

4.2 k-means

Figure 3(a) shows the learning capacity of the non-supervised algorithm for different number of centroids. We can see that 13 centroids maximize the performance of the model. On the other hand, Fig. 3(b) present the sensitivity and specificity of the k-means algorithm for different M values, when the number of centroids is 13. We can see that the sensitivity and specificity are simultaneously maximized when $M = 1.49$.

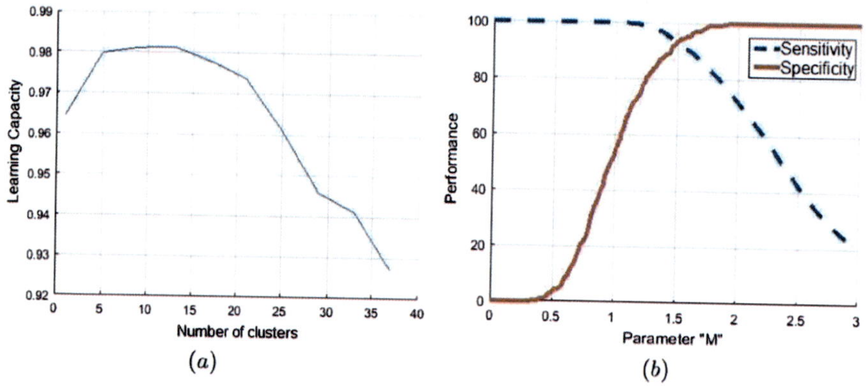

Fig. 3. (a) Learning capacity vs different numbers of clusters with k-means. (b) Sensitivity and specificity vs the parameter M.

In similar way to the neural network, this non-supervised algorithm is evaluated with a k-fold cross-validation ($k = 10$), repeating the process at least 5 times. The average and standard deviation of the obtained results are summarized in Table 7.

Table 7. Summary of the performance reached by k-means.

Parameter	Sensitivity	Specificity
Average	93.29%	92.50%
Standard deviation	0.81%	4.25%

4.3 SOM

The SOM algorithm applied to abnormality detection needs values of the *threshold* and *margin* parameters. The learning capacity of SOM is presented in Fig. 4(a) for 4 different geometries:

- Case A: lineal geometry.
- Case B: $2 \times n$ geometry.
- Case C: $4 \times n$ geometry.
- Case D: $n \times n$ or square geometry.

where n is a number of nodes. From these results, we can establish that $n = 68$ and a linear geometry produce better results. As it is shown in Fig. 4(a), linear and square geometries present the best elasticity, but the first one presents the nearest maximum $n/10$ criteria. Thus, it is recommendable to use a linear geometry with $n/10$ number of clusters.

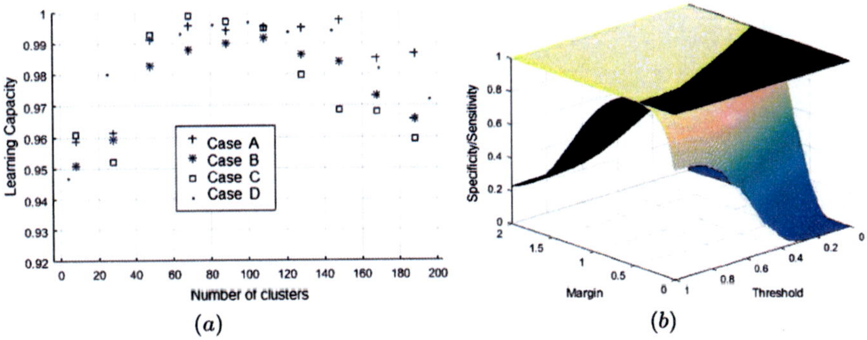

Fig. 4. (a) Learning capacity vs number of clusters for different geometries in SOM. (b) Sensitivity and specificity for different values of threshold and margin.

In addition, the sensitivity and specificity of the model change depending on the *threshold* and *margin* values as shown in Fig. 4(b). We have selected a *threshold* of 0.77 and a *margin* of 0.48 as the best values for these parameters.

In similar way to the previous algorithms, a k-fold cross-validation with $k = 10$ has been repeated at least 5 times to obtain average and standard deviation values. These results are summarized in Table 8.

Table 8. Summary of the performance reached by SOM.

Parameter	Sensitivity	Specificity
Average	99.87%	100%
Standard deviation	0.06%	0%

Finally, Fig. 5 presents two regions: the transient region on the left, where the *threshold* values are effective, and the stationary region on the right, where the *threshold* is too high. All the points in the curves correspond to specificities between 96% and 99% and sensitivity up to 99%. Thus, it should be possible to select, in a general way, the *margin* and *threshold* values without possessing faulty motors.

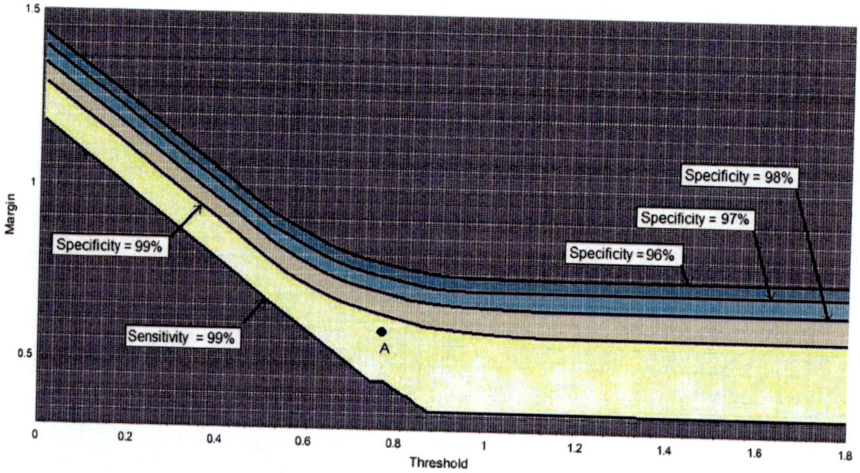

Fig. 5. Margin and threshold values vs performance. Point A shows the parameters tuned with faulty-motor data.

5 Conclusions

The detection of broken bars and faulty bearings in induction motors with a SOM-based anomaly detection method and using electrical current analysis is successful. The performance of the SOM algorithm is similar to the performance obtained by the supervised-learning based method. On the other hand, failure detection based on the k-means algorithm shows lower performance that the other two evaluated algorithms.

Consequently, the construction of a fully designed and trained model based exclusively on healthy-motor data has two main drawbacks when it is going to be implemented in the field. The first drawback is the choice of the mother wavelet, as analyzed in Table 4, since the performance of the model changes dramatically with the choice of different mother wavelets. The second drawback is that faulty-motor data are required to tune the *threshold* and *margin* parameters, although there are some recommendations that could let us to select nearby values to get a high sensitivity and specificity without faulty-motor data, as described previously in Sect. 4.

References

1. Benbouzid, M.E.H.: A review of induction motors signature analysis as a medium for faults detection. IEEE Trans. Ind. Electron. **47**(5), 984–993 (2000)
2. Cusido, J., Romeral, L., Ortega, J.A., Rosero, J., Espinosa, A.G.: Fault detection in induction machines using power spectral density in wavelet decomposition. IEEE Trans. Ind. Electron. **55**(2), 633–643 (2008)
3. Depren, O., Topallar, M., Anarim, E., Ciliz, M.K.: An intelligent intrusion detection system (IDS) for anomaly and misuse detection in computer networks. Expert Syst. Appl. **29**(4), 713–722 (2005). https://doi.org/10.1016/j.eswa.2005.05.002
4. Eren, L., Devaney, M.J.: Bearing damage detection via wavelet packet decomposition of the stator current. IEEE Trans. Instrum. Meas. **53**(2), 431–436 (2004)
5. Faiz, J., Ebrahimi, B.M.: A new pattern for detecting broken rotor bars in induction motors during start-up. IEEE Trans. Magn. **44**(12), 4673–4683 (2008)
6. Guo, A., Yu, D., Du, H., Hu, Y., Yin, Z., Li, H.: Cyber-physical failure detection system: survey and implementation. In: 2016 IEEE International Conference on Cyber Technology in Automation, Control, and Intelligent Systems (CYBER), pp. 428–432. IEEE (2016)
7. He, D., Li, R., Zhu, J.: Plastic bearing fault diagnosis based on a two-step data mining approach. IEEE Trans. Ind. Electron. **60**(8), 3429–3440 (2013). https://doi.org/10.1109/TIE.2012.2192894
8. Ince, T., Kiranyaz, S., Eren, L., Askar, M., Gabbouj, M.: Real-time motor fault detection by 1-d convolutional neural networks. IEEE Trans. Ind. Electron. **63**(11), 7067–7075 (2016)
9. Kang, P., Hwang, S.S., Cho, S.: Continual retraining of keystroke dynamics based authenticator. In: International Conference on Biometrics, pp. 1203–1211. Springer (2007)
10. Kubat, M.: An Introduction to Machine Learning. Springer, Berlin (2017). https://doi.org/10.1007/978-3-030-15729-6
11. Kumar, A., Sinha, M.: A new burg method based approach to mcsa for broken rotor bar detection. In: 2012 IEEE International Conference on Power Electronics, Drives and Energy Systems (PEDES), pp. 1–4. IEEE (2012)
12. Liggins II, M., Hall, D., Llinas, J.: Handbook of Multisensor Data Fusion: Theory and Practice. CRC Press, United State (2017)
13. Sharifi, R., Ebrahimi, M.: Detection of stator winding faults in induction motors using three-phase current monitoring. ISA Trans. **50**(1), 14–20 (2011). https://doi.org/10.1016/j.isatra.2010.10.008
14. Thomson, W., Fenger, M.: Case histories of current signature analysis to detect faults in induction motor drives. In: IEEE International Conference on Electric Machines and Drives, IEMDC'03. vol. 3, pp. 1–4 (2003)
15. Wildi, T.: Electrical machines, drives, and power systems. Pearson, 6 edn. (2014)
16. Zarei, J., Poshtan, J.: Bearing fault detection using wavelet packet transform of induction motor stator current. Tribol. Int. **40**(5), 763–769 (2007)

Physical Activity Classification Using an Artificial Neural Networks Based on the Analysis of Anthropometric Measurements

Antonio J. Alvarez[1(✉)], Erika Severeyn[1,2], Sara Wong[1,3], Héctor Herrera[1,4], Jesús Velásquez[1,2], and Alexandra La Cruz[5]

[1] Applied Biophysics and Bioengineering Group, Simón Bolívar University, Caracas, Venezuela
{severeynerika,swong,haherrera,jmvelasquezf}@usb.ve
[2] Department of Thermodynamics and Transfer Phenomena, Simón Bolívar University, Caracas, Venezuela
[3] Department of Electronics and Circuits, Simón Bolívar University, Caracas, Venezuela
[4] Department of Biological and Biochemical Process Technology, Simón Bolívar University, Caracas, Venezuela
[5] Faculty of Engineering, Ibagué University, Ibagué, Tolima, Colombia
alexandra.lacruz@unibague.edu.co

Abstract. Physical activity is one of the most important factors in leading a healthy life, which has increased the interest in the scientific community to evaluate methods and tools that can help people maintain an exercise routine, such as portable devices that can track the movements of the user and provide an appropriate feedback. Interest has also emerged in assessing the discrimination between physically active and inactive persons through the use of readily available data, which is the aim of this work. In this case, we used an auto-encoder to find the most outstanding characteristics of an anthropometric data set, in order to get the most representative attributes. Then use them to train an Artificial Neural Network (ANN), so that it could learn to identify between a physically active and a sedentary person. The ANN obtained 81% accuracy, 82% precision, 88% recall, 83% F1 score and 0.89 AUC. These results position the ANN as a viable model that could be used as a tool in scenarios such as customer profiling for different interested companies.

Keywords: Classification · Artificial Neural Network · Physical activity

1 Introduction

Physical activity has been shown to be one of the most important factors in a healthy life. The benefits that it produces in the body are: prevention of cardiovascular disease, diabetes, specific types of cancer and osteoporosis [31]. In

M. Botto-Tobar et al. (Eds.): ICCIS 2020, AISC 1273, pp. 60–70, 2021.
https://doi.org/10.1007/978-3-030-59194-6_6

addition, exercise has an impact on mental health, where it has been found to improve symptoms of depression and anxiety [20]. However, according to the World Health Organization (WHO) "Globally, about 31% of adults aged 15 and older were not sufficiently active in 2008 (men 28% and women 34%). Approximately 3.2 million of deaths each year are attributable to insufficient physical activity" [19]. This, together with the fact that physical inactivity is responsible for 6% of deaths, makes it the fourth largest risk factor of death worldwide [16], and one of the major problems of modern society.

Currently, there are efforts to promote physical activity that involve technologies such as clothing and smart phone applications; with a number of factors such as real-time feedback, goal adjustments and rewards, with the purpose of creating a long-lasting habit of physical routine [26]. In fact, with the progress of technology, there has been a proliferation of apparel such as accelerometers with real-time feedback using artificial intelligence for the identification of sedentary behavior and correction of specific postures and exercises [15,17,24,30].

However, all of these methods focus more on real-time feedback, so the user can monitor their performance and/or the effectiveness of the exercise and ultimately you need at least a wearable device to make it work. That leaves room for research into physical activity and inactivity through the use of more readily available data, such as anthropometric variables (height, weight, length and diameter of some muscles and bones), demographic variables (age and gender) and clinical evaluation of chemical composition of blood. This could be particularly useful when there is a need to determine whether a person is physically active or not, without having to rely on their word for it, and there is a limit to the type of information that can be obtained. An example of such a situation is determining whether a potential client of an insurance company is high risk or not; in that case, the insurance company may assume that the potential client will benefit from obtaining the insurance at the lowest possible price.

The aim of this paper is to determine, on the basis of a series of readily available anthropomorphic measurements and demographic variables, a possible distinction between a physically active or inactive person. After this step, we intend to build an automatic learning model that makes such discernment.

The rest of the work is composed of the methodology in Sect. 2, where the data set, machine learning models, metrics of evaluation and the experimental setup are described. The results are presented in Sect. 3 and the corresponding discussion in Sect. 4. Finally, the conclusions are exposed in Sect. 5.

2 Method

2.1 Dataset

The data set used for this study consisted of 4,553 instances with 77 attributes. Each instance is a person whose anthropometric values were taken. The 77 attributes are related to the length and diameter of some muscles and bones, the chemical composition of the blood drawn from a sample, and some general information about the person such as age and sex and whether or not the person was physically active.

Since one of the objectives of this work is to build a model that can discern between physically active and inactive people, the first step in pre-processing was to eliminate the instances that lacked the physical activity attribute, leaving 3,628 instances. The next step was to eliminate the missing attributes, leaving eleven (11) attributes: gender, date of birth, age, weight, height, right arm circumference, left arm circumference, right triceps skin fold, left triceps skin fold, body mass index, and frame size index [6] (calculated from diameter and height of humerus). After this pre-processing, the total number of physically active persons in the data set was 1,380, while the number of physically inactive persons was 2,248.

2.2 Machine Learning Models

Artificial Neural Network (ANN). An ANN is a learning model that can be used in classification problems [12]. ANNs are usually formed of multiple layers of a unit called "neuron" [29]. The neurons in the first layer (a.k.a input layer) receives the input vector, while the neurons in posterior layers are connected to the neurons in previous layers to receive a combination of their output with a designated weight [9,29].

Taking the formulation described in [8] for the first layer and generalizing for each layer, the output of layer i is then given by:

$$\mathbf{h}^i = g^i(\mathbf{W}^i\mathbf{h}^{i-1} + \mathbf{b}^i) \tag{1}$$

where g is an activation function, $\mathbf{W}$ is the matrix of weights, and b is the bias vector. $\mathbf{h}^i = \mathbf{x}$ for $i = 1$ with $\mathbf{x}$ as the input vector. In this work, we used the rectifier linear (ReLU) [7] as the activation function for the hidden layers, where $z = \mathbf{W}^\intercal\mathbf{x} + \mathbf{b}$, noting that $\mathbf{z}$ represents a vector of probabilities:

$$g(\mathbf{z}) = max(\mathbf{0}, \mathbf{z}) \tag{2}$$

For binary classification, an ANN tries to approximate a Bernoulli distribution to the data by means of the logistic function in the output layer [8]:

$$\sigma(z) = \frac{1}{1 + e^{-z}} \tag{3}$$

with argument $z = \mathbf{W}^\intercal\mathbf{x} + b$, where $\mathbf{W}$ is the vector of weights, $\mathbf{x}$ is the input vector, and b is the bias. The cost function used to approximate the optimal weights is the cross-entropy of the Bernoulli distribution detailed in Eq (4) [8], where y is the binary result.

$$J(\mathbf{W}) = -\ln\sigma\left((2y - 1)z\right) \tag{4}$$

For this work, an ANN composed of 3 hidden layers with 200, 100 and 50 nodes was used. The weight update is done using an algorithm called ADAM [10]). The optimization of the parameters for the ANN was done through the *GridsearchCV* module from *Scikit-learn* [22].

Autoencoders. An autoencoder is an ANN that tries to recreate the input vector through the output vector, passing through a hidden layer that serves as an encoder. If we see an autoencoder as a two-component model, then we have an encoder that serves as a function $h = f(x)$, and a decoder that reconstructs with the function $r = g(h)$, with the intention of making $r \approx x$ [8].

The purpose of an autoencoder is to obtain useful information about the data that could be used for the reduction of the dimensionality, features learning [8], or even learn about the distribution of the input space in order to generate new data with the same characteristics, as it is the case with variational autoencoders [11].

For the present work, we want to obtain the most prominent features of the twelve mentioned in Sect. 2.1. In order to accomplish this task, we use an undercomplete autoencoder, which can be defined as an autoencoder where all the hidden layers have smaller dimension than the input layer, this layout forces the autoencoder to identify the most salient features of the training data set [8]. The undercomplete autoencoder used has only one hidden layer with: four (4) nodes, with the rectifier linear unit (ReLU) as the activation function; the optimizer algorithm used is ADAM; the output layer activation function is the identity function; and the cost function is Mean Square Error (MSE) presented in Eq. (5) [8], where m is the number of instances in a batch.

$$J(\mathbf{W}) = \frac{1}{m}\sum_{i=1}^{m} \|z^{(i)} - y^{(i)}\|^2 \tag{5}$$

2.3 Classification Evaluation Metrics

A confusion matrix for 2 classes $(Class_1, Class_2)$ is shown in Table 1: columns correspond to the true classes and rows to the predicted classes. The main diagonal corresponds to the correct classification made by the model.

Table 1. A confusion matrix for 2 classes.

Predicted	True	
	Positive	Negative
Positive	True Positive (TP)	False Negative (FN)
Negative	False Positive (FP)	True Negative (TN)

From the confusion matrix, accuracy, precision, recall, F1 score, ROC curve and the Area Under the Curve (AUC) were computed to assess the classification performance as follows [23,27]:

Accuracy. *Acc*, represents the proportion of correctly classified instances from the total:

$$Acc = \frac{\sum TP + \sum TN}{\sum TP + \sum TN + \sum FP + \sum FN}. \tag{6}$$

Precision. The precision Pre represents the proportion of correctly classified instances that are true positives from the total of classifications that were given to positive label:

$$Pre = \frac{\sum TP}{\sum TP + \sum FP}. \tag{7}$$

Recall. The recall Rec, also known as sensitivity, represents the proportion of correctly classified instances that are true positive from the total of instances that are true positives:

$$Rec = \frac{\sum TP}{\sum TP + \sum FN}. \tag{8}$$

The recall encompasses how many instances of a class were identified.

F1 Score. Represents a weighted average of precision and recall, where an F1 score reaches its best value at 1 and worst at 0:

$$F1 = 2 \times \frac{Pre \times Rec}{Pre + Rec} \tag{9}$$

False Positive Rate. The FPR represents the proportion of false positives from all the instances that are true negatives:

$$FPR = \frac{\sum FP}{\sum FP + \sum TN} \tag{10}$$

ROC Curve and AUC. The ROC curve is a plot made from considering the True Positive Rate (Recall) of the model against the False Positive Rate. According to Ferri et al. (2011), "the AUC estimates the probability that a random positive is ranked before a random negative, without committing to a particular decision threshold" [5].

2.4 Experimental Setup

One of the goals of this work is to find a subset of attributes that can be used to identify whether a given person is physically active or not. Therefore, we use the incomplete autoencoder presented in Sect. 2.2 to determine which are the most outstanding features, which tend to be the ones that are most fatally reconstructed (i.e., the reconstruction loss is smaller).

The next step is fitting the ANN described in Sect. 2.2 with the attributes obtained by the autoencoder. The resulting model will then be measured with the metrics described in Sect. 2.3. This process is repeated ten (10) times to get a average value.

The data set was stratified into a test set representing 20% of the original data set, and a training set representing 80%. In addition, ANN training was done with a 10-fold cross-validation over the training set.

All the experiments were done in an Intel(R) Core(TM) i5-3470 CPU @ 3.20 GHz, 4 processors y 4 GB of RAM. Pytorch 1.4.0 [21] and Scikit-learn 0.19.2 library [22] on Python 3.8.1 were used to perform the tasks. The code for this work can be found in[1].

3 Results

Figures 1, 2, 3 show scatter plots of the autoencoder reconstructions for the most prominent variables. This give us: frame size index, age and height as the most important variables from the eleven (11) mentioned in Sect. 2.1. Furthermore, we added gender through empirical testing, noting that it makes a difference in the accuracy of ANN.

In Table 2 the Accuracy, Precision, Recall and F1 score for the ANN that discriminates between physically active and inactive people are presented.

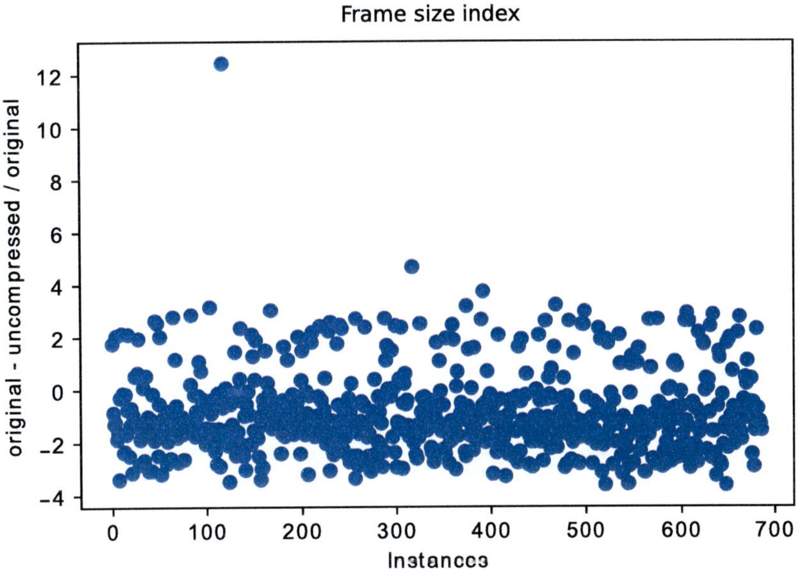

Fig. 1. Scatter plot of frame size index against reconstruction error

Table 2. Validation results for the ANN

Metric	Value
Accuracy	0.81
Precision	0.82
Recall	0.88
F1	0.83

[1] https://github.com/ragnell93/physical-activity/blob/master/GBBA3.ipynb.

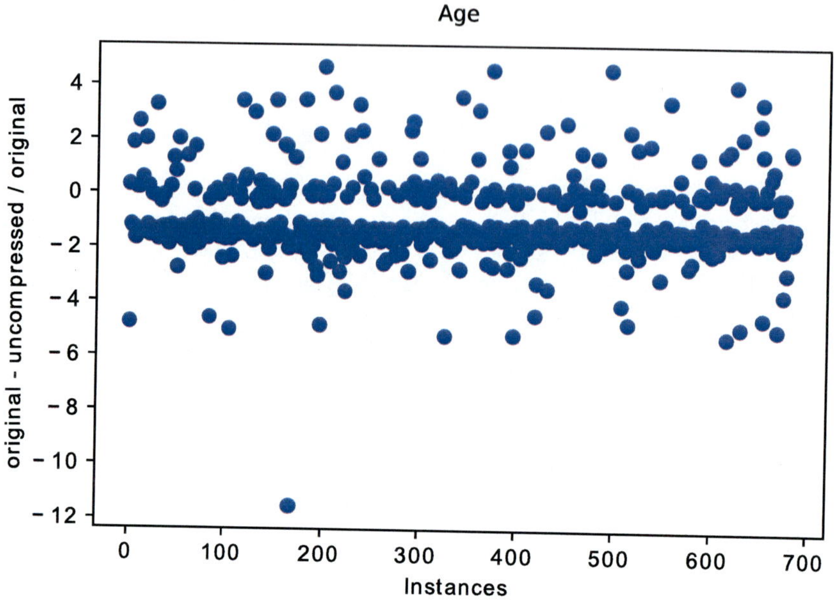

Fig. 2. Scatter plot of age against reconstruction error

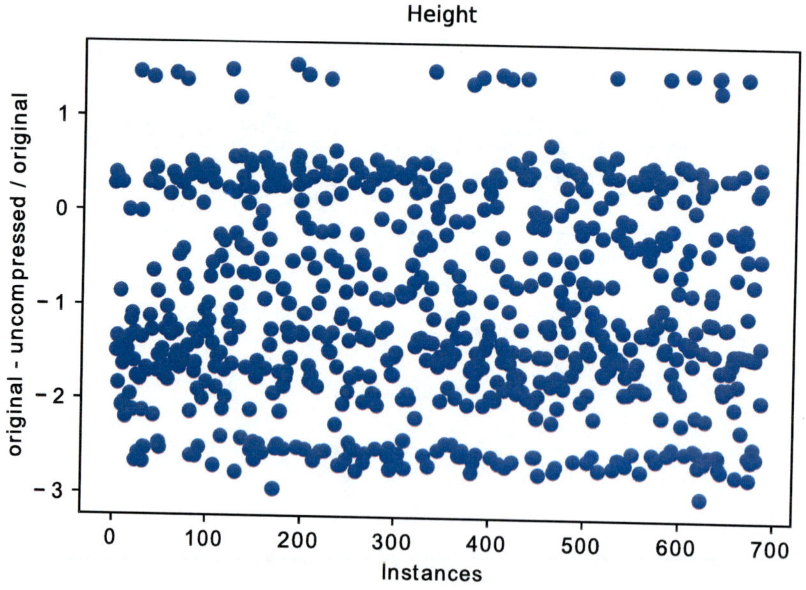

Fig. 3. Scatter plot of height against reconstruction error

In Fig. 4 the ROC curve and the AUC of the ANN are shown.

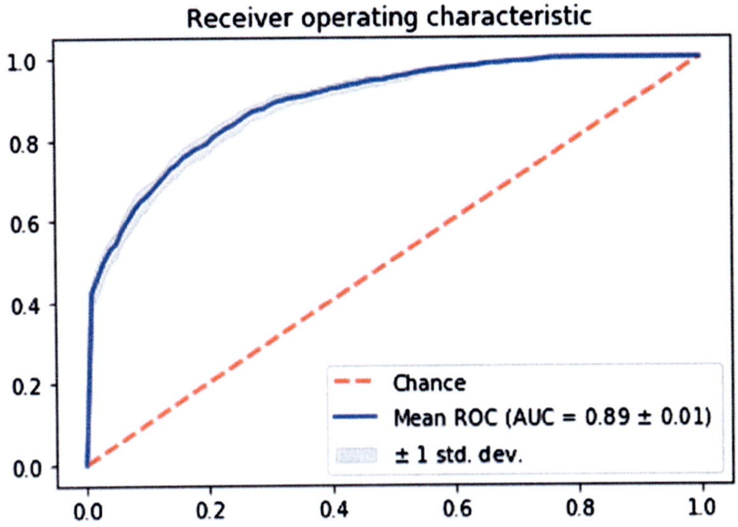

Fig. 4. ROC curve and AUC for the ANN

4 Discussion

The results seen in Figs. 1, 2, 3 show the social and cultural trends related to physical activity. First of all, some studies have found a correlation between a decline in physical activity with aging [3,18], obesity [18] and gender [1,4,13]. The high level of recall obtained by ANN indicates that, in general, physically active people have traits that differentiate them relatively well from sedentary people. This is consistent with the findings in [18], where obese people were found to be more sedentary than those in the correct weight range.

Interestingly, as a factor, one of the correlations found between age and physical activity is that with aging there is a decrease in physical activity. Where the transition between adolescence and youth marks an important moment for the long-term formation of the habit [3].

According to [14], under clinical standards, an AUC between 0.7 and 0.8 is considered acceptable; an AUC between 0.8 and 0.9 is considered excellent and above 0.9 as outstanding. All results of this work fall in the range of 0.7 and 0.8, including the AUC. Thus, under this criterion; ANN trained with gender, age, height and frame size index may be considered an excellent classifier. This means that although it cannot be used as an autonomous discriminator in a clinical setting, it might be used as a tool for a physician or in other areas as a good estimator.

In [25], the authors used a questionnaire form to recall data from 595 participants about their physical activity habits. After the collection, they classified the subjects into 1 of 3 classes: highly sedentary, moderately sedentary and non-sedentary. The next step was determining which questions where most influential to distinguish between highly sedentary and non-sedentary people using Sequential Forward Selection methods (SFS) [28]. Finally, they did ranked and obtained 90.9% accuracy using Linear Discriminant Analysis (LDA) [2].

The main differences between [25] and the present work is that in [25], a selection of features was made to obtain the 10 most prominent, some of them are: "Time on the Internet", "Levels of laziness", "Presence of bone or cartilage diseases", "Grip strength", "Presence of buzz" and "Muscle mass". Clearly, most of the information obtained depends on the honesty and perception of the person interviewed; whereas our work only needs factual information that can be easily obtained, and our model is reduced to only 4 attributes.

The other difference between this work and [25] is that in [25], the instances that are difficult to classify ("moderately sedentary") are eliminated for the classification task while we keep all the subjects regardless of their proximity to the class boundaries. This, and the use of more attributes explains the discrepancy between the 90.9% obtained in [25] and the 81% obtained in the present work.

It is important to highlight that there are some facts that could be biasing the results of this research. Social factors could influence the distribution of the data as all instances come from one country (Venezuela). Besides, many of the attributes were wasted because there are many instances with no data recorded, which effectively leaves only eleven (11) attributes to work on.

5 Conclusion

For the present work, we first identified the subset of attributes that best characterize a physically active person and a physically inactive person through the use of an incomplete autoencoder. This resulted in age, sex, height and body proportions as the most prominent attributes. These results suggest that social and cultural factors play an important role in adherence to a physical activity routine [1,3,4,13,18].

We then built an ANN that can distinguish between physically active people from sedentary people with 81% accuracy and 0.89 AUC. This qualifies the ANN as an excellent model for a clinical environment [14], which could be used as a complementary tool for information retrieval with easily acquired data.

Future work could address the problem of missing data to see if attributes such as blood composition could help in classification. Another possible work could be try to predict whether a person will maintain a physical activity routine over months or years, using automatic learning models on longitudinal data.

Acknowledgment. This work was funded by the Research and Development Deanery of the Simón Bolívar University (DID) and the Research Direction of the Ibagué University. Full acknowledgement is given to authors of "Deep Learning" by Goodfellow et al. (MIT Press) and "Evaluation: From Precision, Recall and F-Factor to ROC, Informedness, Markedness & Correlation" by David Powers (BioInfo Publications™).

References

1. Ács, P., Bergier, J., Salonna, F., Junger, J., Melczer, C., Makai, A.: Gender differences in physical activity among secondary school students in the visegrad countries (v4). Health Probl. Civiliz. **10**(3), 21–29 (2016)
2. Balakrishnama, S., Ganapathiraju, A.: Linear discriminant analysis-a brief tutorial. Inst. Signal Inf. Process. **18**, 1–8 (1998)
3. Corder, K., Winpenny, E., Love, R., Brown, H.E., White, M., Van Sluijs, E.: Change in physical activity from adolescence to early adulthood: a systematic review and meta-analysis of longitudinal cohort studies. Br. J. Sports Med. **53**(8), 496–503 (2019)
4. Ding, D.: Surveillance of global physical activity: progress, evidence, and future directions. Lancet Global Health **6**(10), e1046–e1047 (2018)
5. Ferri, C., Hernández-Orallo, J., Flach, P.A.: A coherent interpretation of AUC as a measure of aggregated classification performance. In: Proceedings of the 28th International Conference on Machine Learning (ICML 2011), pp. 657–664 (2011)
6. Frisancho, R., Flegel, P.: Elbow breadth as a measure of frame size for US males and females. Am. J. Clin. Nutr. **37**(2), 311–314 (1983)
7. Glorot, X., Bordes, A., Bengio, Y.: Deep sparse rectifier neural networks. In: Proceedings of the Fourteenth International Conference on Artificial Intelligence and Statistics, pp. 315–323 (2011)
8. Goodfellow, I., Bengio, Y., Courville, A.: Deep Learning. MIT Press, Cambridge (2016)
9. Haykin, S.S., et al.: Neural Networks and Learning Machines/Simon Haykin. Prentice Hall, New York (2009)
10. Kingma, D.P., Ba, J.: Adam: a method for stochastic optimization. In: Proceedings of the 3rd International Conference on Learning Representations (ICLR) (2015)
11. Kingma, D.P., Welling, M., et al.: An introduction to variational autoencoders. Found. Trends® Mach. Learn. **12**(4), 307–392 (2019)
12. Krogh, A.: What are artificial neural networks? Nat. Biotechnol. **26**(2), 195 (2008)
13. Magoc, D., Tomaka, J., Shamaley, A.G., Bridges, A.: Gender differences in physical activity and related beliefs among Hispanic college students. Hispanic J. Behav. Sci. **38**(2), 279–290 (2016)
14. Mandrekar, J.N.: Receiver operating characteristic curve in diagnostic test assessment. J. Thorac. Oncol. **5**(9), 1315–1316 (2010)
15. Mannini, A., Sabatini, A.M.: Machine learning methods for classifying human physical activity from on-body accelerometers. Sensors **10**(2), 1154–1175 (2010)
16. Mathers, C., Stevens, G., Mascarenhas, M.: Global Health Risks: Mortality and Burden of Disease Attributable to Selected Major Risks. World Health Organization, Geneva (2009)
17. Munguia Tapia, E.: Using machine learning for real-time activity recognition and estimation of energy expenditure. Ph.D. thesis, Massachusetts Institute of Technology (2008)
18. Norman, A., Bellocco, R., Vaida, F., Wolk, A.: Total physical activity in relation to age, body mass, health and other factors in a cohort of Swedish men. Int. J. Obes. **26**(5), 670–675 (2002)
19. World Health Organization: Physical Inactivity: A Global Public Health Problem (2008)
20. Paluska, S.A., Schwenk, T.L.: Physical activity and mental health. Sports Med. **29**(3), 167–180 (2000)

21. Paszke, A., Gross, S., Chintala, S., Chanan, G., Yang, E., DeVito, Z., Lin, Z., Desmaison, A., Antiga, L., Lerer, A.: Automatic differentiation in PyTorch (2017)
22. Pedregosa, F., Varoquaux, G., Gramfort, A., Michel, V., Thirion, B., Grisel, O., Blondel, M., Prettenhofer, P., Weiss, R., Dubourg, V., et al.: Scikit-learn: machine learning in python. J. Mach. Learn. Res. **12**(Oct), 2825–2830 (2011)
23. Powers, D.M.: Evaluation: from precision, recall and F-measure to ROC, informedness, markedness and correlation. J. Mach. Learn. Technol. **2**(1), 37–63 (2011)
24. Ren, X., Ding, W., Crouter, S.E., Mu, Y., Xie, R.: Activity recognition and intensity estimation in youth from accelerometer data aided by machine learning. Appl. Intell. **45**(2), 512–529 (2016)
25. Siirtola, P., Pyky, R., Ahola, R., Koskimäki, H., Jämsä, T., Korpelainen, R., Röning, J.: Detecting and profiling sedentary young men using machine learning algorithms. In: 2014 IEEE Symposium on Computational Intelligence and Data Mining (CIDM), pp. 296–303. IEEE (2014)
26. Sullivan, A.N., Lachman, M.E.: Behavior change with fitness technology in sedentary adults: a review of the evidence for increasing physical activity. Front. Public Health **4**, 289 (2017)
27. Tharwat, A.: Classification assessment methods. Applied Computing and Informatics (2018)
28. Ververidis, D., Kotropoulos, C.: Sequential forward feature selection with low computational cost. In: 2005 13th European Signal Processing Conference, pp. 1–4. IEEE (2005)
29. Wang, S.-C.: Artificial neural network. In: Interdisciplinary Computing in Java Programming, pp. 81–100. Springer, Boston (2003)
30. Wannenburg, J., Malekian, R.: Physical activity recognition from smartphone accelerometer data for user context awareness sensing. IEEE Trans. Syst. Man Cybern. Syst. **47**(12), 3142–3149 (2016)
31. Warburton, D.E., Nicol, C.W., Bredin, S.S.: Health benefits of physical activity: the evidence. CMAJ **174**(6), 801–809 (2006)

Analysis of Receiver Operating Characteristic Curve Using Anthropometric Measurements for Obesity Diagnosis

Erika Severeyn[1(✉)], Jesús Velásquez[1], Héctor Herrera[3], Sara Wong[2], and Alexandra La Cruz[4]

[1] Department of Thermodynamics and Transfer Phenomena, Simón Bolívar University, Caracas, Venezuela
{severeynerika,jmvelasquezf}@usb.ve

[2] Department of Biological and Biochemical Process Technology, Simón Bolívar University, Caracas, Venezuela
swong@usb.ve

[3] Department of Electronics and Circuits, Simón Bolívar University, Caracas, Venezuela
haherrera@usb.ve

[4] Faculty of Engineering, Ibagué University, Ibagué, Tolima, Colombia
alexandra.lacruz@unibague.edu.co

Abstract. Today, obesity is a major public health problem. Obesity increases the risk of diabetes, coronary artery disease, stroke, cancer, premature death and contributes substantially the costs to society. Obesity can be diagnosed with body mass index (BMI). According to the World Health Organization, the diagnosis of overweight is made with a $BMI \geq 25\,\mathrm{Kg/m^2}$, and obesity with a $BMI \geq 30\,\mathrm{kg/m^2}$. The diagnosis of obesity has been made using the abdominal circumference, the hip circumference, the thickness of the skin folds and the percentage of body fat (measured directly or indirectly). Besides, the characteristic operating receiver curves (ROC) have been used to find the optimal cut-off points of hip and waist circumference for the diagnosis of obesity. The aim of this study is to evaluate the ability of anthropometric measures for diagnosing overweight and obesity. A database of 1053 subjects with 26 anthropometric measurements was used. For evaluating the predictive ability of anthropometric measures, the area under the ROC curve (AUC_{ROC}), the sensitivity (SEN), the specificity (SPE), the negative predictive value (NPV) and the positive predictive value (PPV) were calculated. The hip circumference was the anthropometric value that best detected overweight/obese subjects with a $AUC_{ROC} = 0.932$ ($SEN = 0.871$, $SPE = 0.855$, $PPV = 0.536$ and $NPV = 0.972$) and an optimal cut-off point of 97.2 cm for recognition of obesity. The findings reported in this research suggest that the diagnosis of obesity can be made with anthropometric measurements. In the future, machine learning techniques, such as: k-means, neural networks or support vector machines; will be explored for the detection of overweight and obesity.

M. Botto-Tobar et al. (Eds.): ICCIS 2020, AISC 1273, pp. 71–80, 2021.
https://doi.org/10.1007/978-3-030-59194-6_7

Keywords: ROC curves · Anthropometric measurements · Body mass index · Classification

1 Introduction

Nowadays, obesity is a major public health problem [31,32]. Obesity increases the risk of diabetes, coronary artery disease, stroke, cancer, premature death and contributes substantially the costs to society [29]. A person suffering from obesity represents a significant public expense in the future due to the multifactorial nature of the disease [8]. Moreover, obesity is related to the metabolic syndrome, insulin resistance [2] and abnormal insulin and glucose values during oral glucose tolerance test [1,28].

Obesity is diagnosed with the Body Mass Index (BMI). BMI is defined as body mass (weight) divided by the square of body height, and it is universally expressed in units of kg/m^2. BMI is used consistently in epidemiological studies, and has been recommended for individual use in clinical practice to guide recommendations for weight loss and weight control [21,27]. According to the World Health Organization, the diagnosis of overweight is made with a BMI equal to or greater than $25\,kg/m^2$, and the diagnosis of obesity is made with a BMI equal to or greater than $30\,\mathrm{kg/m^2}$ [30]. These values change according to age, adults over 60 are diagnosed with obesity with a BMI equal to or greater than $26\,\mathrm{Kg/m^2}$ [20]. The cut-off points may also change depending on the ethnic group.

Abdominal circumference, hip circumference, skinfold thickness and percentage of body fat (measured directly or indirectly) have been used to diagnose obesity [19,24,25]. Numerous studies have found a directly proportional relationship between skinfold thickness, hip and waist circumference, and BMI [3,16]. Furthermore, high correlations have been found between the accumulation of fat in the abdominal area and high levels of insulin secretion [10,26], as well as a high risk of developing metabolic syndrome, hypertension [15,17] and type 2 diabetes [11,33].

Some methods have been used to find cut-off points of anthropometric variables for the diagnosis of obesity. Such methods include analysis regression and receiver operating characteristics curves (ROC) [22]. The ROC curves have been used to find the optimal cut-off points of hip [12] and waist circumference [4] for the diagnosis of obesity, finding areas under the ROC curves above 0.78. Indicating that anthropometric variables have the ability to detect overweight or obese subjects [7].

The aim of this study is to evaluate the ability of anthropometric variables to diagnose overweight and obesity. A database of 1053 subjects with twenty-eight (28) anthropometric measurements was used; they were: weight, height, arm circumferences, flexed arm circumferences, waist circumference, hip circumference, thigh circumferences, calf circumferences, triceps skinfold thicknesses, subscapular skinfold thicknesses, suprailiac skinfold thicknesses, abdominal skinfold thicknesses, thigh skinfold thicknesses, calf skinfold thicknesses, femoral and humerus diameters. ROC curves were used to evaluate the predictive ability of anthropometric measured variables. In the following section, the methodological procedure will be explained. In section three and four, the results and discussion are presented. And finally, in section five, the conclusions and proposals for future work are exposed.

2 Methodology

2.1 Database

Between 2004 and 2012 [13], 1053 (male = 308) adult men and women from the capital district of Venezuela were recruited in the Nutritional Assessment laboratory of the Simón Bolívar University. Anthropometric measurements were made on each subject; they were: height, weight, arm circumferences, bent arm circumferences, waist circumference, hip circumferences, thigh circumferences, calf circumferences, triceps skin folds, subscapular skin folds, supra-iliac skin folds, abdominal skin folds, thigh skin folds, calf skin folds, humerus and femur epicondyle diameters.

The diagnosis of overweight was made using the WHO guidelines which state that an overweight person has a BMI greater than or equal to 25. From the group of overweight subjects, 23 participants had a BMI greater than or equal to 30, indicating that they suffer from obesity [30]. Both overweight and obese subjects were placed in the same group for this study since we want to classify subjects with dysfunctional weight values.

All the procedures carried out in the study were in accordance with the ethical standards of the Bioethics Committee of the Simón Bolívar University and the 1964 Declaration of Helsinki and its subsequent amendments or comparable ethical standards. All subjects accepted the study by signing an informed consent form. Table 1 shows the characteristics of the database used for control subjects and subjects with overweight or obesity.

2.2 Performance Measurements

Twenty-six (26) anthropometric parameters were used in this research for the diagnosis of overweight/obesity, height and weight were not used as they were applied in the calculation of BMI. The ROC curves and the area under the ROC curve (AUC_{ROC}) were estimated to evaluate the ability of the anthropometric parameters for the diagnosis of overweight/obesity. Each anthropometric parameter was defined as a parameter vector (PV) for these tasks. The minimum and maximum of the PV were found and a diagnostic vector (DV) was then constructed with numbers equidistant between the minimum and maximum values of the PV.

Each DV value was taken as a cut-off point, and evaluated against the diagnostic vector of overweight/obesity. For the evaluation of each cut-off point of the DV vector, the true positives (TP), false positives (FP), true negatives (TN) and false negatives (FN) were obtained. Then, the sensitivity (SEN), the specificity (SPE), the positive predictive value (PPV) and the negative predictive value (NPV) were found for each value of the DV vector. Equations (1), (2), (3) and (4) describe performance measurements used in this study.

Table 1. Anthropometrics variables characteristics.

Anthropometrics variables	Control Male = 246, n = 883	Overweight/Obesity[c] Male = 62, n = 170
Age [years]	20.940 ± 2.773^b	21.130 ± 2.919
Weight[a] [Kg]	56.125 ± 8.304	73.582 ± 10.174
Height [cm]	162.831 ± 8.521	163.274 ± 9.171
Right arm circumference[a] [cm]	25.762 ± 2.493	30.461 ± 2.373
Left arm circumference[a] [cm]	25.650 ± 2.425	30.432 ± 2.427
Right flexed arm circumference[a] [cm]	26.818 ± 2.717	31.381 ± 2.642
Left flexed arm circumference[a] [cm]	26.581 ± 3.109	31.119 ± 2.679
Waist circumference[a] [cm]	70.301 ± 7.095	84.824 ± 9.271
Hip circumference[a] [cm]	91.806 ± 5.471	103.539 ± 6.237
Right thigh circumference[a] [cm]	45.236 ± 2.370	49.594 ± 2.532
Left thigh circumference[a] [cm]	44.699 ± 2.565	49.093 ± 2.536
Right calf circumference[a] [cm]	33.736 ± 2.370	38.094 ± 2.532
Left calf circumference[a] [cm]	33.699 ± 2.565	38.093 ± 2.536
Right triceps skinfold[a] [mm]	13.593 ± 4.852	19.012 ± 5.634
Left triceps skinfold[a] [mm]	13.384 ± 4.788	18.861 ± 5.664
Right subscapular skinfold[a] [mm]	12.657 ± 3.991	20.512 ± 5.616
Left subscapular skinfold[a] [mm]	12.768 ± 4.005	20.658 ± 5.595
Right suprailiac skinfold[a] [mm]	11.837 ± 5.073	20.151 ± 6.451
Left suprailiac skinfold[a] [mm]	11.869 ± 5.102	20.216 ± 6.522
Right abdominal skinfold[a] [mm]	22.037 ± 5.073	30.351 ± 6.451
Left abdominal skinfold[a] [mm]	22.869 ± 5.102	31.216 ± 6.522
Right thigh skinfold[a] [mm]	19.899 ± 5.136	25.660 ± 6.019
Left thigh skinfold[a] [mm]	20.705 ± 5.219	26.282 ± 5.799
Right calf skinfold[a] [mm]	13.099 ± 5.136	18.860 ± 6.019
Left calf skinfold[a] [mm]	13.505 ± 5.219	19.082 ± 5.799
Right humerus diameter epicondylar[a] [cm]	5.973 ± 0.640	6.296 ± 0.634
Left humerus diameter epicondylar[a] [cm]	5.976 ± 0.635	6.301 ± 0.635
Right femur diameter epicondylar[a] [cm]	8.831 ± 0.661	9.588 ± 0.703
Left femur diameter epicondylar[a] [cm]	8.828 ± 0.660	9.595 ± 0.720
Body mass index $[Kg/m^2]^a$	21.106 ± 2.092	27.525 ± 2.349

[a]Statistically significant difference (*p-value* < 0.05) between control and overweight/obesity.

[b]Average and standard deviation.

[c]The database has 23 subjects with obesity.

$$SEN = \frac{TP}{TP + FN} \tag{1}$$

$$ESP = \frac{TN}{TN + FP} \tag{2}$$

$$PPV = \frac{TP}{TP + FP} \tag{3}$$

$$NPV = \frac{TN}{TN + FN} \tag{4}$$

The AUC_{ROC} were calculated using the trapeze method. An $AUC_{ROC} \geq 0.70$ was considered an acceptable predictive value [23]. The optimal cut-off point was defined as the shortest distance between the ROC curve and the coordinate point (0, 1) of the ROC curve trace.

2.3 Statistical Analysis

Non-paired samples were treated with a different distribution than normal, thus determining the differences between groups of two. The non-parametric Mann-Whitney U test was used, a *p-value* less than 5% was considered statistically significant [18]. The data in the text and in the Table 1 are presented as values of mean and standard deviation.

3 Results

Table 1 reports the anthropometrics values of control and overweight/obese subjects. The database consists of 1053 subjects, 83.86% belongs to the control group and 16.14% endure overweigh/obesity. The classification were made according to WHO, all the subjects with $BMI \geq 25$ were classified as overweight. The 13.5% of the subjects that belongs to overweight/obesity group have obesity since their $BMI \geq 30$. Figure 1 shows the ROC curves of the twenty-six (26) anthropometrics parameters for overweight/obesity diagnosis. Table 2 shows the AUC_{ROC} as well as the sensitivity, specificity, positive predictive value and negative predictive value for the optimal detection cut-off point of the twenty-six (26) anthropometric parameters for the diagnosis of overweight/obesity.

4 Discussion

Overweight/obese subjects (see Table 1) have higher anthropometric values (with a statistically significant difference $p < 0.05$) than control subjects, which corroborates the fact that anthropometric measurement values such as skin circumferences and skinfolds have a directly proportional relationship to BMI [9]. Otherwise, the abdominal circumference of the overweight/obesity group is within the normal values managed in the diagnosis of metabolic syndrome [5], despite

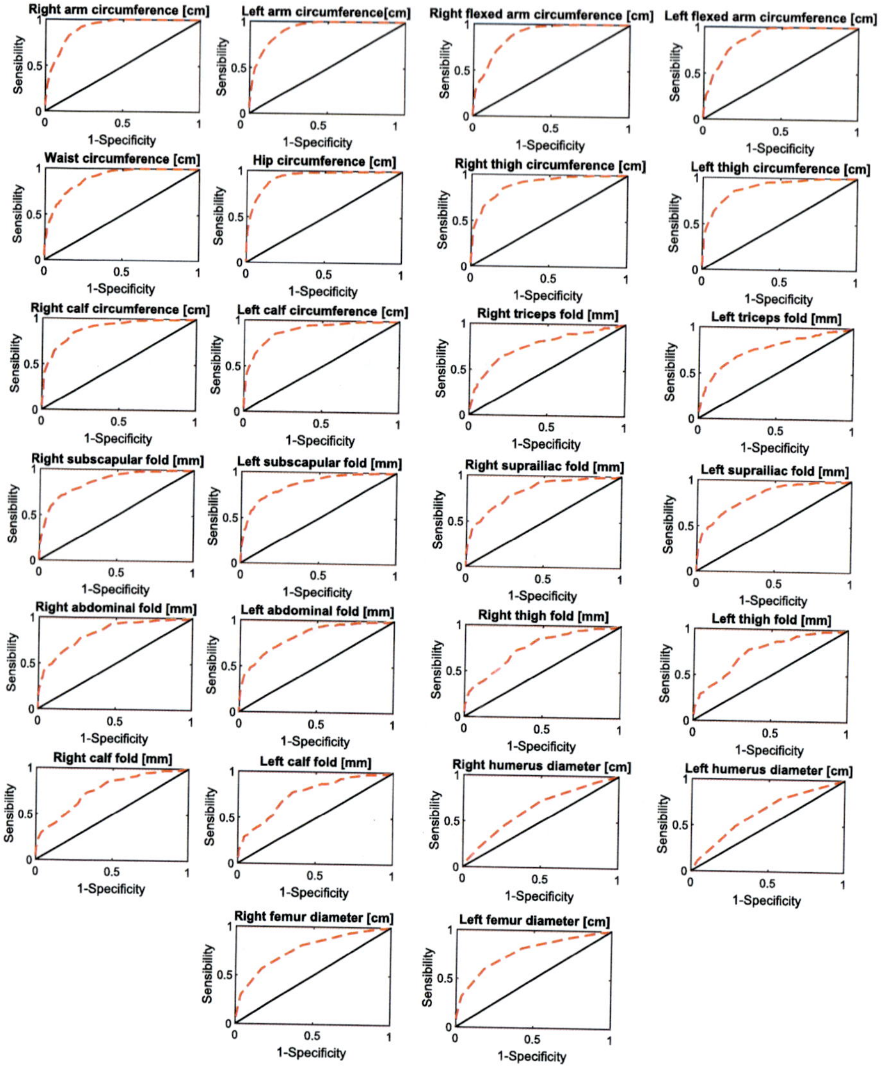

Fig. 1. Area under the ROC curves for the overweight and obesity predictability.

there being significant differences between the control group and the overweight/obesity group. This could indicate that the group of overweight/obese subjects do not have an accumulation of fat in the abdominal area, therefore, their cardiometabolic risk is lower [6].

The epicondylar diameters of the humerus and femur bones of overweight/obese subjects have higher values than control subjects ($p < 0.01$), which is due to the fact that the way of measuring the bones is above the skin and the measurement is affected because obese subjects tend to have a thicker adipose panicle in that area [9].

Based on the analysis of the ROC curves (Table 2, Fig. 1), it can be verified that the anthropometric measurements that can best detect subjects with dysfunctional BMI were: arm circumference, flexed arm circumference, waist circumference, hip circumference, and thigh and calf circumference with a $AUC_{ROC} > 0.890$ ($SEN > 0.820$, $SPE > 0.790$). Having a high probability of detecting subjects of normal weight ($NPV > 0.956$).

Finally, hip circumference was the anthropometric value that best detected overweight/obese subjects with an $AUC_{ROC} = 0.932$ ($SEN = 0.871$, $SPE = 0.855$, $PPV = 0.536$, $NPV = 0.972$) and an optimal cut-off point of 97.2 cm for

Table 2. SEN, SPE, PPV, NPV, AUC_{ROC} and optimal cut-off point for the evaluation of anthropometric variables for overweight and obesity diagnosis.

Variable	AUC_{ROC}	SEN	SPE	PPV	NPV	Optimal cut off point
Right arm circumference [cm]	0.914	0.824	0.849	0.513	0.962	28.400
Left arm circumference [cm]	0.919	0.847	0.818	0.472	0.965	27.900
Right flexed arm circumference [cm]	0.886	0.871	0.760	0.411	0.968	28.500
Left flexed arm circumference [cm]	0.888	0.794	0.815	0.453	0.954	28.700
Waist circumference [cm]	0.897	0.806	0.803	0.441	0.956	76.000
Hip circumference [cm]	0.932	0.871	0.855	0.536	0.972	97.200
Right thigh circumference [cm]	0.896	0.853	0.795	0.445	0.966	47.000
Left thigh circumference [cm]	0.897	0.859	0.796	0.448	0.967	46.500
Right calf circumference [cm]	0.896	0.853	0.795	0.445	0.966	35.500
Left calf circumference [cm]	0.897	0.859	0.796	0.448	0.967	35.500
Right triceps fold [mm]	0.769	0.653	0.793	0.378	0.922	17.600
Left triceps fold [mm]	0.772	0.682	0.761	0.355	0.926	16.800
Rignt subscapular fold [mm]	0.872	0.753	0.824	0.452	0.945	16.300
Left subscapular fold [mm]	0.869	0.765	0.818	0.447	0.948	16.500
Rignt suprailiac fold [mm]	0.838	0.782	0.729	0.358	0.946	14.500
Left suprailiac fold [mm]	0.836	0.765	0.717	0.342	0.941	14.500
Rignt abdominal fold [mm]	0.838	0.782	0.729	0.358	0.946	24.700
Left abdominal fold [mm]	0.836	0.765	0.717	0.342	0.941	25.500
Right thigh fold [mm]	0.761	0.735	0.670	0.300	0.929	22.300
Left thigh fold [mm]	0.757	0.782	0.647	0.299	0.939	22.600
Right calf fold [mm]	0.761	0.735	0.670	0.300	0.929	15.500
Left calf fold [mm]	0.757	0.782	0.647	0.299	0.939	15.400
Right humerus diameter epicondylar [cm]	0.642	0.735	0.497	0.220	0.907	5.800
Left humerus diameter epicondylar [cm]	0.643	0.494	0.712	0.249	0.880	6.200
Right femur diameter epicondylar [cm]	0.774	0.818	0.576	0.271	0.943	8.900
Left femur diameter epicondylar [cm]	0.778	0.600	0.820	0.391	0.914	9.400

the detection of obesity. This corroborates the work that has documented that hip circumference has a direct proportional relationship with BMI [14].

5 Conclusions

The findings reported in this research strongly suggest that the diagnosis of obesity can be made from anthropometric variables. In this study, the anthropometric measurement that best detected overweight subjects was the hip circumference, with a high probability of detecting normal weight subjects. Because hip circumference is an easy measure to acquire, it could be used in the diagnosis of obesity, especially in low-income countries where the availability of equipment is scarce.

As future work, tests with automatic learning techniques such as k-means, neural networks or support vector machines for the detection of overweight and obese subjects will be carried out.

Acknowledgment. This work was funded by the Research and Development Deanery of the Simón Bolívar University (DID) and the Research Direction of the Ibagué University. Full acknowledgement is given to David Powers, author of "Evaluation: From Precision, Recall and F-Factor to ROC, Informedness, Markedness & Correlation" (BioInfo Publications™).

References

1. Altuve, M., Severeyn, E., Wong, S.: Adaptation of five indirect insulin sensitivity evaluation methods to three populations: Metabolic syndrome, athletic and normal subjects. In: 2014 36th Annual International Conference of the IEEE Engineering in Medicine and Biology Society, pp. 4555–4558 (2014)
2. Altuve, M., Severeyn, E., Wong, S.: Unsupervised subjects classification using insulin and glucose data for insulin resistance assessment. In: 2015 20th Symposium on Signal Processing, Images and Computer Vision (STSIVA), pp. 1–7 (2015)
3. Anoop, S., Misra, A., Bhatt, S.P., Gulati, S., Mahajan, H., Prabakaran, G.: High plasma glucagon levels correlate with waist-to-hip ratio, suprailiac skinfold thickness, and deep subcutaneous abdominal and intraperitoneal adipose tissue depots in nonobese Asian Indian males with type 2 diabetes in North India. J. Diab. Res. **2017** (2017)
4. Anusruti, A., Jansen, E.H., Gao, X.Y., Xuan, H.B., Schoettker, B.: Longitudinal associations of body mass index, waist circumference, and waist-to-hip ratio with biomarkers of oxidative stress in older adults: results of a large cohort study. Obes. Facts **13**(1), 66–76 (2020)
5. Aschner, P., Buendía, R., Brajkovich, I., Gonzalez, A., Figueredo, R., Juarez, X.E., Uriza, F., Gomez, A.M., Ponte, C.I.: Determination of the cutoff point for waist circumference that establishes the presence of abdominal obesity in Latin American men and women. Diabet. Res. Clin. Pract. **93**(2), 243–247 (2011)
6. Ashwell, M., Gunn, P., Gibson, S.: Waist-to-height ratio is a better screening tool than waist circumference and bmi for adult cardiometabolic risk factors: systematic review and meta-analysis. Obes. Rev. **13**(3), 275–286 (2012)

7. Bener, A., Yousafzai, M.T., Darwish, S., Al-Hamaq, A.O., Nasralla, E.A., Abdul-Ghani, M.: Obesity index that better predict metabolic syndrome: body mass index, waist circumference, waist hip ratio, or waist height ratio. J. Obes. **2013** (2013)
8. Bomberg, E., Birch, L., Endenburg, N., German, A., Neilson, J., Seligman, H., Takashima, G., Day, M.: The financial costs, behaviour and psychology of obesity: a one health analysis. J. Comp. Pathol. **156**(4), 310–325 (2017)
9. Bratke, H., Bruserud, I.S., Brannsether, B., Abmus, J., Bjerknes, R., Roelants, M., Juliusson, P.: Timing of menarche in norwegian girls: associations with body mass index, waist circumference and skinfold thickness. BMC Pediatr. **17**(1), 138 (2017)
10. Castro, A.V.B., Kolka, C.M., Kim, S.P., Bergman, R.N.: Obesity, insulin resistance and comorbidities? mechanisms of association. Arq. Bras. Endocrinol. Metabologia **58**(6), 600–609 (2014)
11. Cheng, C.H., Ho, C.C., Yang, C.F., Huang, Y.C., Lai, C.H., Liaw, Y.P.: Waist-to-hip ratio is a better anthropometric index than body mass index for predicting the risk of type 2 diabetes in Taiwanese population. Nutr. Res. **30**(9), 585–593 (2010)
12. Cheng, Y.H., Tsao, Y.C., Tzeng, I.S., Chuang, H.H., Li, W.C., Tung, T.H., Chen., J.Y.: Body mass index and waist circumference are better predictors of insulin resistance than total body fat percentage in middle-aged and elderly Taiwanese. Medicine **96**(39), e8126 (2017)
13. Herrera, H., Rebato, E., Arechabaleta, G., Lagrange, H., Salces, I., Susanne, C.: Body mass index and energy intake in venezuelan university students. Nutr. Res. **23**(3), 389–400 (2003)
14. Huxley, R., Mendis, S., Zheleznyakov, E., Reddy, S., Chan., J.: Body fat assessed from total body density and its estimation from skinfold thickness: measurements on 481 men and women aged from 16 to 72 years. Euro. J. Clin. Nutr. **64**(1), 16–22 (2010)
15. Imai, A., Komatsu, S., Ohara, T., Kamata, T., Yoshida, J., Miyaji, K., Takewa, M., Kodama, K.: Visceral abdominal fat accumulation predicts the progression of noncalcified coronary plaque. Atherosclerosis **222**(2), 524–529 (2012)
16. Kriemler, S., Puder, J., Zahner, L., Roth, R., Meyer, U., Bedogni, G.: Estimation of percentage body fat in 6-to 13-year-old children by skinfold thickness, body mass index and waist circumference. Br. J. Nutr. **104**(10), 1565–1572 (2010)
17. Lê, K.A., Ventura, E.E., Fisher, J.Q., Davis, J.N., Weigensberg, M.J., Punyanitya, M., Hu, H.H., Nayak, K.S., Goran, M.I.: Ethnic differences in pancreatic fat accumulation and its relationship with other fat depots and inflammatory markers. Diab. Care **34**(2), 485–490 (2011)
18. Marusteri, M., Bacarea, V.: Comparing groups for statistical differences: how to choose the right statistical test? Biochem. Medic **20**(1), 15–32 (2010)
19. Mirmiran, P.: Body mass index as a measure of percentage body fat prediction and excess adiposity diagnosis among iranian adolescents. Arch. Iran. Med. **17**(6), 400 (2014)
20. Misra, A., Chowbey, P., Makkar, B.M., Vikram, N.K., Wasir, J.S., Chadha, D., Joshi, S.R.: Consensus statement for diagnosis of obesity, abdominal obesity and the metabolic syndrome for Asian Indians and recommendations for physical activity, medical and surgical management. J. Assoc. Phys. India **57**(2), 163–170 (2009)
21. Ng, C., Elliott, M., Riosmenna, F., Cunningham, S.: Beyond recent BMI: BMI exposure metrics and their relationship to health. SSM-Population Health, p. 100547 (2020)

22. de Oliveira, C.M.C., Kubrusly, M., Mota, R.S., Choukroun, G., Neto, J.B., da Silva, C.A.B.: Adductor pollicis muscle thickness: a promising anthropometric parameter for patients with chronic renal failure. J. Renal Nutr. **22**(3), 307–316 (2012)
23. Reitsma, J.B., Glas, A.S., Rutjes, A.W., Scholten, R.J., Bossuyt, P.M., Zwinderman, A.H.: Bivariate analysis of sensitivity and specificity produces informative summary measures in diagnostic reviews. J. Clin. Epidemiol. **58**(10), 982–990 (2005)
24. Ronnecke, E., Vogel, M., Bussler, S., Grafe, N., Jurkutat, A., Schlingmann, M., Koerner, A., Kiess, W.: Age-and sex-related percentiles of skinfold thickness, waist and hip circumference, waist-to-hip ratio and waist-to-height ratio: results from a population-based pediatric cohort in germany (life child). Obes. Facts **12**(1), 25–40 (2019)
25. Selcuk, A., Bulucu, F., Kalafat, F., Cakar, M., Demirbas, S., Karaman, M., Ay, S.A., Saglam, K., Balta, S., Demirkol, S., Arslan, E.: Skinfold thickness as a predictor of arterial stiffness: obesity and fatness linked to higher stiffness measurements in hypertensive patients. Clin. Exp. Hypertens. **35**(6), 459–464 (2013)
26. Shin, H.Y., Lee, D.C., Chu, S.H., Jeon, J.Y., Lee, M.K., Im, J.A., Lee, J.W.: Chemerin levels are positively correlated with abdominal visceral fat accumulation. Clin. Endocrinol. **77**(1), 47–50 (2012)
27. Trijsburg, L., Geelen, A., Hollman, P.C., Hulshof, P.J., Feskens, E.J., van't Veer, P., Boshuizen, H.C., de Vries, J.H.: BMI was found to be a consistent determinant related to misreporting of energy, protein and potassium intake using self-report and duplicate portion methods. Public Health Nutr. **20**(4), 598–607 (2017)
28. Velásquez, J., Wong, S., Encalada, L., Herrera, H., Severeyn, E.: Lipid-anthropometric index optimization for insulin sensitivity estimation. In: Romero, E., Lepore, N., García-Arteaga, J.D., Brieva, J. (eds.) 11th International Symposium on Medical Information Processing and Analysis, vol. 9681, pp. 195–204. International Society for Optics and Photonics, SPIE (2015)
29. Wang, Y., McPherson, K., Marsh, T., Gortmaker, S., Brown, M.: Health and economic burden of the projected obesity trends in the USA and the UK. Lancet (London, Engl.) **378**(9793), 815–825 (2011)
30. Who, E.C.: Appropriate body-mass index for Asian populations and its implications for policy and intervention strategies. Lancet (London, Engl.) **363**(9403), 157 (2004)
31. Williams, E., Mesidor, M., Winters, K., Dubbert, P., Wyatt, S.: Overweight and obesity: prevalence, consequences, and causes of a growing public health problem. Curr. Obes. Rep. **4**, 363–370 (2015)
32. Wyatt, S., Winters, K., Dubbent, P.: Overweight and obesity: prevalence, consequences, and causes of a growing public health problem. Am. J. Med. Sci. **331**(4), 166–174 (2006)
33. Yki-Järvinen, H.: Liver fat in the pathogenesis of insulin resistance and type 2 diabetes. Dig. Dis. **28**(1), 203–209 (2010)

Using Low-Frequency EEG Signals to Classify Movement Stages in Grab-and-Lift Tasks

V. Diego Orellana[1(✉)], Beatriz Macas[2], Marco Suing[2], Sandra Mejia[3], G. Pedro Vizcaya[1], and Catalina Alvarado Rojas[1]

[1] Department of Electronics, Pontificia Universidad Javeriana, Bogotá 11001000, Colombia
{diego.orellana,pvizcaya,catalina_alvarado}@javeriana.edu.co

[2] Department of Electronics, Universidad Nacional de Loja, Loja 110110, Ecuador
{beatriz.macas,marco.suing}@unl.edu.ec

[3] Faculty of Human Health, Universidad Nacional de Loja, Loja 110110, Ecuador
sandra.mejia@unl.edu.ec

Abstract. Nowadays, Brain-Computer Interface (BCI) systems are considered a tool with enormous potential to establish communication alternatives, restore functions, and provide rehabilitation processes to patients with neuromotor impairment. A wide variety of invasive and non-invasive methods has been studied to control BCI systems, especially with electroencephalography (EEG) signals. However, despite numerous studies in this field, much work remains to be done to understand the underlying neural mechanisms and to develop versatile and reliable BCI systems. Typically, BCI systems oriented to motion decoding are based on information extracted from sensorimotor rhythms, which correspond to the EEG signal in the mu (8–12 Hz) and beta (18–30 Hz) bands. In this work, we focus on the search for information in low-frequency bands (0.1–7 Hz). To accomplish this goal, we work on the classification of six stages of gripping and lifting movements of an object. The features of the signals were extracted applying Discrete Wavelet Transform (DWT) and Empirical Mode Decomposition (EMD). Our results suggest that, for this case, the most significant amount of discriminant information is within the (0–4 Hz) band (maximum accuracy of $89.22 \pm 0.81\%$). Another remarkable result is the high similarity observed between the waveforms belonging to the same stage between different subjects. This result is especially motivating since numerous studies have demonstrated that the EEG signals present a high inter-subject and inter-session variability.

Keywords: Brain-Computer Interface (BCI) · EEG · Discrete Wavelet Transform (DWT) · Empirical Mode Decomposition (EMD) · Dynamic Time Warping (DTW) · Motor imagery · Random Subspace Method

1 Introduction

A Brain-Computer Interface (BCI) is a system that acquires, processes, and translates brain signals into real-time output commands [1, 2]. It consists of four stages: signal

M. Botto-Tobar et al. (Eds.): ICCIS 2020, AISC 1273, pp. 81–93, 2021.
https://doi.org/10.1007/978-3-030-59194-6_8

acquisition, feature extraction, automatic classification, and execution [3]. Although BCI research was initiated more than two decades ago, there are still many challenges before it can be widely used outside of research settings. Potentially, this kind of system offers a wide variety of applications; for instance, in the medical field, research is currently being conducted on three types of systems, namely silent speech decoding to support people with communication disabilities [4], movement decoding to restore, rehabilitate, or improve motor skills [3], and measurement of clinical markers for the study of psychological illnesses and cognitive processes [5, 6]. Among the applications to improve motor impairments is the so-called "Motor Imagery" (MI), which is the mental rehearsal of physical movement tasks. In recent years, this type of system has received considerable attention in the scientific community and is now used as a tool for the rehabilitation and improvement of motor skills. Numerous studies using medical images [7] have demonstrated that there are brain regions that are activated similarly during the real execution of movements (ME) and their imagination [7]. Based on these similarities, a valid approach to study MI consists of first analyzing patterns in the electroencephalography (EEG) signals produced by the actual ME and then extending the procedures to EEG signals generated by their imagination.

Usually, studies on motor imagery BCI (MI-BCI) have focused on decoding movement based on the analysis of so-called "sensorimotor rhythms" (SMR) [8, 9]. They are neurophysiological activities recorded on the sensorimotor cortex and modulated by ME or MI. Specifically, SMRs are characterized by a decrease in the mu (8–12 Hz) and beta (14–24 Hz) bands, accompanied by an increase in the gamma frequency band (> 30 Hz) [10]. Unlike traditional approaches, the purpose of this work is to analyze the significance of EEG information at low frequencies (0–7 Hz) during the execution of complex movements. To carry out this analysis, we classified six stages of the grip and elevation movement of an object executed with the right hand. The process implemented included the following stages: First, we conducted a visual and quantitative analysis of the similarity between the waveforms belonging to the movement stages applying the Dynamic Time Warping algorithm (DyTW). Second, to evaluate the relevance of the information in the band of interest and contiguous bands, we performed a feature extraction in 0–4 Hz, 4–7 Hz, and 7–15 Hz bands, applying two successful techniques in the field of signal processing, Discrete Wavelet Transform (DWT) and Empirical Mode Decomposition (EMD). Third, we classified the six stages using the Random Subspace Method (RSM); the discussion of the results is mainly based on the evaluation metrics (accuracy and F1-score) of the classifiers and the possible contributions of the work performed in the improvement of the BCI systems. Unlike most of the studies, where the signals originate from movements of the right and left hand, in this database, the movements were executed only with the right hand, which increases the difficulty of the analysis since there is no contralateral activation.

2 Materials and Methods

2.1 Database

We worked with the WAY-EEG-GAL database. This dataset contains scalp EEG recordings from grasping and elevation tasks in humans and has been used to test different techniques for decoding the intention, sensation, and execution of movements [11]. For each trial, the participant was asked to reach for the object, grasp it with thumb and forefinger, lift it, hold it for a couple of seconds, place it back on the support surface, and, finally, release it and return the hand to the selected resting position, as shown in Fig. 1. The experiment was divided into series, with each series containing approximately 32 runs of the grab-and-lift movement. In this study, we used the first eight series, with a total of 260 runs per participant. Only right-handed people were selected as participants. The study included 12 subjects, 8 females and 4 males, aged 19–35 years. For more details about the database, please see [11].

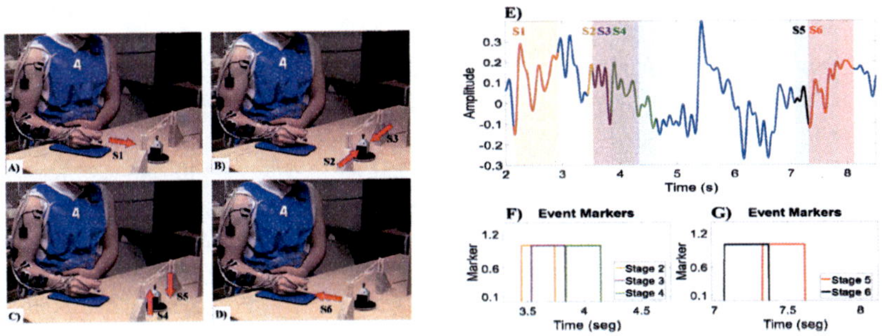

Fig. 1. a) S1 = forward movement, b) S2 and S3 correspond to the grip movement, S2 = contact of the first digit, and S3 = contact of the second digit, c) S4 = lifting the object and S5 = lowering the object to its original position. d) S6 = release of the object, e) EEG channel C4, f and g) Stage start markers. Images were taken from [11].

2.2 Analysis of EEG Signals

As already mentioned, the features were extracted by applying DWT and EMD. The DWT has been widely employed in the development of BCI systems and in practically all fields involving time-frequency analysis [12]. Empirical Mode Decomposition is a relatively new and effective tool for analyzing non-linear and non-stationary signals. In recent years, EMD has gained wide acceptance in image and video processing and is increasingly being explored in new fields [13, 14]. The proposed similarity analysis was implemented by applying the DyTW, which is one of the most popular techniques for measuring the similarity between time series, and its application extends to all types of data sequences[15, 16].

Discrete Wavelet Transform is a method for the decomposition of a signal $\boldsymbol{x[n]}$ using a scaled and shifted version of a mother wavelet function (in this case, $\boldsymbol{x[n]}$ corresponds to the EEG signal) [12]. The DWT of a signal $\boldsymbol{x[n]}$ is extracted by applying a bank of filters. First, the original signal is passed through a low-pass filter with impulse response $\boldsymbol{g}$, and in parallel through a high-pass filter $\boldsymbol{h}$.

$$y_{low}[n] = x[n] * g[n] = \sum_{k=-\infty}^{\infty} x[k]g[n-k] \quad (1)$$

$$y_{high}[n] = x[n] * h[n] = \sum_{k=-\infty}^{\infty} x[k]h[n-k] \quad (2)$$

The output from the high-pass filter constitutes the detail coefficients, and the output from the low-pass filter corresponds to the approximation coefficients. Since half of the signal frequencies have already been removed, half the samples can be discarded according to Nyquist's rule. The output of the low-pass filter is then followed by a dyadic decimation (down-sampling by 2) and further processed by successively passing it through a new low- and a high-pass filter with half of the cut-off frequency of the previous stage.

Focused on capturing the information contained in our band of interest, we performed the decomposition by DWT in six levels of detail plus one level of approximation, which, with a sampling frequency of 500 Hz and a mother function Daubechies of 1, corresponds to the following bands: 0–3.9 Hz, 3.9–7.8 Hz, 7.8–15.6 Hz, 15.6–31.2 Hz, 31.2–62.5 Hz, 62.5–125 Hz, and 125–250 Hz. The feature space was formed by collecting the approximation coefficients (8 values) of each channel and concatenating them. Our database contained 1,560 events with 32–channel EEG signals. Therefore, our features matrix had a dimension of 1,560 * 256. To assess the contribution of higher frequency bands, we repeated this process by adding the coefficients of the 4–7 Hz and 7–15 Hz bands.

Empirical Mode Decomposition is a signal decomposition method to decompose a non-linear and non-stationary signal into its oscillatory modes, called "Intrinsic Mode Functions" (IMFs); every IMF satisfies two conditions [17]:

- P1.-The difference between the total number of extrema (local minima and local maxima) and the total number of zero-crossings is zero or one.
- P2.- Over the entire signal, the mean value of the envelope defined by the local maxima and local minima is zero or close to zero [12].

Fulfillment of these conditions guarantees the existence of narrow-band IMFs (single scale) and allows each IMF to contain physically relevant information. The algorithm for performing EMD is described elsewhere [12]. The EMD is entirely data-driven and does not rely on any *a priori* basis function in contrast to the wavelet transform. These two properties make EMD an ideal tool for the analysis of non-linear and non-stationary signals. However, it suffers from the significant drawback of mode mixing. Mode mixing is defined as the presence of a single-component signal in several IMFs [18]. To reduce this problem, some variants of the EMD algorithm have been developed. The most popular alternatives are Multivariate Empirical Mode Decomposition (MEMD) [19] and Adaptive-Projection Intrinsically Transformed MEMD (APIT-MEMD) [20].

In our case, to extract characteristics, we first applied EMD to each channel of the EEG signal and then reconstructed the signal using only the IMFs that were within our range of interest. Then, to reduce the dimensionality and keep the relevant information, we performed a subsampling according to Nyquist's theorem, that is, for the 0–4 Hz band, we applied a sampling frequency of 8 Hz, which is a 1- sec analysis window and represents eight values for each channel. Therefore, the feature vector had a length of

256. To evaluate the contribution of contiguous bands, this process was repeated to cover 15 Hz.

Dynamic Time Warping. Generally, the degree of similarity between time series is represented by defining a specific distance between two series, the so-called "similarity distance". The smaller the distance, the higher the similarity. The DyTW is a measure of similarity, with a high tolerance to peak variability and to the variability of time or speed. This feature makes it suitable for assessing the similarity of complex series in which there is displacement, compression, or time scaling, such as EEG signals. In our case, the objective was to analyze the similarity between the signals belonging to each stage of the movement and coming from different subjects. The algorithm can be checked at [21].

2.3 Classification Algorithm

Since EEG signals are multichannel, a recurrent problem in this type of work is the high number of attributes in relation to the number of events. For this reason and based on the ability to deal with problems where the number of features is much greater than the number of training points, we chose the RSM, which is one of the algorithms that has demonstrated excellent results in a wide variety of applications [22]. In machine learning, the RSM [23], also named attribute bagging [24] or feature bagging, is an ensemble learning approach that strives to reduce the correlation between estimators in the ensemble by training them with subsets of randomly selected features rather than the entire feature set. Algorithm 1 explains the operation of this classifier array.

Algorithm 1: Random Subspace Method

Input: Dataset $\{(x_1, y_1), (x_2, y_2), \ldots, (x_n, y_n)\}$
Initialization: Base learning classifier F
Number of random subspace rate $k<N$
Number of learning rounds T
for t=1 to T
$D_t = RS(D, k)$ %Random generate a subspace sample from D
$f_t = F(D_t)$ %Train a base(weak) classifier f_t
end
Output: $F(x) = argmax_{y \in Y} \sum_{i:f_{t(x)=y}}^{T} 1$

3 Results

3.1 Similarity Analysis

Figure 1e depicts channel C4 of the recorded EEG signal for an entire grab-and-lift movement. The shaded areas represent the analysis window, which has a duration of one second. In Fig. 1f and Fig. 1g, it is possible to clearly notice that stages 2–3 and 5–6 are fairly close temporarily. From Fig. 1, we know that the gripping is performed with the thumb and index fingers. Stage 2 corresponds to the touch of the object with the first finger (it is not specified which finger), and stage 3 corresponds to the touch of the second digit and the beginning of the elevation. Therefore, it is reasonable that this temporary difference is small (approx. 30 ms), and in some cases, practically null (a simultaneous touch of the two fingers). Something similar occurs between stages 5 and 6, corresponding to replacing the object and releasing it.

To perform a preliminary visual inspection, in Fig. 2, we separately display the EEG segments of the six stages that compose the movement. These waveforms have a length of 1,000 ms, extracted 200 ms before the beginning of the stage and 800 ms after. To visualize the information in our band of interest, we filtered these signals using a bandpass Finite Impulse Response (FIR) of 20th order filter, a cut-off frequency 0.5–7 Hz, and a Kaiser window with a shape parameter of 0.5. Subsequently, we obtained the average of all the events for each stage. In all subjects, the waveforms belonging to stages 2 and 3 were highly similar; this is reasonable considering what has been explained in the previous paragraph. Similarly, the waveforms of stages 5 and 6 were similar, although to a lesser extent. An interesting fact is the remarkable similarity of the plots between different subjects. The most noticeable similarity was observed in stage 1, with a characteristic peak at approximately 0.5 s and two local minima at 0.3 and 0.6 s.

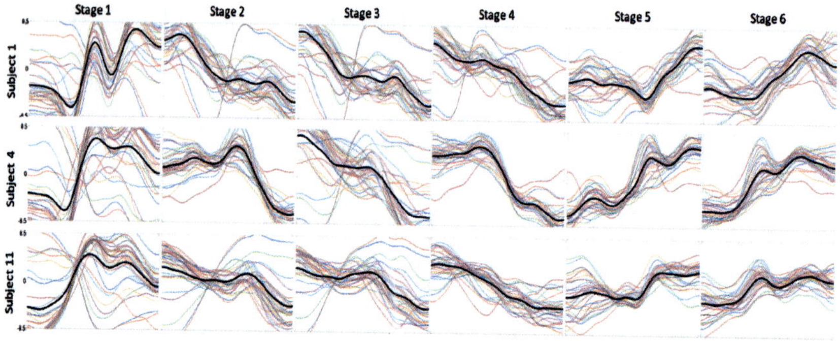

Fig. 2. Waveforms of subjects 1, 4, and 11. The columns represent the stages and the rows the subjects. The colored lines depict the 32 channels, and the thickest black line displays the average of all channels.

To obtain a quantitative measure of the similarity between waveforms, we applied the DyTW measure. Since similarity metrics were performed between pairs of signals, before running DyTW, all events belonging to the same class were averaged. Table 1 lists the similarity measurements between participants 1 and 9 (the remaining participants showed a similar pattern). The values in Table 1 are consistent with the plots in Fig. 2. The diagonal values, which represent the distance between the same stages from different subjects, tended to be lower, which implies that these waveforms present high similarity.

3.2 Classification Using Discrete Wavelet Transform

The feature matrices (1560 * 256) of the 12 subjects obtained by DWT were fed into the RSM classifier and evaluated by applying 10 × 10 k-fold cross validation, (almost

Table 1. Similarity measures applying Dynamic Time Warping algorithm.

		Subject 9					
		Stage 1	Stage 2	Stage 3	Stage 4	Stage 5	Stage 6
Subject 1	Stage 1	**41.7**	172.7	163.4	186.7	110.0	59.9
	Stage 2	76.9	**53.6**	41.4	40.5	117.5	99.5
	Stage 3	108.2	**46.3**	**40.9**	**30.9**	143.9	144.9
	Stage 4	147.2	34.3	28.3	**30.7**	169.0	198.1
	Stage 5	89.8	74.2	77.59	106.9	**0.32**	61.19
	Stage 6	38.1	189.9	178.4	199.7	128.2	**36.8**

Measures of similarity between subjects 1 and 9. The values are dimensionless.

identical results were obtained using 70% of the data for the classifier training and 30% for the test stage). The results are presented in Table 2. The results are quite promising; using only information from the 0–4 Hz band, a maximum of 89.2% and an average accuracy of 77.3% were achieved. With the proposed method, the performance in the classification decreased as the levels increased. This behavior is not usual, since it would be expected that other frequency bands contribute to discriminatory information. In our opinion, this may occur due to the way the characteristics vector is created, since, when adding another band, the number of attributes increases exponentially. Therefore, the size of the training set was probably insufficient. To alleviate this problem, we additionally applied dimensional reduction by means of Principal Components Analysis (PCA) and Linear Discriminant Analysis (LDA). Even though with this alternative, we tried to include information up to 15 Hz, the classification results did not improve.

Table 2. Classification accuracy using Discrete Wavelet Transform.

Subj	1	2	3	4	5	6	7	8	9	10	11	12	Av
Ac	89.2	69.2	82.6	82.6	71.0	77.9	76.6	75.8	79.4	78.8	73.0	71.9	**77.3**
Ac + 1d	88.5	67.6	80.0	80.9	70.6	76.9	76.5	75.6	79.3	78.9	72.2	70.9	**76.4**
Ac + 2d	85.6	59.0	78.4	76.9	62.3	72.4	66.4	66.7	75.8	68.5	63.8	60.8	**69.7**

Ac = Approximation coefficients, 1d = Detail Coefficients 1, 2d = Detail Coefficients 2, Av = Average

The F1-score can be interpreted as a weighted average of the precision and recall. Therefore, this score takes both false positives and false negatives into account and allows us to obtain a better idea of the classifier's performance. Table 3 presents the results of the F1-score accuracy (one vs. all) for the 12 subjects analyzed. Figure 3 shows the confusion matrices of subjects 1, 4, and 12. These subjects correspond to the best case, the average case, and the worst case, respectively.

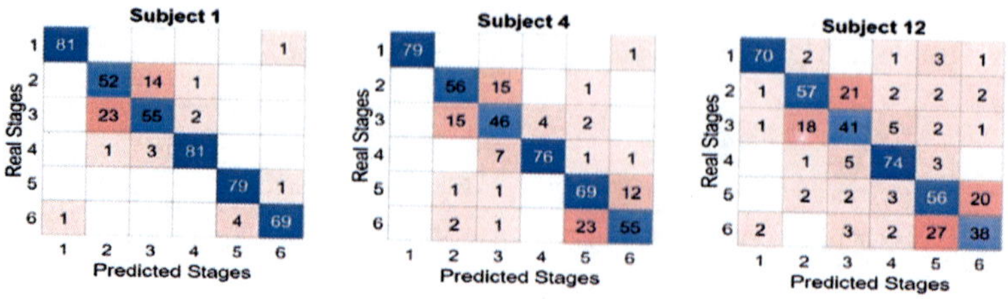

Fig. 3. Confusion matrix for the best case (Subject 1), for a normal result (Subject 4), and for the worst case (Subject 12).

From Table 3 and Fig. 3, we can see that for all participants, the easiest stages to classify were stages 1 and 4. For subjects 1, 3, 5, 8, 9, 10, and 11, the most difficult stages are 2 and 3, and for subjects 2, 4, 6, 7, and 12, the most complicated ones were stages 5 and 6. In general, taking into account that stages 2 and 3 correspond to the grip of the object (touching the object with the thumb and index fingers) and that stages 5 and 6 correspond to releasing the object, the performance metrics suggested that the extracted characteristics and the classification algorithm applied are adequate to discriminate stages of the movement temporarily distant and that it is necessary to explore additional alternatives to classify stages that are temporarily near.

Table 3. Classification results using DWT.

S	Acc	F-S1	F-S2	F-S3	F-S4	F-S5	F-S6
1	89.22 ± 0.81	98.8	72.7	72.4	95.9	96.9	95.2
2	69.21 ± 1.87	78.8	73.9	70.8	77.8	63.3	56.3
3	82.60 ± 2.05	98.2	68.1	64.6	87.2	87.6	89.2
4	82.64 ± 2.01	99.4	76.7	67.2	92.1	77.1	73.3
5	71.03 ± 1.72	87.1	61.8	56.7	86.6	73.9	70.6
6	77.91 ± 1.31	98.6	77.7	73.1	88.6	71.5	68.1
7	76.61 ± 2.56	88.9	80.0	75.2	81.4	67.1	66.2
8	75.74 ± 1.95	89.7	65.3	58.3	80.0	83.2	84.7
9	79.41 ± 2.30	93.4	67.1	65.8	91.8	83.2	75.3
10	78.79 ± 2.18	96.6	56.6	59.5	83.8	76.0	84.4
11	73.02 ± 1.31	80.0	56.9	60.7	81.9	83.1	81.6
12	71.91 ± 1.69	92.7	69.1	58.6	87.1	63.6	56.7
Avg	**77.34 ± 1.81**	**91.85**	**68.8**	**65.24**	**86.1**	**77.2**	**75.13**

Values are expressed as a percentage (%). S = Subject, Avg = Average, Acc = Accuracy, F-S = F1-Score

3.3 Classification Using Empirical Mode Decomposition

Figure 4a depicts the decomposition of an EEG channel into seven IMFs, and Fig. 4b illustrates the power spectral density estimate for each of the IMFs obtained. It can be seen that the EMD decomposition operates as a filter bank whose cutoff frequencies vary slightly according to the information content of the signal under analysis. It should be noted that when only working with the 7, 6, and 5 IMFs, our band of interest was already covered. By adding one more component (fourth IMF), it was possible to reach up to 17 Hz.

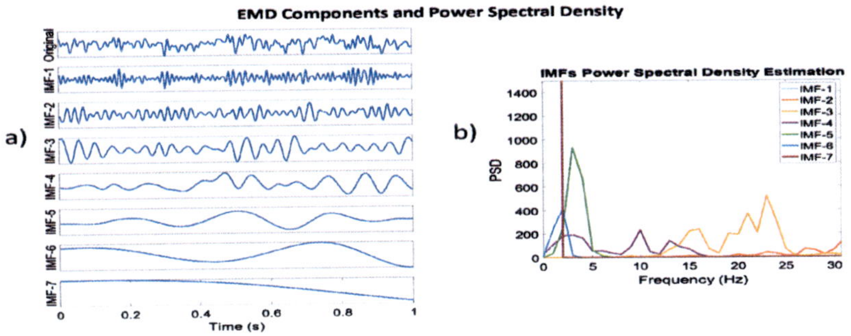

Fig. 4. a) Decomposition of an electroencephalography (EEG) channel in seven Intrinsic Mode Functions (IMFs) applying Empirical Mode Decomposition (EMD), b) estimation of the power spectral density of each IMF. PSD = Power Spectral Density (Watts/Hz)

Figure 5 describes the results obtained with this approach; the results obtained with EMD are quite similar to ones observed with the DWT approach. In addition, it is notorious that a large amount of discriminant information is contained in the low frequency; only with two IMFs (0–4 Hz), a maximum accuracy of approximately 85% was achieved. With three IMFs (0–7 Hz), we achieved a maximum accuracy of 88.95%. Finally, it is interesting to note that the contribution of the band up to 17 Hz (four IMFs), in this case, was insignificant, and there was an improvement of only 0.09%.

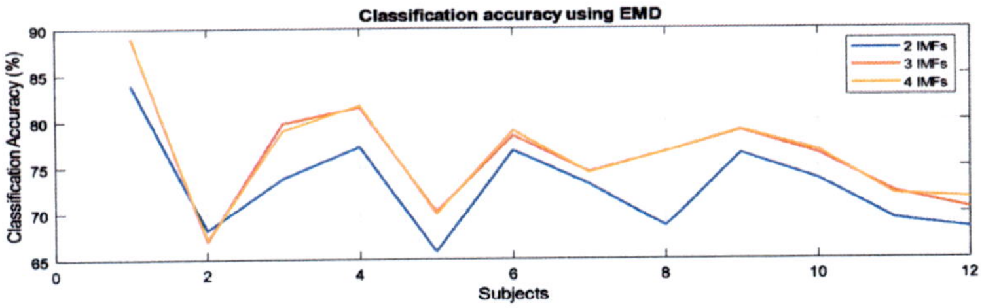

Fig. 5. Classification accuracy using Empirical Mode Decomposition (EMD). The blue line represents the accuracy achieved using 2 Intrinsic Mode Functions (IMFs) to reconstruct the signal, the red line using 3 IMFs and the yellow line using 4.

3.4 Classification Using a Single Classifier

Based on the similarity between waveforms observed in Fig. 2, we tried to use a single classifier for all subjects. For that purpose, we concatenated the characteristic arrays obtained by DWT from all subjects, representing a set of 18,720 events with 256 attributes each, and then fed the RSM algorithm with these data. Figure 6 presents the confusion matrix obtained with this procedure. Considering that BCI systems are usually customized according to the particularities of each subject, in our opinion, the results are quite encouraging. We achieved an accuracy of 64.71% (48.04% over chance level). As observed in the individual classification, the most considerable difficulty for discrimination occurred between the stages that were temporarily closest, i.e., stages 2–3 and 5–6.

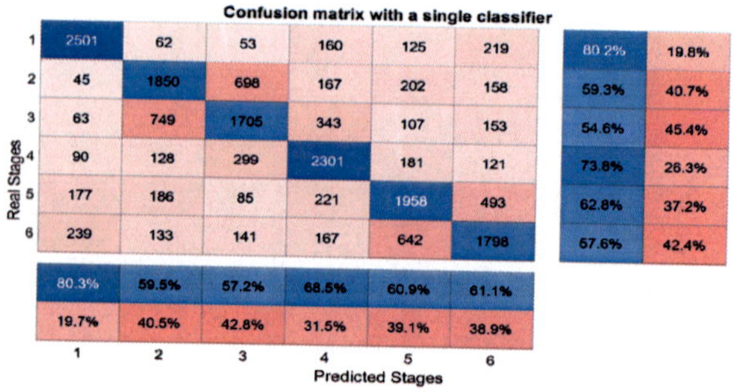

Fig. 6. Confusion matrix by using a single classifier for all participants.

4 Discussion

Several studies have demonstrated inter-subject and inter-session variability in EEG signals [25, 26]. This means that the EEG signals that belong to the same type of movement but were generated by distinct subjects are different. Moreover, this variability is observed in signals of the same person but recorded in different sessions. This fact causes the BCI systems based on EEG to be customized, implying that the stages of the system must be adjusted to the particularities of each user. Additionally, this factor also prevents the creation of massive databases in which it is possible to apply complex classification algorithms without suffering the "dimensionality curse." The low-frequency waveforms shown in Fig. 2 depict an unusual similarity in EEG signals, and that could contribute to the solution of the problems as mentioned above. The results of the classification using a single classifier for all users (Fig. 6) support this hypothesis, since, in our opinion, 64.71% accuracy is a promising result considering that we did not employ any technique of domain adaptation, synchronization, or Transfer Learning.

The two proposed methods are uncomplicated and efficient; we reached a maximum value of 89.22%. Based on the inspection of the confusion matrices, we discovered that the proposed method can satisfactorily discriminate the temporarily distant stages of the movement and that the most challenging stages are those that are close (approximately 30 ms). This result suggests the exploration of complementary processes focused solely on the classification of these stages. For example, a hierarchical classification could be implemented, in which in the first level, the classes 2–3, are merged and a classifier of five classes is applied, and in the second level, a classifier focused on discriminating only these stages by means of characteristics extracted with methods oriented to maximize the variance in binary problems, such as the Filter Bank Common Spatial Patterns (FBCSP), is used. It is important to emphasize that, so far, BCI systems aimed at decoding movements through scalp EEG have achieved reliable decoding of only two movements (left hand and right hand) [3]. The analysis performed in this work differs from the rest since the movements of this experiment are different from those commonly analyzed. The movements of this database are executed by the same hand; that is, there is no contralateral activation that facilitates detection and, besides, the movements are executed at normal speed.

5 Conclusions

In our opinion, the most relevant result is the remarkable similarity observed between the waveforms coming from different subjects. This similarity was verified through the application of a single classifier for all users, which achieved an accuracy of 64.71%, which is 48% more than the chance probability. This fact suggests the existence of specific patterns in low-frequency EEG signals that present robust features, i.e., patterns that remain somewhat invariant despite the high variability reported intra and inter subjects in EEG signals.

Our results are consistent with other work focused on low frequencies, such as [27]. Using information obtained only from the 0–4 Hz band, we achieved a maximum precision of 89.22% (chance probability 16.67%). This data demonstrates that for this type of problem, the most relevant information is concentrated in low frequencies. From the visual analysis and the confusion matrices, we identified that the problems in classification are because the stages are almost completely overlapped. The database used corresponds to EEG signals of executed movements. However, considering that numerous medical imaging studies show that there are regions of the brain that are activated in a similar way during ME and MI [7], in our opinion, the work done could be useful in the development of MI-BCI systems.

Data Availability
The database can be downloaded at https://www.nature.com/articles/sdata201447.

Conflicts of Interest
The authors declare that they have no known competing for financial interests or personal relationships that could have appeared to influence the work reported in this paper.

Funding Statement
This research was supported by the Ecuadorian Government *"Instituto de Fomento al Talento Humano"* (IFTH) and the National University of Loja.

References

1. He, B., Gao, S., Yuan, H., Wolpaw, J.R.: Brain–computer interfaces. In: He, B. (ed.) Neural Engineering. Second Edition, Boston, MA, US, pp. 87–151. Springer (2013)
2. Graimann, B., Allison, B., Pfurtscheller, G.: Brain–computer interfaces: a gentle introduction. In: Graimann, B., Pfurtscheller, G., Allison, B., (eds.) Brain-Computer Interfaces: Revolutionizing Human-Computer Interaction, pp. 1–27. Springer, Heidelberg (2009)
3. Lotte, F., et al.: A review of classification algorithms for EEG-based brain-computer interfaces: a 10-year update. J. Neural Eng. **15**(3), 031005 (2018)
4. Meltzner, G.S., Heaton, J.T., Deng, Y., De Luca, G., Roy, S.H., Kline, J.C.: Silent speech recognition as an alternative communication device for persons with Laryngectomy. IEEE/ACM Trans. Audio Speech Lang. Process. **25**(12), 2386–2398 (2017)
5. Han, S.H., Jeong, H.B., Park, K.Y., Shin, H.W., Youn, Y.C., Kim, S.Y.: P300 latency changes in patients with mild cognitive impairment after taking Choline Alphoscerate: a preliminary study. eNeurologicalSci **11**, 5–8 (2018)
6. Diez, A., Juárez, A.L., Cañas, B., Alvarez-Astorga, A., Nevado, A., Molina, V.: T181 Abnormal frontal and parietal synaptic gain responses in first episode schizophrenia during a P300 target detection task. Schizophr Bull. **44**(suppl_1), S186–S186 (2018)
7. Guillot, A., Di Renzo, F., Christian, C.: The neurofunctional architecture of motor imagery. Adv. Brain Neuroimaging Top. Heal. Dis. Meth. Appl. **3**, 433–456 (2017)
8. McFarland, D.J., Sarnacki, W.A., Wolpaw, J.R.: Effects of training pre-movement sensorimotor rhythms on behavioral performance. J. Neural Eng. **12**(6), 066021 (2015)
9. León, M., Orellana, D., Chuquimarca, L., Acaro, X.: Study of feature extraction methods for BCI applications. In: Botto-Tobar, M., León-Acurio, J., Díaz Cadena, A. (ed.) Advances in Emerging Trends and Technologies, pp. 13–23. Springer, Cham (2020)
10. Cheyne, D.O.: MEG studies of sensorimotor rhythms: a review. Exp. Neurol. **245**, 27–39 (2013)
11. Luciw, M.D., Jarocka, E., Edin, B.B.: Multi-channel EEG recordings during 3,936 grasp and lift trials with varying weight and friction. Sci. Data **1**, 140047 (2014)
12. Alickovic, E., Kevric, J., Subasi, A.: Performance evaluation of empirical mode decomposition, discrete wavelet transform, and wavelet packed decomposition for automated epileptic seizure detection and prediction. Biomed. Signal Process. Control **39**, 94–102 (2018)
13. Nunes, J.C., Deléchelle, É.: Empirical mode decomposition: applications on signal and image processing. Adv. Adapt. Data Anal. **1**(1), 125–175 (2009)
14. Maheshwari, S., Kumar, A.: Empirical mode decomposition: theory and applications. Int. J. Electron. Electr. Eng. **7**(8), 873–878 (2014)
15. Izakian, H., Pedrycz, W., Jamal, I.: Fuzzy clustering of time series data using dynamic time warping distance. Eng. Appl. Artif. Intell. **39**, 235–244 (2015)
16. Varatharajan, R., Manogaran, G., Priyan, M.K., Sundarasekar, R.: Wearable sensor devices for early detection of Alzheimer disease using dynamic time warping algorithm. Cluster Comput, 1–10. Springer, New York (2017)
17. Huang, N.E. et al.: The empirical mode decomposition and the Hubert spectrum for nonlinear and non-stationary time series analysis. Proc. R. Soc. A Math. Phys. Eng. Sci. **454**(1971), 903–995 (1998)
18. Kedadouche, M., Thomas, M., Tahan, A.: Monitoring machines by using a hybrid method combining MED, EMD, and TKEO. Adv. Acoust. Vib. **2014**, 1–10 (2014)
19. Rehman, N., Mandic, D.P.: Multivariate empirical mode decomposition. Proc. R. Soc. A Math. Phys. Eng. Sci. **466**(2117), 1291–1302 (2010)
20. Hemakom, A., Goverdovsky, V., Looney, D., Mandic, D.P.: Adaptive-projection intrinsically transformed multivariate empirical mode decomposition in cooperative brain-computer

interface applications. Philos. Trans. R. Soc. A Math. Phys. Eng. Sci. **374**(2065), 20150199 (2016)
21. Petitjean, F., Forestier, G., Webb, G.I., Nicholson, A.E., Chen, Y., Keogh, E.: Faster and more accurate classification of time series by exploiting a novel dynamic time warping averaging algorithm. Knowl. Inf. Syst. **47**(1), 1–26 (2016)
22. Hosseini, M.P., Pompili, D., Elisevich, K., Soltanian-Zadeh, H.: Random ensemble learning for EEG classification. Artif. Intell. Med. **84**, 146–158 (2018)
23. Ho, T.K.: The random subspace method for constructing decision forests. IEEE Trans. Pattern Anal. Mach. Intell. **20**(8), 832–844 (1998)
24. Bryll, R., Gutierrez-Osuna, R., Quek, F.: Attribute bagging: Improving accuracy of classifier ensembles by using random feature subsets. Pattern Recognit. **36**(6), 1291–1302 (2003)
25. Chatterjee, S., Bhattacharyya, S., Khasnobish, A., Konar, A., Tibarewala, D.N., Janarthanan, R.: Study of inter-session variability of long-term memory and complexity of EEG signals. In: Proceedings of the 2012 3rd International Conference on Emerging Applications of Information Technology. (EAIT 2012), pp. 106–109 (2012)
26. Espinosa, R.: Increased signal to noise ratio in P300 potentials by the method of coherent self-averaging in BCI systems. Int. J. Medical, Pharm. Sci. Eng. **7**(11), 698–702 (2013)
27. Ofner, P., Schwarz, A., Pereira, J., Müller-Putz, G.R.: Upper limb movements can be decoded from the time-domain of low-frequency EEG. PLoS ONE **12**(8), e0182578 (2017)

Fuzzy Control of Temperature on SACI Based on the Emotion Recognition

Maria Sol Soria, Violeta Maldonado, Danilo Chavez(✉), Kleber Patiño(✉), and Oscar Camacho

Escuela Politécnica Nacional, Quito, Ecuador
{maria.soria01,violeta.maldonado,danilo.chavez,kle-ber.patino, oscar.camacho}@epn.edu.ec

Abstract. This article presents a new alternative for the design of the thermal comfort system for a classroom, based on an experimental and empirical fuzzy model, since it is currently important to provide solutions on topics such as health, well-being and learning performance in students within an academic institution, and because in the scientific literature there are no mathematical models of human behavior that help to design a control algorithm for the management of comfort applications, the present work proposes an alternative of modeling and control algorithms focused on studies that establish the relationship between emotions, temperature and academic performance of students, this being essential in the technological, educational and psychological fields, in such a way that simulations of the proposal were carried out using Matlab, which allowed determining relationships between the student's emotion and the environmental factors, in addition to showing that a temperature control can influence the emotions of a group of students.

Keywords: Thermal comfort · Fuzzy control · Emotions · Temperature

1 Introduction

Nowadays, with the use of technology, it is more common to implement systems that provide environmental comfort to human beings through the control systems, for example: lighting, temperature, air, etc. In addition, this type of application is important within a classroom since a high level of environmental quality may considerably improve attention, concentration, learning, hearing, and performances of students [1].

The first scientific studies about the effects of the thermal environment quality in classrooms began around the middle of 1950 s. Since this period, investigations have been carried out, such as effects of different classroom temperatures or autonomic cardiac control and cognitive performance in undergraduate students, [2] where there is evidence of how the temperature affects the health of students. Other works are acoustic, thermal and luminous comfort in classroom [3]. The effects of classroom air temperature and outdoor air supply rate on the performance of schoolwork by children [4], among others,

M. Botto-Tobar et al. (Eds.): ICCIS 2020, AISC 1273, pp. 94–106, 2021.
https://doi.org/10.1007/978-3-030-59194-6_9

that analyze the influence of temperature on the student, showing that the control of temperature in the classroom influences in health, concentration, learning, etc. Over time, various thermal comfort indices have been established for the analysis of indoor climates in which it can be seen that different factors intervene, such as physical (temperature, humidity, wind, etc.), physiological (age, sex, health, etc.) emotional (sadness, joy, etc.), social (fashion in dress) and among other aspects. However, in the case of a young person and mentally healthy, the feeling comfort is caused by the thermal environmental conditions. Therefore, that person's comfort is said to be given by the mental condition of satisfaction that the environment causes on him [5].

The use of better controller algorithms to manage the indoor environments has been increasing with the growth of automation, as has the use of fuzzy theory to control different systems such as HVAC air condition systems [6]. An example of this is the study that was carried out, [7] where the objective was to find a fuzzy mathematical model to analyze the health of agricultural workers, related to thermal and acoustic environment, where the output variable was the human well-being index. Another example is the Temperature Control inside a Room using Fuzzy logic method, this work use a Fuzzy logic controller for room temperature controller, the input for this controller are six ambient parameters and the output is the temperature [8]. The use of controllers based on especially fuzzy rules allows the implementation of a multi-criteria control used mainly when the literature does not present mathematical models of the system to be studied, therefore it is essential to incorporate knowledge about fuzzy sets, functions of membership, blurred operations, inference rules, defuzzification necessary for decision-making of a control system with uncertainties as is the case in this work [6].

In this document, the design of a fuzzy model is proposed considering the analysis between classroom temperature, body temperature and student emotion. For the control, four basic emotions will be used, which are sadness, anger, tranquility and happiness, and important factors such as the type of mechanical effort that the student performs in the classroom, the type of clothing that the student uses and the thermal sensation according to the temporality of the city.

2 Thermal Comfort at SACI

Thermal comfort is a determining aspect of interiors and more in a classroom because the students spend a great part of time there. Over time, different models of thermal comfort have been produced in interior spaces, giving rise to multiple norms and standards [9]. In this work, the norm INEN – ISO 7730 detailed below is used.

2.1 Importance of Thermal Comfort

The thermal comfort at classroom is important because in agreement with the temperature produces a reaction on the students for example: if the temperature is hot the students could fall asleep or get distracted on the other hand, if the classroom is cold the students get distracted because they tried to warm up [8]. The ideal temperature in static classroom is between 20°C and 23.6°C. The temperature in a classroom where students are in constant motion is between 18°C and 20°C [4] However, this also depends on the ventilation in the classroom.

2.2 Factors that Influence Thermal Comfort

According to the norm INEN – ISO 7730 the thermal sensation by a human is related to the overall thermal balance of body. This balance depends on physical activity, clothing, and environmental parameters. The thermal sensation can be estimated by PMV (predicted mean vote) [10]. The environment factors that influences the thermal sensation are air temperature, average radiant temperature, RH, and airspeed. While the physical factors are age, sex, gender, weight, metabolic rate, heat exchange, and residence time. Also there are sociocultural and psychological factors such as type of dress, an expectation of comfort, and eye contact with the outside [5, 10].

3 Impact of Emotions on Temperature Bodily

Emotions in the human being not only cause a change in the person's attitude but also produce a variation in body heat in some body areas. This variation will be can increase or decrease the body heat in a person, for this reason is important to analyze emotion and according to the body heat it produces, determine if the temperature of the classroom should increase or decreases to later produce a variation in the student emotion. According to a study on Finland to 700 people, [11, 12] there exists a relation between the emotion and body temperature in a person, it shows that each emotion produced more or less body heats variating the body temperature of people. According this study, the emotions are classifying in simple (anger, fear, disgust, happiness, sadness and neutral) and complex (anxiety, love, depression, contempt, pride, shame, and envy).

Figure 1 shows the four emotions that will be analyzed. If the emotion is sadness the temperature of the people decreases and so the room temperature should be increased. If the person is angry or happy, their temperature increased, and the room temperature should be decreased. In a classroom there are many students and a social set point is used based on the percentage of people with an emotion.

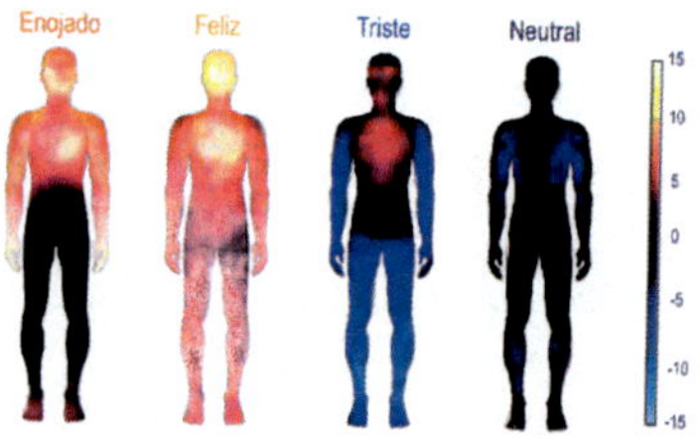

Fig. 1. Body temperature of four emotions

4 Control Loop Proposal

In order to establish a model that does not require several expensive hardware resources, such as humidity sensors, thermal insulation sensors, infrared human temperature sensors, and because there is no mathematical model in the literature that relates the thermal sensation and emotions of the human being with the external temperature of the environment, the following control model is proposed applying fuzzy control criteria.

4.1 Importance of Thermal Comfort

Based on the ISO 7730 standard [10] which deals with the ergonomics of the environment for thermal well-being, the thermal comfort model for humans proposed in this project involves several factors that must be considered, which are: Classroom temperature, Type of activity carried out in the classroom, Type of insulation of student clothing. When relating the factors mentioned above, a criterion can be obtained under the fuzzy control structure of the thermal sensation for the student within a classroom in such a way that it is first necessary to establish which are the inputs and outputs of the model, at the same time, indicate what type of fuzzy control membership functions are used as shown below:

Where the membership functions for each input and output were chosen as follows:

- "type of clothing" is classified according to the climatic season in which the city of Quito is located, these were chosen because it is much simpler to program an algorithm that establishes what type of clothing the students commonly use according to the time zone and climatic station in which they are located, therefore since it is a city where there are only two seasons (winter and summer), it has been divided into three triangular type membership functions as shown in Fig. 2.

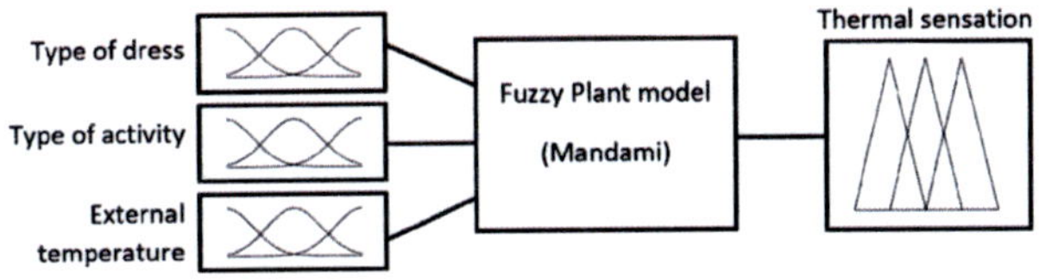

Fig. 2. Inputs and outputs of the fuzzy plant model.

- "type of activity" is classified into three triangular membership functions according to the type of academic activity that is frequently carried out in the university, including working with machines such as engines, sensors, robots, etc., working at the software level with computers, and work clearly in classes where the teacher is the only one who moves and the students remain seated.
- "External temperature" is classified into five triangular membership functions according to the temperature values established and suitable for the thermal comfort of students so that from 18 to 20 degrees Celsius is an optimum temperature for students who are in motion or making some mechanical effort, and from 20 to 22 degrees Celsius it is considered an optimum temperature for students who are sitting listening to the teacher or working on the computer.
- "Thermal sensation" is classified into four functions of triangular membership according to the thermal sensations that generally exist in the city of Quito considering that it belongs to a tropical zone, such as slightly cold, comfortable, slightly hot and hot (Fig. 3).

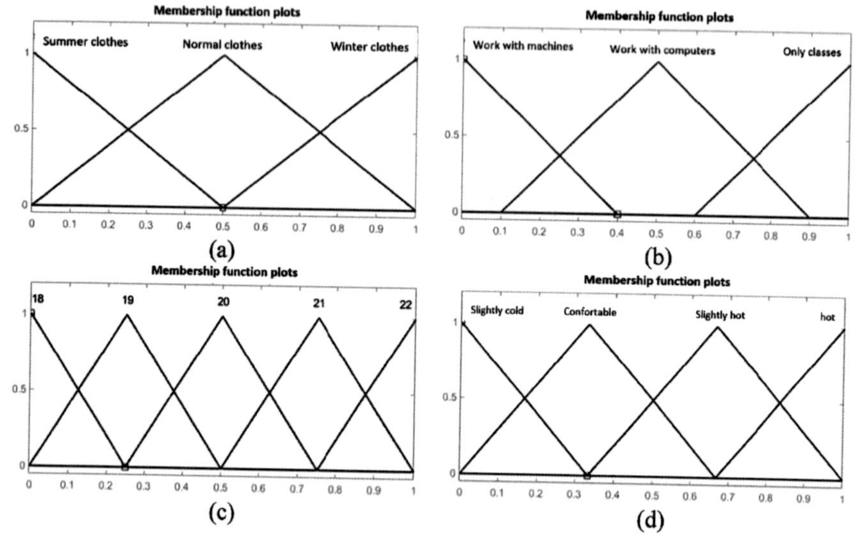

Fig. 3. (.a) Input variable "Type of dress", (b) Input variable "Type of activity", (c) Input variable "External temperature", (d) Output variable "Thermal sensation".

Where the 45 rules of the model are stated as follows, considering that in the type of activity "A" means working with machines, "B" means working with computers and "C" means only classes (Table 1).

Table 1. Rules of the model.

if	And	And	Then
Type of dress is	Type of activity is	External temperature is	Thermal sensation is
Summer clothes	A, B, C	18	Slightly cold
Normal clothes	A	18	Comfortable
Normal clothes	B, C	18	Slightly cold
Winter clothes	A, B	18	Comfortable
Winter clothes	C	18	Slightly cold
Summer clothes	A	19	Comfortable
Summer clothes	B, C	19	Slightly cold
Normal clothes	A	19	Comfortable
Normal clothes	A, C	19	Slightly cold
Winter clothes	A	19	Slightly hot
Winter clothes	A, C	19	Comfortable
Summer clothes	A, B, C	20	Comfortable

(*continued*)

Table 1. *(continued)*

if	And	And	Then
Normal clothes	A	20	Slightly hot
Normal clothes	B	20	Comfortable
Normal clothes	C	20	Slightly cold
Winter clothes	A	20	Slightly hot
Winter clothes	B, C	20	Comfortable
Summer clothes	A	21	Slightly hot
Summer clothes	B	21	Comfortable
Summer clothes	C	21	Slightly hot
Normal clothes	A, B, C	21	Comfortable
Winter clothes	A	21	Hot
Winter clothes	B, C	21	Slightly hot
Summer clothes	A, B	22	Comfortable
Summer clothes	C	22	Slightly hot
Normal clothes	A	22	Slightly hot
Normal clothes	B	22	Comfortable
Normal clothes	C	22	Slightly hot
Winter clothes	A, B, C	22	Hot

Once the diffuse model that relates the thermal sensation to the ambient temperature has been established, another fuzzy model must be created that establishes a relationship between the emotion and the thermal sensation of the students. This model is based on experimental temperature data that were carried out with ten students of the National Polytechnic School during three weeks in the month of January of the year 2020 of which two of them were weeks of exams where the students were subjected to academic pressures that are regularly experienced within the campus. The experimental tests consisted of measuring the internal temperature of the student at the same time as the classroom temperature and subsequently storing the data on mood, thermal sensation, type of dress, and state of concentration. These data can be summarized in the following table.

According to Table 2 the emotion depends on some factors and the people personality but if the temperature is cold the person feels tired and if the temperature is hot the person tends to get distracted and feels happy or angry. Also, how can be observed the thermal sensation directly influences the emotion regardless of the internal temperature at which the students are found since there are also variants such as the type of clothing they use and the ambient temperature at which the classroom is located, so that the following model includes the input of disturbances considering that at any time the thermal sensation in the classroom can vary, either by an increase or decrease of people with different thermal sensations, so the following relationship is proposed (Fig. 4).

Table 2. Average data of experimental tests performed on students at the University

External temperature [°C]	Internal student temperature [°C]	Emotion	Thermal sensation	Mood	Type of dress
17.17	36.20	Sad	Slightly cold	Tired	Winter clothes
17.33	36.56	Sad	Slightly cold	Hungry	Winter clothes
18.21	36.18	Normal	Slightly cold	Concentrated	Winter clothes
18.33	36.25	Normal	Slightly cold	Concentrated	Winter clothes
18.43	36.50	Normal	Comfortable	Concentrated	Normal clothes
18.81	36.60	Normal	Comfortable	Concentrated	Normal clothes
19.15	36.70	Normal	Comfortable	Concentrated	Normal clothes
19.51	36.70	Happy	Comfortable	Distracted	Normal clothes
20.17	36.40	Happy	Comfortable	Distracted	Summer clothes
20.43	36.80	Happy	Slightly hot	Distracted	Normal clothes
20.55	36.70	Happy	Slightly hot	Distracted	Normal clothes
20.73	36.50	Angry	Slightly hot	Distracted	Summer clothes
20.81	37.12	Angry	Slightly hot	Frustrated	Normal clothes
21.12	37.23	Angry	Hot	Distracted	Summer clothes
21.47	37.15	Angry	Hot	Stressed	Summer clothes
21.76	36.85	Angry	Hot	Stressed	Summer clothes

Where the membership functions for each input and output were chosen as follows:

- "thermal sensation" is classified into four triangular membership functions which coincide with the membership functions of the output of the fuzzy plant model.

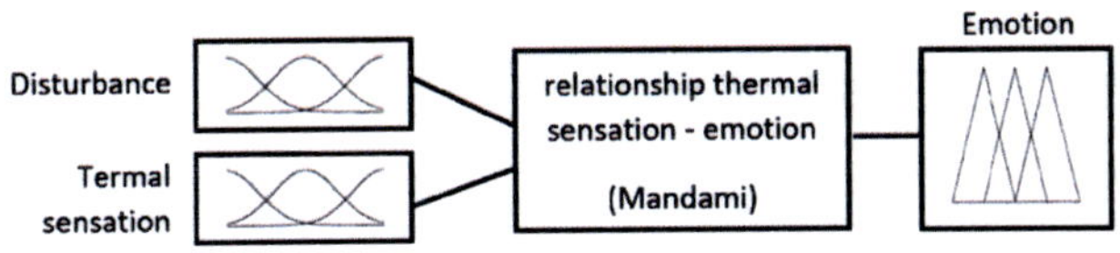

Fig. 4. Inputs and outputs of the relationship between thermal sensation and emotion.

- "disturbances" are classified into four triangular membership functions which are equal to those of thermal sensation.
- "Emotion" is classified into four functions of triangular membership according to the emotions that typically university students, considering that they are under academic pressure, such as: sad, normal, happy, angry (Fig. 5).

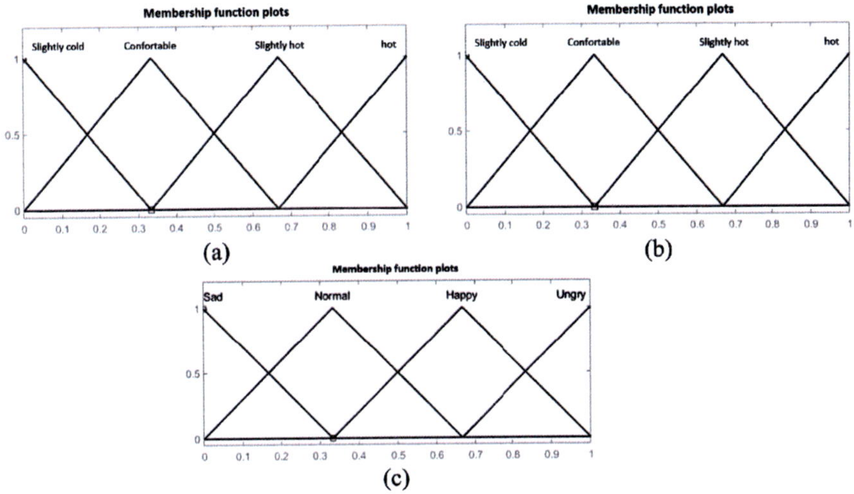

Fig. 5. (.a) Input variable "Thermal sensation", (b) Input variable "Disturbances", (c) Output variable "Emotion".

Where the 16 rules of the model are raised as follows (Table 3):

4.2 Control Proposal

Since the plant is created under the criteria of a fuzzy control system, the most logical thing is to use a controller that uses the same logic, so as it is known that the working plant is the emotion of the students based on the data of the thermal sensation thereof, the following control loop structure is taken into account (Fig. 6).

Table 3. Rules of the relationship between thermal sensation and emotion.

if	And	Then
Thermal sensation is	Disturbances is	Emotion is
Slightly cold	Slightly cold, Comfortable	Sad
Slightly cold	Slightly hot	Normal
Slightly cold	Hot	Happy
Comfortable	Slightly cold	Sad
Comfortable	Comfortable	Normal
Comfortable	Slightly hot	Happy
Comfortable	Hot	Angry
Slightly hot	Slightly cold	Normal
Slightly hot	Comfortable	Happy
Slightly hot	Slightly hot, Hot	Angry
Hot	Slightly cold	Normal
Hot	Comfortable	Happy
Hot	Slightly hot, Hot	Angry

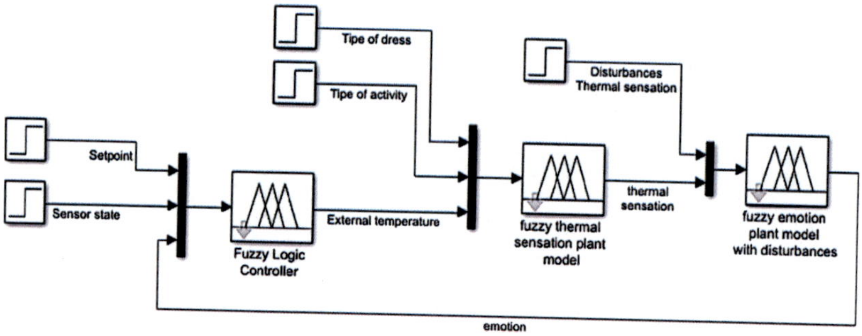

Fig. 6. Control loop proposal.

Where fuzzy controller inputs and outputs come from (Fig. 7):

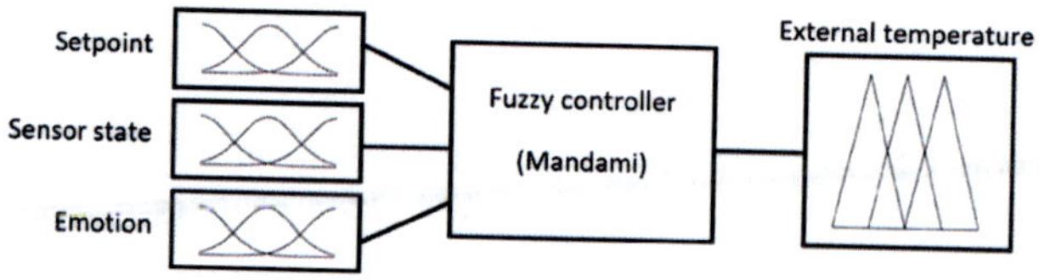

Fig. 7. Inputs and outputs of the fuzzy controller.

Where the membership functions for each input and output were chosen as follows:

- "Setpoint" corresponds to a trapezoidal type membership function corresponding to the desired state of emotion, in this case, the normal state is the most appropriate state for the proposed control loop because with the experiments it has been concluded that students have a higher concentration when they have a normal or neutral emotion, that is to say, that the student is not sad, not happy or not angry.
- "Sensor status" corresponds to 2 trapezoidal membership functions belonging to the state of the emotion sensor, in this case, the sensor can be a bed that uses facial recognition to detect the emotions of each student in the classroom, thus the data that are sent will be the result of an algorithm that establishes a single social emotion.
- "Emotion" corresponds to 4 triangular membership functions that belong to the emotions measured by the sensor, those that will serve as feedback of the control system of this study.
- "External temperature" corresponds to 5 triangular type membership functions that belong to the desired temperature value according to the emotions detected, at this point, it is important to mention that the system actuator would be the HVAC conditioned air system but in this case, it is consider that the fuzzy controller for this system has already been studied in publications such as [6], with good results, therefore it is taken into account taking into account the desired temperature as a value already controlled and processed by the HVAC.

Where the 5 rules of the model are raised as follows (Table 4):

Table 4. Rules of the relationship between thermal sensation and emotion.

if	and	and	Then
Setpoint is	Emotion is	Sensor state is	External temperature is
Normal	Sad	on	22
Normal	Normal	on	20
Normal	Happy	on	19
Normal	Angry	on	18
Normal	none	off	21

5 Results

To show the results, a case of conditions is chosen which are presented in the following table, these data will be observed in Fig. 8 and Fig. 9 (Table 5).

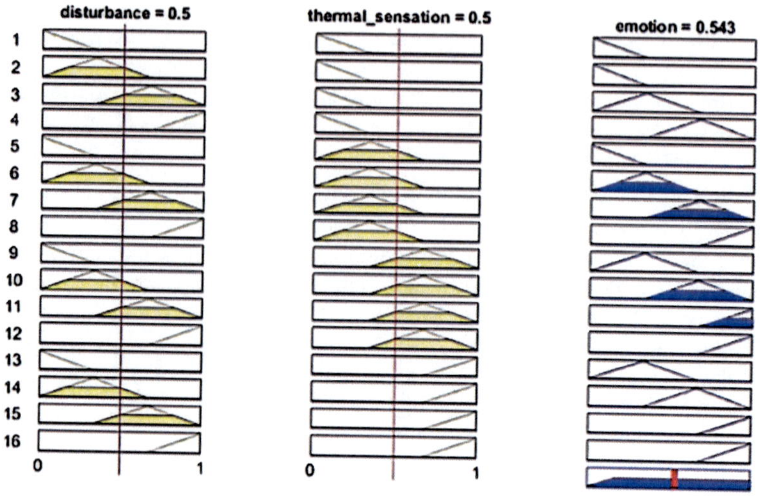

Fig. 8. Output signal.

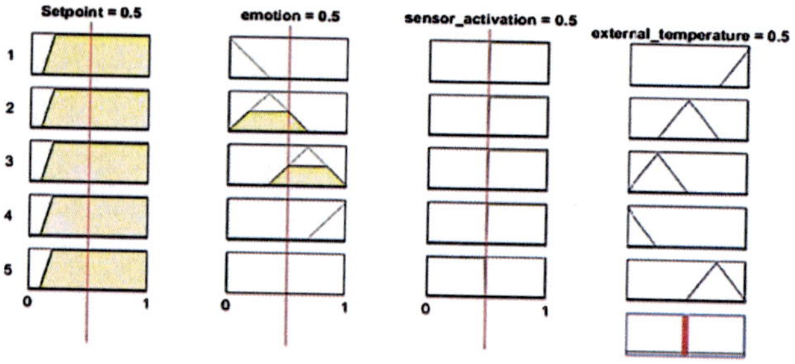

Fig. 9. Control signal.

Table 5. Conditions of case.

Condition	Setpoint	Sensor state	Type of dress	Type of activity
Value of Membership function	Normal	on	Summer clothes	Work with machines
	0.1	0.3	0.1	0.1

With these conditions the results are:

Figure 8 shows that the value of the output signal is 0.543 which corresponds to a "normal" membership function in the emotion, so the control loop is working properly.

6 Discussion

There are some studies that show the benefits of thermal comfort for human beings in the development of their activities, as shown in the article by P. Mondal [8], where a room temperature control is performed using fuzzy logic based on external temperature, humidity and pressure. While in the present work, this approach is used not only to establish a control algorithm since it also presents an alternative modeling of a system where the fundamental variables are the thermal sensation and the emotions of the human being taking into account conditions environmental, clothing and the activity that will be carried out in a classroom, differentiating itself by establishing better conditions of comfort to increase the performance of attention and concentration of students in their academic activities. On the other hand, there is a study by C. Beall [2], which analyzes the effect of ambient temperature on health and cognitive performance in students. In it, it is observed that a thermal discomfort of the student affects the cognitive performance showing sympathetic cardiac hyperactivity, therefore, under this analysis, the present work includes the effect of the student's emotions in the diffuse modeling of a system of control, making it possible to know the level of thermal comfort so that a temperature control of the environment is subsequently executed, achieving better conditions so that the student does not lose attention and concentration.

The results obtained from the implementation of this proposal show great viability of using diffuse control techniques for modeling and controlling a classroom temperature based on the emotions of the students, but, it is important to clarify that these data are for a specific case study which is a group of students from the National Polytechnic School where the environmental conditions, clothing and psychological depend on the climate of the city of Quito, however, this study serves as a guide for future analysis and implementation work of thermal comfort for the institution's classrooms.

7 Conclusions

- The Temperature Fuzzy Control allows for establishing ranges on which a person can work based on the different factors that influence thermal comfort
- In a classroom there are some people and each one has a different emotion, for this reason a social setpoint must be established analyzing the percentage of emotions, same with which they Fuzzy Control will be performed.
- It concludes that Fuzzy control works according to the different conditions that influence the thermal sensation of the student.

The PIJ 17-10 project from Escuela Politécnica Nacional is thanked for providing us with full availability of equipment and technical assurance in order to carry out the validation tests of the project and to implement it properly.

References

1. Corgnati, S.P., Filippi, M., Viazzo, S.: Perception of the thermal environment in high school and university classrooms: subjective preferences and thermal comfort. Build. Environ. **42**(2), 951–959 (2007)
2. Beall,C., Piipari, K., Al-qassab, H., Smith, M.A.: Effects of differen classroom temperatures on cardiac autonomic control and cognitive performances in undergraduate students. Biochem. J. (May), 0–14 (2010)

3. Kruger, E.L., Zannin, P.H.T.: Acoustic, thermal and luminous comfort in classrooms. Bulid. Environ. **39**, 1055–1063 (2004)
4. Alle, N.K.: The effects of classroom air temperature and outdoor air supply rate on the performance of school work by children, pp. 368–372 (2005)
5. Cervantes Pérez, J., Barradas Miranda, V.: Ajuste de escalas de sensación térmica para Xalapa, Veracruz, México. Investig. y Cienc. la Univ. Autónoma Aguascalientes (48), 30–37 (2010)
6. Alcalá, R., Benítez, J.M., Casillas, J., Cordón, O., Pérez, R.: Fuzzy control of HVAC systems optimized by genetic algorithms. Appl. Intell. **18**(2), 155–177 (2003)
7. Schiassi, L., Melo, N.S.M., Tavares, G.F., De Souza, ÍP., Araújo, H.B., Della Giustina, C.: Modelagem fuzzy Em Parâmetros De Bem-Estar Humano fuzzy modeling in human well-beingparameters. Nativa **01**(15), 8–12 (2013)
8. Mondal, P., Mondal, M.: Temperature control inside a room using fuzzy logic method, pp. 7679–7685 (2018)
9. Arballo, B., Kuchen, E., Scientific, N., Naranjo, Y.A.: Evaluación de modelos de confort térmico para interiores. In: VIII Congreso Regional de Tecnología de la Arquitectura-CRETA, Desarro. tecnológico Reg. sustentable, no. October, p. 10 (2016)
10. NTE INEN-ISO 7730, "Ergonomía Del Ambiente Térmico. Determinación Analítica E Interpretación Del Bienestar Térmico Mediante El Cálculo De Los Índices Pmv Y Ppd Y Los Criterios De Bienestar Térmico Local (Iso 7730:2005, Idt)," pp. 5–25 (2014)
11. Nummenmaa, L., Hari, R., Hietanen, J.K., Glerean, E.: Maps of subjective feelings. Proc. Natl. Acad. Sci. U. S. A. **115**(37), 9198–9203 (2018)
12. Gunes, H., Piccardi, M.: Bi-modal emotion recognition from expressive face and body gestures. J. Netw. Comput. Appl. **30**(4), 1334–1345 (2007)

Implementation of Recognition System of People and Computers in a Classroom Through Artificial Vision

Lesly Cadena, Xavier David Rógel, Danilo Chavez(✉), Kleber Patiño(✉), and Jackeline Abad

Escuela Politécnica Nacional, Quito, Ecuador
{lesly.cadena,xavier.rogel,danilo.chavez,kleber.patino, jackeline.abad}@epn.edu.ec

Abstract. This work encompasses the implementation of an autonomous system that recognizes people and object in a classroom to later develop a smart classroom. This system uses artificial intelligent, artificial vision and deep learning technics to process images, sent by a camera that was previously installed in the class. The information is processed through a YOLO-type convolutional neural network, which is open source and is already trained and validated for real-time object detection applications, this information will contribute to the creation of an intelligent classroom. The aim of the system is getting the number of people and laptops that are inside a defined work area. The smart classroom goal is to influence the student's behavior to: improve their academy performance, cognitive skills, and most importantly create a dynamic and interactive learning environment where the relationship between classmates and teacher improves.

Keywords: Neural network · Machine learning · Deep learning · Yolo · Layers

1 Introduction

Nowadays, the equipment of public education centers does not have the appropriate characteristics to generate the best learning environment. Poor lighting, unpleasant room temperature also limited didactic technological resources are some aspects that produce a faulty school environment. These aspects generate loss of concentration, tiredness and eye strain. Because of long learning period in education centers, the students' performance reduces gradually.

The technological development lets student getting better learning tools; like websites or online platforms that make easy students learning. These tools not only allow students get any kind of information but also, they get data to generate a better learning environment. For this reason, developing a smart class allows us to construct a general profile of students to scale the classroom depending on the user's needs. The aim of a smart class is to include new technologies in a classic scholar environment, which

M. Botto-Tobar et al. (Eds.): ICCIS 2020, AISC 1273, pp. 107–120, 2021.
https://doi.org/10.1007/978-3-030-59194-6_10

eventually will develop learning methods to reach useful experiences like improving the relationship between classmates, teachers and new technology.

Smart classroom (SACI) is a place where students and teachers have a good relationship: they could easily talk about any topic because they feel comfortable and have all technological devices to learn fast. Developing a smart classroom produce the following advantages:

- Interactive and dynamic learning experience
- Distance learning
- Safe environments
- Adequate energy management
- Communication improvement between student and teacher
- Development of cognitive skills

The aim of this work is to record the number of people and laptops that are inside the classroom. This information will be used during the creation and development of a social set point related to the number of people in the classroom in order to modify the lighting and temperature inside the classroom. It is important to know that this information is carried out to a subsequent process: emotion recognition. This new phase uses these results to adapt the comfort parameters of the work environment and influence over control actions that allow getting an adequate learning environment.

The detection of people and objects is done by artificial vision algorithms. Unlike current system, this method uses images processing by probabilistic mechanism, which are trustworthy and efficient. A pre-trained convolutional neural network called YOLO (You Only Look once) was implemented in this project. This type of network is used in artificial vision applications. The aim of YOLO is to extract the most important features from the input image, and then perform a prediction process based on the objects that need to identify. The system is composed of: a camera to capture information of the classroom in real time; a computer to process the camera's information and show the results in a graphic interphase.

This paper is organized as follow: In the Sect. 2, we present a brief summary about the purpose of the work and the current mechanisms to identify people and objects in specify work area. In Sect. 3, information about the algorithms and tools, used to develop the present work, is introduced. Then, in Sect. 4, we show the results of this work. Finally, section five provides a brief conclusion and suggesting for future works.

2 State of Art

The aim of the general project is to provide the basis for the smart class (SACI) developing, whose purpose is to improve the learning environment of education centers. A suitable study environment uses new learning methods related with continuous technological development that promotes the students' cognitive skills and avoid tiredness due to log hours of study.

To develop a suitable classroom environment, it is necessary to know some characteristics of the work area that helps to motivate, give comfort for the students and get an efficiently manage energy. Each one of these features are evaluated before processing and decoding the information about the work area in order to take control actions that solve the students' needs. Specifically, this work records the number of people and laptops that are inside the classroom in order to modify the temperature and lighting of work area. For this reason, it is necessary to know some identification and recognition mechanisms currently used.

2.1 Current Identification Techniques

The current technics to record people mostly use electronic sensors, which detect the presence of any object that cross its measurement area. These devices form a big electronic system to increase the detection area and improve the capacity of detecting the presence or absence of any people in the class. There are some identification and record methods frequently used:

Sensor Kinect. The sensor Kinect of Microsoft is a useful tool in some applications like people motion detection in real time. This sensor is used in Xbox platform to interact in video games using augmented reality. It has the capacity of discriminating the position and the motion of each human joint that are identify by the Kinect. This sensor has a depth system named RGB/Depth, which divides the image in three parts or segments, these layers are translated in Cartesians coordinates that has the position of each joints detected in the capture image.

Figure 1 shows the developing interface in Labview. This picture shows each of the detected joints by the sensor. This information is saved on a matrix of pixels and used to get the range of movement of the detected person [1].

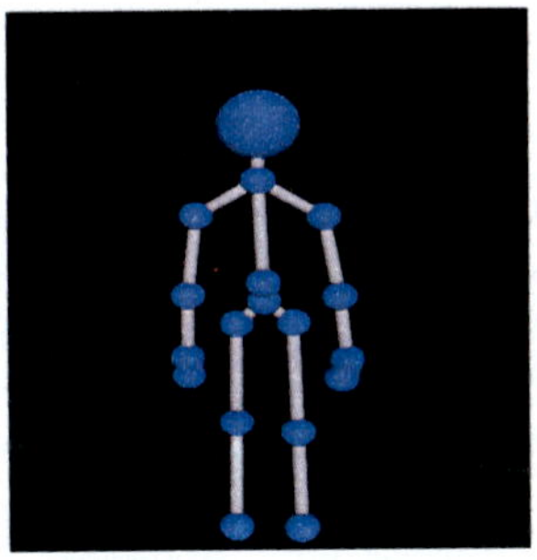

Fig. 1. People recognition interface of sensor Kinect in LabView

The interface development uses libraries of sensor Kinect; however, this method shows some difficulties during the testing process due to the small number of people that Kinect can identify in one capture. Light is another aspect that influence adversely in the image capture. However, the most relevant disadvantage is the software: it has some limitations during the programming process, and it has a high cost for the commercialization of the final product. These are the reasons why this method has been discarded [1].

People Quantifier with Electronic Sensors. It is one of the most widely used detection mechanisms. Although it is easy to implement, this method needs a complex hardware system to fulfill the aims of this project. This kind of sensors has some advantages related to a fast data processing and signals interpretation. However, it does not have the capacity to distinguish the variety of different objects. They only detect the presence or absence of an object. The most used electronic sensors are: [2].

- Inductive sensor
- Photoelectric Sensor
- Optical Sensor
- Capacitive Sensor

Furthermore, the information acquired by these sensors is limited and requires more software processing to cover the needs of the SACI. In addition, it requires more development of the tools and maintenance of the hardware, which increases the final product cost. For that reason, this option is also discarded.

Computers Quantifier Using IP Addresses. This method uses IP addresses of the computers to identify the number of devices that are connected in the classroom. However, it means that the system monitors only the internet network inside the classroom. Nevertheless, being a system of easy installation and monitoring, depends directly on the distribution of the network. For this reason, if there is a fault in the network all the information will be lost. In addition, it is necessary that all computers are always connected to this network, otherwise they are not counted. For that reason, this method is not the most suitable for this project [3].

2.2 Detection and Identification Methods of People and Objects Using Artificial

The machine vision mechanisms, which use trained convolutional neural networks, are based on probabilistic algorithms suitable for classification, location, detection and even segmentation of different objects. The type and quantity of objects that the neural network can identify will depend directly on its architecture and training process.

It is important to note that this identification mechanism can quickly classify objects or people according to the database, which they were trained. This method can be implemented using free programming software such as Python, which are currently widely used since they are multi-platform and multi-paradigm [4].

Unlike using several electronic sensors to count the number of people, this method only needs a webcam to cover the work area, together with a computer where the information captured by the camera is processed and stored.

After evaluating some characteristic aspects of current detection techniques and identification mechanisms using artificial vision, the latter is selected as the most appropriate method to be implemented in an intelligent classroom due to programming easiness, hardware installation, and above all the information acquired from the work environment. These data will not only allow the counting of people and laptops, which is the main objective of this project, but also this information will be useful to cover some needs of the SACI, previously mentioned such as the creation of a social setpoint that allows the manipulation of lighting and temperature of the student environment [4].

3 Methodology

There are several artificial vision algorithms used to identify and recognize specific characteristic, patterns within images. Currently, due to the great technological development and especially the advance of computational processing, the neural networks have had an exponential advance in this field.

Neural networks are a subset of Machine Learning. They are characterized by their ability to learn and they are specialists in pattern recognition. For the creation of a neural network, mathematical and probabilistic algorithms are used such as decision trees so that the network learns automatically without the need to explicitly program each neuron [5].

3.1 Convolutional Neural Networks [4]

They are a type of artificial network where neurons join in receiving fields in a similar way to brain neurons in the primary cortex of the visual field. Therefore, they are very efficient in artificial vision applications, whose main function is the classification and recognition of images or patterns. These networks are made up of several sets of layers, which perform specific functions from image acquisition, application of filters to classification processes and detection of objects. The term convolutional network refers to the main characteristics of the entered image, such as the representation of a face, the most relevant parts being the nose, eyes, and mouth.

Convolutional networks have 5 sets of main layers.

- **Input Layer:** The layer that receives raw or unprocessed images.
- **Convolutional Layers:** Characteristic layers of this network, are responsible for applying filters to the input image to be processed later, so that certain characteristics of the image become more dominant and stand out above the rest.
- **Activation Layers:** An activation function such as Relu, Sigmoid Tangent Hyperbolic, or any other depending on the application is applied to these layers. These functions are applied to the output of the convolutional layers to represent each neuron as a non-linear model and to find a specific model that fits my detection problem.
- **Pooling Layers:** The main function of this process is to reduce the size of the image characteristic map and obtain the most relevant information from it. It can capture the most representative parts of the image and store them in a matrix of pixels. This information is useful for carrying out a data analysis that provides the most representative characteristics of the input image as a final result. The pooling function divides the image's pixel map by regions from which it obtains a unique value. The best known filters are: Max Pooling, Average Pooling and Min Pooling; but among these the most used is the one that obtains the maximum value of each region, that is, the Max Pooling filter. This type of filter has the function of dividing the entire image into "n" matrices, from which the highest number is chosen to subsequently form a final matrix as shown in the following Fig. 2.

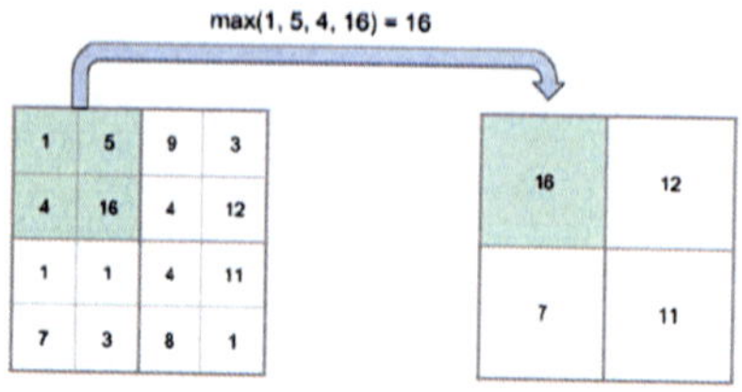

Fig. 2. Filter type Max pooling [4]

- By reducing the number of parameters on the image pixel map, the training process is performed in less time. In addition, the network does not vary when the input image makes small changes of translation or rotation since it always takes the maximum value of a region.
- **Classifying Layer (Full Connected Layer).** This is a traditional neural network placed at the end of the convolutional network, so that it is in charge of giving the final result of classification and detection of the object in the image. They are neural networks with layers hidden inside, known as FULLY CONECTED LAYER. Also, they have an architecture that detect patterns of objects in images or real time detection in video. The following graph shows the architecture of a traditional neural network. [3] (Fig. 3).

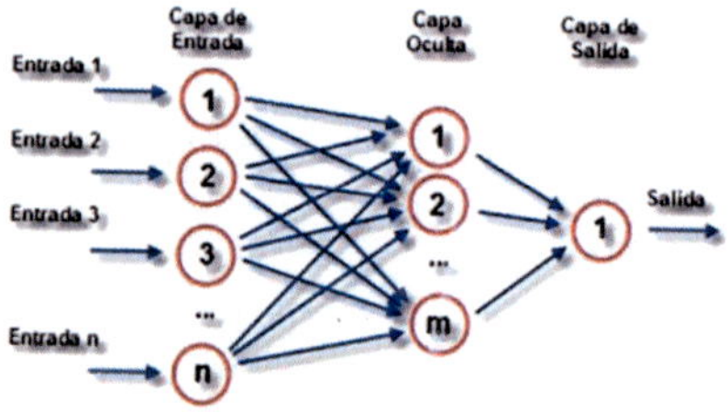

Fig. 3. Traditional architecture of a neural network (Feed Forward) [4]

3.2 Network Implemented (YOLO) [4, 5]

The main function of the network is classification and detection of objects. This network uses a deep learning technique called one-phase algorithm, which divides the image into regions, with the aim of predicting identification and probability tables. These regions go through a convolutional network, later they are modified and grouped considering the prediction obtained. This technique is used for real-time applications since image processing is faster compared to other two-phase deep learning techniques. Because all the processing is carried out in a single phase, the pressure decreases, but the result is still adequate, additionally, computational resources are saved.

YOLO is different from traditional detection methods because it has a different approach, this algorithm looks at the image only once, not like traditional classifiers that reuse information, hence its name (You Only Look once). This algorithm works by placing a type of grid on the entire input image. Then, it is divided into a matrix (n1 x n2) and a location algorithm is applied to each of these cells. Therefore, it has the ability to identify where is located the main object within an image, getting the center of the object and placing a rectangle around it. The classification algorithms, in addition to finding the main object, can classify it in an array of "n" possible answers. To classify, a prediction sub algorithm known as One/Hot encoding is used. It works when an object belongs to a category (i), it is represented with a vector whose component (i) is 1 and the rest is zero, this means that belongs to that category.

A problem of this technique is that a cell in the grid can only detect one object. Then, if we have objects overlapping on the same cell, they cannot be detected. To solve this problem, the algorithm uses the anchor box technique. The main function is to define rectangles of different shapes for each object and associate predictions to each detected object. Figure 4 shows the classification process of YOLO network.

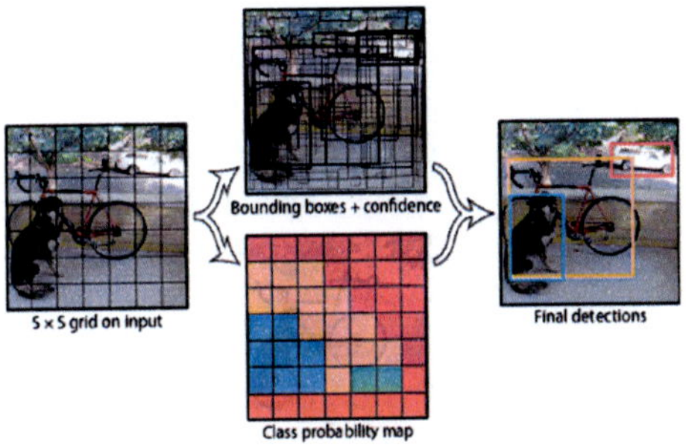

Fig. 4. Yolo network classification stages [4].

The implemented network was YOLOv3, which has the detection and location function in the same algorithm. YOLO network is made up of 24 convolutional layers whose main function is to get the most important characteristics of the image. In addition, it has 2 complete connection layers, which are responsible for predicting the probability of each identified box or cells in the input image.

The dynamics of the YOLOv3 network algorithm are described below. [4]

The steps used in the algorithm are mentioned below:

- The input image in the network is divided into an S x S grid. If the center of an object falls into a grid cell, that grid cell is responsible for detecting that object.
- Each cell in the grid predicts bounding boxes B and confidence scores for those boxes. If there is not object in that cell, the confidence scores must be zero.
- Each bounding box consists of 5 predictions: x, y, w, h and reliability. The coordinates (x, y) represent the center of the box in relation to the limits of the grid cell. Width and height are predicted relative to the whole image.
- Each cell in the grid also predicts conditional class probabilities. These probabilities are conditioned by the cell in the grid that contains an object.

The following image presents the Yolo network architecture (Fig. 5).

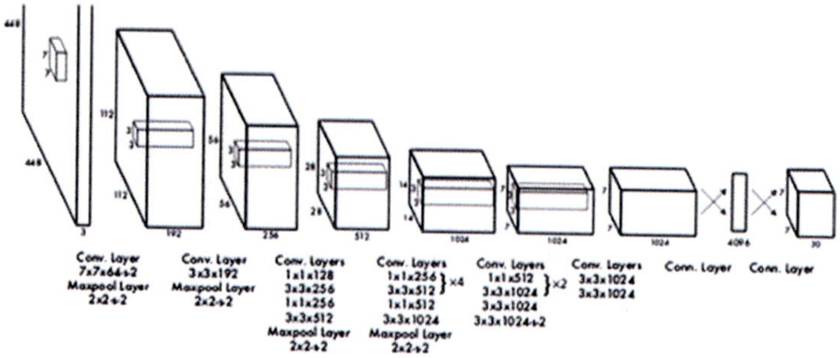

Fig. 5. Image processing on the Yolo network [4]

After obtaining the objects identified from the work environment through the implementation of the YOLOv3 network, the next step is to record the number of people and laptops by designing a small counting algorithm in Python language. It uses the resulting vector of the network as input, this vector specifies the type of object, the probability prediction, and its location in the image.

With this information, the algorithm creates an array that stores the object detected and the number of times they appeared in the image. To avoid counting the same object twice, the location of the object is considered. Finally, the image that enters on the neural network, the number of laptops and people found, and each of the identified objects are enclosed in a rectangle, which is shown in the graphic interface. Something important to note is that the YOLOv3 neural network model used in this project is trained for the CoCo Dataset, therefore, the network already can make predictions about 80 different objects, within this database. This dataset has images of people and laptops, for this reason there is no need to retrain the network (Fig. 6).

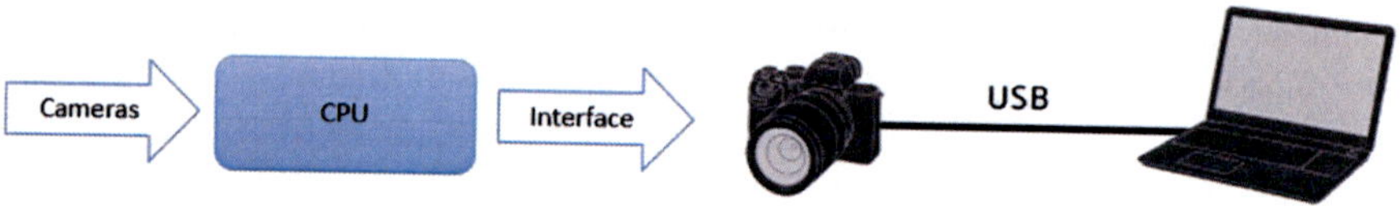

Fig. 6. System architecture

3.3 Hardware Development

The hardware implemented to acquire the information of environment only uses a webcam with the following characteristics.

- Resolution 720p–1080p/30fps
- Visual field (78°)
- Autofocus
- Connection USB

- Compatible in OS such as Windows 7. Windows 8, Windows® 10 or higher.

This camera captures images of the environment and sends the information to the computer. As a suggestion, the computer should have the following characteristics.

- Processor Intel Core i7 2 Ghz
- OS such as Windows 7. Windows 8, Windows® 10 or higher
- RAM Memory 12Ghz

The camera is connected to the computer through a USB port, the position of the camera in the classroom depends on the viewing angle of the camera, and in this case the camera is placed in front of the classroom (Fig. 7).

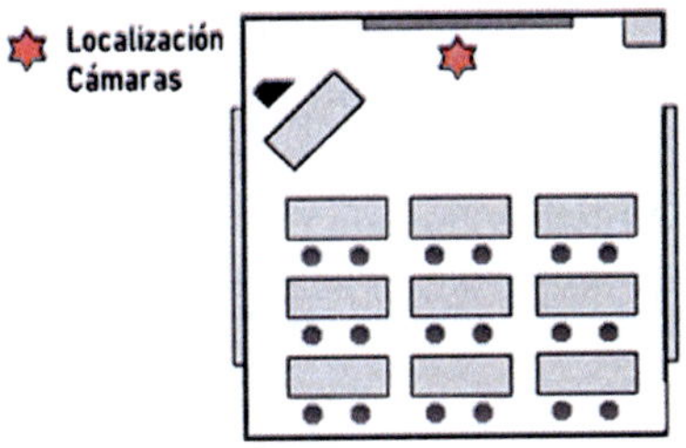

Fig. 7. Position of the camera in the classroom

The camera used in the testing stage has a viewing angle of 78°, this characteristic avoids getting an adequate image of the classroom and hence incomplete information of the environment. For this reason, it is recommended to use a camera that has a viewing angle of 180°, which can capture a complete image of the visualization area. All the information sent by the camera is processed in the computer, the results of this process are displayed in a small interface developed on the same computer (Fig. 8).

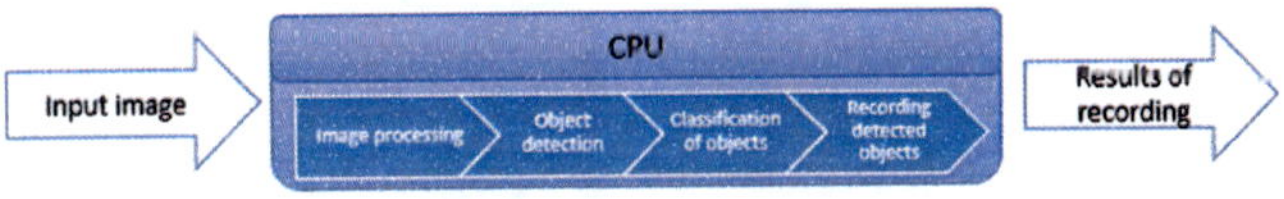

Fig. 8. System architecture

3.4 Software Development

The program implemented is made up of YOLOv3convolutional neural network which uses an image captured by the camera through a series of filters, to later process the information, choosing the most important parts of the image and comparing them with a database that has the objects that the system wants to recognize, in this case people and laptops. The network results send probabilistic information of the kind of object that was detected in the input image.

The result of network is stored in an array which contains the type of object found, the coordinates in pixels of the object's location, and the probability prediction of each identified cell. With this information, the system can record the number of people and computers there are in the work area. Finally, results are showed in the graphic interface developed in the free software Python.

4 Results

The validation process of the designed system was made in two phases. The first test uses downloaded images from the internet, which described a daily classroom environment. The algorithm analyzes a set of thirty images which had different number of people and laptops inside of the work area. Thus, you could know the system efficiency contrasting measure results with real developer results.

4.1 Test Performed with Images Downloaded from the Internet

In this case the input images were download from the internet, to analyze the efficiency of the system, they were chosen according to the number of people and laptops that the algorithm could find. The next figures present some results obtained by the algorithm explained on section three. It identifies and records the number of people and laptops found inside the work area. Then the algorithm draws a rectangle surrounding the object or person detected. Each of figures have a table where you find the real and measure number of people and laptops found inside the work area. These tables show the efficiency of the system in different environments (Figs. 9, 10, 11 and 12).

Image 1	People	Laptops
Measure	5	4
Real	6	5
Efficiency	83.33%	80%

Fig. 9. System input N° 3 [6]

Image 2	People	Laptops
Measure	9	7
Real	9	7
Efficiency	100%	100%

Fig. 10. System input N° 6 [7]

Image 3	People	Laptops
Measure	15	0
Real	18	0
Efficiency	83.3%	100%

Fig. 11. System input N° 9 [8]

Image 4	People	Laptops
Measure	13	7
Real	15	7
Efficiency	86.6%	100%

Fig. 12. System input N° 11 [9]

4.2 Results Review

The Fig. 13 and Fig. 14 show the algorithm results, this image presents a higher efficiency of network when the input image has a number of people less than fifteen; however, if the number of people increases, the system efficiency decreases.

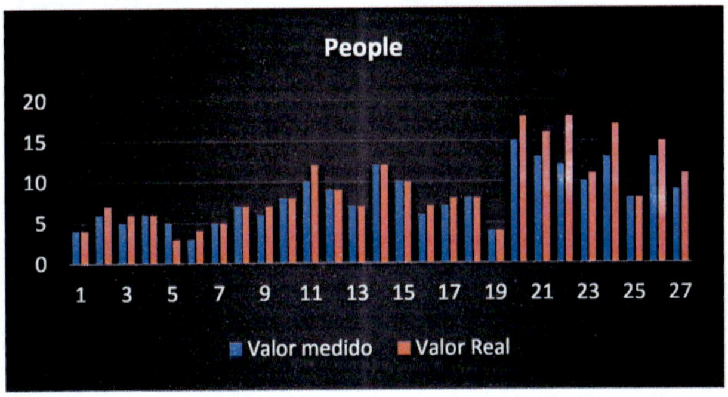

Fig. 13. Comparative graph of real results vs measure results, in the case of recognition of persons.

The right operation of the system depends on the position of camera inside the smart classroom. If camera is set in a wrong place, the vision angle could be twisted. In this case, the first test was made with downloaded images thus its impossible know the vision angle of the camera and the maximum or minimum distance when the algorithm can identify people and laptops in the classroom. However, these characteristics will be considered in the next test when the algorithm uses captured images of a daily learning environment in real time.

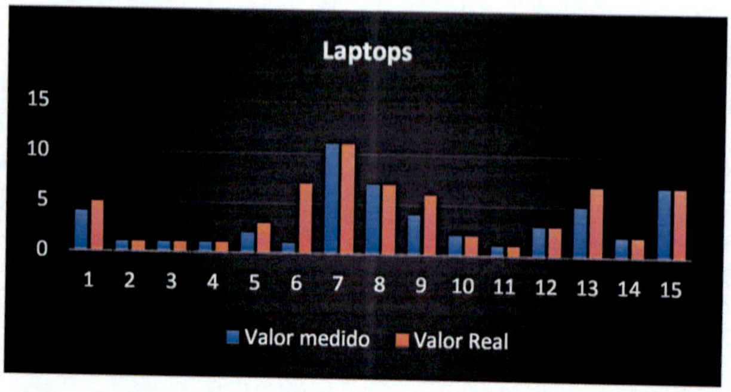

Fig. 14. Comparative graph of real results vs measure results, in the case of recognition of laptops.

4.3 Tests Carried Out in Different Classrooms of the National Polytechnic School

The next figures present the results of the system. In this case, the input images are captured by a camera installed in the Laboratory "Unidad de Mantenimiento Electrónico" of "Escuela Politecnica Nacional". The dynamic of the algorithm is the same (Figs. 15, 16, 17 and 18).

Image1	People	Laptops
Measure	3	1
Real	3	3
Efficiency	100%	33.33%

Fig. 15. System input N° 1.

Image 2	People	Laptops
Measure	3	2
Real	3	3
Efficiency	100%	66.67%

Fig. 16. System input N° 2

4.4 Results Review

According to the results, camera position is an important point to improve the system because of the laptops shape. The laptop could be confused during classification process unlike people, which could be identified in any position and despite the algorithm gives a suitable result. Humans have some relevant characteristics which are easily recognized, unlike some objects that are easily confused ones with others.

In this phase, the algorithm is applied in real time. The web cam is placed on a window and connected to the computer by a USB port. The computer has following characteristics:

Image 3	People	Laptops
Measure	3	2
Real	3	2
Efficiency	100%	100%

Fig. 17. System input N° 3.

Image 4	People	Laptops
Measure	3	2
Real	3	2
Efficiency	100%	100%

Fig. 18. System input N° 4

- Windows 10, 64-bit operating systems
- Intel Core i7 2 Ghz processor
- 12 GB RAM memory

The system processor speed is close to 2FPS. This speed depends on characteristic of the computer. This project does not need a high speed to process the camera information since the control actions to change variables, like temperature, have a slow response inside the work area. However, you can increase the speed using a computer with better characteristic or with a video graphics card.

5 Conclusions

- The developed system to identify people and laptops using artificial vision mechanisms provides greater efficiency and reliability than the classic electronic systems based on presence sensors. According to the results obtained in this work, this type of systems is able to get more information about the work area that could hardly be acquired with the commercial electronic sensors currently used for the same purposes.
- The use of new techniques, based on artificial intelligence, helps to simplify the physical and electronic hardware for applications such as object recognition and counting people in open and close space.
- The stored information of this work helps with the development of a future project named smart class. This information will be used for processing a social Set Point, which will control certain aspects of internal environments such as lighting, temperature, noise, etc. with the aim have an optimal learning environment in classrooms.
- The results obtained in the present work show the efficiency and versatility of this type of systems based on artificial vision mechanisms; however, it is proposed for future works to couple a mechanism for detecting the position of objects in three dimensions,

in order to obtain a social set point that fits easily in the work area and reduces the inconvenience of the angle of vision produced by the camera position inside the work area.

The PIJ 17-10 project from Escuela Politécnica Nacional is thanked for providing us with full availability of equipment and technical assurance in order to carry out the validation tests of the project and to implement it properly.

References

1. Franco, Ò., Perez-Sala, X., Angulo, C.: Identificación y seguimiento de personas usando kinect por parte de un robot. Fondo Europeo de Desarrollo Regiona, Catalunya España (2009)
2. Fegemu Automatismos S.L.: fegemu solutions (2018). https://www.fegaut.com/es/productos/sensores-de-presencia/2/. Accessed 07 Mar 2020
3. Adrian, E.C.: Protocolos TCP/IP de Internet. Revista Digital Universitaria **5**(8), 4 (2004)
4. Angela, C.: Biblioteca Unirioja (2018). https://biblioteca.unirioja.es/tfe_e/TFE004595.pdf. Accessed 05 Mar 2020
5. Zhao, Z.-Q.: Mystery of Object Detection (2019). https://dudeperf3ct.github.io/object/detection/2019/01/07/Mystery-of-Object-Detection/. Accessed 05 Mar 2020
6. EFE: TVN Noticias (2016). https://www.tvn-2.com/tecnologia/internet-computadoras-llegan-rural-Peru_0_4636036369.html. Accessed 07 Mar 2020
7. Frame Pool (2019). https://footage.framepool.com/es/shot/522115082-pizarra-alumno-de-primaria-deberes-para-el-hogar-escolar. Accessed 07 Mar 2020
8. Proyetco451 (2017). http://www.proyecto451.com/apple-y-google-en-la-batalla-por-conquistar-las-aulas/. Accessed 07 Mar 2020
9. Alamy (2016). https://www.alamy.es/foto-los-estudiantes-que-toman-un-test-de-laptops-en-un-aula-de-la-escuela-secundaria-la-prueba-es-la-prueba-de-parcc-nacional-104931704.html. Accessed 07 Mar 2020

Development of an Algorithm Capable of Classifying the Starting, Gear Change and Engine Brake Variables of a Vehicle by Analyzing OBD II Signals

Rivera Campoverde Néstor Diego(✉), Paúl Andrés Molina Campoverde, Gina Pamela Quirola Novillo, and Andrea Karina Naula Bermeo

Universidad Politécnica Salesiana, Cuenca, Ecuador
nrivera@ups.edu.ec

Abstract. The present project shows an algorithm capable of visualizing the behavior of the PID signals of a vehicle by identifying common driving maneuvers such as: starting, changing gear and engine braking during a road test. The data of this study were acquired through the OBD-II connector of a vehicle, by means of a data logger device capable of collecting the information and writing it to a data file. Later the data is post processed in MATLAB® using the unsupervised learning method k-means. Finally, the results obtained by the algorithm are able to discriminate the erroneous values of the signals by means of a multivariate statistical analysis that classifies the study variables: start, gear change and engine brake.

Keywords: OBD II · k-means · Start · Gear change · Engine brake

1 Introduction

Pollutant emissions produced by vehicles are worrisome due to the negative effects they cause on human health and environment [1]. Therefore, searching for new ways to reduce fuel consumption in vehicles powered by gasoline internal combustion engines is essential for reducing polluting emissions [2]. This study aims to identify characteristics associated with driving styles that are not visualized with the use of conventional automotive diagnostic tools, due to the fact that the driver is an influential factor, but is often unnoticed in vehicle performance [3]. However, the use of strategies such as eco driving and data analysis are low-cost tools that generate immediate improvements in fuel consumption of up to 45% [4], in addition to being controllable factors [5]. Regarding fuel consumption, it can be seen that the speed of the vehicle and the synchronization in the gear change allow a fuel saving of around 1.5 L/100 km. This shows the importance of this factor in fuel economy and polluting emissions, which is why the study of these driving techniques is necessary [6].

M. Botto-Tobar et al. (Eds.): ICCIS 2020, AISC 1273, pp. 121–133, 2021.
https://doi.org/10.1007/978-3-030-59194-6_11

In the following study sections, the characteristics of the signals obtained during the normal operation of an internal combustion engine will be determined. Subsequently, these signals are analyzed graphically and statistically, verifying the influence of the driver on the study variables: start, gear change and engine brake.

2 State of Art

In order to describe the characteristics associated with driving style, data is acquired through the OBD II communications protocol [7] which is a fundamental tool for the vehicle diagnosis and the information collection. A commonly used device for collecting vehicles real-time data is the Portable emissions measurement system (PEM's) which is characterized by generating a considerable increase in the mass of light vehicles, meanwhile a data logger device does not adhere mass to test conditions and stores data under normal driving conditions [3]. Many authors based on the obtained data propose different methodologies for their study. Door et al. [8] propose a fuzzy logic control in which categorizes the speed in various thresholds and assigns a driving style coefficient between one and three, being the value of one a comfortable driving style and three a sporty look. Meseguer et al. [9] uses a neural network for a classification problem, using as input data the speed, acceleration and revolutions per minute of the engine, obtaining as outputs types of road and driving style. The inputs values for each parameter were normalized between zero and one, considering the range of possible values. Results are obtained using a three-layer neural network, capable of cataloging the type of road in: urban, suburban and highway, while the driving style divides it between calm, normal and aggressive. Milesich et al. [10] develops a model from neural networks, based on this model develops an optimized driving cycle that searches for the most important parameters along a route. Driving includes a route through the city, out of the city and along a highway, data for the neural network is obtained from the CAN Bus network and a GPS sensor. Following the built model and the route given, the most efficient route in terms of fuel consumption was obtained.

Y. Bao and W. Chen [11] propose a route search method, which considers the driver's driving style, road traffic, and geographic information. They classify driving styles using a neural network (LVQ) with inputs such as vehicle speed, acceleration, and engine speed. Results show that the proposed route search method can achieve a driving style prediction precision of 82.17%, and reduce fuel consumption by 16%. J Molina [12] uses a model based on the Machine learning architecture, using a Random Forest technique achieves a precision of 0.98 and explains the behavior of the PID's signals in front of a person's driving mode, according to engine specific consumption and polluting emissions.

3 Materials and Methods

3.1 Methodology

For the model characterization, this work is based on the proposal of Rivera et al [13], where the most influential variables during driving are determined and the input, study, noise and controllable variables are classified as shown in Table 1.

Table 1. Incidence factors

Variables	Factors	Symbol	Unit
Input variables	Longitudinal acceleration	A	m/s^2
	Vehicle speed	VSS	m/s
	Intake manifold absolute pressure	MAP	kPa
	Engine rotation speed	H	rpm
	Throttle position	TPS	%
Controllable	Coolant temperature	$T_{Coolant}$	°C
	Rolling temperature	$T_{Rolling}$	°C
Noise	Lubricant temperature	$T_{Lubricant}$	°C
	Oxygen sensor	O_2	V
	Air temperature	T_{Air}	%
	Fuel temperature	T_{Fuel}	°C
	Atmospheric pressure	P_{Atm}	Pa
	Air relative humidity	H_{Rel}	%
Study	Power	P	kW
	Specific consumption	C_{esp}	g/kWh

3.2 Data Acquisition Through OBD II by Means of Data Logger Device

Data acquisition is carried out using the OBD II protocol, with the use of a Data Logguer "Freematics ONE +" device that records the information from the vehicle's sensors and actuators and writes this data in a text file. The variables used for the study are shown in Table 2.

Table 2. Stored variables

Variables	Nomenclature	Units
Longitudinal acceleration	Acceleration	m/s^2
Speed vehicle	VSS	m/s
Coolant engine temperature	ECT	°C
Air intake temperature	IAT	°C
Intake manifold absolute pressure	MAP	KPa
Oxygen sensor	O_2	V
Engine speed	RPM	rpm
Throttle	TPS	%
Long-term fuel trim	LTFT	%
Short-term fuel trim	STFT	%

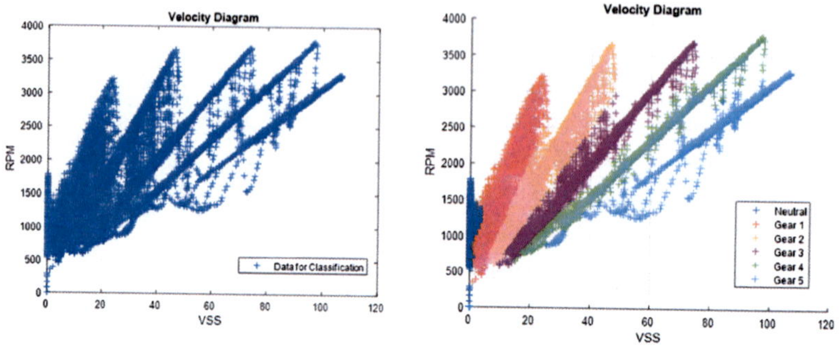

(a) Real velocity Diagram

(b) Data classification using k/means algorithm

Fig. 1. Velocity diagram

4 Development

4.1 Application of the K-Means Classification Method Applied to the Velocity Diagram

By using the K-means algorithm, an attempt is made to represent the different speed steps for each gear of the vehicle, obtained from the relationship between vehicle speed and engine revolutions per minute. In Fig. 1a can be seen that the actual classification diagram differs from the theoretical one due to the large number of values represented in the graph. In addition to this study, variables such as: neutral gear, upshifting, downshifting, engine brake, clutch, disengaged, etc. are involved, therefore, it makes difficult the interpretation of the slopes in each march obtained based on the experimental test. It is then that the k-means classification identifies the gears applied in the experimental samples as observed in Fig. 1b.

4.2 Vehicle Starting Study from Absolute Rest.

Figure 2 shows the speed sensor (VSS) behavior of at start-up, this classification being still imprecise, due to the fact that unwanted information intervened, such as neutral gear values, braking and gear changes at certain points. The classification still needs adjustment so as not to involve misinformation.

In this section, the data of the neutral and one gear category are considered influential in the study of vehicle starting, so it is necessary to filter information, the best option being to implement discrimination by the vehicle speed vector (VSS), where a speed of $0\,Km/h$ is considered absolute rest as can be seen in Fig. 3. Once the correction has been made, the readjustment results are shown as it is a more appropriate way to analyze the driving maneuver called "vehicle start" as seen in Fig. 4.

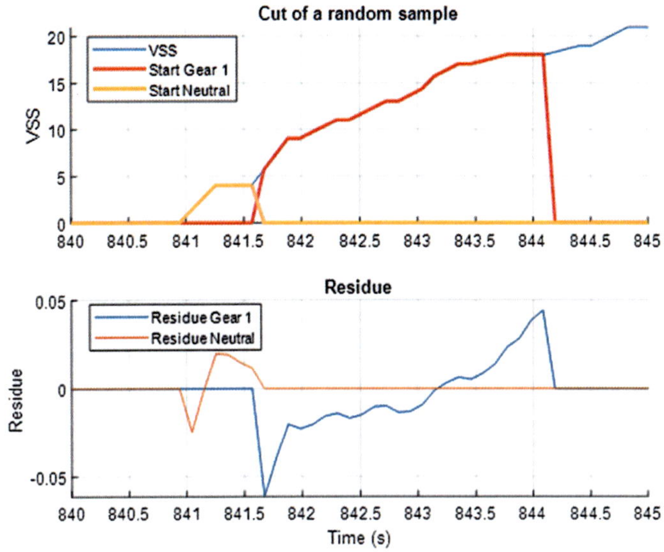

Fig. 2. Misinterpretation of waste based on wrong rating.

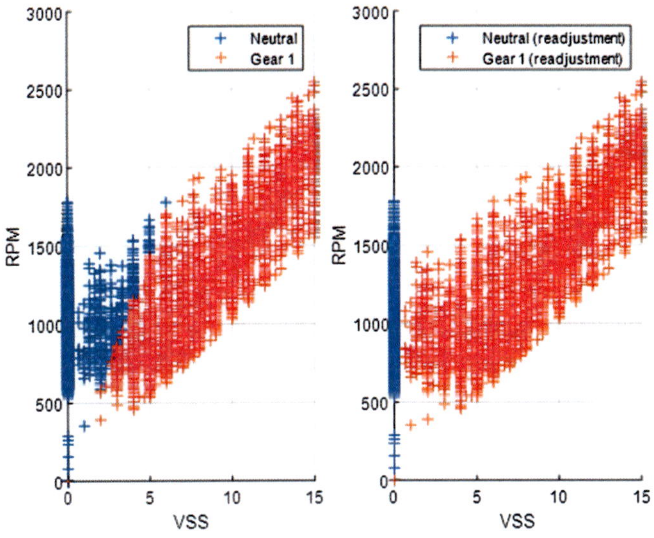

Fig. 3. Readjustment k/means on the basis of information discrimination

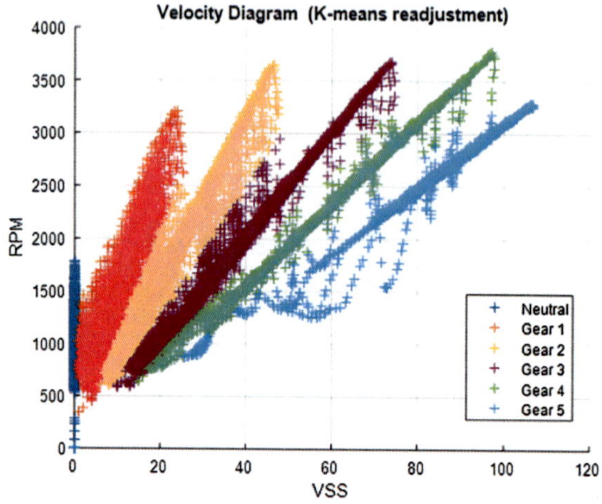

Fig. 4. Readjustment display in k/means data classification.

Once the organization that comes closest to reality has been determined, the determination of the vehicle's start is made, through a discrimination of the VSS sensor, with which a reference value of $0\,Km/h$ is sought, as this is the identifying value of the vehicle at rest absolute. After this, it is evident that the value of the speed begins to decrease, otherwise the influence of the first gear is detected as shown in Fig. 5. A recurring pattern in the residuals when leaving absolute rest is the display of a valley formed by the clutch pedal as the ratio (VSS/RPM) is broken allowing debris to drift away from other data. At this moment, the vehicle exit is identified until the residuals follow a trend line understood as a constant gear, and it is observed that the residuals do not stray too far from the reference point 0.

Study Focused on Motor Braking Applied to an Experimental Test. Conditions of this category are based on the discrimination of the data captured by the TPS, MAP, RPM and VSS sensors, without the application of the accelerator pedal. In this condition, the intake manifold pressure is below idle with these two being different. Furthermore, the speed will not always be different from zero and there will be circumstances where the engine brake acts as a pedal brake and in other conditions, the vehicle will remain in a state of absolute rest. As shown in Fig. 6 residues during the application of an engine brake are distinguish by two patterns. In the former, the speed is different from 0, therefore most of the data is not dispersed until the user applies a disengagement while driving. The second pattern is identified, when the speed reaches 0 by the application of the normal brake, because this element takes the vehicle from a speed other than zero until it reaches absolute rest. In this instance, the reduction of the gears by the driver to the neutral position where the application of the clutch is necessary for the gear change is common. By this action, residues

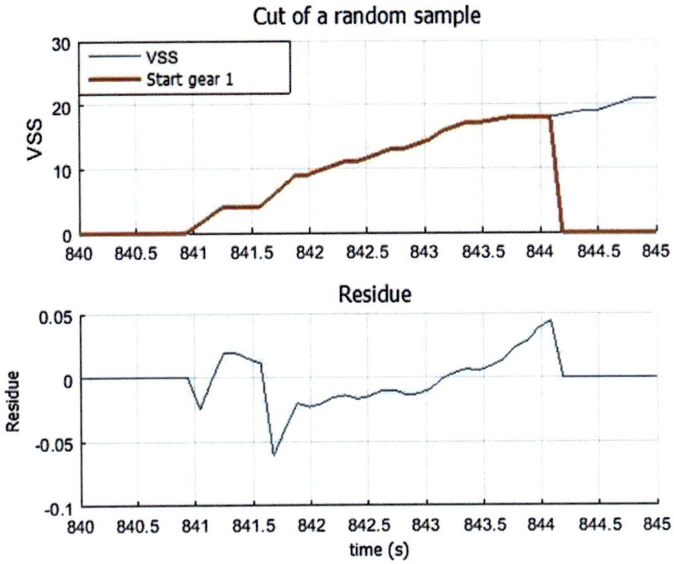

Fig. 5. Real behavior of residues when starting the vehicle.

are disperse from the rest of the data, due to the change gearshift over time. The motor brake applied during the experimental test can be seen in the speed diagram (Fig. 7) where the application of the motor brake at different speeds and engine revolutions is evidenced.

Study Focused on the Search for Gear Changes in an Experimental Test . For the study of the gear change during an experimental test, the analysis of the VSS/RPM relationship is performed, where it is inferred that during each gear change there is an agglomeration of data when it remains constant in certain time intervals as shown in Fig. 8a. However, it is interesting to analyze the produced data during the gear change that is of interest within this study. Due to the diversity of data found during each march, it can be determined that the standard deviation identifies the border where each march takes place, while the other data are produced by irregular maneuvers during the driving process as can be seen in the Fig. 8b.

Another way to visualize the aforementioned is to work on the (VSS/RPM), apply a residue analysis to verify the existence of data far from the media that are interpreted as the intervention of another event (Fig. 9). The events shown in the experimental test are more susceptible to distortions as the different gears are executed, thus facilitating the search for changes made.

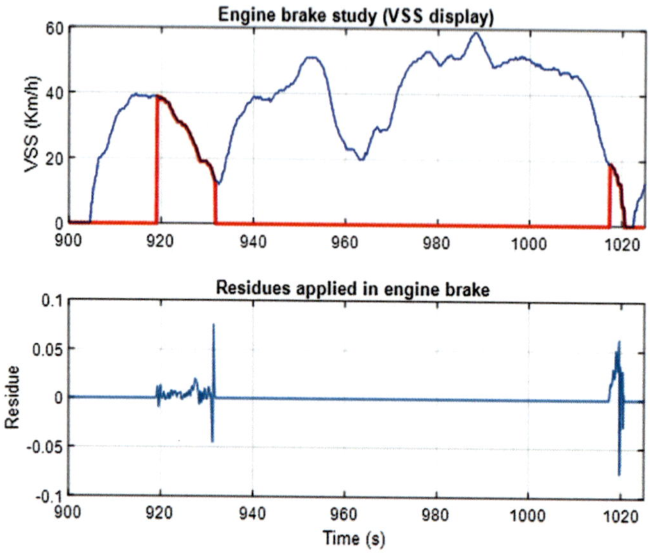

Fig. 6. Search and display of engine brake according to the vehicle speed

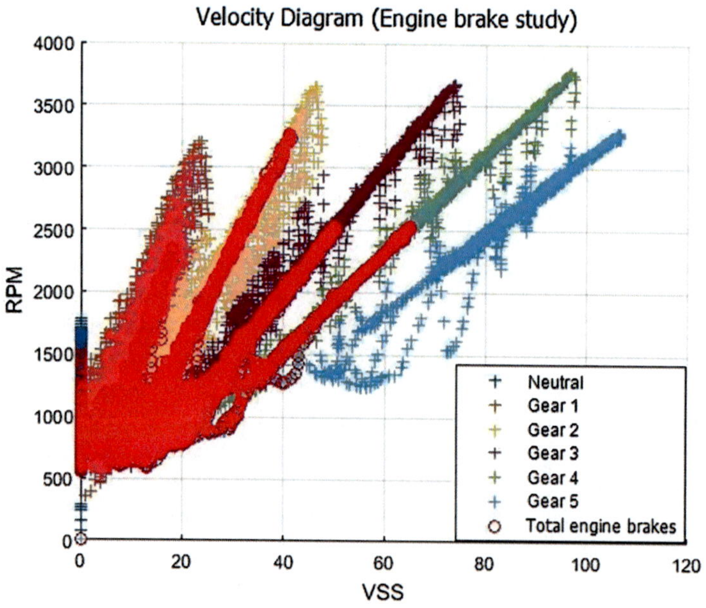

Fig. 7. Introduction of all the engine brakes found in the experimental test.

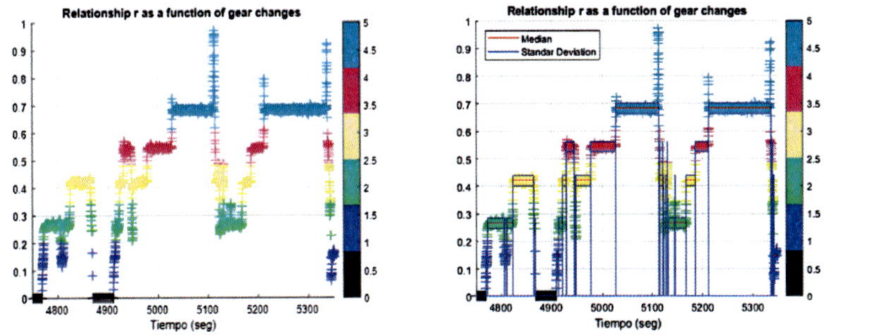

(a) Gears as a function of the relationship (VSS /RPM).

(b) Standard deviation in gears as a function of the relationship (VSS / RPM)

Fig. 8. Relation depending on gear changes

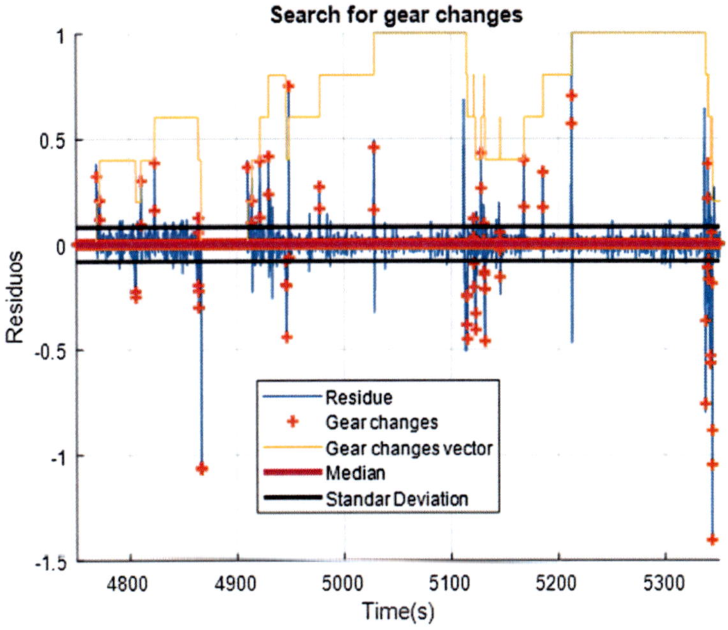

Fig. 9. Interpretation of gear changes made in the vehicle with help of applied residuals in the ratio (VSS/RPM) and data deviation for data discrimination.

5 Discussions

Once the algorithm has been performed and the vectors that define each of the proposed maneuvers have been obtained, it is possible to visualize in this case the vehicle start, engine brake and gear changes carried out during the experimental tests as shown in the Figure 10.

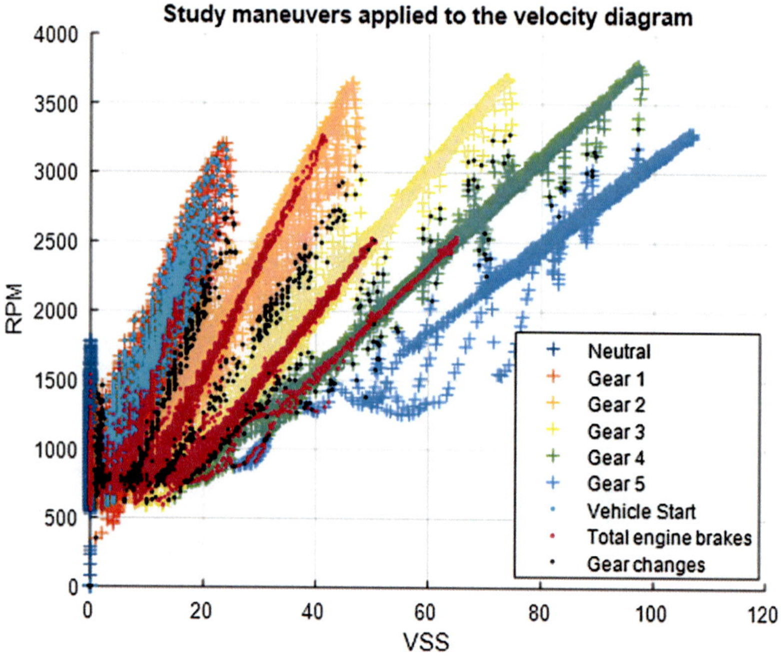

Fig. 10. Interpretation of gear changes made in the vehicle with help of applied residuals in the ratio (VSS / RPM) and data deviation for data discrimination.

5.1 Application of the Algorithm by Changing Input Variables for Adaptability Validation in Different Vehicles

The importance of this study lies in the fact that the algorithm carried out can be executed without problems in any kind of vehicle during experimental tests. It should be noted that mathematically and statistically, no model is the same as another, that is, if the vehicle is the same and only the driver is changed, there will be no similarity between the two analyzes as shown in Fig. 11.

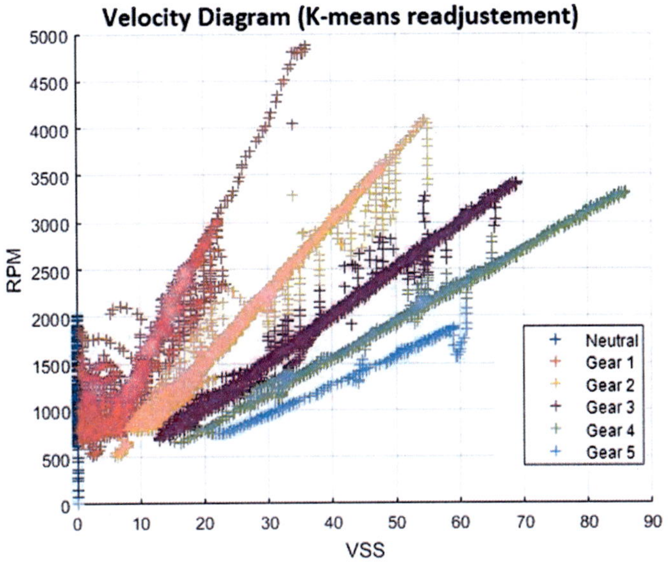

Fig. 11. Speed diagram based on a different vehicle.

6 Conclusions and Futures Works

The work shows an algorithm capable of classifying the starting, gear shifting and engine brake maneuvers of a vehicle through the application of unsupervised K-means learning through the signals obtained by OBD II during a random road test.

The K-means model presents a better correlation of results in relation to vector support machines (SVM) due to the difficulty of knowing with certainty the labels of the model, reaching values of 60 % accuracy and on other occasions causing overtraining. For the vehicle-starting category, it was necessary to use classification readjustment criteria, due to gears that are difficult to interpret, such as first gear and neutral gear because they directly affect the waste that discriminates the information.

Clutch is a factor that generates dispersion in the data during the experimental tests and the relationship (VSS/RPM) is no longer linear, which is why it is considered a descriptive statistical method (standard deviation), in order to identify the corresponding data to another category.

Driving maneuvers executed by the user generate unusual behaviors in the waste, causing scattered values caused by the bad driving habits during clutch application. This fact is evident in little everyday driving styles.

Unsupervised K-means learning, limits the categories in which the user can appreciate an experimental test, for this study 6 categories were needed (neutral and 5 speed gears). In those databases where there is no 5-speed information, this algorithm will take 4-speed data as if it were 5-speed, thus being considered a failed test. The algorithm will search for 6 centroids when 5 are actually needed (neutral and 4 gear speeds). Therefore, adjustments in the algorithm are necessary to compensate for these types of cases in future works.

Vehicles are big producers of green house emissions therefore in future works this paper will be complemented with pollutant emissions estimation according to every action such as brake, accelerate and gear shifting. The aim will be identify efficient parameters to every state in order to reduce emissions.

References

1. Gao, J., Ma, C., Xing, S., Sun, L., Huang, L.: Nanostructure analysis of particulate matter emitted from a diesel engine equipped with a NTP reactor. Fuel **192**, 35–44 (2017)
2. Zavalko, A.: Applying energy approach in the evaluation of eco-driving skill and eco-driving training of truck drivers. Transp. Res. Part D **62**, 672–684 (2018)
3. Huanga, Y., Ng, E.C.Y., Zhoua, J.L., Surawskia, N.C., Chan, E.F.C., Hong, G.: Eco-driving technology for sustainable road transport: a review. Renew. Sustain. Energy Rev. **93**, 596–609 (2018)
4. Sivak, M., Schoettle, B.: Eco-driving: strategic, tactical, and operational decisions of the driver that influence vehicle fuel economy. Transp. Policy **22**, 96–9 (2012)
5. Xu, Y., Li, H., Liu, H., Rodgers, M.O., Guensler, R.L.: Eco-driving for transit: an effective strategy to conserve fuel and emissions. Appl. Energy **194**, 784–97 (2017)
6. Jeffreys, I., Graves, G., Roth, M.: Evaluation of eco-driving training for vehicle fuel use and emission reduction: a case study in Australia. Transp. Res. Part D: Transp. Environ. **60**, 85–91 (2018)
7. International Organization for Standardization: Road vehicles. Diagnostic systems. Keyword Protocol 2000 (1999)
8. Dorr, D., Grabengiesser, D., Gauterin, F.: Online driving style recognition using fuzzy logic. In: 2014 IEEE 17th International Conference on Intelligent Transportation Systems (ITSC), 8-11 October (2014)
9. Meseguer, J., Toh, C.K., Calafate, C.T., Cano, J., Manzoni, P.: Drivingstyles: a mobile platform for driving styles and fuel consumption characterization. J. Commun. Netw. **19**(2), 162–168 (2017)
10. Milesich, T., Bucha, J., Gulan, L., Danko, J.: The possibility of applying neural networks to influence vehicle energy consumption by eco driving. In: Březina, T., Jabłoński, R. (eds.) Mechatronics 2017. MECHATRONICS 2017. Advances in Intelligent Systems and Computing, vol. 644. Springer, Cham (2018). https://doi.org/10.1007/978-3-319-65960-2_46
11. Bao, Y., Chen, W.: A personalized route search method based on joint driving and vehicular behavior recognition. In: 2016 IEEE MTT-S International Wireless Symposium (IWS), Shanghai, pp. 1–6 (2016)

12. Molina Campoverde, J.J.: (2020) Driving mode estimation model based in machine learning through PID's signals analysis obtained from OBD II. In: Botto-Tobar, M., Vizuete, M.Z., Torres-Carrión, P., Montes León, S., Vásquez, G.P., Durakovic, B. (eds.) Applied Technologies. ICAT 2019. Communications in Computer and Information Science, vol. 1194. Springer, Cham (2019). https://doi.org/10.1007/978-3-030-42520-3_7
13. Rivera, N., Chica, J., Zambrano, I., García, C.: Estudio Del Comportamiento De Un Motor Ciclo Otto De Inyección Electrónica Respecto De La Estequiometría De La Mezcla Y Del Adelanto Al Encendido Para La Ciudad De Cuenca. Revista Politécnica. **40**(1), 59–67 (2017)

Characterization of Braking and Clutching Events of a Vehicle Through OBD II Signals

Paúl Andrés Molina Campoverde(✉), Néstor Diego Rivera Campoverde, Gina Pamela Novillo Quirola, and Andrea Karina Bermeo Naula

Universidad Politécnica Salesiana, Cuenca, Ecuador
pmolinac1@ups.edu.ec

Abstract. This work presents an algorithm capable of identifying two habitual driving maneuvers such as braking to decrease speed and disengaging to produce a gear change in the vehicle by studying the PID's (Identification Parameters) signals from the electronic control unit. These signals are acquired through the OBD II connector through a data logger device capable of storing information during the engine operation. The obtained data is prost-processed using K-mean unsupervised learning algorithm. The algorithm is capable of identifying braking and clutch events in addition to classifying whether the vehicle has a motorized or mechanical body acceleration system. With the used algorithm, it is possible to determine the pilot's driving style and the most frequently used change during a route. For this research, four vehicles from different years and manufacturers were used to verify the functionality of the algorithm.

Keywords: K-means · Eco-driving · Clutching · Braking · OBD II · PID's signals

1 Introduction

Vehicles powered by fossil fuels allow the mobility of people and products around the world, but are responsible for the detriment of air quality. One way to reduce pollution from vehicle-dependent drivers is to introduce environmentally friendly driving techniques [1]. Eco-driving consists of a series of parameters intended to reduce polluting emissions and decrease fuel consumption [2], these rules must be applicable to both new and old vehicles [3]. Over the years, many efforts have been made to reduce emissions and improve fuel efficiency, such as the implementation of international standards [4] (e.g. Euro 6/VI standards), fuel improvements, etc. However, an unnoticed factor that can generate significant long-term savings is the driver, through the use of eco-driving [5].

Eco-driving is a broad concept that ranges from the selection and maintenance of the car, choice of routes and driver behavior (being classified as: aggressive, conservative or shy) [6]. Several studies agree that fuel consumption can be

M. Botto-Tobar et al. (Eds.): ICCIS 2020, AISC 1273, pp. 134–143, 2021.
https://doi.org/10.1007/978-3-030-59194-6_12

reduced from 6 to 15% [7–9] with the use of these techniques. Strategies include avoiding sudden accelerations and stops. Furthermore, a correct synchronization in the gear change can mean a fuel saving of around 1.5 L/100 km [10]. That is why it is important to identify various aspects of driving and to be able to describe how the braking and disengaging process of a vehicle takes place on a random route.

2 Lirerature Review

Eco-Driving is a term used to describe efficient use of energy in vehicles, based on decisions and behaviors adopted by the driver [11]. These characteristics associated with driving style could be acquired through the OBD II communications protocol [12]. Data logger devices allow information to be collected under normal handling conditions without sticking mass as is the case with the Portable emissions measurement system (PEM's) [5]. From the samples collected, many authors propose different methodologies for their study. Yanzhi et al. [13] use an algorithm designed to optimize fuel consumption by limiting the specific power and maintaining the average vehicle speed. Milesich et al. [14] develops a model using neural networks and identifies influential parameters along a route. J Molina [15] by means of a Machine learning algorithm achieves a 98% precision on the behavior of the PID's signals compared to a person's driving mode, describing a relationship between specific fuel consumption and engine power. Kang [16] designs a strategy for the optimal performance of the gear change from a stochastic model, using data obtained from traffic to improve fuel consumption. Yeh [17] uses Big Data to relate the behavior of bus drivers with respect to the consumption of fuel and highlights engine torque, rpm, speed, engine temperature and travel distance as the most influential variables in fuel consumption.

Orfila et al. [18] develop an android application to provide real time evaluation of the driving behavior, through OBD II interface. Acquired information advice the driver about consumption and improve the model by means driver's own driving statistics. Chrenko [19] designs a driving cycle based on the road information and driver behavior (acceleration, deceleration and maximum speed), pointing out increase in acceleration and decrease in acceleration show a lineal influence in fuel economy. Ferreira et al. [20], collect data from a fleet of Lisbon's public transportation buses trough GPS identifying the best driving practices and styles. The research focuses in demonstrate how eco-driving techniques such as clutch, idle and control in rpm can reduce fuel consumption $4l/100$ km in average. From another point of view Maamria et al. [21] propose two methodologies (time method and space methods) to estimate fuel consumption in vehicles equipped with internal combustion engines for a fixed road, finding the best profile velocity and power consumption.

3 Materials and Methods

3.1 Methodology

For the characterization of the model, this work is based on the proposal of Rivera et al. [22], where the most influential variables during driving are determined and the input, study, noise and controllable variables are classified as shown in Table 1. An experimental run is proposed where data from the ECU is collected through the Data logger, during normal driving cycles. Drivers of the test vehicles do not have any instruction about how to drive with which the tests randomness is guaranteed. Driving cycles were carried out in the urban, rural and highway parts of Cuenca city, in Ecuador. The carried out driving maneuvers were performed at the drivers discretion to through the automatic learning find patterns that describe conduction variables behavior during braking and clutching , specially is analyzed the behavior of VSS/RPM relation.

Table 1. Incidence factors

Variables	Factors	Symbol	Unit
Input variables	Longitudinal acceleration	A	m/s^2
	Vehicle speed	VSS	m/s
	Intake manifold absolute pressure	MAP	kPa
	Engine rotation speed	H	rpm
	Throttle position	TPS	%
Controllable	Coolant temperature	$T_{Coolant}$	°C
	Rolling temperature	$T_{Rolling}$	°C
Noise	Lubricant temperature	$T_{Lubricant}$	°C
	Oxygen sensor	O_2	V
	Air temperature	T_{Air}	%
	Fuel temperature	T_{Fuel}	°C
	Atmospheric pressure	P_{Atm}	Pa
	Air relative humidity	H_{Rel}	%
Study	Power	P	kW
	Specific consumption	C_{esp}	g/kWh

3.2 Data Acquisition Through OBD II by Means of Data Logger Device

Data acquisition is carried out using the OBD II protocol, with the use of a Data Logguer "Freematics ONE +" device that records the information from the vehicle's sensors and actuators and writes this data to a text file. The variables used to the study will be seen in Table 2.

4 Development

K-means algorithm makes it possible to group the points generated in the speed diagram by means of the signals acquired by the speed sensor (VSS) and the engine revolutions. As can be seen in Fig. 1.

The algorithm can classify by color each of the gears. Based on this grouping, it can be apply a residue analysis to know the behavior of variables analyzed. The residuals presented in the gears are the difference between the values observed or experimentally acquired against the values predicted from a regression line, as shown in Fig. 2. Applied these criteria to braking and actuation events of the clutch pedal during gear shifting.

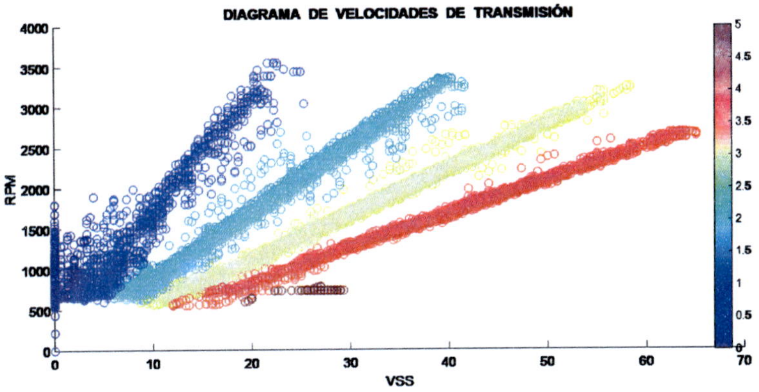

Fig. 1. Transmission speed diagram

Table 2. Stored variables

Variables	Nomenclature	Units
Longitudinal acceleration	Acceleration	m/s^2
Speed vehicle	VSS	m/s
Coolant engine temperature	ECT	°C
Air intake temperature	IAT	°C
Intake manifold absolute pressure	MAP	KPa
Oxygen sensor	O_2	V
Engine speed	RPM	rpm
Throttle	TPS	%
Long-term fuel trim	LTFT	%
Short-term fuel trim	STFT	%

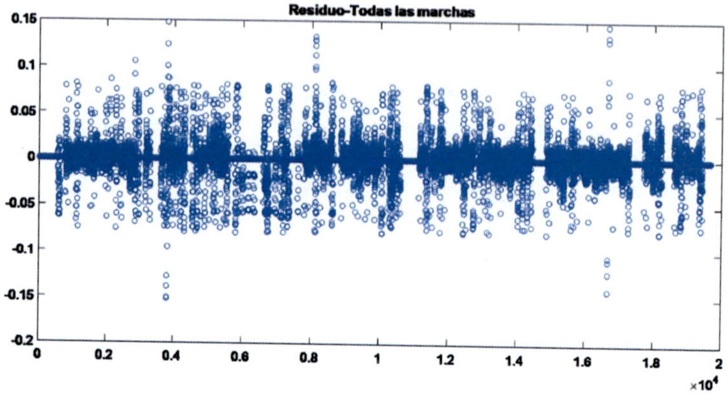

Fig. 2. Total waste from the gears

4.1 Braking and Clutching Process

By means of the algorithm, it is possible to obtain those points that only correspond to braking events, placing emphasis on the moments that the vehicle is

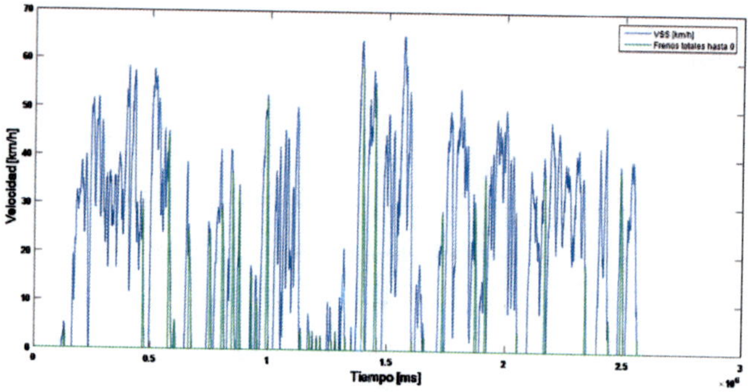

(a)Totalbrakestozero

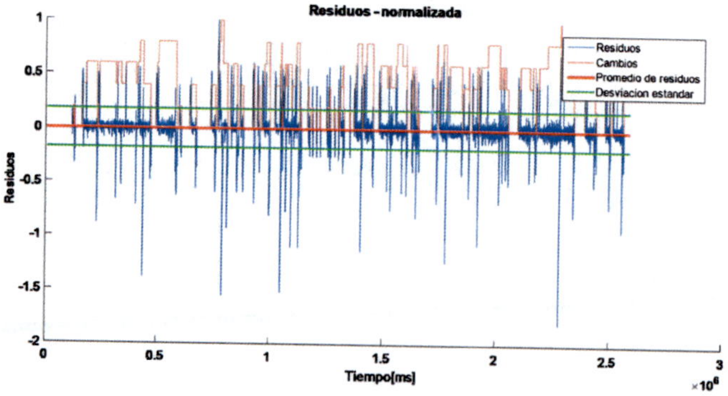

(b)Clutchedanddisengaged

Fig. 3. Analysis cases

completely stopped, as indicated in Fig. 3a. It can be distinguish in green the moments where the vehicle comes to a complete stop.

For the vehicle clutch and disengagement process, the standard deviation and the value of the residuals are used, obtaining the values corresponding to the process, as shown in Fig. 3b. Those values corresponding to the residuals (blue color) that protrude from the graph (green color) are obtained and are considered as moments of disconnection of the motor with the wheels, in addition the red line represents the gear number.

5 Results

5.1 Braking Process

During the braking process the residues produce a formation of peaks due to the action of the clutch coinciding with the graph of gear change. The MAP sensor

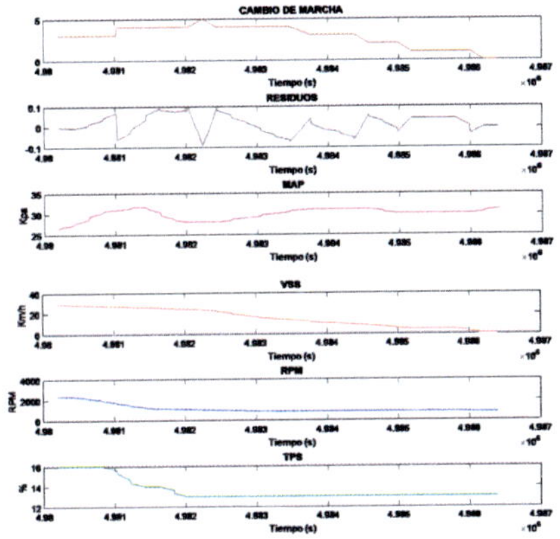

(a) Motorized body throttle signal behavior

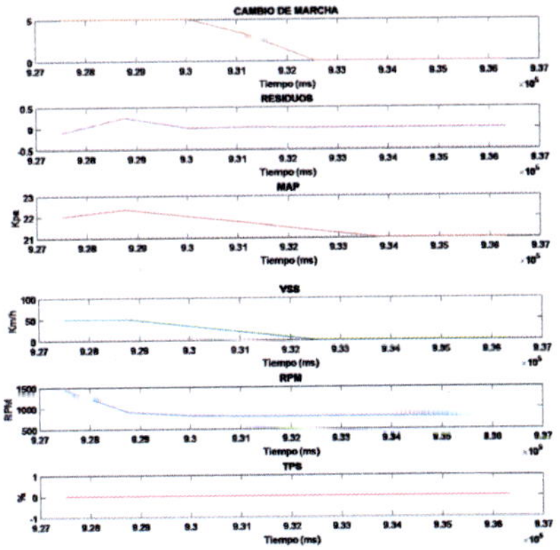

(b) Behavior of conventional throttle signals

Fig. 4. Obtained signals during the braking process

signal has a graph with a smooth curvature during the braking event, increasing and decreasing its value by actuating the clutch pedal, as seen in Fig. 4a. For vehicles that have a conventional throttle mechanism, the MAP sensor has an almost linear behavior, very similar to the behavior of the residuals with a single peak at the moment that a fifth gear change to neutral is made. Figure 4 shows the behavior of the other signals.

5.2 Clutch Process

During this event a particular area is chosen and with reference to the standard deviation, the behavior of the residues in the disengagement can be observed and

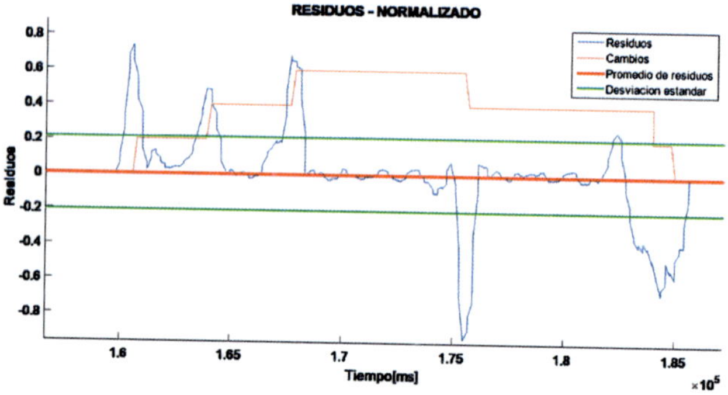

Fig. 5. Waste behaviour

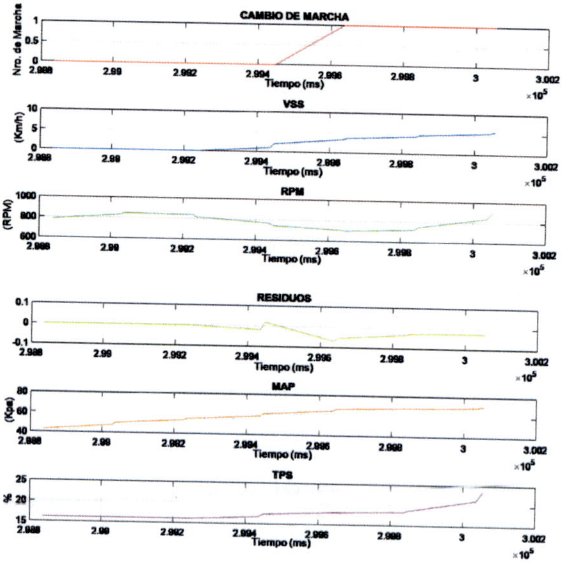

Fig. 6. Upshifting from neutral to first gear

clutch for each gear, as shown in Fig. 5. The red line represents the current gear of the vehicle, while the blue line represents the behavior of the waste during the clutch and disengagement process (Figs. 6 and 7).

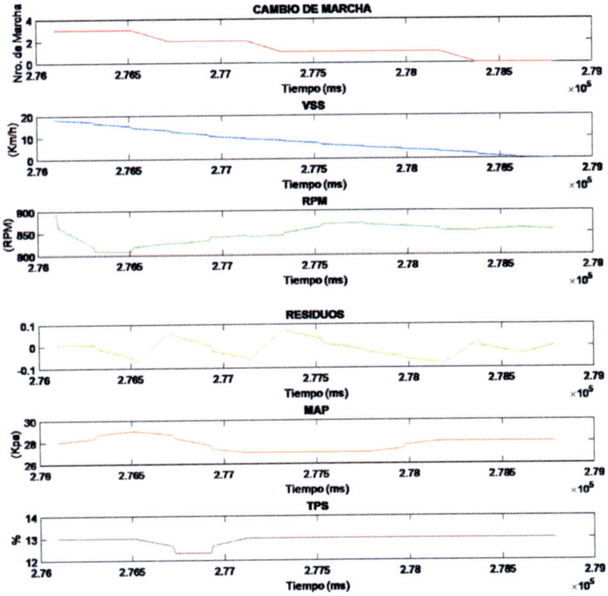

Fig. 7. Downshifting from neutral to first gear

6 Conclusions and Futures Works

The study focuses on PID's signals behavior during the braking and clutching process necessary to adapt the power and torque of the engine in different regimes and driving conditions. Based on obtained data, it can be seen that during the braking, disengagement and clutch process, acceleration system with motorized body allows to establish patterns that identify the description of the process. For the classification of the speed diagram, K-means algorithm presents a high percentage of precision, allowing to demonstrate the driving style and the most used gear during a journey.

During the study of the braking process of vehicles equipped with a non-motorized acceleration body, a non-linear behavior of the MAP sensor signals and the residual graph is evidenced. In addition, these signals have abrupt changes while a gear change occurs. After the event, the signals stabilize. Meanwhile for vehicles equipped with motorized acceleration body signals with smooth transitions are presented in their graphs as observed in Fig. 4b.

In the clutch process during an upshift in vehicles with motorized acceleration bodies, progressive and staggered changes are formed, at the moment of clutch, also being for downshifting. For the clutch process at the instant where the downshifts are carried out randomly, peaks occur according to the number of gear to which it has been decreased, and a correction is noted by the electronic injection system of the engine , the same events are generated in the engine speed and the TPS sensor.

In vehicles equipped with non-motorized acceleration bodies (mechanical systems), it is observed that during the gear raising at the moment when the clutch process is performed, the PID's signals have gradual and progressive changes. However, the signals from the MAP and TPS sensors do not generate repetitive patterns. Also in this system, during the clutch process, no corrections are evidenced by the electronic injection system.

The algorithm is capable of describing the behavior and driving style of different vehicles and pilots, it can also identify the instants of time of use of each gear so that in future works the phenomena studied and their influence on specific consumption could be related of engine fuel.

References

1. Nègre, J., Delhommea, P.: Drivers' self-perceptions about being an eco-driver according to their concern for the environment, beliefs on eco-driving, and driving behavior. Transp. Res. Part F: Traffic Psychol. Behav. **46**(Pt. A), 96–110 (2017). Eslevier
2. Mensing, F., Bideaux, E., Trigui, R., Ribet, J., Jeanneret, B.: Eco-driving: an economic or ecologic driving style? Transp. Res. Part C Emerg. Technol. **38**, 110–121 (2014)
3. Rionda, A., Paneda, X.G., García, R., Díaz, G., Martínez, D., Mitre, M., Arbesu, D., Marín, I.: Blended learning system for efficient professional driving. Comput. Educ. **78**, 124–139 (2014)
4. European Commission Amending regulation (EC) No 7152007 of the European Parliament and of the Council and Commission regulation (EC) No 6922008 as regards emissions from light passenger and commercial vehicles (Euro 6). Off. J. Eur. Union **142** , 16–24 (2012)
5. Huanga, Y., Ng, E.C.Y., Zhoua, J.L., Surawskia, N.C., Chan, E.F.C., Hong, G.: Eco-driving technology for sustainable road transport: a review. Renew. Sustain. Energy Rev. **93**, 596–609 (2018)
6. Sivak, M., Schoettle, B.: Eco-driving: strategic, tactical, and operational decisions of the driver that influence vehicle fuel economy. Transp. Policy **22**, 96–99 (2012)
7. Barth, M., Boriboonsomsin, K.: Energy and emissions impacts of a freeway-based dynamic eco-driving system. Transp. Res. Part D: Transp. Environ. **14**(6), 400–410 (2009)
8. Beusena, B., Broekxa, S., Denysa, T., Beckxa, C., Degraeuwea, B., Gijsbersa, M., Scheepersa, K., Govaertsa, L., Torfsa, R., Panisa, L.I.: Using on-board logging devices to study the longer-term impact of an eco-driving course. Transp. Res. Part D: Transp. Environ. **14**(7), 514–520 (2009)
9. Rutty, M., Matthews, L., Andrey, J., Del Matto, T.: Eco-driver training within the City of Calgary's municipal fleet: monitoring the impact. Transp. Res. Part D: Transp. Environ. **24**, 44–51 (2013)

10. USDoE, Driving more efficiently. http://www.fueleconomy.gov/feg/driveHabits.jsp. Accessed 12 April 2018
11. Ayyildiz, K., Cavallaro, F., Nocera, S., Willenbrock, R.: Reducing fuel consumption and carbon emissions through eco-drive training. Transp. Res. Part F: Traffic Psychol. Behav. **46**(Part A), 96–110 (2017)
12. International Organization for Standardization, 1999: Road vehicles, Diagnostic systems, Keyword Protocol 2000 (1999)
13. Xu, Y., Li, H., Liu, H., Rodgers, M.O., Guensler, R.L.: Eco-driving for transit: an effective strategy to conserve fuel and emissions. Appl. Energy (2016). http://dx.doi.org/10.1016/j.apenergy.2016.09.101
14. Milesich,, T., Bucha, J., Gulan, L., Danko, J.: The possibility of applying neural networks to influence vehicle energy consumption by eco driving. In: Březina, T., Jabłoński, R. (eds.) MECHATRONICS 2017. AISC, vol. 644. Springer, Cham (2018)
15. Molina Campoverde, J.J.: Driving mode estimation model based in machine learning through PID's signals analysis obtained from OBD II. In: Botto-Tobar, M., Zambrano Vizuete, M., Torres-Carrión, P., Montes León, S., Pizarro Vásquez, G., Durakovic, B. (eds.) Applied Technologies. ICAT. CCIS, vol. 1194. Springer, Cham (2019)
16. Kang, M., Gao, J.: Design of an eco-gearshift control strategy under a logic system framework. Front. Inform. Technol. Electron. Eng. **21**, 340–350 (2020). https://doi.org/10.1631/FITEE.1900459
17. Yeh, C.F., Lin, L.T., Wu, P.J., Huang, C.C.: Using on-board diagnostics data to analyze driving behavior and fuel consumption. In: Zhao, Y., Wu, T.Y., Chang, T.H., Pan, J.S., Jain, L. (eds.) Advances in Smart Vehicular Technology, Transportation, Communication and Applications. VTCA 2018. SIST, vol. 128. Springer, Cham (2018)
18. Orfila, O., Saint Pierre, G., Messias, M.: An android based ecodriving assistance system to improve safety and efficiency of internal combustion engine passenger cars. Transp. Res. Part C: Emerg. Technol. **58**(PP), 772–782 (2015). https://doi.org/10.1016/j.trc.2015.04.026
19. Chrenko, D.: Influence of hybridisation on eco-driving habits using realistic driving cycles. IET Intell. Transp. Syst. **9**(5), 498–504 (2015). https://doi.org/10.1049/iet-its.2015.0005
20. Ferreira, J.C., De Almeida, J., Da Silva, A.R.: The impact of driving styles on fuel consumption. a data-warehouse-and-data-mining-based discovery process. IEEE Trans. Intell. Transport. Syst. **16**(5), 2653–2662 (2015). https://doi.org/10.1109/TITS.2015.2414663
21. Maamria, D., Gillet, K., Colin, G., Chamaillard, Y., Nouillant, C.: Which methodology is more appropriate to solve eco-driving Optimal Control Problem for conventional vehicles?, pp. 1262–1267 (2016). https://doi.org/10.1109/cca.2016.7587980
22. Rivera, N., Chica, J., Zambrano, I., García, C.: Estudio del comportamiento de un motor ciclo otto de inyección electrónica respecto de la estequiometría de la mezcla y del adelanto al encendido para la ciudad de cuenca. Rev. Politécnica **40**(1) (2017)

PESTEL Analysis as a Baseline to Support Decision-Making in the Local Textile Industry

Erik Sigcha[1], Andrés Martinez-Moscoso[2,3], Lorena Siguenza-Guzman[1,4], and Diana Jadan[5](✉)

[1] Department of Computer Sciences, Faculty of Engineering, Universidad de Cuenca, Cuenca, Ecuador
{erik.sigchaq,lorena.siguenza}@ucuenca.edu.ec
[2] Faculty of Economics and Business Administration, Universidad de Cuenca, Cuenca, Ecuador
andres.martinez@ucuenca.edu.ec
[3] School of Law, Universidad de Cuenca, Cuenca, Ecuador
[4] Research Centre Accountancy, Faculty of Economics and Business, KU Leuven, Louvain, Belgium
[5] Department of Applied Chemistry and Systems of Production, Faculty of Chemical Sciences, Universidad de Cuenca, Cuenca, Ecuador
diana.jadan@ucuenca.edu.ec

Abstract. The concept of sustainability has several possible meanings in the Ecuadorian context. This is especially true when it comes to the business sector because of the complexity related to implementing the Good Living concept and honoring the constitutional Rights of Nature. In Ecuador, most of the business sector is represented by micro, small, and medium-sized enterprises (MSMEs). Thus, their role in the development of national economies is crucial. This sector has difficulty adapting its organizations to the way of sustainable development. In fact, the decision to incorporate the principle of sustainability in this transition period is a big challenge for MSMEs due to their limited resources and the competitive environment they face. This work studies the contextual aspects influencing the regular operations of the Ecuadorian MSMEs, with particular emphasis on the textile industry. To do so, a comprehensive literature review was performed by applying a political, economic, social, technological, environmental, and legal (PESTEL) analysis. As a result, factors and implications were identified and explained in context, comprehensively, not individually. The information gathered yielded an overview of the sector, which serves as a baseline for the definition of the required models and tools for supporting sustainable decision-making in the local context.

Keywords: Decision Support Systems · Sustainability · Textile MSMEs · Context analysis · Strategic management

1 Introduction

Sustainable development is a principle that has attracted the interest of industry, academia, and policymakers. This concept seeks to find a balanced integration of economic performance, social inclusiveness, and environmental resilience to benefit current

M. Botto-Tobar et al. (Eds.): ICCIS 2020, AISC 1273, pp. 144–156, 2021.
https://doi.org/10.1007/978-3-030-59194-6_13

and future generations [1]. The introduction of sustainable practices in management and business operations involves considering many aspects, such as social investment, improving working conditions, rational use of materials, and reducing environmental impacts [2]. Lately, Latin American (LA) policymakers are encouraging sustainable development by addressing principles of social welfare and the Rights of Nature in local regulations for persons and organizations. One essential category of organizations involved in this trend should be the micro, small, and medium-sized enterprises (MSMEs) since they play a crucial role in the development of national economies [3]. In Ecuador, MSMEs constitute 99% of formally established enterprises, and their sales are 28.66% of the country's total sales [4]. Nevertheless, applying the principle of sustainability in MSMEs is a big challenge due to their limited resources and the competitive business environment. In this context, the need to provide simple tools that could help local MSMEs to gain an understanding of the impacts and benefits of incorporating sustainability, in anticipation of the transition to a more circular economy, constitutes an interesting research opportunity.

Currently, MSMEs often have low productivity levels due to inefficient management of activity and resources [5]. This inefficiency causes their primary concern to be profit maximization, leaving aside additional considerations of sustainability [6]. Nevertheless, management models supporting sustainability can be a powerful tool to integrate sustainable practices in the daily operation of businesses. Different approaches to incorporating sustainability decisions into management models have emerged with varying levels of ambition [7]. In Ecuador, the number of companies involved in sustainability decisions is still quite small, and generally correspond to medium and large enterprises. However, regulations and policies are increasingly addressing the negative impacts of unsustainable management found in various productive sectors, such as the textile and clothing industry. Even though the textile industry is paramount in Ecuador [8], companies in this sector are continually seeking innovative ways to foster social awareness without decreasing their efficiency, due to severe shortcomings in the industry. Thus, the development of an easily implementable sustainable management model on a local scale for textile MSMEs is a timely research topic that merits more investigation.

To address the construction of a sustainable management model for the Ecuadorian industry, it is first essential to make a structured examination of contextual aspects influencing the regular operations of MSMEs. This article describes an analysis of the Ecuadorian MSME's current context, with an emphasis on the textile industry. To accomplish this aim, a study based on the political, economic, social, technological, environmental, and legal (PESTEL) structure [9] was carried out, which allowed the development of the main characteristics of the textile MSMEs in Ecuador. This baseline identifies the main factors with consequences in business operations that can influence the selection of theories, models, and indicators for financial/economic accountability, social responsibility, and environmental leadership in the local textile MSME sector. A complete overview of the business sector is essential to choose the most appropriate tools and data models for sustainable decision-making and the best use of optimization and Big Data techniques.

2 Methods

In most of the cases studied, the researchers have focused on various specific aspects, but looking at the decision-making context facilitated understanding phenomena from several different perspectives. A wide variety of characteristics need to be examined to analyze the context of the Ecuadorian textile MSMEs sector comprehensively. These characteristics usually fall under dimensions, such as legal, economic, cultural, or organizational, which highlights the need to carefully select context dimensions that will be considered in this analysis. For the present study, the use of the PESTEL analysis was proposed. This approach addresses six categories of external environmental factors (Political, Economic, Social, Technological, Environmental, and Legal), so it allows an understanding of the problem from multiple perspectives and categories. PESTEL analysis is used for two purposes in studying the business sector: a) allowing identification of the environment within which a business operates, and b) providing information to help predict situations and circumstances that can influence business operations [10].

The first phase of this study consisted of an examination of relevant questions in each category, after a careful analysis of the pertinent bibliography (primary sources). Initially, these questions were intended to generate a general perspective of the business context in Ecuador, identifying external and internal factors that favor or discourage sustainable development. Also, the questions were focused on identifying the role of the textile MSMEs sector in the economy and general society. An original set of questions was presented to the project team; each member suggested adding or deleting questions to refine the research.

In the second phase, to address the questions mentioned above, a first selection of databases and websites was made to find papers that suited each factor. The main articles were by Scholar, Scopus, Scielo, and Dialnet. The information gathered comes from sources published from 2006 to 2020. Also, an essential source of information was the National Institute of Statistics and Censuses (INEC), which provides relevant information regarding gains or losses in different business sectors in Ecuador. Moreover, to present a clear overview of the regulations that currently govern Ecuadorian companies' operation according to each factor, specific aspects of the Constitution of the Republic of Ecuador and seven different laws derived from it were reviewed, as were regulations corresponding to these laws. Also, a review of the Statistical Yearbook of the Economic Commission for Latin America and the Caribbean (ECLAC) was included.

3 Results

Information presented in this section summarizes the main aspects influencing the operations of local textiles MSMEs for the PESTEL dimensions. The resulting scheme of the factors reviewed can be observed in Fig. 1.

3.1 Political Scope

During the two most recent decades, Ecuador has changed its political vision from neoliberalism to Andean socialism (XXI Century Socialism). In 2008, the country modified its Constitution and incorporated an Andean Political cosmovision called Sumak

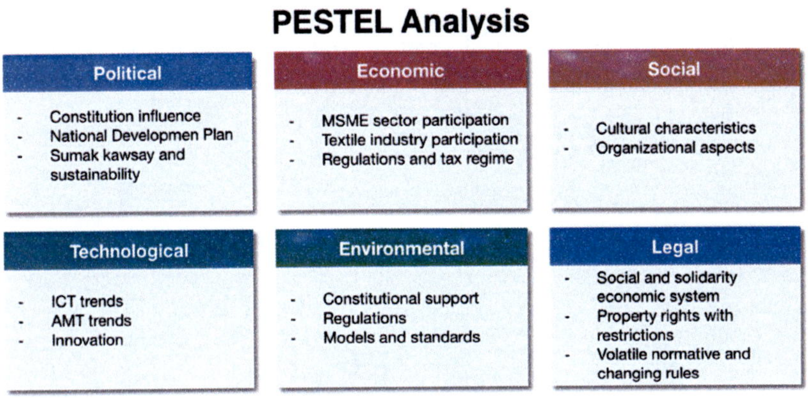

Fig. 1. PESTEL factors reviewed.

Kawsay (Good Living), which became the principal objective of its development plan [11]. This law is seen as distinct from constitutionally guaranteed Rights of Nature, as well as the freedom to work as a social right, which forms the base of the economy. Sumak Kawsay can be informed by some of the Corporate Social Responsibility (CSR) principles since it says: a) Ecuador is a culturally diverse country, where constitutional rights should be applied without prejudice, and be practiced with justice, solidarity and respect for diversity; b) the economic system should be built on the principles of justice, democratic relationships, productivity, and producing goods and services with social and environmental responsibility; and, c) nature rights and the right to a healthy environment will be developed through environmental principles, preventive principles, and objective responsibility for impacts on nature or damaged caused.

Additionally, one of the most important related amendments to Ecuador's Constitution is the establishment of a route to achieving the goals of Sumak Kawsay, through the Development Regime, which is the primary instrument to put into practice the social and solidarity economic system [12]. Since this plan is mandatory in the public sector and facultative for the private sector, its structure could be seen as centralized.

3.2 Economic Scope

In Ecuador, the MSME sector is defined by the size classification of companies, which is based on parameters established by the Andean Community of Nations (CAN). The criteria for categorizing an Enterprise as micro, small, or medium is: a) the number of workers lower than 200, b) the amount of sales less than \$ 5 million, and c) the value of assets must not exceed \$ 4 million. The participation of this sector in the national economy has been measured by the National Institute of Statistics and Census (INEC) [4]. For the year 2018, MSMEs represented 99% of formally established companies in Ecuador. They contributed more than 25% of the country's non-oil GDP, and their generation of labor comprises nearly 70% of the Economically Active Population (EAP). According to data from the Superintendent of Companies, in 2015, MSMEs reported incomes accounted for 26% of GDP (\$ 25,962.4 million), generating \$ 1,366.9 million in profit [13]. Nevertheless, when comparing sales and employment generated, they are surpassed by large companies, which create more than 70% of national annual sales. Figure 2a presents a summary of their participation in the national economy.

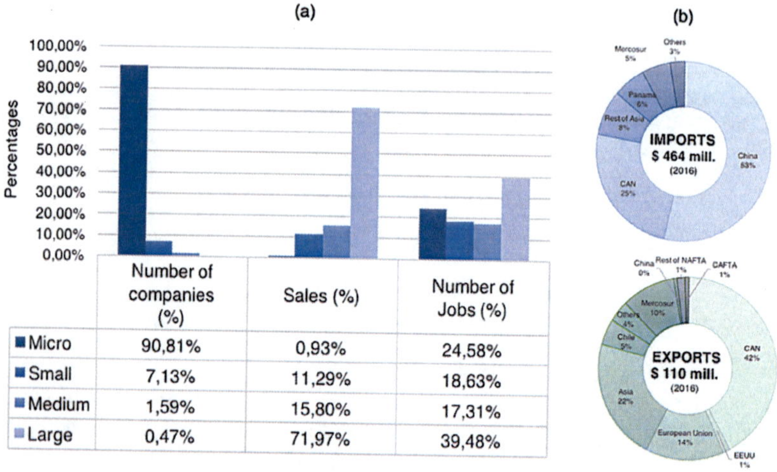

Fig. 2. (a) Participation in the economy by size, and (b) Textile imports and exports in 2016

According to data from the Textile Industrial Association of Ecuador (AITE) [8], this sector represents 18% of the employment generated by all manufacturing companies, and 85% of the textile actors are MSMEs [16]. The average share of the textile industry in the national GDP is 1.31%. Figure 2b summarizes import and export behavior in the textile sector in 2016 (based on [17]). AITE reports that their main destinations are the CAN, European Union (EU), and Asia. Also, local studies indicate that domestically produced items are among the most commonly used raw materials, including cotton fabrics, manufactured continuous filament yarn fabrics, and wool fabrics. Flax fabrics and other manufactured continuous filament yarn fabrics are also used but mostly imported [18].

Regarding policies, foreign trade is an essential issue in Ecuador's economic policies. Currently, agreements with the CAN and the EU establish and maintain preferential relationships regarding exports. The existence of these agreements, which aim to reduce tariffs, has made these countries Ecuador's main export destinations for industries such as textiles. However, the arrangements also generate disadvantages for local companies. In the CAN sector, for example, Ecuador's dollarized economy does not allow its products to be competitive since consumers tend to buy cheaper products from Colombia or Peru. Also, in the case of the EU, there is a deadline for reducing tariffs to 0% on products from these countries. Thus, European companies will compete strongly against local companies [19].

Concerning economic regulations, the Committee of Foreign Trade (COMEX) is in charge of regulating and approving national trade policies. For a considerable period, COMEX has been a regulatory entity protecting the national industry. For example, during the 2007-2016 period, COMEX issued a balance of payments safeguard measures meant to protect the Ecuadorian economy from negative consequences of monetary devaluation in Colombia and Peru. The textile sector was affected by the import tariff surcharges imposed in March 2015. These tariffs decreased this sector's growth outlook by making imports of raw materials, supplies, and machinery more expensive.

3.3 Social Scope

The social dimension contains a large number of characteristics that are very difficult to generalize. There are specific characteristics of each company, such as organizational characteristics and cultural aspects of its members, which influence the daily activities of organizations.

On the one hand, an analysis of cultural aspects in the local MSME sector can entail the study of a wide range of characteristics related to the region, type of business, and other organizational features. Information about cultural aspects, defined in [14], has been considered to obtain an initial working framework for organizational cultural aspects for Latin American countries. The authors propose a set of five core cultural characteristics (i.e., power distribution, social relationships, environmental relationships, time/work patterns, and uncertainty and social control), and a regional analysis based on data from countries belonging to some country clusters, like Anglo, Eastern Europe, East/Southeast Asia, LA, or Latin European.

Concerning LA's cultural aspects, *Power Distribution* is characterized as moderately hierarchical due to a regional tendency to concentrate power in an authority figure and a vertical organization with autocratic or centralized decision making. *Social Relationships* are seen as moderately collectivistic since members share the belief that group goals are more important than individual goals. Regarding *Environmental Relationships*, LA organizations are characterized as moderately harmony-oriented since this society seeks to live in relative harmony with the surrounding environment instead of dominating it. The authors noticed an emphasis on striving for social progress, quality of life, and welfare. *Time/Work Patterns* is an aspect that is strongly polychromic, which means that they are more flexible about time and address several problems simultaneously, being oblivious to time and resisting firm deadlines. Finally, *Uncertainty and Social Control* is characterized as strongly relationship-based since, to some degree, these cultures tend to emphasize developing mutually beneficial interpersonal relationships and trust as a substitute for rules and procedures.

On the other hand, some organizational characteristics, such as property structure, workplace customs, relationships, seniority, size, and design, can be examined as part of the context analysis. Determining an ultimate set of organizational aspects to evaluate is an ambitious task. On property structure, results in [15] show that, in the local textile MSME sector, most microenterprises are familial or unipersonal. Small enterprises have a family structure, and medium enterprises have similar percentages of familial or non-familial structures (44% and 41%, respectively).

Organizational innovation is another subject of study in the literature reviewed, referring to the introduction of new organizational methods in the business practices, workplace organization, or external relations of companies. In [16], an analysis is performed, quantifying levels of innovation related to revising organizational practices, such as scheduling, incentives, staff training, and quality management. The study reports that more than 50% of Ecuadorian MSMEs tend to innovate in areas like business practices, workplace improvements, and relationships with external stakeholders.

3.4 Technological Scope

Determining the technological level in the MSMEs sector requires reviewing two technologies: a) Information and Communication Technologies (ICT) and b) production technologies. For the ICT trend analysis, several documents reviewed discuss the most used tools in this sector in LA (see Appendix A). These tools were classified into a rubric based on [17], including 1) basic, 2) internal integration, 3) external integration, 4) decision support, and 5) infrastructure. As can be seen in Fig. 3, the ICT tools with the highest level of use fall into the *Basic* and *Infrastructure* categories. These categories are related because basic hardware and infrastructure networks are required for operating basic tools, such as Office software or e-mail. *External integration* tools are used at an intermediate level since this category is comprised of software, such as Web sites, software for financial and bank transactions, e-commerce, and points of sale. MSMEs easily access these tools because they are configured before purchase and do not require specialized hardware or infrastructure. *Internal integration* tools and Decision Support Systems (DSS) are used at low levels in MSMEs.

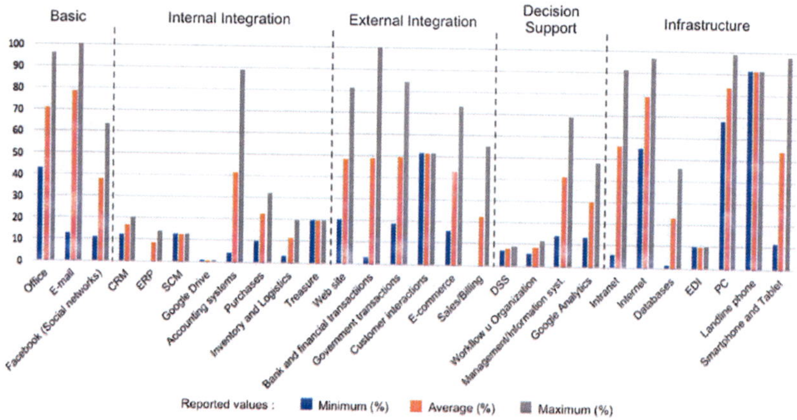

Fig. 3. Levels of ICT use in the MSME sector

Regarding technology for production, the application of Advanced Manufacturing Technologies (AMT) in MSMEs was considered. In [18], four categories are mentioned: 1) product design technologies, 2) process technologies, 3) transactional and logistic applications, and 4) communication and integration application systems. The results show that the most prevalent AMT in manufacturing MSMEs are production inspection and control systems (67.2%), computer-based inventory management (66.6%), computer-aided drawing (64.1%), and quality control systems (61.9%). These results demonstrate that MSMEs tend to adopt technologies for logistics and monitoring operations. The least prevalent is Material Requirement Planning (MRP-II) (17.9%), flexible manufacturing systems (18.7%), and automated handling of materials (20%). This implies that AMT for integration and flexibility have low rates (less than 30%) of adoption, except for Enterprise Resource Planning (ERP) systems.

In the case of technological innovation, according to INEC, machinery-equipment acquisition (31.22%) and training (20.18%) are the most applied innovation activities. Less attention is paid to the research/development of software, hardware, and technical support (13% average) [19]. Furthermore, [15] reports that product innovation is a common trend in MSMEs in the textile sector, with a frequency greater than 60%. Other business areas where innovation is applied are process, organizational, and marketing. Nonetheless, less attention is paid to these areas in the MSMEs textile sector than in other sectors of the economy. In Ecuador, government institutions promote laws and regulations to support innovation. Some of the existing laws are COPCI [20] and INGENIOS [21]. The COPCI code, also called "Código Orgánico de la Producción, Comercio e Inversiones" spells out the role of the state, funding, and tax deductions for innovation investments, all of which are relevant to the entire production sector. Similarly, the INGENIOS code "Código Orgánico de la Economía Social de los Conocimientos, la Creatividad y la Innovación" defines responsible research, innovation, research freedom, construction of a model of knowledge management, financing, tax incentives, and participation of the public, private, and mixed sectors with the popular and solidarity economy.

3.5 Environmental Scope

Concerning environmental aspects, Ecuador has regulations aiming to control possible environmental impacts produced by industries, including MSMEs. These regulations vary depending on the type of project and location, among other parameters. At the national level, protection of the environment has been a recurring concern ever since the mid-1990s, when the population's right to live in a healthy and ecologically balanced environment was recognized. This right was codified in the current Constitution (2008), which also guarantees the Rights of Nature [22]. A decade after, these rights were validated in the Constitution, the Organic Environmental Code (COA) was issued as an instrument to "... *guarantee the right of people to live in a healthy and ecologically friendly environment balanced, as well as protecting the rights of nature...*"

In addition, COA defines categories of environmental damage as *not significant, low, medium, or high.* The Unique Environmental Information System (SUIA) establishes requirements related to environmental permits granted for every level of impact. For *not significant* impacts, organizations have no obligation to get a license. COA also indicates that project operators, works, or activities causing a *low environmental impact*, are required to submit a specific environmental management plan in accordance with secondary regulations issued for this purpose. Likewise, a COA regulation dictates that the Competent Environmental Authority, through the SUIA, will grant environmental administrative authorization for works, projects, or activities of *medium or high impact*, as outlined in terms of an environmental license.

Regarding environmental management models, the one that is most commonly acknowledged is the ISO 14001 Environmental Management [23]. For companies, the ISO 14000 family is a strategic tool that creates opportunities to improve efficiency, reduce costs, and increase productivity. This standard requires companies to have the following: a) an environmental policy according to their needs, b) documentation of environmental impacts that occur when preparing products or services, c) records of the fulfillment of legal requirements, and d) a statement of environmental objectives and goals. In 2018, the number of companies in Ecuador with this certification was 130. Companies with this certification address Environmental Management by conducting audits to evaluate their environmental policies and processes, their products must adhere to rules for labeling, and waste generated must be handled through life cycle assessment (LCA) [24].

3.6 Legal Scope

The Ecuadorian legal structure plays an important role in the development of business activity since it sets the rules for fair market competition. The local social and solidarity economic system addresses public, private, mixed, popular, and solidarity organizations, as well as other economic organizations (e.g., families, domestic work, and autonomous productive organizations) [11]. These parts of the economy function according to the following principles: quality, sustainability, systemic productivity, work valuation, and economic and social efficiency. Organizations in the popular and solidarity economy receive government benefits and are characterized by the particular relationships and purposes of its members [25]. Their activities are related to social solidarity, as opposed to commercial or civil acts [26].

Regarding property rights, Ecuador's Constitution recognizes different forms of property (i.e., public, private, community-based, state-owned, associative, cooperative, or a mixture of these forms), but with some restrictions related to social and environmental functions [11]. Similarly, Ecuador's Companies Law recognizes six different types of companies [27]. In the case of MSMEs, the most commonly used business forms are the Limited Liability Company (LLC) and the Limited Partnership (LP). In most cases, companies in the textile industry use the first form because of their structure (limited, for example, by a family nexus) and their responsibility (partners are shielded from personal liability for the debts or other obligations of the company).

For legal requirements, textile industries, like other sectors, need a legal structure to operate in Ecuador. The problem is that the requirements and processes are too slow and with too much bureaucracy. To begin operations, companies first choose the type of company: LLC or/and LP. Next, a name must be chosen. Then, under the advice of an attorney, company by-laws must be written. This document must be notarized and approved by the State Authority (SC). Later, a local registration process must be performed (allowing for municipal taxes and authorizations), along with registering the company and submitting documentation of the election of the CEO and President to the Commercial Registry. The last step establishes fiscal affiliation with the Tributary Administration (Ecuador's Internal Revenue Service, SRI), which assigns a fiscal number (known as a 'RUC'). By completing such procedures, companies can operate in compliance with local legal requirements.

4 Conclusions

Although a substantial amount of information was gathered for the present work, some topics were omitted or not covered in depth. This section discusses the results, implications, and omitted information.

Regarding the *political* sphere, how the state promotes sustainable development through the Constitution has been analyzed. In Ecuador, this is done by applying the Sumak Kawsay principle. In this way, all laws, statutes, or policies promoted by government entities (i.e., National Development Plan) must function in accordance with this principle. From an *economic* perspective, topics such as MSME and textile industry participation in the national economy were reviewed. This dimension allowed identifying the factors that influence and are influenced by companies in the textile MSME industry. Unfortunately, the most recent statistical data is not available; thereby, information from 2015–2018 was used. Additionally, although several government institutions regulate business participation in the national economy, a few examples of economic regulation institutions were mentioned. Since *organizations and culture* are broad subjects, information from regional studies was considered. It is hoped that the cultural core set of characteristics reviewed can serve as an initial framework to identify relevant cultural features in local companies. Regarding organizational issues, a sample related to analyzing organizational aspects was selected. The information analyzed is limited to a business sector context analysis, but it can serve as an initial compendium of factors to be reviewed in more detail in subsequent studies.

The *technological* dimension of a business sector can vary widely over time, which represents one of the limitations faced in this work. Also, most of the information gathered is related to LA and is not specific to the Ecuadorian textile industry. It is assumed that there are no unique characteristics that could differentiate these organizations by country. By comparing the literature about Ecuador with research about Taiwan and China, it is clear that the technology and innovation found in Ecuadorian MSMEs contrast sharply with technology and innovation in Taiwan and China [28, 29]. In these companies, applications like Big Data Analytics and the Internet of Things (IoT) are addressed, which require an ICT infrastructure level not available in the local context (see Fig. 3). Referring to AMT technology trends, studies of specific AMT for the textile industry were not available for this study. Hence, the present study includes results of AMT trends for the MSME sector in general. Concerning the *environmental* dimension, the current study presents a summary of the legal requirements mandated by governmental and regulatory environmental authorities. The environmental issue is promoted by the Constitution, as reported in Sect. 3.5. However, there are other ways to encourage proper management of environmental impacts by private or public entities that were not reviewed in this article. For the *legal* dimension, the information presented shows that, despite its small size, Ecuador has many legal requirements that influence the business sector. Despite constitutional provisions driving national entrepreneurship, many requirements and regulations complicate startup, operation, development, and decision-making processes. Also, the central government is continuously present during business operations and acts as a tutor for local organizations.

The current article presents a structured review of the environment in which Ecuadorian textile MSMEs operate, envisioning possible events that would influence their activities. The PESTEL analysis yielded a contextual which allows a deeper understanding of each characteristic of the local context. Thus, these dimensions were analyzed holistically, avoiding an isolated analysis that gives rise to incorrect interpretations of the current reality. This holistic analysis of the current local context shows that MSMEs are important actors in the economic development of Ecuador. However, local companies face a complex environment where changing rules governing Economic, Political, and Legal dimensions destabilize the business sector. This discourages investment and local entrepreneurship. Likewise, although the national context presents interesting opportunities for innovation and sustainable development, this advantage is reduced by the many requirements and regulations that exist in Ecuador compared to other countries in the region.

This study's results can be applied as a basis for other investigations, but, due to the high variability and instability of some issues in the LA, periodic systematic research updates are recommended. Understanding the contextual factors influencing companies allows recognizing external entities (i.e., government support initiatives), specific requirements (legal), or internal characteristics (e.g., technological change and innovative social structures) that must be considered when using DSS architectures or implementing other technological tools to support sustainable decision-making.

Acknowledgments. This study is part of the research project "Incorporating Sustainability concepts to management models of textile Micro, Small and Medium Enterprises (SUMA)", supported by the Flemish Interuniversity Council (VLIR) and the Research Department of the University of Cuenca (DIUC).

Appendix A - List of References on ICT Levels of Use

The list of works reviewed to determine the levels of use of ICT tools can be found at https://imagineresearch.org/wp-content/uploads/2020/05/PESTEL-Appendix-A.pdf.

References

1. Geissdoerfer, M., Savaget, P., Bocken, N.M.P., Hultink, E.J.: The circular economy – a new sustainability paradigm? J. Clean. Prod. **143**, 757–768 (2017). https://doi.org/10.1016/j.jclepro.2016.12.048
2. Lyra, F.R., Barbosa De Souza, M.J., Verdinelli, M.A., Lana, J.: Corporate social responsibility: comparing different models. Soc. Responsib. J. **13**, 728–742 (2017). https://doi.org/10.1108/SRJ-02-2017-0036
3. Westman, L., Luederitz, C., Kundurpi, A., Mercado, A.J., Weber, O., Burch, S.L.: Conceptualizing businesses as social actors: a framework for understanding sustainability actions in small- and medium-sized enterprises. Bus. Strategy Environ. **28**, 388–402 (2019). https://doi.org/10.1002/bse.2256
4. Instituto Nacional de Estadísticas y Censos: Boletín Técnico - Directorio de Empresas y Establecimientos, 2018 (2019)

5. Forth, J., Bryson, A.: The Impact of Management Practices on SME Performance. Social Science Research Network, Rochester (2018)
6. Bridge, S., O'Neill, K.: Understanding Enterprise: Entrepreneurship and Small Business. Macmillan International Higher Education, Basingstoke (2012)
7. Jørgensen, T.H., Remmen, A., Mellado, M.D.: Integrated management systems – three different levels of integration. J. Clean. Prod. **14**, 713–722 (2006). https://doi.org/10.1016/j.jclepro.2005.04.005
8. Asociación de industrias textiles del Ecuador: AITE - Asociación de industrias textiles del Ecuador. https://www.aite.com.ec/
9. Johnson, G., Angwin, D.N., Whittington, R., Regner, P., Scholes, K.: Fundamentals of Strategy. Pearson, London (2015)
10. Yüksel, I.: Developing a multi-criteria decision making model for PESTEL analysis. IJBM **7**, 52 (2012). https://doi.org/10.5539/ijbm.v7n24p52
11. Asamblea Constituyente: Constitución de la República del Ecuador (2008)
12. Consejo Nacional de Planificación: Plan Nacional de Desarrollo 2017–2021 (2017)
13. Zabala, V., Muñoz, M., Naranjo, C., Guerrero, M.: Especial pymes 2016. Revista Ekos **269**, 44–46 (2016)
14. Steers, R.M., Sanchez-Runde, C.J., Nardon, L.: Management across Cultures: Challenges and Strategies. Cambridge University Press, New York (2010)
15. Zaruma, J.R., Santos, J.L.S., Escudero, D.V., Vivanco, M.O.: Innovation and technology in the ecuadorian textile industry. In: 2017 12th Iberian Conference on Information Systems and Technologies (CISTI), pp. 1–7 (2017)
16. Astudillo, S., Carpio, X., Cordero, F., Pozo, S.: El efecto de la innovación en las mipymes de Cuenca, Ecuador. Estudio observacional descriptivo transversal. Maskana **4**, 35–46 (2013). https://doi.org/10.18537/mskn.04.02.03
17. Siqueira, É.S., de Souza, C.A., Reinhard, N.: Digital divide in small and medium business: an analysis of sector differences in Brazilian companies based on data from a nationwide survey, vol. 13 (2015)
18. Uwizeyemungu, S., Poba-Nzaou, P., St-Pierre, J.: Assimilation patterns in the use of advanced manufacturing technologies in SMEs: exploring their effects on product innovation performance. JISTEM **12**, 271–288 (2015). https://doi.org/10.4301/S1807-17752015000200005
19. Falcón, L.M.M., Guajala, M.E.R., Abril, C.M.M., Córdova, A.G.V.: La innovación tecnológica de las PYMES manufactureras del Cantón Ambato. Revista Científica Hermes 3–17 (2017)
20. Asamblea Nacional: Código Orgánico de la Producción, Comercio e Inversiones - COPCI (2010)
21. Asamblea Nacional: Código Orgánico de la Economía Social de los Conocimientos - INGENIOS (2016)
22. Martínez-Moscoso, A.: El nuevo marco jurídico en materia ambiental en Ecuador: estudio sobre el Código Orgánico del Ambiente. Actualidad Jurídica Ambiental 14–46 (2019)
23. International Organization for Standarization: ISO 14001 - Environmental Management Systems (2015). https://www.iso.org/standard/60857.html
24. Comisión Económica para América Latina y el Caribe: Anuario Estadístico de América Latina y el Caribe 2019. CEPAL (2020)
25. Asamblea Nacional: Ley Orgánica de Economía Popular y Solidaria y del Sector Financiero Popular y Solidario. Registro Oficial **444** (2011)
26. Asamblea Nacional: Reglamento de la Ley Orgánica de Economía Popular y Solidaria y del Sector Financiero Popular y Solidario. Registro Oficial Suplemento **648** (2012)
27. Superintendencia de Compañías del Ecuador: Ley de compañías. Registro Oficial **312**(20), 20 (1999)

28. Chen, C.-L.: Value creation by SMEs participating in global value chains under industry 4.0 trend: case study of textile industry in Taiwan. J. Global Inf. Technol. Manage. **22**, 120–145 (2019). https://doi.org/10.1080/1097198X.2019.1603512
29. Saleem H., LI Y., Ali Z., Mehreen A., Mansoor MS.: An empirical investigation on how big data analytics influence China SMEs performance: do product and process innovation matter? Asia Pac. Bus. Rev. 1–26 (2020). https://doi.org/10.1080/13602381.2020.1759300

Occupational Health and Safety for Decision-Making in the Framework of Corporate Social Responsibility: Models, Guidelines, and Indicators

Byron Calle[1], Erik Sigcha[2], Rodrigo Guaman[3], and Lorena Siguenza-Guzman[2,4](✉)

[1] Faculty of Chemical Sciences, Universidad de Cuenca, Cuenca, Ecuador
byron.calle1001@ucuenca.edu.ec

[2] Department of Computer Sciences, Faculty of Engineering, Universidad de Cuenca, Cuenca, Ecuador
{erik.sigchaq,lorena.siguenza}@ucuenca.edu.ec

[3] Department of Applied Chemistry and Systems of Production, Faculty of Chemical Sciences, Universidad de Cuenca, Cuenca, Ecuador
rodrigo.guaman@ucuenca.edu.ec

[4] Research Centre Accountancy, Faculty of Economics and Business, KU Leuven, Louvain, Belgium

Abstract. Occupational Safety and Health (OSH) is a primary concern in the Corporate Social Responsibility (CSR) discussion since the workers' welfare is one of the main aspects of operating in a socially responsible manner. When searching for information about OSH management models, a large amount of content is obtained; however, they have not been compiled or synthesized for their analysis. The purpose of this study is to define the main models, guidelines, and indicators that allow building a conceptual framework for decision-making regarding the topic of OSH as a component for CSR. A literature review methodology was applied to select relevant information in the period 2009–2018 about OSH in manufacturing industries. This literature review allowed analyzing a final sample of 96 articles. As a result, this paper identifies the most applied OSH models, their benefits and challenges, as well as the main guidelines and indicators to be included as part of an OSH-based management model.

Keywords: Occupational Safety and Health · Management model · Literature review · Indicators

1 Introduction

In previous years, companies were seen as organizations with the sole purpose of maximizing shareholder value, without taking into account secondary factors affecting the company [1]. Later on time, this manner of thinking has changed within enterprises due

M. Botto-Tobar et al. (Eds.): ICCIS 2020, AISC 1273, pp. 157–169, 2021.
https://doi.org/10.1007/978-3-030-59194-6_14

to a growth of interest in integrating social and environmental aspects to provide value-added and to generate sustainable development [1]. Corporate Social Responsibility (CSR) is known as a voluntary commitment that companies make to achieve sustainable development, taking into consideration economic, social, and environmental matters [2]. CSR provides benefits to companies, such as improving its reputation, generating social welfare, protection of human and labor rights, preservation of the environment, and establishment of anti-corruption practices [3]. One of the main benefits pursued by CSR is the welfare of the members of a company; thus, the topic of Occupational Safety and Health (OSH) is considered as a relevant part of CSR, especially in the industrial sector. The application of OSH practices acquires great importance, as it allows reducing accident rates during production. The International Labour Organization (ILO) estimated that globally, from 2010 to 2015, 153 accidents occur at work every 15 s and at least one death at work. The economic costs associated with workplace accidents are estimated at 4% of the world's gross domestic product, which has an impact on countries' economies, business productivity and is a public health problem [4].

In the literature, a wide variety of works and guides is found on CSR and OSH implementation for all types of organizations. For instance, the work performed by Cobos describes an ergonomic risk assessment process for a refrigerator assembly company [5]. Oña and Bermeo describe the implementation of a safety management model based on OHSAS 18001 and an environmental management system based on the ISO 14001 standard for a body assembly process [6]. However, prior works are focused on a particular set of industrial activities and are based on a specific standard of safety. Unfortunately, little work is observed on the application of OSH management models within production companies, specifically in manufacturing companies, nor an overview of the aspects that need to be addressed to implement OSH practices as part of socially responsible management. Besides, tools and resources available to implement socially responsible practices can be adjusted to fit the specific features of the different types of companies. Hence, implementing CSR and OSH requires sophisticated methods to support this process and to generate tools and indicators for managers and decision-makers. Similarly to the use of software tools to control and monitor production, other tools and indicators can be created to control the fulfillment of tasks in a socially responsible manner within industrial companies. For all these reasons, there is a need to investigate the subject of OSH management that allows recognizing its guidelines, models, principles, and indicators aligned with principles of CSR. The resulting insights can be applied as part of a management model in assembly companies, which leads to the development of automated tools for monitoring and supporting decision making.

Therefore, the aim of this study is the analysis of OSH principles, guidelines, and indicators, which can be later incorporated into a conceptual framework to build a CSR-based management model and the data models required for the development of decision support systems. To this end, a literature review has been performed to identify a set of models for OSH management, its benefits, limitations, guidelines, and indicators for the implementation of OSH as a key component of a management model.

2 Methodology

This literature review applies Fink's methodology for conducting research literature reviews to analyze completed and recorded papers that have been produced by researchers and practitioners [7]. Fink's procedure allows a systematic, structured, and focused review for research purposes. This methodology consists of seven main steps. 1) Formulation of research questions. 2) Selection of source databases. 3) Selection of research terms. 4) Application of practical screening criteria. 5) Use of methodological screening criteria. 6) Revision of the documents. Finally, 7) summary of the results.

The initial step, for the development of the literature review, is the formulation of research questions to be answered by the analysis of the information; therefore, it was necessary to establish the approach and scope of the investigation. In this step, it was determined that the identification of the management models or guides for CSR and OSH was a key aspect to be considered. Also, it was required to obtain knowledge on the main guidelines and principles on which these models are based, along with methods or strategies and indicators for evaluating OSH management performance. Finally, it was seen useful to recognize the main benefits and challenges that companies face when performing management based on CSR and OSH. Hence, the following research questions were formulated: 1) What are the most applied OSH management models in the manufacturing industry? 2) What benefits and challenges do manufacturing companies obtain when incorporating OSH management models? 3) How to evaluate the performance of OSH models (indicators) within the processes of companies? Also, 4) What are the main OSH guidelines to be incorporated into a management model?.

The second step corresponds to the selection of digital databases to be consulted. This step is done through a bibliographic search and considers resources such as books, scientific journals, websites, and digital databases. The current study focused on digital resources in the following databases: Scielo, Scopus, ScienceDirect, Taylor and Francis, Ebsco, ProQuest, RRAE, and Google Scholar. Later, as a third step, it was necessary to determine the keywords, which, for this investigation, were mainly derived from the following basic terms: "Occupational Safety and Health", "Management Model", "Safety and Health Business". The detail of the bibliographic databases, search terms, and the number of results obtained is shown in Table 1.

The fourth step was the application of the practical screening criteria since, in the preliminary searches, a large number of articles were obtained. In this case, 1046 items were initially gathered when using the search terms[1] (English and Spanish), many of which did not contain relevant information to this investigation. A large number of papers omitted were related to OSH application in other types of organizations, such as educational or hospitals. In addition, the selected information was considered for the revision if it was published in the period from 2009 to 2018. Likewise, only articles published in Spanish and English were selected. The Spanish language allowed finding principles, guidelines, and indicators applied in local industries. As for English, it is the universal scientific language [8]; thus, it is considered fundamental in all researches.

[1] For Science Direct, Proquest, and Scielo, a filter corresponding to the year of publication was applied, taking as a reference a period of ten years.

Table 1. Search terms utilized for the literature review.

Database	Query	Spanish results	English results
Google Scholar	"modelos de gestión de seguridad y salud ocupacional"	1	
	"Directrices de seguridad y salud ocupacional"	4	
	"seguridad y salud ocupacional industrial"	11	
	"Safety and Health occupational"	12	
	"occupational safety and health" + "assembly" + "management" + "csr"	4	564
	"seguridad y salud ocupacional empresarial"	1	
	"principios de seguridad y salud ocupacional"	20	
	"seguridad y salud laboral en la industria"	7	
Springer	"seguridad y salud laboral"		6
	seguridad y salud ocupacional	20	
	"Industrial Occupational Safety and health"		3
	"Safety and Health guidelines"		12
Taylor and Francis	"seguridad y salud ocupacional"		1
	"Occupational Safety and Health guidelines"		17
	"Occupational Health and safety system"		16
Science Direct	seguridad y salud ocupacional en la manufactura	19	
	"seguridad y salud laboral"	25	
	"modelos de gestión de seguridad"	1	
ProQuest	"modelo de gestión de seguridad"	6	
	"seguridad y salud ocupacional"	51	
RRAE	seguridad y salud ocupacional	4	
Scopus	"Occupational safety and health model"		138
Scielo	responsabilidad social empresarial	99	
	seguridad y salud ocupacional en la manufactura	2	
	TOTAL	**287**	**761**

After applying the practical filter, a total of 243 literary references about the application of OSH practices in manufacturing companies were identified.

Once the information was selected, the fifth step is the application of the methodological screening criteria. It required the use of a "Findings Matrix", which has the function of listing the important data after the implementation of a practical filter. This matrix contains information such as the year of publication, author name, document title, content information, and the abstract. It is used to perform a quick analysis of the content to determine if the articles contain relevant information for the research. After the application of the practical and methodological screening criteria, a final sample containing 96 articles for the literary analysis was determined (Appendix A).

The sixth step corresponds to the study of the selected literature. For this, papers were read to identify the information that allows answering the questions formulated in the first step. To this end, the Atlas.ti[2] software tool was used, which helps systematically discovering and analyzing hidden phenomena in unstructured data, by coding and labeling findings in the bibliographic material. For the use of this software, a set of 38 codes or tags was structured (Appendix B). They were organized as follows: demographics (author, year, country, title, magazine), structures of the management model, benefits, challenges, components of management models, types of manufacturing companies, accident rates, and performance assessment methods. Therefore, text excerpts that the reader considered necessary were marked through these tags.

Finally, the seventh step formulates the results; hence, the data was exported to a spreadsheet to perform metadata and content analysis. For metadata analysis, a statistical study was carried out through the frequencies of information, such as the year of publication, applications of management models, and the amount of data selected in each country. The content analysis identified the guidelines, benefits, indicators, and aspects considered relevant to the reader through tags in the software.

3 Results and Discussion

This section describes the main findings obtained after the systematic literature review using the qualitative data analysis software. Results allowed gaining insights into the OSH state of the art.

3.1 Metadata Analysis

The demographic criteria provided additional statistics on the revised articles. The country of publication criterion allowed identifying the number of studies that have been carried out per country concerning OSH. Figure 1 shows that developed countries such as the USA and Spain have a significant contribution to the topic of OSH with 11 published articles. These findings may be due to the growth in accident incidence rates, which has generated greater interest in safety management model applications [9, 10]. Also, it can be noted that countries such as Korea, Canada, Australia, Iran, and the UK have also contributed to works on the subject. In addition, this analysis has shown that Latin American countries have provided a low amount of information on OSH.

Another selection criterion was the year of publication, which indicates that documents published over ten years (2009–2018) were considered. In this manner, a trend analysis could be performed overtime on the investigations about OSH. Figure 2 shows that the years with the most publications were 2016, 2017, and 2018, where 16 articles

[2] https://atlasti.com/.

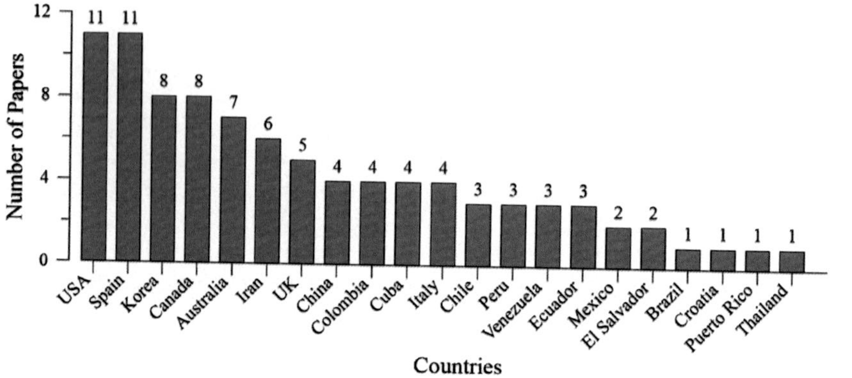

Fig. 1. Amount of OSH information selected by countries.

were published on digital databases. This figure also shows that, from 2009 to 2015, few papers were issued related to the topic. Therefore, the number of publications has steadily grown over the last 5-yr period.

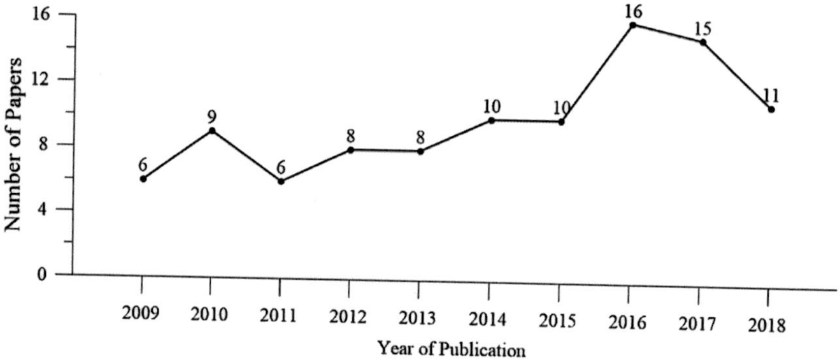

Fig. 2. The number of OSH publications by year.

Since the type of bibliographic material analyzed corresponds to articles, an analysis of the top scientific journals that have published content about OSH has been carried out. Table 2 shows that the "International Journal of Occupational and Environmental Health" has published a higher number of articles. In addition, the journals "International Journal of Occupational Safety and Ergonomics" and "Safety and health at work" have contributed to five scientific papers related to safety and ergonomics management models at the workplace. To simplify the reading of the table, journals that have only published an article related to the topic were not included.

Table 2. The number of OSH publications by journals.

Journal	# of Articles
International Journal of Occupational and Environmental Health	6
International Journal of Occupational Safety and Ergonomics	5
Safety and Health at Work	5
Ergonomía	3
Revista Internacional de Investigación de Producción	2
Responsabilidad Social y Gestión de la Seguridad y la Salud en el Trabajo	2

3.2 Content Analysis

For the content analysis, the research questions established in Sect. 2 are answered. Concerning the question: *1) What are the most applied OSH management models in the manufacturing industry?* The articles describing implementations or application of practices based on a particular model within production or manufacturing companies were previously located. In this manner, the identification of the primary standards, models, or guidelines for OSH management was carried out. Thus, it was determined that the OHSAS 18001 model was the most standard mentioned in the literature analyzed [11]. OHSAS focuses on aspects related to worker safety and is designed for companies of different sizes and geographical conditions [10]. Another guideline identified was the ISO 26000 standard for CSR, which addresses the issue of Occupational Safety and other work practices [12]. However, since it is a CSR model, it encompasses several additional aspects and does not specialize in OSH. For this reason, it can be assumed that its frequency of appearance in the analyzed literature is much lower compared to the OHSAS model.

Another element identified was the Deming management model, which focuses on the continuous improvement of companies. The Deming model is mostly related to quality management and has been considered part of this work since it addresses OSH as a constant improvement component [13, 14]. In addition, the ISO 45001 standard has been identified, which guides making work areas safe and meeting various occupational safety and health criteria [15]. The ISO 45001 standard, despite being promoted by the International Organization for Standardization, does not have much-related research yet because it was recently published. Finally, applications of other management models, such as LOPCYMAT [16], MODEL ECUADOR [17], ERGOPAR [18], and HSEQ [19], have been identified in specific countries, e.g., Venezuela, Ecuador, and Spain, respectively. Figure 3 details the list of models identified in the review with the number of mentions.

Regarding the research question, *2) What benefits and challenges do manufacturing companies obtain when incorporating OSH management models?* For their determination, the frequency of mention of each identified benefit or challenge was analyzed. On the one hand, the advantages with the most remarks were found to be: 1) the reduction of occupational injuries and diseases, 2) the improvement of climate, performance, and business image, and 3) the generation of more sources of employment for society.

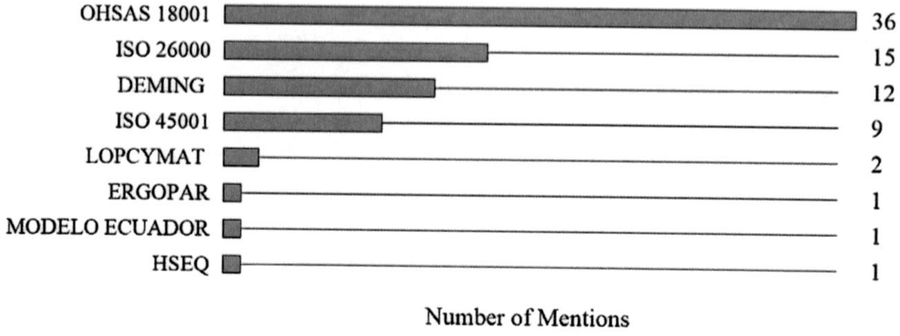

Fig. 3. Number of mentions per OSH model or guide

On the other hand, about the identified challenges, it was determined that cultural change is the challenge that appears the most. Another problem identified was the selection of techniques for specific risk assessments. Finally, worker collaboration and participation that allows for a cultural change is another challenge that was considered significant. Table 3 shows a list of benefits and challenges found in the literature review.

Table 3. Security management benefits and challenges mentioned in the articles.

Benefits	Mentions	Challenges	Mentions
Reduction of occupational injuries and diseases	27	Change or improve the company's culture	23
Improve the image of the company	11	Select appropriate techniques for specific risk assessment	8
Give all rights to workers	9	Voluntary collaboration of the worker	8
Improve worker performance	7	Development of personalized occupational health and safety programs	6
Improve the working climate	6	Flexibility to cultural change	5
Cost reduction, which involves hiring	5	The responsible staff should have a vast knowledge of the company's activities	4
More sources of employment	5	Extensive knowledge in the area of occupational safety and health	3
Welfare in society	4	Participation in management groups	3
Avoid fines and penalties	3	High costs in the application of the management model	3

Concerning the question, *3) How to evaluate the performance of OSH models (indicators) within the processes of companies?* The use of indicators is one of the most widely used strategies. In the field of occupational safety within production companies, the indicators to be used depend on the type of analysis required, in which two types of applied studies have been identified, productivity and accident. About productivity analysis, the assessment of positive or negative changes in the company after using a management model is essential. Within this type of study, the Efficiency Index is the indicator that is most mentioned in the literature, and it serves to measure the degree of achievement of the objectives. Likewise, it has been known that the Effectiveness Index, which relates the number of final products obtained from certain resources, is another indicator that appears in the articles. For accident analysis, the indices mentioned in the literature are frequency index, incidence rate, and severity index. In several articles, these three indicators are used in case studies to check the frequency of accidents within a period. Table 4 shows the productivity and accident indicators found in the literary review.

Table 4. Summary of indicators found in the literature review

Indexes	# of mentions
Productivity	
Efficiency index	29
Gross Domestic Product (PIB)	9
Effectiveness index	5
Index of risks not controlled by the worker	2
Rate of improvement of risk conditions in the work areas	2
Accident Rate	
Frequency index	10
Incidence rate	8
Severity index	8
Average Casualty Duration Rate	2

Ultimately, about the research question: *4) What are the main OSH guidelines to be incorporated into a management model?* As a preliminary step, a set of guidelines has been taken as a reference based on the content of the main OSH management models. Therefore, it was possible to unify its content to analyze and identify the most mentioned guidelines at the literature level. Figure 4 shows the number of mentions by guideline, where Assessment has 33 remarks and refers to the process of reviewing occupational safety and health risks. Assessment is based on the identified hazards, taking into account the effectiveness of existing controls [15]. Planning is the second most frequently mentioned guideline and corresponds to ensuring that the occupational health and safety management system can achieve the expected results and generate continuous improvement. Thirdly, the Improvement has been identified as a primary objective, which guides managing incidents and non-conformities to react on time [15]. Other guidelines identified are worker leadership, knowledge of the context, management of needs and expectations, and understanding of internal and external aspects of staff. Therefore, the importance of correct target formulation for OSH management systems can be noted. Finally, the guidelines on operation, results and support have been identified, which relate to the support required for the development of the activities proposed for OSH management.

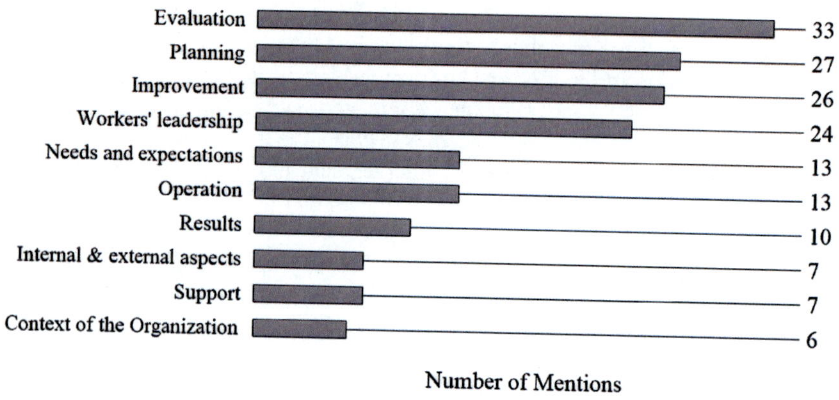

Fig. 4. Number of mentions per OSH principles

Table 5. Guidelines components found in the literature review

Guideline	Frequency	Component	Authors
Evaluation	21	Monitoring, measurement, scan and performance assessment	[20]
	12	Compliance assessment	[21]
	7	Internal auditoriums	[20, 22]
Planning	16	Actions for arc risks and opportunities	[19, 22]
	11	Hazard identification and risk and opportunity assessment	[23, 24]
	8	Action planning	[25]
Improvement	26	Continuous improvement	[26, 27]
Leadership	24	Leadership and commitment Encouraging OST policies	[28]
Needs and expectations	13	Understanding the needs and expectations of workers and other parties	[18]

In the current analysis, the guidelines have a more detailed level and can be defined as the components that allow its management. To facilitate their presentation, the elements found in the literature review were grouped and summarized because they have differences between the articles, the results can be seen in Table 5. It is recommended to complete these supplements with the components of existing models and standards for occupational health and safety management.

4 Conclusions

This article describes a systematic literature review on Occupational Health and Safety, which is intended to generate an approach to management models oriented to CSR. In this manner, some research questions have been formulated to allow obtaining relevant information, such as the most applied guidelines, the indicators used for the evaluation of the management, and the most used OSH guides and models in the manufacturing industry. For the development of this work, Fink's literature review methodology was applied, based on a definition of research questions that allowed identifying the resources related to the subject investigated [7]. In the first part of the literature review, a search for information was developed in some selected digital databases. Later, applying practical and methodological screening criteria, a sample of 96 articles of scientific content was established to be analyzed. Through the development of this review, it was possible to determine that the OHSAS 18001 model has been the most applied tool for manufacturing companies.

Moreover, reducing occupational injuries and diseases, improving the climate and worker performance are the main perceived benefits of implementing an OSH management. This fact implies that, when implementing occupational safety practices, companies address the issues of social welfare and quality as a component of CSR. Additionally, the cultural change of the whole company is the most critical challenge that has been encountered in the literary review since it is necessary to include the worker in the planning, execution, and evaluation of management. Concerning the evaluation methods of OSH management, indicators such as frequency, incidence, and severity indices have been determined to assess the performance of OSH practices. The identified indicators are intended to be analyzed for decision-making and incorporated in a software platform for process management, allowing to link SSO indicators as new variables for a CSR-based Business Process Management. Finally, ten guidelines were found for the OSH management implementation. They were prioritized according to the number of applications or mentions in the literature.

This work may be seen as an initial conceptual framework for the construction of a process management model based on CSR aspects, which considers the OSH dimension as a critical aspect of the manufacturing sector. As future work, it is planned to develop a detailed study of guidelines and identifiers found, which will be validated in the local context. This will allow identifying a more significant number of indicators and determining the most relevant aspects of OSH to the local industry.

Acknowledgments. This study is part of the research project "Modelo de gestión basado en variables de Calidad y Responsabilidad Social Empresarial para la optimización de procesos de Ensamblaje", supported by the Research Department of the University of Cuenca (DIUC).

Appendix A – Full List of References Read in the Literature Review

The list of sources used in the literature review can be found online at https://imagineresearch.org/wp-content/uploads/2020/02/OSH-Appendix-A.pdf.

Appendix B – List of Codes Utilized in the Literature Review

The list of codes used in Atlas.ti for the literature review can be found online at https://imagineresearch.org/wp-content/uploads/2020/05/OSH-Appendix-B.pdf.

References

1. Huerta, P., Rivera, C., Torres, A.: Anàlisis de la relaciòn entre responsabilidad social corporativa y los resultados empresariales: un estudio por estrategia corporativa 30 (2015)
2. Simó, M.J.M., Padilla, R.A.A., Rey, J.M.: La seguridad y salud en el trabajo en el marco de la responsabilidad social de la empresa. Rev Fom Soc. 19 (2007)
3. Ibarra Padilla, A.M.: Principles of corporate social responsibility in the Colombian law. Rev. Derecho Uninorte **41**, pp. 51–82 (2014)
4. Gómez García, A.R., Algora Buenafé, A.F., Suasnavas Bermúdez, P.R., Silva Peñaherrera, M.G., Vilaret Serpa, A.: Notificación de Accidentes de Trabajo y Posibles Enfermedades Profesionales en Ecuador, 2010–2015. Cienc Trab. **18**(57), 166–172 (2016)
5. Cobos, M.E.L.: Evaluacíon de los factores de riesgo ergonomico de la linea de ensamblaje de refrigeradoras Indurama 84 (2017)
6. Oña, I.F.B., Bermeo, S.A.F.: Diseño de un Sistema de Seguridad Industrial para el Edificio de Ingeniería Civil y Ambiental de la Escuela Politécnica Nacional 579 (2016)
7. Fink, A.: Conducting Research Literature Reviews: From the Internet to Paper, 4th edn., p. 257. SAGE, Thousand Oaks (2014)
8. Ramírez, M.G., Falcón, I.A.L.: Importancia del lenguaje en el contexto de la aldea global. Horizontes Educacionales **15**, 14 (2010)
9. Arocena, P., Núñez, I.: An empirical analysis of the effectiveness of occupational health and safety management systems in SMEs. Int. Small Bus. J. Res. Entrep. **28**(4), 398–419 (2010)
10. Origen y evolución de OHSAS 18001 [Internet]. Nueva ISO 45001 (2014). https://www.nueva-iso-45001.com/2014/06/origen-y-evolucion-de-ohsas-18001/
11. OHSAS Project Group: Occupational health and safety management systems - requirements: OHSAS 18001. S.L.: OHSAS Project Group (2007)
12. ISO 26000: Normas RSE **1**(1), 126 (2012)
13. Gupta, A., McDaniel, J.C., Kanthi, H.S.: Quality management in service firms: sustaining structures of total quality service. Manag. Serv. Qual. Int. J. **15**(4), 389–402 (2005)
14. Ghahramani, A.: An investigation of safety climate in OHSAS 18001-certified and non-certified organizations. Int. J. Occup. Saf. Ergon. **22**(3), 414–421 (2016)
15. ISO TMG: ISO 45001. Psicogente 60 (2018)
16. Caraballo-Arias, Y.: Occupational safety and health in Venezuela. Ann. Glob. Health **81**(4), 512 (2015)
17. Mora, L.: MODELO ECUADOR 1 (2015)
18. García, A.M., Boix, P.G., Benavides, F., Gadea, R., Rodrigo, F., Serra, C.: Participación para mejorar las condiciones de trabajo: evidencias y experiencias. Gac. Sanit. **30**, 87–92 (2016)
19. Riveros, A.V.: Formulación de una Propuesta de Mejoramiento, Basado en la Aplicación de las Normas Iso 9001:2015, Iso 14001:2015 E Iso 18001:2007, en la Empresa Jc Construcciones & Mantenimientos S.A.S, en la Ciudad de Bogotá 139 (2018)

20. Nordin, M., Dul, J.: Summary and prospects: occupational health and safety. Ergonomics **33**(3), 387–389 (1990)
21. Yepes González, N., López Castaño, E., Quintero Guzmán, C., González Bulla, J.J.: Factores que impactan en la selección e implantación del sistema de gestión ISO/TS 16949: Caso del sector industrial de autopartes de la ciudad de Bogotá. Cienc E Ing Neogranadina **24**(1), 143 (2014)
22. Molina, M.: Caracterización de la gestión de calidad bajo la Norma OHSAS 18001, en las micro y pequeñas empresas del sector manufactura **1**, 68 (2018)
23. Vega Pérez, J.G., Vargas Ramos, M.F., Amores Guevara, P.D.R., Arias Tapia, S.A.: Riesgos psicosociales y la seguridad industrial en las lavanderías textiles del Cantón Pelileo. Rev. Comun. SEECI (43), 135 (2017)
24. Ulutasdemir, N., Kilic, M., Zeki, Ö., Begendi, F.: Effects of occupational health and safety on healthy lifestyle behaviors of workers employed in a private company in Turkey. Ann. Glob. Health **81**(4), 503 (2015)
25. Kim, Y., Park, J., Park, M.: Creating a culture of prevention in occupational safety and health practice. Saf. Health Work. **7**(2), 89–96 (2016)
26. Pheng, L.S., Kwang, G.K.: ISO 9001, ISO 14001 and OHSAS 18001 management systems: integration, costs and benefits for construction companies. Archit. Sci. Rev. **48**(2), 145–151 (2005)
27. Marín, C.V.L.: Sistema de calidad integrado con la seguridad y salud ocupacional en la empresa Dim-Moble 159 (2016)
28. De Oliveira Matias, J.C., Coelho, D.A.: The integration of the standards systems of quality management, environmental management and occupational health and safety management. Int. J. Prod. Res. **40**(15), 3857–3866 (2002)

Model of Classification of University Services Using the Polarity of Opinions Taken from Social Networks

Javier Sánchez-Guerrero[1](✉), Silvia Acosta-Bones[1], Rosario Haro-Velastegui[1], and Marco V. Guachimboza-Villalva[2]

[1] Facultad de Ciencias Humanas y de la Educación, Universidad Técnica de Ambato, Av. Los Chasquis entre Payamino y Guayllabamba, Ambato, Ecuador
{jsanchez,silviabacostab,charytoharo}@uta.edu.ec
[2] Facultad de Contabilidad y Auditoría, Universidad Técnica de Ambato, Av. Los Chasquis entre Payamino y Guayllabamba, Ambato, Ecuador
marcovguachimboza@uta.edu.ec

Abstract. The objective of this research is to use the Big Data generated through the opinions of the social networks of Higher Education Institutions, which are free from bias at the time of expressing themselves about the IT services offered by these institutions, providing a data bank that serves as raw material to perform a data and text mining that is used to guide the ITIL process in its phase of Service Strategies. The mining process uses the methodology known as CRISP - DM that obtains values that determine the quality of the service studied, giving an input for decision making that will maintain, improve or eliminate the IT service analyzed. Finally, it is concluded that using social networks to know the opinions of the IT services offered by the institutions and at the same time apply a mining process to guide the adaptation of ITIL best practices is a reliable process because users feel free to express their opinion about the service received.

Keywords: Big data · ITIL · Social networks · IT services · CRISP-DM

1 Introduction

The growing influence in the daily activities of the human being, supported by the use of information technology, networks and technologies, involves the generation of a massification of information derived from them. According to [1], new forms of information processing and processing are required that can handle a large amount of very diverse data at high speeds, with the objective of obtaining knowledge that allows for making appropriate decisions.

The digital transformation at a global level is undeniable, for companies and organizations this support in ICT has become an opportunity to scale up and increase their business model and take advantage in a globalized and too competitive environment.

Today we are part of a society that is based on information and knowledge supported by technology giving rise to the information society and as this phenomenon has grown

M. Botto-Tobar et al. (Eds.): ICCIS 2020, AISC 1273, pp. 170–186, 2021.
https://doi.org/10.1007/978-3-030-59194-6_15

it transforms from a technological line to a more relevant one based on the user and the content he handles in consumption and generation of information, generating profound changes in different contexts [2].

It is necessary to take into account the imperative need to align the rules of business or institutional business to the primary use of IT by optimizing technology at various levels of the organization in order to maintain efficient processes to keep costs reasonable and predictable. The great problem at present in the provision of services is to ensure that the provision of the same, is in accordance with the needs of the customer without the need to incur greater costs or efforts [3].

In universities, teaching, administrative and student activities are marked by the generation of information that leads to knowledge and decision making. It is extremely important to know how and where information is organized so that it can be easily interpreted [4].

In this way, higher education institutions have been developing academic services through IT Information Technologies, which allow them to manage and administer the activities of the different university actors in an appropriate manner. In addition, the constant use of social networks gives way to generate opinions on various topics of interest within the community, which serve to capture opinions frankly and without bias of the functionality of the services offered [5].

On the one hand, those in charge of hardware and software, technicians or administrators, and on the other hand, those who need these services, that is, clients who often may not feel satisfied with the services provided even when those in charge indicate that high standards of the service offered are maintained. The goal must be to achieve an efficient combination of people, processes and technology to provide better management of IT services, optimizing the use of resources and continuous improvement of service levels [6].

In order to effectively achieve the above, the work of companies and institutions must be aligned to have good practices that lead to the achievement of the objectives pursued by the institutions and therefore their authorities. This means that if, in spite of making high investments in technology, hardware and software, if it is not administered properly, it simply does not work and does not provide the expected results.

In order to have a guide to administer, manage, evaluate and, above all, a process of continuous improvement, it is very important to rely on a known and proven regulatory framework that helps this process and serves as a reference and has documentation and validity in the IT world, such as ITIL, which has a series of good practices that serve as a guide in the administration or management of IT services.

In order to detect the best criteria that guide this articulation of IT management with a methodological framework of best practices, it is necessary to know the criteria of the clients and others involved in such a way that they resort to tools such as social networks that give them the freedom to express themselves and to express their opinion around a certain topic, in our particular case the use of technological services that higher education centers have. These opinions or comments issued by users of the services can be thousands, which forms the Big Data of opinions on a particular topic.

[7], the social network Facebook has become part of our social imaginary, increasing its participation and notoriety in the new era of Big Data; taking as a premise that the term

Big Data does not yet have a defined concept, since it has been described as a "relative term" [8], as well as the excessive growth of information as well as data storage, that is to say, that the term "large data" has three categories in its definition: (i) when the focus of its product is qualitative in the size of the data, (ii) when data collection, processing and use processes are involved and finally (iii) in relation to the human being with cognitive capacities in particular; allowing to integrate both the focus and the scope of focus.

This work analyzes the opinions and obtaining their polarity (positive, negative), this process allows services to be classified according to the polarity assigned to each of them. The proposal is to identify the polarity of the text and then try to interpret this as part of a positive or negative emotion. It is not intended to ensure the polarity of a text, because even this depends on the mood of the person, hence our results are only a guide for decision making by a human expert.

The technique used for a part of this work is based on data mining and Point Mutual Information (PMI) allowing to obtain the semantic orientation (SO) of the phrases that compose an opinion. Phrase selection is carried out following the part of speech patterns of unsupervised classification proposed by [30].

The overall objective of the study is to classify IT services according to the polarity of opinions obtained from social networks, for effective decision-making.

However, this should be framed in a methodological process such as ITIL, in such a way that it will have specific objectives to reach the general one:

- Validate the fulfillment of the objectives of the processes carried out by the institution in each phase of the service life cycle based on ITIL V3.
- Design a big data structure to capture opinions about IT services and their quality.
- To design a technique that allows, through the polarity of opinions on IT services, to improve management decision making.

2 State of the Art

IT Information Technology

As indicated by [9], Information Technology or IT refers to the set of elements, computers, optical fiber, cameras, robots and some other devices that, together with computer-assisted applications and systems, have the fundamental task in the company of processing, storing, evaluating and communicating information. There are three areas that integrate what we can point out as IT: information technology, telecommunications and office automation, generating an infrastructure of machines and systems that allow the relationship of the different activities of the company or institution, and between the company and its customers and suppliers.

[10] mention that IT or Information Technologies are responsible for the generation of ideas, knowledge and information in the business and corporate environment so that its management plays a very important role in institutions. The proper functioning of IT resides in balancing the real demands of project development within the expected resources and deadlines [11].

IT appeared more than four decades ago and although they represent a great support in institutions, in many of them they still represent a costly problem to solve; this is given because senior management is not yet effectively involved in adapting its services to the use of technologies, limiting their operational capacity and quality [12].

The integration of IT into the business model must have a systematic approach, based on three phases:

- A long-term IT renewal process that is strongly linked to the company's strategy.
- Have a technology platform that is simple and unifying, that is framed in an open architecture with horizontal design and serves the company as a whole.
- Functional, service-focused, high-performance IT management that works as a team according to corporate standards.

Services

IT services are a way of delivering value to customers, delivering results that they want to obtain without having to assume specific costs and risks derived from the transaction [13].

Processes

They are referred to as a set of systematized actions that combine and implement resources and skills in order to achieve a result, and that directly or indirectly provide value for the user or client [13].

Value is understood as the combination of utility plus warranty, with the utility being what the customer wants or requires, while the warranty is the assurance that the services or products supplied are available. Any service is protected by one or more processes, and this notion is fundamental in order to know what IT Services and their management are [14].

Services From You

Refers to a group of actions that respond to one or more needs of a client by changing the condition of computer assets, enhancing their value and reducing the inherent risk of the system. An IT Service is based on the use of Information Technologies and supports the customer's business processes, is composed of a combination of people, processes and technologies and should be defined in a Service Level Agreement.

Service Management of You

For Abad, IT Service Management is a set of management skills specialized in providing value to the client in the form of services, following a process-based regime, aimed at aligning IT services, provided with the needs of the company or institution, with special emphasis on the benefits that the final client will receive.

Services of You in the Institutions of Higher Education

The infrastructure of the so-called Information Technologies is of fundamental importance for organizations to build their information system [15]. This infrastructure is made up of the physical equipment (hardware) as well as the programs that work on that hardware (software), this means that there is a combination between storage media, communication and programs, in addition to printers, cameras and other necessary hardware.

Within the organizations are also involved the Higher Education Institutions, which handle large amounts of documentation and at the same time IT services, which are also called Academic IT Services such as the Enrollment System, Libraries grouped in a global information system that is at the service of their customers who are mainly thousands of students [16].

Process Management

It consists of coexistence with functional administration, in which people are assigned to key processes and of greater interest, giving the possibility of creating an inter-functional management that generates added value for its clients, which means their satisfaction, giving indicators of the processes to be changed or improved, creating priorities and maintaining improvement plans that allow objectives to be reached, showing strengths and weaknesses.

Best Practices for the Management of Services You

According to [17], best practices are defined as a coherent set of actions that have led to success in a given context and that are expected to have the same results in similar environments.

Reference Frameworks for it Service Management

In order for organizations not to maintain a vision based on the use of technology and on their own organizations, the continuous improvement and quality of the services they offer must be considered and the focus must be on the relationships with their clients.

Higher Education institutions are not outside these objectives and IT Service Management should be based on the promotion and support of the application of best practices, as well as reference frameworks and other internationally accepted patterns, from which we can cite: ITIL, COBIT, ISO.

ITIL

[18] as indicated in Office of Government Commerce, ITIL is the framework of good practices in the administration of the service, which allows to establish and improve its capabilities, in addition to ITIL offers a body of knowledge useful to reach this standard.

ITIL comes from the acronym IT Infrastructure Library, which translated into Spanish would be IT Infrastructure Library, offers a set of knowledge useful to achieve high standards. ITIL is composed of: core and complementary publications; being that the core has the group of best practices that can be applied in any type of organization that provides IT services. The ITIL complementary publications have a set of industry-specific publications, operating models and other technological architectures.

There are 5 books that make up the core of ITIL and at the same time form its life cycle:

- Service Strategy
- Service design
- Service Transition

- Operation of the service
- Continuous improvement of the service

Areas of Knowledge of ITIL

ITIL in its V3 version is composed of five publications or books that are briefly described in Table 1.

In the case of this research, the ITIL Service Strategy will be based on the Polarity of the opinions in Social Networks applied to the classification of university services, so that the Big Data generated in social networks will be taken from the opinions of the services provided in a Higher Education Institution.

Big Data

Describes Big Data as the trend resulting from technological progress, which has given way to a new paradigm of understanding and timely decision-making that originates in the large amount of data obtained in an unstructured, structured and semi-structured way, which would be very demanding to analyze through a relational database demanding time and money.

It is a computational model designed to process and deal with huge amounts of data from different and varied sources which allows it to be always available, traceable, visible and calculable [20]. In order to understand Big Data, one must first understand that it is the data and where they are, in such a way that it mentions that they are found everywhere and that human beings are the ones who generate them; through the most varied forms taken by sensors, social networks, heartbeats, respiration, meters, which in passing also open the way to the Internet of Things.

Data Mining

It is also known as KDD Knowledge Discovery in Databases and represents an area of knowledge that allows discovering new potentially useful information collected from large data sources [22].

The great growth of data creates the need to have meaningful information that is extracted from these large volumes of data, thus, this field of data mining allows to extract information and at the same time improve the quality of the processes [23] taking into consideration that data mining techniques provide a set of several very powerful tools that allow to generate several classification systems that are driven by previous data, this is how these techniques can play a too important role in the refining of rules [24].

In data mining a classification system that is based on rules is very important since it allows making decisions that are mostly assertive, a decision tree can be used in the generation of rules that are driven by the data, but for the size of a sample these rules tend to describe several outliers that do not necessarily generalize to a larger data set [25].

Table 1. IT service management

Book	Description
Service strategy	We seek to improve strategic impact through the design, development, implementation and practice of Service Management Strategy generation Demand management Service portfolio management IT financial management
Service design	Provides guidance on how to design IT services and how to evaluate interactions between solutions and their business environments. It includes the processes of: 1. Service catalogue management 2. Service level management 3. Capacity management 4. Availability management 5. IT Service continuity management 6. IT Security management
Service transition	Provides guidance on how to deliver a service for operational use It includes the processes of: 1. Transition planning and support. 2. Change management Management of the configuration and assets of the service Management of versions and deployments 5. Service validation and testing 6. Evaluation 7. Service knowledge management
Service operation	Provides guidance on achieving agreed levels of quality for the service delivered. It includes the processes of: 1. Event management 2. Incident management 3. Request management 4. Problem management 5. Access management 6. IT operations
Continuous service improvement	Provides guidance on how to adapt current services to new customer needs. Includes the process of: 1. Continuous service improvement process

Source: [19]

Text Mining

While the volume of data increases considerably, foresight and its methods will benefit from these same data, in the same way as in text mining, which is based on the heterogeneity of textual data, the same that address a new field of research in integration [26], allowing access to and examination of data that is aggregated in a largely automated manner.

Thus, by exploiting new data sources such as Twitter, Facebook, Instagram, among others (web mining) [27], more actors are being integrated and at the same time points of view with more emphasis on the analysis of social changes; at present, it is intended to exploit the potential of text mining for foresight considering different data sources, such as social networks, foresight methods and the text mining approach [28].

The purpose of text mining is to facilitate the detection and examination of emerging issues and technologies by broadening the previous knowledge base, i.e. text mining provides a solid basis for reflection on possible futures.

Text mining techniques include categorizing texts, detecting topics, extracting concepts, searching, grouping and retrieving documents, and at the same time finding some non-trivial information from a collection of documents with the aim of developing recommendation systems [29], through a set of evaluation criteria that measure the effectiveness of these techniques with the intention of facilitating the selection of an appropriate technique.

3 Methodology

Once the Big Data of the opinions downloaded from Facebook about the Registration Service has been built, the following methodology is established.

The investigation is framed in the Qualitative - Quantitative approach, these two approaches help the investigation, with the qualitative we focus to the users to realize consultations of opinion in relation to the subject to investigate and quantitative we focus in the two variables to analyze the obtained results to determinate the relation and the impact that exists between them. Having identified the requirements in the construction of the lexicon of opinions, the CRISP-DM methodology is proposed for usability.

CRISP-DM is an open, free, flexible methodology that can be easily customized to create models that adapt to the needs of the project, facilitating the understanding and management of the different interactions between the stages involved in the process. It should also be noted that one of the main advantages of the proposed methodology is that it does not necessarily require highly specialized personnel for its application.

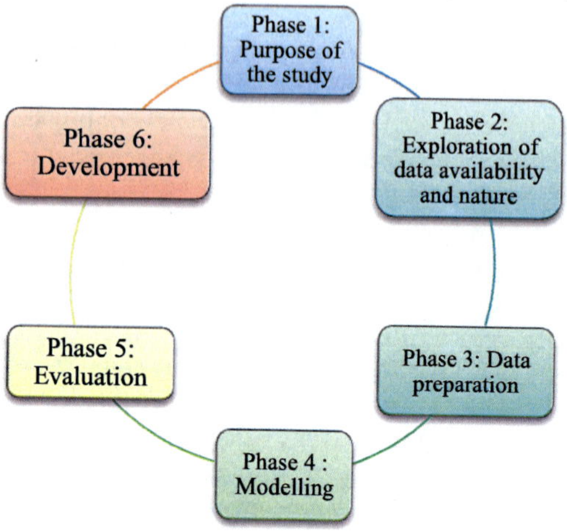

Phase 1: Purpose of the Study
It proposes to analyze opinions and obtain their polarity (positive, negative). This process allows services to be classified according to the polarity assigned to each of them.

Phase 2: Exploring the Availability and Nature of Data
This phase explains in more detail the process that has been followed to collect the information. Essentially, two sources have served as a starting point for the analysis of social networks. The method of data extraction is produced by means of the interaction.

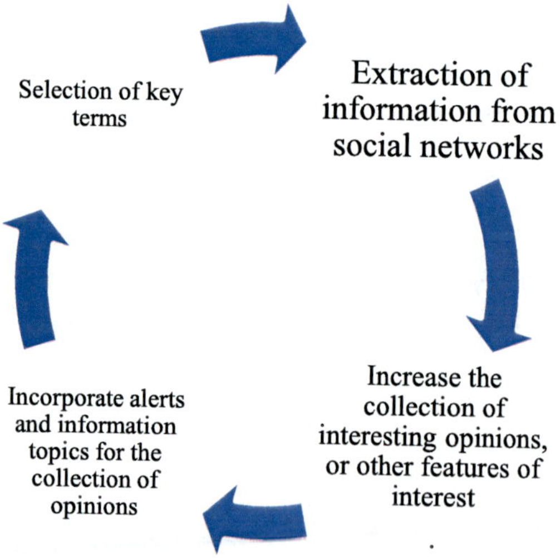

Phase 3: Data Preparation
The reliability of the selected data is improved. This includes clarifying the data, eliminating empty values and non-significant data. Depending on the nature of the data and the study, this step may involve complex statistical methods and even predictive techniques, and an initial cleaning of the data is necessary in order to work with them.

Phase 4: Modeling
The process of determining the emotional tone behind a series of words is used to understand the attitudes, opinions and emotions expressed in an online citation. The feeling analysis is extremely useful in the monitoring of social networks as it allows us to get an idea of general public opinion on certain issues.

For the modeling different packages will be applied to know the opinion about university services in social networks such as facebook or twitter using R. For this will show how to connect from R to Twitter. We will look for the associated opinions of the users associated to the social network facebook, where the emotions and polarity of the opinions will be classified to then visualize the evolution of the opinion and the clouds of words, where the negative opinions will be detected and why the comments of the public are due towards the university services, allowing this way to improve the services that they offer.

The clustering algorithm will be applied in order to analyze what kind of services can be considered as positive and which ones are considered negative for users of this Facebook page.

Phase 5: Evaluation
The patterns discovered are evaluated and interpreted in relation to the opinions that the users have towards the university services established at the beginning of the process. This step focuses on the understanding and usefulness of the induced model. For the evaluation we select the page and get comments, filter the opinions, the analysis of feelings and then analyze the results of the Classification.

Phase 6: Development
Finally, the knowledge discovered in the process is incorporated. The idea is to incorporate it into the environment, apply it and measure its impact.

4 Analysis of Results

Once the service to be analyzed has been selected, users generate opinions about it on social networks such as Facebook, freely giving their opinion about the quality of the service presented, from which a file is formed with the Big Data of the opinions through a file with extension CSV type Excel, which is known as Corpus.

The opinions of Corpus CVS must be filtered before starting to execute the polarity prior to classifying academic services by cluster. The entire process is described:

a. Filtering Opinions

The opinions that are disseminated on the subject of university services are validated beforehand by an expert in the subject, selecting in this way the opinions that are going to form part of the set that will be analysed through Point wise mutual information (PMI) and the analysis of feelings.

1. Selection of the theme of university services.

The official website of the UTA was accessed and the enrolment service for university leveling and transport young people was selected (Figs. 1, 2, 3 and 4).

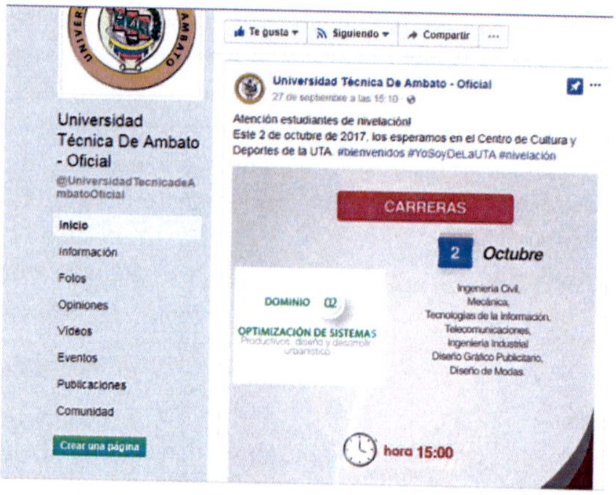

Fig. 1. Case study selection: levelling enrollment

By applying the whole process described above, the comments of the transport service are obtained in JSON format. Click on "Send" and the document with the opinions is generated.

To convert the opinions obtained in JSON format into CSV format, the Konklone application was used: https://konklone.io/json/, the comments obtained with the Facebook API are copied.

The opinions of the CSV file should be filtered by the expert before executing the polarity and analyzing the feeling prior to classification with the clustering algorithm.

Having said this, we proceed to use the software R, elaborated previously, and execute the comments in csv.

There are more positive than negative comments about the UTA Public Transport service, as can be seen in Fig. 5. There is the probability that the possible terms included in the opinions of the UTA registration services are the right terms and I can with a high probability between 0.16 and 0.25 in the opinions of the users of the UTA registration service. We worked with training data, with a current accuracy of 0.60.

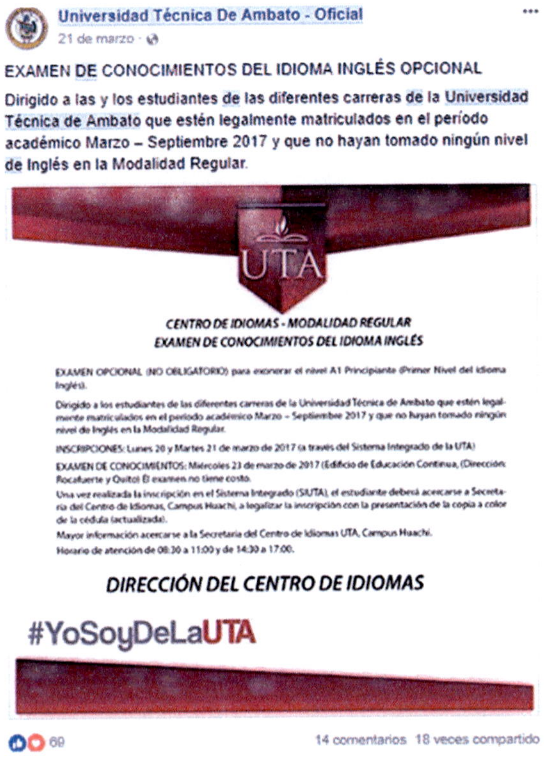

Fig. 2. Case study selection: enrollment university test english language.

Fig. 3. Obtaining comments from the registration service to take university English language.

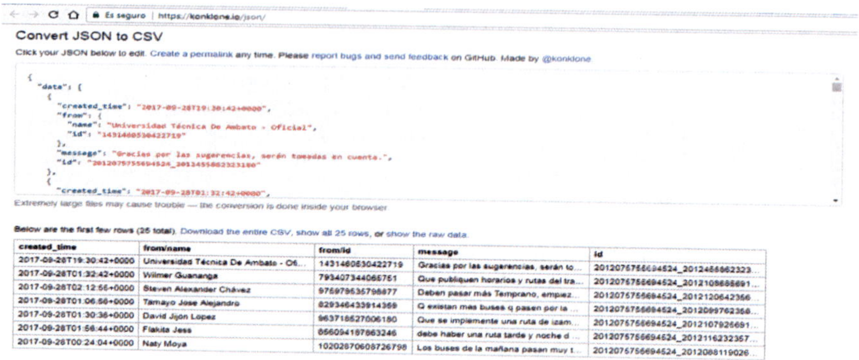

Fig. 4. Conversion from JSON to CSV with topic of study

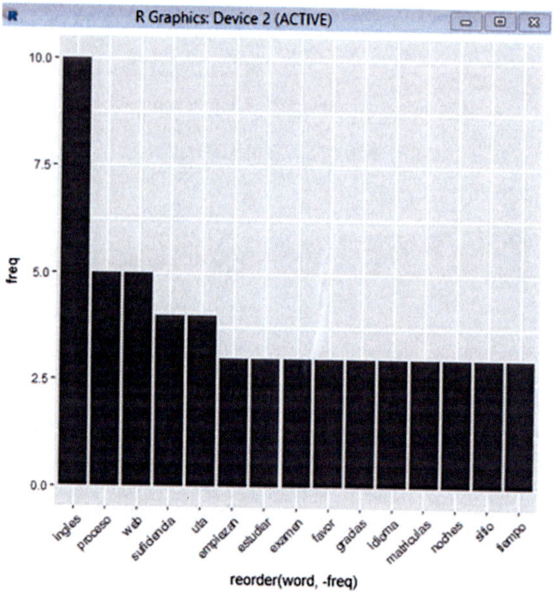

Fig. 5. Results of the study topic, enrollment to take English language.

Graph 8 shows the results of the words that appear at least 25 times in the opinions about the transport service, the words that stand out in the graph show a greater presence in the comments of the users and Graph 9 shows the 100 words with a greater occurrence in the comments of the UTA Transport services that may serve in the future to form a new file or Corpus of words in Spanish (Figs. 6 and 7):

In Fig. 8, to get a better idea of where the groups are in the dendrogram, with R it was possible to identify the groups of words that are related to each other; five groups can be observed, as indicated in the red boxes. If you want to highlight a different number of groups, then you can change the code to get another group variance.

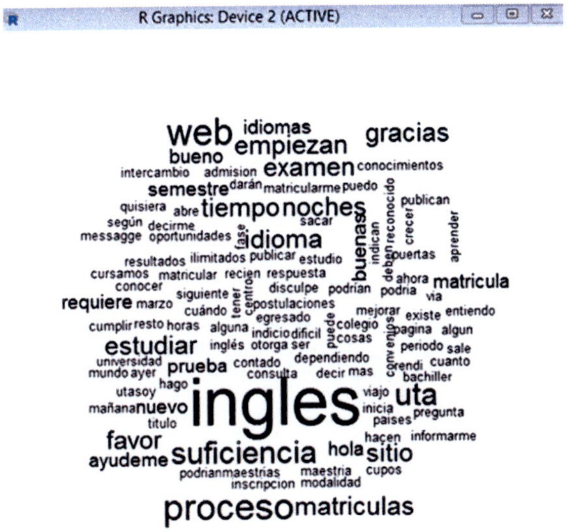

Fig. 6 Terms of Frequency at least 25 times for UTA English language enrolment services.

Fig. 7. Terms of Frequency of the 100 most used words in the registration services to take English language UTA.

In Fig. 9 we obtain a clustering with the Kmeans algorithm in which it allowed us to group a set of observations of closer words in the comments.

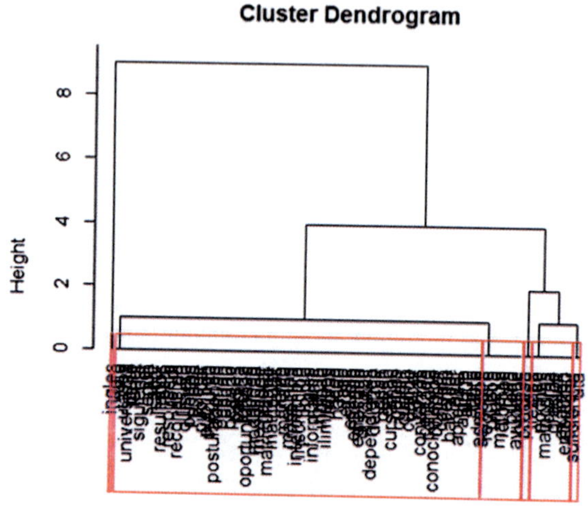

Fig. 8. Clustering of the study theme by Dendogram

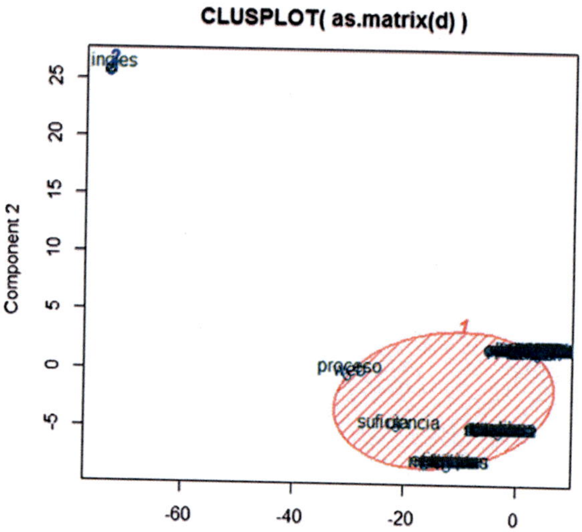

Fig. 9. Clustering of the study theme by Kmeans

5 Conclusions

The accuracy obtained is greater than 65%, it is considered as valid for the system, since it starts from free context opinions on the quality of service.

The decisions can be taken efficiently through the results obtained from the application of the CRISP-DM methodology that generates indices of confidence, because it is based on the analysis of the opinions freely issued on a particular subject by users, allowing decisions to be made tailored to the customer's need adapted to the best practices of ITIL in each IT service offered in the institution.

The social networks were an effective source of data collection to build a Big Data in a Corpus file which served to generate text mining and then data mining that issued reliable results for decision making.

In higher education institutions, the inclusion of ITIL standards and good practices becomes a good option to properly manage the use of technology, in order to improve service quality and customer satisfaction, thus fulfilling the function of Service Level Management, within the ITIL framework, whose benefits are: alignment between IT services and business needs, and improve IT relationships with customers (business), in this case university students.

The present research work was limited by the requirement of an expert person who initially analyzed the opinions to classify in: positive or negative in their polarity and also requires a data cleaning in the opinions entered in social networks because they use inappropriate words and also the use of emoticons.

As for the use of free software R, it requires management through the use of its con-sola command line, when error messages are displayed is complex interpretation of them for novice users.

The CRISP-DM Methodology requires the operator to have a broad knowledge of both the Business and the Data with which it works.

The results shown in this document, open opportunities for future work, to conduct studies and experiments on various topics such as education, infrastructure, economy, society, among others. The application of the methodology presented is encouraged in order to take advantage of the massive use of technologies and, as part of this, that of social networks.

References

1. Paredes, A.: Big Data: Estado de la cuestión de. Int. J. Inf. Syst. Softw. Eng. Big Co. **2**(1), 38–59 (2015)
2. Fernández, Á.: Sistemas de mobile learning para alumnado con necesidades especiales. Universidad de Granada, Granada (2014)
3. Cárdenas, Y.C.M., Bautista, D.W.R.: Mejores prácticas de gestión para la calidad de los Servicios en Tecnologías de la Información. Gerencia, Tecnología e Información. **11**(29), 47–57 (2011)
4. Dixson, Y.R., Maturel, L.N.: La inteligencia de negocio como apoyo a la toma de decisiones en el ámbito académico. GECONTEC: Revista Internacional de Gestión del Conocimiento y la Tecnología. **3**(2), 63–73 (2015)
5. Díaz, V.: Mitos y Realidades de las Redes Sociales. Prisma Soc. 1–26 (2011)
6. Acevedo, H.: Magazcitum, 16 junio 2010. [En línea]. http://www.magazcitum.com.mx/?p=50#.Wa39f_jyjIV
7. Parra, P., Gordo, A.J., D'Antonio, S.A.: La investigación social aplicada en redes sociales. Una innovación metodológica para el análisis de los Me gusta en Facebook. Revista Latina de Comunicación **69**, 195–212 (2014)
8. Minelli, M., Chambers, M., Dhiraj, A.: Big Data, p. 9. Big Analytics. Wiley, Hoboken (2013)
9. Liu, B.: Sentiment Analysis and Opinion Mining. Morgan and Claypool Publishers, San Rafael (2012)
10. Huerta, E., Larraza-Kintana, M.: Tecnologías de la información de la información e innovación organizativa en la empresa española. Productividad y organización **340**, 51–60 (2001)

11. Lohmann, G.G., Pizzinatto, A.K., Pizzinatto, N.K., Ponchio, M.C., Belli, H.C.: Gestión de Proyectos Internos de Ti Un Estudio Sobre Las 100 Empresas Más Importantes De Brasil, Invenio **15**, 81–93 (2012)
12. Chappel, R., Paliwal, K.: An educational platform to demonstrate speech processing techniques on Android based smart phones and tablets. Speech Comm. **57**, 13–38 (2014)
13. Feld, S. Stoddard, D.: TI a la manera correcta. Harvard Bus. Rev. 64–71 (2004)
14. Abad, W.: Gestión de Servicios de Tecnologías de la Información, 18 Febrero 2013. [En línea]. http://gsticperu.blogspot.com/2013/02/que-es-servicio-de-ti.html
15. Hassanlou, K., Fathian, M., Akhavan, P., Azari, A.: Information technology policy trends in the world. Technol. Soc. **31**(2), 125–132 (2009)
16. Laudon, K., Laudon, J.P.: Sistemas de informação gerenciais: administrando a empresa digital. Prentice Hall, Sao Paulo (2004)
17. Mitić, S., Nikolić, M., Jankov, J., Vukonjanski, J., Terek, E.: The impact of information technologies on communication satisfaction and organizational learning in companies in Serbia. Comput. Hum. Behav. **76**, 87–101 (2017)
18. López, P.: Fundamentos de gestión de servicios de tecnologías de información, 28 Noviembre 2013. [En línea]. http://itilunidad5.blogspot.mx/2013/
19. Loayza, A.: Modelo de gestión de incidentes para una entidad estatal. Interfases **9**, 221–254 (2016)
20. Guzmán, A., García, A.: ITIL v3 -Gestión de Servicios de TI. Revista Ecorfan, 801–806 (2012)
21. Rennó, R.: Smart cities e big data: o cidadão produtor de dados URBS. Revista de Estudios Urbanos y Ciencias Sociales, 13–24 (2014)
22. Paniagua, S.: Un mundo de sensores. De los datos al Big Data, Revista TELOS (Cuadernos de Comunicación e Innovación), 1–3 (2013)
23. Terán, H.E.E., Alcivar, M.: Aplicaciones de Mineria de Datos en Marketing. Revista Publicando **3**(8), 503–512 (2016)
24. Asif, R., Merceron, A., Ali, S.A., Haider, N.G.: Analyzing undergraduate students' performance using educational data mining. Comput. Educ. **113**, 177–194 (2017)
25. Peral, J., Maté, A., Marco, M.: Application of data mining techniques to identify relevant key performance indicators. Comput. Stand. Interfaces **54**, 76–85 (2017)
26. Singaravelana, S., Aruna, R., Arunshunmugama, D., Jerina, S., Muruganb, D.: inner interruption discovery and defense system by using data mining. J. King Saud Univ. Comput. Inf. Sci. **32**(5), 592–598 (2017)
27. Hussein, H., Alaaeldin, H., Hassan, M.: Selection criteria for text mining approaches. Comput. Hum. Behav. **51**, 729–733 (2015)
28. Truyens, M., Van Eecke, P.: Legal aspects of text mining. Comput. Law Secur. Rev. **30**(2), 153–170 (2014)
29. Kotu, V., Deshpande, B.: Chapter 9 – Text Mining, Predictive Analytics and Data Mining, pp. 275–303 (2015)

Clustering Analysis of Electricity Consumption of Municipalities in the Province of Pichincha-Ecuador Using the K-Means Algorithm

Oscar F. Núñez-Barrionuevo(✉), Edilberto A. Llanes-Cedeño, Javier Martinez-Gomez, Jorge I. Guachimboza-Davalos, and Jesús Lopez-Villada

Universidad Internacional SEK, Quito, Ecuador
ofnunez.mee@uisek.edu.ec

Abstract. This work shows an analysis of electricity consumption by parishes in the province of Pichincha to define consumption affinities between them. To do this, categorical variables are encoded to be used with the unsupervised machine learning algorithm k-means. As relevant results, four types of consumption are established and they are analyzed in relation to population density while algorithm represents 100% of the variability of the data, and municipalities with different consumption trends are identified.

Keywords: Power consumption · Clustering · K-means

1 Introducción

Global warming has become one of the biggest concerns of governments around the world, so efforts are increasingly being made to counteract its effects and mitigate its impact on the population. In virtue of this, numerous studies have been carried out where it has been shown that the polluting gases generated by industries, transport and many of the habits of the population have been the cause of this phenomenon. Adverse events such as fires, floods and droughts are only part of the effects of climate change on the planet [1]. One of the main causes is emissions of polluting gases, including CO_2. These emissions must be controlled immediately before the consequences are irreversible for the population, so both the issuance of public policies and the search for new renewable energy sources must be proposed by governments for their mandatory application in the areas of the industry [2]. Furthermore, these effects of global warming are related to the modification of the electricity consumption of the population suffering extreme temperatures, causing an increase in the use of heating systems [3]. Therefore, one of the ways of acting to limit and prevent its environmental, social and economic consequences, related to the increase in temperature, is to reduce energy

M. Botto-Tobar et al. (Eds.): ICCIS 2020, AISC 1273, pp. 187–195, 2021.
https://doi.org/10.1007/978-3-030-59194-6_16

consumption. Since the current generation, transport and consumption model is absolutely dependent on fossil fuels. In this way it is unsustainable as a consequence of the climate change that it implies.

In Latin America, there is an average of 0.65 tons of CO_2 for each MWh generated [4,5]. In Ecuador, this result is 397.5 g of CO_2 per KWh, where the highest value corresponds to 2010 since the generation of energy from non-renewable sources was (52.2%) [6]. However, with the change in the productive matrix, 60.85% of electricity production is based on renewable energy where 58.53% is related to Hydroelectric sources [7]. This production is oriented 30.93% to residential consumption and 26.01% to the Industrial sector [6]. However, the generation of electricity through power plants causes changes in ecosystems, the disappearance of species, soil erosion, among others. For this reason, it is necessary to carry out a process of recognition of consumption parameters that allow discovering people's habits. This process is carried out in order to have information for all market participants to cooperate effectively in saving electricity consumption. In addition, it provides information to power generation plants to carry out proper planning and avoid overloading their equipment [8].

With the constant increase in household electricity demand, the analysis of electrical energy consumption (EEC) is increasingly important. Therefore, the acquisition and analysis of data has become a widely used tool these days since it allows to determine the behavior of EEC in a precise way. In this sense, the use of user tags (identifiers) and the appropriate technology can provide a more intuitive and concise expression to EEC analysis compared to traditional analysis [9]. For this reason, the use of machine learning algorithms allows us to recognize more relevant patterns of the phenomenon by studying based on complex mathematical calculations that seek to emulate certain functionality of the human brain, such as grouping into sets of objects in relation to their most characteristic evident. In this sense, by having information on electricity consumption by cantons and parishes, the grouping criterion allows them to be organized in such a way that subgroups with attributes similar to each other, but different from others, are formed. This process is framed within unsupervised learning since there is only one set of input data that we must obtain information about the structure of the output domain, which is not available [10]. With this, an EEC can be established by sectors of the country and defined electricity consumption policies. However, the data acquisition process can be a challenging task. This is because most government databases have mostly categorical information. Consequently, it significantly limits data analysis [11]. For this, there are techniques that allow encoding variables to convert them to a numerical system. However, it is necessary to be careful in the use of these attributes since they can oversize the machine learning algorithm and provide erroneous information when generating the model. It must be emphasized that the geopolitical distribution is divided by provinces, cantons, and municipality [6].

For this reason, the present work exposes the development of clusters within the energy consumption of Ecuador in the province of Pichincha to group them into population areas with similar consumption trends related to their canton and municipality. For this, a data preprocessing stage is performed to encode the categorical variables and eliminate those that do not present relevant information to the machine learning algorithms. Subsequently, technical criteria are used to determine the number of groups to be carried out and finally establish electricity consumption trends. As a result, this study presents an acceptance of 100% that explains the variability of each group.

The rest of the document is structured as follows: Sect. 2 presents the related works. Section 3 indicates the methodological scheme of data analysis. Section 4 presents the cluster analysis to provide the grouping of electricity consumption. Section 5 shows the results obtained with the different error analyses. Finally, section6 presents the results and future work.

2 Related Works

Works such as [8–10] have presented important contributions related to the grouping in the behavior of electricity consumption of residential customers, in order to understand the personalized demands of the user and provide them with specific services. However, consumption patterns can be significantly different in relation to geographic location, population density, consumption habits, among others. In addition, most of the works use a supervised analysis criterion where the output variables are known. In this case, the aim is to find non-existent grouping parameters at a glance. That is why there are pending problems, mainly oriented in the Ecuadorian population since there is a lack of these studies and above all, they do not use machine learning algorithms.

3 Materials and Methods

This section shows, on the one hand, data acquisition (Sect. 3.1), normalization (Sect. 3.2), the selected grouping algorithm (Sect. 2.3) and the data analysis scheme (Sect. 3.3).

3.1 Data Acquisition

The data collection is through the information presented by the Ministry of Electricity and Renewable Energy in its monthly reports during the every year. In these reports, the company that distributes electricity to the provinces, cantons, and municipalities is explained in detail. In this specific case, the consumption information for the years 2019 and 2018 is taken. All the data is stored in the array $\mathbf{Y} \in \mathbb{R}^{m \times n}$ where m represents the number of samples and n the attributes of the database. In this case, $m = 2{,}550$ and $n = 11$. Attributes refer to: month, canton, municipality, type of equipment with a voltage at 220, number of customers, energy billed, increase in consumption, residential consumption, subsidy energy generated, billing of the service and value of the subsidy.

3.2 Database Normalization

Matrix normalization means adjusting the measured values on different scales with respect to one in common. This process is carried out prior to making mathematical models [7]. With this, it is avoided that there are variables that, according to their nature of the data, contribute in a large weighted amount to the model as they have very high values in relation to the rest. In addition, it prevents the use of categorical variables from being diminished due to its low scale (0 to 1). That is why there are different methods and forms of standardization. In this case, the standard normalization given by the following formula is used:

$$\frac{X - \mu}{\sigma} \tag{1}$$

3.3 Clustering

One of the important computational challenges is having the ability to recognize characteristics to group similar elements. The `K-means` method aims to partition a set of m observations into k groups. Where each value of m belongs to a group of k whose mean value of the distance is the closest. To carry out this process, the following process is established:

Algorithm 1. K-means pseudo-code

Input: Dataset
Output: each m assigned to a group k

1: Choose the number of k fo clusters
2: Select at random K points the centroids
3: Assign each data point to the closest centroid (That forms k clusters)
4: Compute and place the new centroid of each cluster
5: Reassign each data point to the new closest centroid
6: **if** any reassignment took place **then**
7: go to step 4
8: **else**
9: go to Fin
10: **return:** Each m on k cluster

However, the randomness of the values of k can cause different forms of grouping. Therefore, it is necessary to properly define its value in order to group only the data that contains the largest number of similar attributes [12]. Consequently, the `K-means ++` version allows us to eliminate this problem by analyzing a set of observations $(x_1, x_2, ..., x_p)$, where each observation is a real vector $d - dimensional$, the grouping of $K - means$ aims to divide the p observations into $k(<= p)$ sets $S = S_1, S_2, ..., S_k$ to minimize the sum of squares within the group (WCSS) (that is, the variance). This can be seen in the next equation:

$$\arg\min_{\mathbf{S}} \sum_{i=1}^{k} \sum_{\mathbf{x}\in S_i} \|\mathbf{x} - \boldsymbol{\mu}_i\|^2 = \arg\min_{\mathbf{S}} \sum_{i=1}^{k} |S_i| \operatorname{Var} S_i \quad (2)$$

where ν_i is the point average in S_i.

3.4 Data Analysis Scheme

The present work focuses on grouping the municipalities of the Pichincha province by electricity consumption habits. For this, a data analysis model is required that represents everything previously discussed, this is shown in Fig. 1.

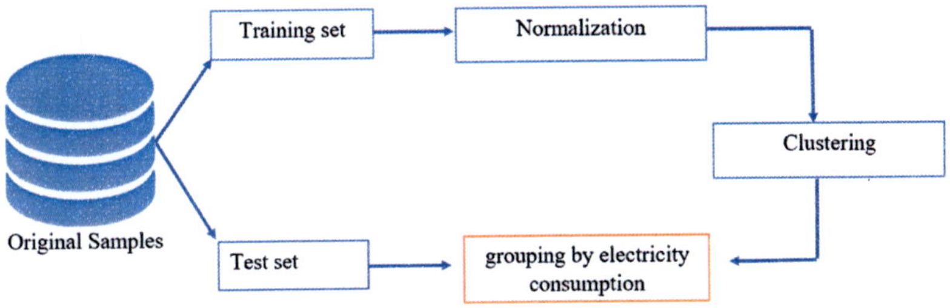

Fig. 1. Data analysis scheme proposed

4 Results

Analysis (WCSS) is performed with a value of $p = 10$ and determine the appropriate value of k. Based on the variance, $k = 4$ is determined to be ideal relative to the attributes of the **Y** database. This can be seen in Fig. 2.

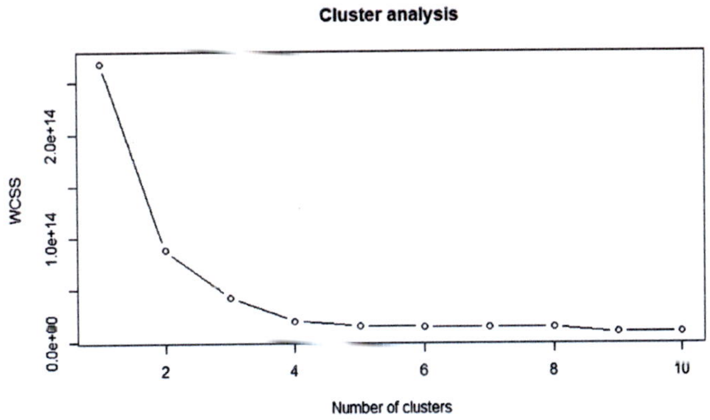

Fig. 2. WCSS analysis to select k value

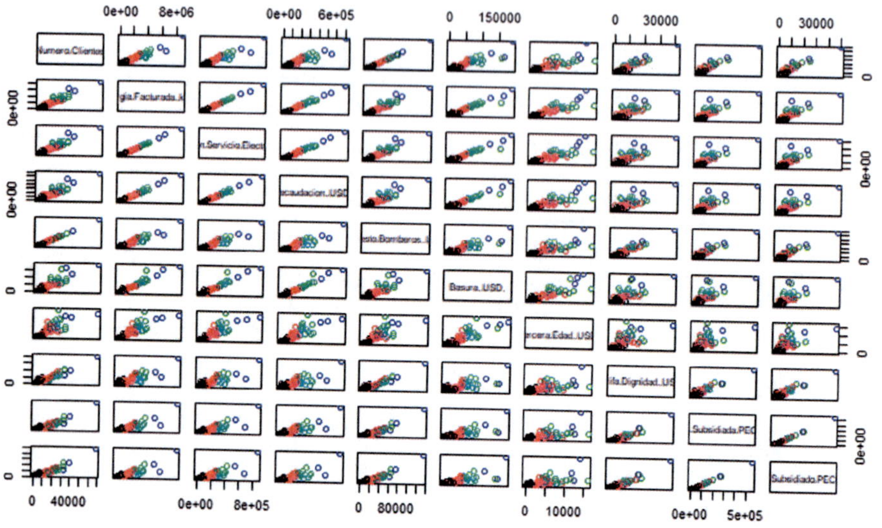

Fig. 3. Data variance analysis to represents the fit the model

In order to know the percentage of acceptance of the model in relation to the distribution of the data. It is done in the analysis of variance. This is together with the implementation of the WCSS model. As a result, there is a normal distribution of the data to the model with an acceptance of 99.9%. The multi-dimensional summary of the analysis of variance can be seen in Fig. 3.

The implementation of the algorithm allows grouping by cantons in relation to consumption in the province of Pichincha-Ecuador. As main information, the centroid values of each cluster are established. Where the electric consumption has the values of: Excessive consumption: 7037116.0 Kwh, high: 3458488.1 Kwh, average: 1984133.6 Kwh and low: 295444.3 Kwh. In order to graphically observe the organization of the clusters, it is carried out by reducing the dimensionality to two dimensions by means of the Principal Component Analysis (PCA) algorithm Fig. 4.

With the machine learning algorithm developed, the municipalities of each canton are grouped in relation to the type of consumption previously established. A summary can be made by cantons of the province of Pichincha. This can be seen in the Table 1.

The Metropolitan District of Quito has the highest electricity consumption in the province. However, its municipalities have great variability in consumption. With the aim of having a geographic analysis, the form of consumption can be established with colors: Excess consumption (red), high (tomato), average (yellow) and low (green). This can be seen in Fig. 5. It can be seen that the parishes of Calderón, Conocoto, and Iñaquito have higher consumption in relation to the rest of the municipalities (around 7037116.0 Kwh). This is due to their high population density and they are much higher than the rest. However,

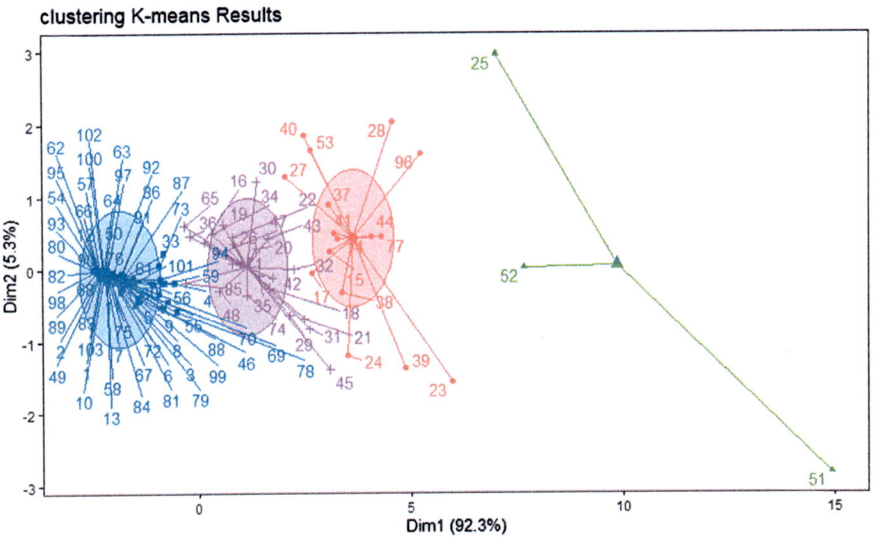

Fig. 4. K-means results for power consumption in Pichincha

Table 1. Summary of grouping of municipalities by type of electricity consumption

Power consumption clustering	Municipalities
EXCESSIVE	IÑAQUITO(25), CALDERON (CARAPUNGO)(51), CONOCOTO(52)
HIGH	BELISARIO QUEVEDO(14), CARCELÉN(15), COCHAPAMBA(17), EL CONDADO(23) GUAMANÍ(24), JIPIJAPA(27), KENNEDY(28), PONCEANO(37), PUENGASÍ(38), QUITUMBE(39), RUMIPAMBA(40), SAN BARTOLO(41), SOLANDA(44), CUMBAYA(53), TUMBACO(77), SANGOLQUI(96)
MODERATE	CENTRO HISTóRICO(16), COMITÉ DEL PUEBLO(18), COTOCOLLAO(19), CHILIBULO(20), CHILLOGALLO(21), CHIMBACALLE(22), ITCHIMBIA(26), LA ARGELIA(29), LA CONCEPCIóN(30), LA ECUATORIANA(31), LA FERROVIARIA(32), LA MAGDALENA(34), LA MENA(35), MARISCAL, SUCRE(36), SAN ISIDRO DEL INCA(42), SAN JUAN(43), TURUBAMBA(45), ALANGASI(47) AMAGUAÑA(48), NAYON(65),POMASQUI(71), SAN ANTONIO(74), MACHACHI(85)
LOW	TANDAPI(1), SAN MIGUEL DE LOS BANCOS(2), PUERTO QUITO(3), CAYAMBE(4) JUAN MONTALVO(5), CANGAHUA(6), OLMEDO (PESILLO)(7) SAN JOSE DE AYORA(8), TABACUNDO(9), LA ESPERANZA(10) MALCHINGUI(11), TOCACHI(12), TUPIGACHI(13), LA LIBERTAD(33), QUITO(46), ATAHUALPA (HABASPAMBA)(49), CALACALI(50), CHAVEZPAMBA(54), CHECA (CHILPA)(55), EL QUINCHE(56), GUALEA(57), GUANGOPOLO(58), GUAYLLABAMBA(59), LA MERCED(60), LLANO CHICO(61), LLOA(62), NANEGAL(63), NANEGALITO(64), NONO(66), PACTO(67), PERUCHO(68), PIFO(69), PINTAG(70), PUELLARO(72) PUEMBO(73), SAN JOSE DE MINAS(75), TABABELA(76), YARUQUI(78), ZAMBIZA(79), CAYAMBE(80), ASCAZUBI(81), CANGAHUA(82), OTON(83), SANTA ROSA DE CUZUBAMBA(84), ALOAG(86), ALOASI(87), CUTUGLAHUA(88), EL CHAUPI(89), SAN RAFAEL(95),(TANDAPI)(90), TAMBILLO(91), UYUMBICHO(92), MALCHINGUI(93), SANGOLQUÍ(94), COTOGCHOA(97), RUMIPAMBA(98), SAN MIGUEL DE LOS BANCOS(99), PEDRO VICENTE MALDONADO(101),PUERTO QUITO(102), MINDO(100)

Iñaquito has an average population density (around 18 thousand people) but with high electricity consumption. On the other hand, parishes such as El Condado or San Antonio have a high population density (25,000 and 30,000 people respectively) do not have excessive electricity consumption. This happens in the same way with the municipalities of El Quinche and Guayllabamba, which has around 17 thousand people and have low electricity consumption (average of 295444.3 Kwh).

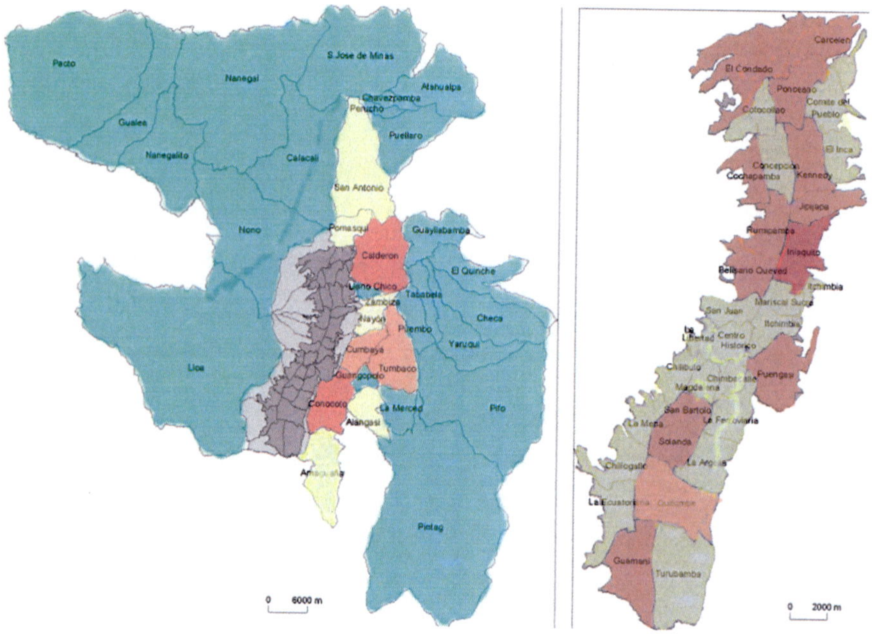

Fig. 5. Cluster distribution of municipalities in Pichincha providence. Excess consumption (red), high (tomato), average (yellow) and low (green).

5 Conclusions and Future Works

This process of grouping by electricity consumption in contrast to the population density can be deduced that parishes with 15 thousand inhabitants have a consumption of more than 17,000 Kwh per month. However, some of them do not present this similar trend and it is necessary to carry out a more in-depth analysis of the characteristics of electricity consumption.

It can be seen in the graph that sectors with similar geographical locations and population density can change the habits of electricity consumption. This can be seen specifically in the high and medium consumption sets. With this, better government policies can be presented for better planning of electricity production.

As future work, it is proposed to make use of a decision-making support tool to have an adequate interface on consumer behavior with monthly and annual reports.

References

1. Huang, X., Wang, S.: Prediction of bottom-hole flow pressure in coalbed gas wells based on GA optimization SVM. In: *Proceedings of 2018 IEEE 3rd Advanced Information Technology, Electronic and Automation Control Conference, IAEAC 2018*. Institute of Electrical and Electronics Engineers Inc, pp. 138–141 December 2018
2. He, L., Song, Q., Shen, J.:K-NN numeric prediction using bagging and instance-relevant combination. In: *Proceedings - 2nd International Symposium on Data, Privacy, and E-Commerce, ISDPE 2010*, pp. 3–8 (2010)
3. Bose, S., Goel, A., Shankar, T., Mageshvaran, R., Rajesh, A.: Energy efficient heterogeneous network with daily load variation. In: *2017 Innovations in Power and Advanced Computing Technologies, i-PACT 2017*, vol. 2017-January. Institute of Electrical and Electronics Engineers Inc, pp. 1–4 January 2018
4. Zhang, X.M., Grolinger, K., Capretz, M.A., Seewald, L.: Forecasting residential energy consumption: single household perspective. In: *Proceedings - 17th IEEE International Conference on Machine Learning and Applications, ICMLA 2018*. Institute of Electrical and Electronics Engineers Inc, pp. 110–117 January 2019
5. Diao, L., Sun, Y., Chen, Z., Chen, J.: Modeling energy consumption in residential buildings: a bottom-up analysis based on occupant behavior pattern clustering and stochastic simulation. Energ. Build. **147**, 47–66 (2017)
6. Ministerio de Electricidad y Energía Renovable – Ente rector del Sector Eléctrico Ecuatoriano. http://historico.energia.gob.ec/
7. Parra Narváez, R.: Factor de emisión de CO₂ debido a la generación de electricidad en el Ecuador durante el periodo 2001–2014, *Avances en Ciencias e Ingeniería*, vol. 7, no. 2, December 2015
8. Wang, Y., Chen, Z., Xu, Z., Gang, G., Lu, J.: User electricity consumption pattern optimal clustering method for smart gird. In: *2018 14th IEEE International Conference on Signal Processing (ICSP)*, pp. 567–570 August 2018
9. Zhong, C., Shao, J., Zheng, F., Zhang, K., Lv, H., Li, K.: Research on electricity consumption behavior of electric power users based on tag technology and clustering algorithm. In: *2018 5th International Conference on Information Science and Control Engineering (ICISCE)*, pp. 459–462 July 2018
10. Kangping, L., Wang, F., Zhao, Z., Zengqiang, M., Hongbin, S., Chun, L., Bo, W., Jing, L.: Analysis on residential electricity consumption behavior using improved k-means based on simulated annealing algorithm. In: *2016 IEEE Power and Energy Conference at Illinois (PECI)*, pp. 1–6 February 2016
11. Toapanta-Lema, A., Gallegos, W., Rubio-Aguilar, J., Llanes-Cedeño, E., Carrascal-García, J., García-López, L., Rosero-Montalvo, P.D.: Regression models comparison for efficiency in electricity consumption in ecuadorian schools: a case of study. In: *Applied Technologies*, Botto-Tobar, M., Zambrano Vizuete, M., Torres-Carrión, P., Montes León, S., Pizarro Vásquez, G., Durakovic, B., Eds. Cham: Springer International Publishing, pp. 363–371 (2020)
12. Rezaei, S., Sharghi, A., Motalebi, G.: A framework for analysis affecting behavioral factors of residential buildings' occupant in energy consumption, *J. Sustain. Archi. Urban Des.* **5**(2), 39–58 (2018). http://jsaud.sru.ac.ir/article_895.html%0Ahttp:/jsaud.sru.ac.ir/pdf_895_9f6e3441f7399851f2185e37696ed98e.html

Behavioral Signal Processing with Machine Learning Based on FPGA

Víctor Asanza[1(✉)], Galo Sanchez[1], Ricardo Cajo[1,2,3], and Enrique Peláez[1]

[1] Facultad de Ingeniería en Electricidad y Computación, FIEC, Escuela Superior Politécnica del Litoral, ESPOL, Campus Gustavo Galindo Km 30.5 Vía Perimetral, P.O. Box 09-01-5863, 090150 Guayaquil, Ecuador
{vasanza,galosanc,rcajo,epelaez}@espol.edu.ec

[2] Research Group of Dynamical Systems and Control, Department of Electromechanical, Systems and Metal Engineering, Ghent University, 9052 Ghent, Belgium
RicardoAlfredo.CajoDiaz@Ugent.be

[3] EEDT—Core Lab on Decision and Control, Flanders Make Consortium, Tech Lane Science Park 131, 9052 Ghent, Belgium

Abstract. This paper focuses on analyzing health problems derived from a sedentary lifestyle. Studies seeking to improve physical activity have shown that a good incentive to increase physical activity requires social feedback, allowing the subject to keep motivated and competitive, along with a feedback of number of steps at the end of the day. This work describes the training and implementation of a neural network as an artificial intelligence model to predict the behavior of an individual, taking advantage of the flexibility provided by Field Programmable Gate Arrays (FPGAs). We propose the design of an edge computing system, analyzing the efficiency on power, area and computational performance. The results are presented through a display, making a comparison of the predicted and expected steps.

Keywords: Behavioral Signal Processing (BSP) · Social Cognitive Theory (SCT) · Edge computing · Field-Programmable Gate Array (FPGA) · Neural Network (NN) · Machine Learning (ML).

1 Introduction

Currently, one of the main causes of death is due to problems related to lack of physical activity, since it derives in conditions such as diabetes, hypertension, colon cancer, obesity and poor circulation [1]. In order to reduce the possibility of contracting these diseases, physical activity must be accompanied by a

Supported by Escuela Superior Politécnica del Litoral (ESPOL) and National Secretariat of Higher Education, Science, Technology and Innovation of Ecuador (SENESCYT).

M. Botto-Tobar et al. (Eds.): ICCIS 2020, AISC 1273, pp. 196–207, 2021.
https://doi.org/10.1007/978-3-030-59194-6_17

performance history that allows to visualize the evolution of the individual. Furthermore, motivation becomes relevant to obtain satisfactory results.

Recent applications of systems identification focused on control theory seek to intervene in the behavior of Physical Activity (PA), whose objective is to improve health by reducing the risk of contracting chronic diseases [2]. This PA can often be done with 30–60 min of daily walking [3].

Any distance people walk needs energy. That is why, in the long term, walking allows weight control [4]. In addition, it allows us to go from one place to another, walking produces benefits for our health, which over time these results are notorious in our daily life. However, human behavior has complex characteristics and is also multi-faceted. An individual's behavior and physical activity can include a reflection of their psychological, neuronal, and cognitive state that involves an interaction of the mind and body. This behavior can also be affected by the environment in which a person develops day by day [5].

A sedentary lifestyle is identified as a risk factor for general mortality. It is estimated to be the main cause of approximately 21–25% of breast and colon cancer, 27% of diabetes [6]. A major factor in wasting the health and wellness potential caused by physical inactivity is due to low levels of walking. A middle-aged person should have the ability to walk 1.6 km comfortably at a speed of 6.4 km/h [4]. Pedometer-based walking programs have been developed, which are inexpensive and small devices that count the steps per day. In this way, an overweight person or specific diets can control the number of minimum steps that must be taken to stay healthy [7].

Based on the principles of control systems engineering, behavior change theory can be used to develop models and decision frameworks for interventions that promote PA among sedentary individuals. An example is the Social Cognitive Theory (SCT) was proposed by Bandura [8]. Additionally, the work of Martín et al. [9] established a dynamic systems fluid analogy model that captures key SCT concepts, which basically consists of learning motor skills through two factors: observation to create a mental model and feedback to correct what has been learned [9].

Behavioral Signal Processing (BSP) refers to computer methods or techniques that involve three main steps: data acquisition, analysis and modeling. This outcome may include high-level descriptions or in quantitative or predictive models for scientific use [5]. In this aspect, the concept of Machine Learning (ML) is introduced. It can be understood that this term refers to an artificial intelligence style (e.g Neural Networks) that is based on obtaining patterns of large amounts of data through software training [10].

Data prediction is important to identify the behavior of the individual based on the established conditions. Thus, it can incentivize and motivate the subject to perform the estimated number of steps [7]. This work presents the implementation of neural networks on a digital system capable of predicting the number of steps of a person.

The rest of the paper is organized as follows. Section 2 provides an extensive overview of the related literature. Sections 3 describes the experimental setup performed in the data acquisition process. Section 4 explains the experimental methodology carried out in the preprocessing of the data, the learning process of the ML algorithm and its implementation in a FPGA. Section 5 presents the results obtained in the classification and performance of the FPGA. Finally, the discussion of the results obtained is presented in Sect. 6, followed by the conclusions.

2 Related Works

Nowadays, it is estimated that at least a quarter of the world's population leads a sedentary life, leading to health problems. Studies have been carried out that seek to relate the behavior of human health and mobile intervention. From a mobile application, a count of the steps taken by the user is performed in real time [11]. L. Vachhani et al. [12] proposed different hypotheses on the step count, which resulted in social feedback and feedback on the progress of the individual every day. This helped to increase the number of steps that the person can take in the coming days compared to people who received no comments.

In BSP three basic technology-related operations involving data collection, development engineering and behavioral modeling can be employed. One of the models that has gained great interest in fields such as robotics and motion detection is multimodal sensing for human behavior [10]. The prediction data process involves various analysis and calculation techniques. One of the methods used is the combination of statistical techniques with their different models in one model (hyper-ensemble), which is a generic tool normally used in geographic applications [13]. CARIMA-type models have also been applied, which are based on autoregressive methods that consider the average of the data series used to predict future values. Here, deep neural networks are more efficient for modeling and allow continuous decision making with the help of Internet of Things (IoT) technologies [14]. In multimodal analysis to achive correct results, the implementation of the activity description vector is intended. It contains the relevant parameters in the study process and this focuses on the creation of a data network. In this way, the digital system is able to make a prediction even using part of the data that make up the entered vector [15].

On the other hand, SCT is implemented in the study of the behavior of an individual for management of well-being and health. It consists of a dynamic system that can adapt to each subject in real time based on certain specific variables of their habitual behavior [9]. Similarly, it has been found that dynamic modeling for human behavior has obtained efficient results, where human actions are best expressed as a sequence of control blocks. For instance, a precise model for the behavior of people during driving was recognized and the derivation of modeling for other dynamic systems of both people and machines is also proposed [16].

Freigoun et al. [3] implemented a smartphone application called "Just Walk" for adaptive walking intervention for sedentary and overweight adults. This development was mainly designed as a tool that allows efficient behavioral modeling, as well as modeling the system for a PA intervention (e.g, walking). However, incorporating all measured disturbances to estimate a SCT behavior model (particularly the construction of the environmental context) can be computationally demanding with identifiable challenges [3,7].

Khan et al. [17] proposed a physical activity recognition system based on machine learning decision tree algorithm with data from 3D acceleration sensors. The results revealed that placing this sensor in the upper part of the spine improved the detection of physical activities (sitting, standing, lying down, walking, running, cycling, climbing and descending stairs). Regarding location-dependent activities, Tirkaz et al. [18] proposed methods to classify people's daily activities, through adequate preprocessing that includes filtering data from 15 motion sensors and 14 door sensors, extraction of characteristics for subsequent classification with the Support Vector Machine (SVM) and Conditional Random Fields (CRFs). The results shown to be successful in correlating a location with activities such as: outdoor, sleeping, moving and undetermined.

Regarding activities of the march, Santhiranayagam et al. [19] worked on the classification of gait in subjects using SVM algorithms. The results affirmed better precision for three gait conditions, including walking with subjects, normal walking and walking with a glass of water. The study carried out by Pentland et al. [16] was the first to successfully implement an automatic detector of a person's emotional state in an FPGA and display the results on a screen. Two techniques were taken as fundamental pillars: K-Nearest Neighbors (K-NN) and Local Binary Pattern (LBP) methods. Then, both techniques were joined and tests were performed with the Matlab and Simulink tools. The database used were videos recorded by 5 participants who watched the same video twice.

Another experiment involving FPGA implementation and prediction is the article written by Upegui et al. [20] that seeks autonomy of navigation in robots. A mobile robot with 8 ultrasonic sensors is presented, it is subject to an unknown geometric place with multiple obstacles. Voronoi theory or better known as Generalized Voronoi Diagrams (GVD) in which his strategy is based on prediction and correction was used for the study.

Neural networks in charge of machine learning are used for the implementation of prediction systems and the development of modeling systems. For instance, it requires flexibility in the hardware to be used, where modifications to the processing system are allowed. Microprocessor implementations have been made, but the need and characteristic of parallel behavior make this a somewhat inefficient application. The FPGA application allows to take advantage of the effectiveness of its parallel computing and synchronism with the neural network. Neural networks of pulsating neurons have been used, but other different models can be used taking advantage of the low economic cost of using FPGAs [20]. It can also be mentioned that works on the FPGA represent advantages such as the use of data paths and control through programmable cells in common logical functions. They are also capable of taking advantage of the memories present on the same chip, due to their high degree of parallel connection, which naturally fits with the "Feed-Forward" characteristic of most of the neural networks used. It is worth mentioning that even certain FPGAs currently allow to modify a certain part of their hardware while it is in operation, reprogramming only a certain section of its structure according to the application required [21].

3 Dataset

In the present work, three test subjects participated, who volunteered for the data acquisition experiment over a period of 50 days. Subsequently, test subjects were asked to fill out an informed consent that clearly explained the scope of the acquisition experiment and the anonymity of the data collected. The test subjects were healthy people with no history of cardiovascular disease. Two of them are male and one female with an average age of 25 ± 1 years.

The experimental methodology consisted of two stages. The first stage was to install two mobile applications on the volunteers' smartphones. The first one helps to record the number of steps taken by a particular individual per hour, allowing him to observe the accumulated detail of the steps taken for the whole day. The second one allows to measure and record every hour the current temperature in the city, where the individual is located. Both applications were configured to carry out the data acquisition process during the hours of 9 am to 8 pm (11 h) per day, given that the greatest step activity of the subjects is concentrated between that period of time. In addition, these applications allow the variables of interest to be recorded with details of time, day, month and year.

In the second stage, test subjects were explained that they should perform their daily activities with absolute normality, while always carrying their smartphone with them and avoiding turning it off. Finally, after the 50-day acquisition period had elapsed, data was collected from each of the test subjects' smartphones.

The data set obtained in the acquisition process was exported in three files with the extension ".xlsx". Each of these files have 6 columns that show (in order from left to right) the date of the day, month, day of the week represented by numbers from 1 to 7 (with 1 = Monday). The fourth column represents the time, when the data is taken (from 8 am to 8 pm). The fifth column indicates the average temperature in degrees centigrade in each time interval and the last column contains the number of total steps taken during each hour. These data are presented in Table. 1. However, the last column (number of steps taken in the current time) of this Table was replaced by the total number of steps taken in the previous hour. This allows to perform the neural network training with the ability to predict the number of steps the test subject will take during the next hour. Consequently, this Dataset with the extension ".xlsx" was imported from the Matlab Workspace and stored as a ".mat" file with the features shown in Table. 2.

The amount of activity of steps taken by each individual throughout a single day presents variations depending on the time when these data were acquired. Figure 1 shows the average steps taken by the test subjects, during the daily time range from 8 a.m. to 8 p.m. for the 50 days of acquisition, as well as data values that were considered aberrant in reference to the normal dispersion values of steps taken by the test subjects.

Table 1. Variables of interest in the dataset.

Date	Month	Day	Schedule	Temperature	Steps
15	11	5	8	23	91
15	11	5	9	24	123
15	11	5	10	25	53
15	11	5	11	26	72
.	.	.	.	.	.
26	12	4	20	24	1036

Table 2. Features of the data used in the training of the algorithms.

Variables	Values
Date	Day of the month: 1–31
Day	Day of the week: 1 - Monday 2 - Tuesday, 3 - Wednesday 4 - Thursday, 5 - Friday 6 - Saturday, 7 - Sunday
Month	Month of the year: 1–12
Schedule	Hours range: 09:00–20:00
Temperature	Average temperature for hour: $\sim$ 20–32
Previous steps	Steps taken the previous hour

4 Methodology

In this section, we present the data acquisition and analysis techniques, the neural network selected and training tool chosen, software and hardware used to develop the behavioral prediction system to predict the number of steps a subject could execute under given conditions. We used the mobile health app "Samsung Health", to acquire the necessary data (e.g. step number, time and weather conditions) for building our prediction model. Our model is based on a "fitting-type" neural network, ideal for numerical data management. It has been selected for its reliability and accurate results obtained in comparison with other neural network models tested. We used MatLab neural network toolbox, as it is a high productivity tool for developing and testing different types of neural networks.

Various architectural approaches were applied during this work by using the Altera's DE10 standard development board, which packages the Cyclone V FPGA with a rich variety of peripherals. This makes it suitable for many research purposes. FPGAs incorporate a lot of resources such as logical elements, DSP blocks, on-chip memory and many other into a single chip and a high productivity environment and flexibility in comparison with ASICs. The architecture presented in this work uses a multicore approach made out of NIOS

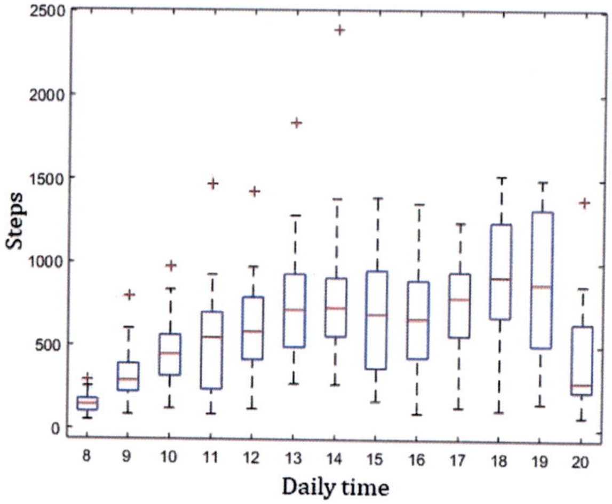

Fig. 1. Average hourly steps taken by test subjects during the 50 days of acquisition.

II soft-processors, to take advantage of the Thread Level Parallelism (TLP), as the system must tackle different tasks in a time efficient manner. This approach sets a baseline for later works as many different custom hardware designs can act as a coprocessors and boost the system performance.

Matlab 2019 software version was used to perform the analysis of the registered data, in which a single file with a .CSV extension is loaded. This file is separated and stored in two different documents. Input data (i.e. the data corresponding to the day, month, day of the week and temperature) will be in a file that represents the system inputs (IN 481 x 5), while the steps recorded each day will represent the system output (OUT 481 x 1). Subsequently, as a system input, we need the above steps, add an additional column and create new arrays InNN (479 x 6) and OutNN (479 x 1).

Based on the new input and output data tables, we obtain the arrangements NInNN (60 x 6) and NoutNN (60 x 1), so that an average value of the first 4 h is obtained (hours from 8 am to 12 pm, from 13 pm to 17 pm and 18 pm to 22 pm) for each of the corresponding columns. Hence, we reduced the amount of data per day in 3 schedules with temperature values and average steps. Because we want to predict the number of steps in the next 4 h, we decided to use the neural network adjustment tool, which trains with the NInNN (inputs) and NoutNN (targets) matrices. Besides, the neural network with 10 neurons was chosen since the precision percentage was higher and the error was almost insignificant compared to neural networks with more or less number of neurons.

The neural network has a data entry base, two layers with feed-forward control, where the hidden and exit layer are found. After tests carried out, 10 neurons were used in the hidden layer, since it presented better matches and the amount of memory available for its implementation was also considered. During the training of this network, 30% percent of the data was destined for validation and testing and 70% percent of these for training, following the values recommended by the Matlab tool. Figure 2 illustrates the neural network model used for this work. It consists of 6 input variables and one output that represents the prediction of steps that will be taken.

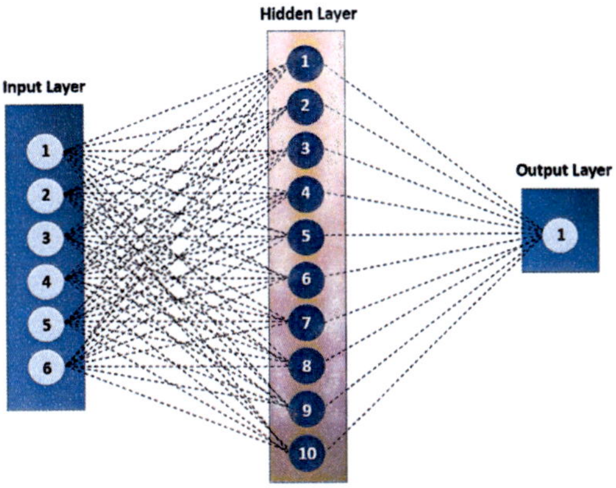

Fig. 2. MLP neural network used in the FPGA for step prediction.

Once the development of the prediction network was completed, we proceed with the hardware implementation in the Altera DE10-Standard FPGA development board. For this, a two-core computer was used to fulfill the required computational tasks and present the results.

5 Results

The neural network model with 6 inputs and 1 output proved to be able to predict the number of steps of a person with an average accuracy of 93%, in which data were taken every 4 h in a time range of 12 h per day. Figure 3a shows the curve of the actual steps taken during the months of October, November and December, and the output prediction of multiple neural network model using different sets of neurons. The dependence of the number of neurons and prediction accuracy in the neural network can be seen in Fig. 3b. This test was performed to determine the number of neurons required in the network to obtain the best fit. Finally, the number of neurons selected in our network is 10 neurons in the hidden layer, which provides better accuracy compared to other models.

Figure 4 shows the trade-offs performed by selecting 10 neurons in the hidden layer in comparison with the closets sets related to power consumption, logic elements used, accuracy and memory usage. Hardware related parameters do not show a major change between sets of neurons, thus selecting the most accurate set of neuron does not affect the overall resource utilization. Figure 5 depicts the architecture designed in the FPGA and the process to be followed by blocks. The

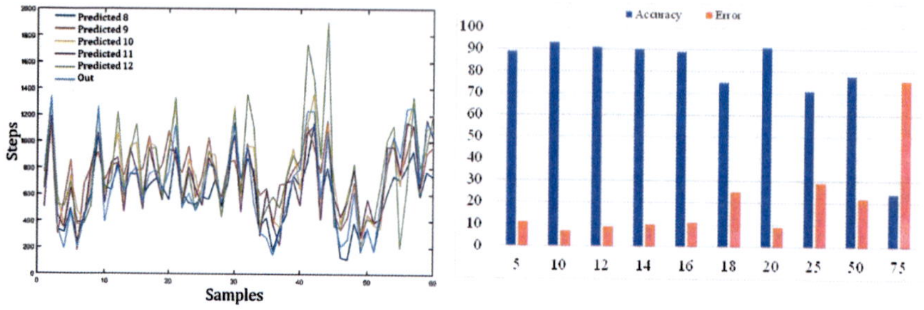

(a) (b)

Fig. 3. (a) Steps performance prediction using MLP with di erent neurons in the hidden layer, (b) accuracy vs. number of neurons in the hidden layer.

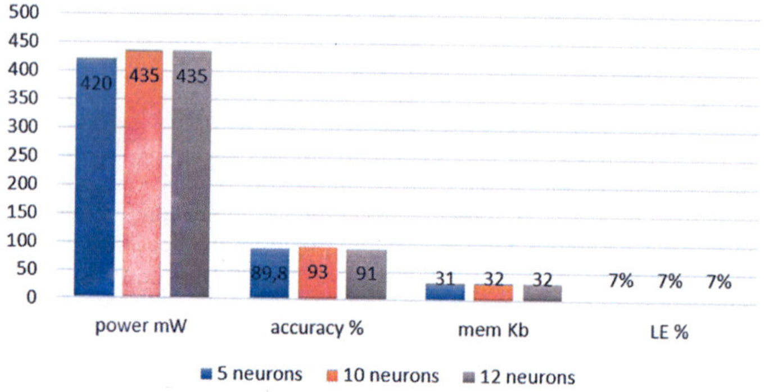

Fig. 4. Results obtained while testing different set of neurons in the hidden layer, and the resources used to tackle the respective workload.

illustration shows how data entry is performed in the FPGA and the printing of the result obtained.

The data used to make the prediction of the steps are read from an SD card by means of the Eclipse console. It can be seen in Fig. 3b, where the data corresponding to the current date and time will be requested, followed by the temperature value and the steps taken in the previous period. In this way, the system will store the step prediction data and its corresponding real value in a vector for subsequent analysis. In addition, this information will be used to implement the comparison graph between the real and predicted data as data is entered.

Figure 3a helped us to select a neural network with 10 neurons. Here, it can be seen that for a larger number of neurons, the prediction precision decreased and the error increased. The precision of the trained neural network is 93% with an error of approximately 7%. Finally, Table 3 summarize the resources used by the FPGA for the implementation of the neural network and the visualization of the results from a VGA port.

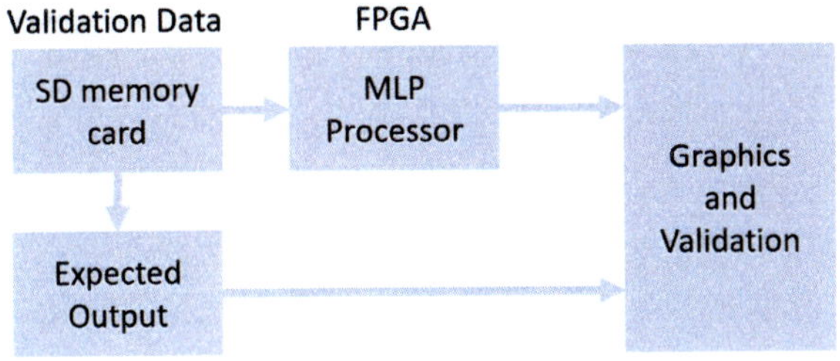

Fig. 5. Overview of our proposed architecture.

Table 3. Resources used by FPGA

Resources	Values
Total pins	135/499 (27%)
Total block memory bits	2167296/5662720 (38%)
Logic utilization (in alms)	2829/41910 (7%)
Power consumption	350 mW

6 Discussion and Conclusions

The SCT model analyzes human behavior, in which from a step count feedback, it will encourage the individual to take more steps in the coming days. The data collected for the analysis of the article was obtained through an application on a Smartphone during the months of October to December 2019.

In addition, the features used in the dataset allowed us to contextualize the behavior of individuals based on the variables of the environment. In our work, the temperature plays an important role in predicting the number of steps each subject will take. It is because the ambient temperature directly impacts the performance of each individual as it is affected by the conditions of the surrounding environment.

The precision and mean square error evaluation was performed to establish the number of neurons with which a better prediction of the output signal is obtained. It was implemented using an FPGA development card that provides great flexibility in its structuring. In this way, with the application of the chosen neural network, it was possible to obtain an accuracy of 93.0% in the prediction of the data. In other words, there was an error of 7.0% and despite having better results. This was chosen due to the CPU resources and C programming libraries available.

The implementation of a multiprocessor computer in the FPGA development card, by using the Qsys graphical programming tool and the elements of the Intel library allowed a greater integration in the development of applications without extra components. With this, it was possible to run the neural network on one of the processors created on the card together with a memory for data storage.

The graphical representation of the predicted data compared to the actual steps, better evidences the precision of the trained and implemented neural network.

References

1. World Health Organization, WHO.: Physical activity, February 2018. https://www.who.int/es/news-room/fact-sheets/detail/physical-activity. Accessed 6 Apr 2020
2. Bauer, U.E., Briss, P.A., Goodman, R.A., Bowman, B.A.: Prevention of chronic disease in the 21st century: elimination of the leading preventable causes of premature death and disability in the USA. Lancet **384**(9937), 45–52 (2014). https://doi.org/10.1016/S0140-6736(14)60648-6
3. Freigoun, M.T., Martín, C.A., Magann, A.B., Rivera, D.E., Phatak, S.S., Korinek, E.V., Hekler, E.B.: System identification of just walk: a behavioral mHealth intervention for promoting physical activity. In: 2017 American Control Conference (ACC), pp. 116-121 (2017). https://doi.org/10.23919/ACC.2017.7962940
4. Richardson, C.R., Newton, T.L., Abraham, J.J., Sen, A., Jimbo, M., Swartz, A.M.: A meta-analysis of pedometer-based walking interventions and weight loss. Ann. Fam. Med. **6**, 69–77 (2008). https://doi.org/10.1370/afm.761
5. Narayanan, S., Georgiou, P.: Behavioral signal processing: deriving human behavioral informatics from speech and language. In: Proceedings of the IEEE Institute of Electrical and Electronics Engineers, vol. 101(5), pp. 1203-1205 (2013). https://doi.org/10.1109/JPROC.2012.2236291
6. Fujiki, Y., Kazakos, K., Puri, C., Starren, J., Pavlidis, I., Levine, J.: NEAT-o-games: ubiquitous activity-based gaming. In: Proceedings of the 2007 ACM Conference on Human Factors in Computing Systems (CHI), pp. 2369-2374. ACM Press (2007). https://doi.org/10.1145/1240866.1241009
7. Asanza, V., Martin, C.A., Eslambolchilar, P., van Woerden, H., Cajo, R., Salazar, C.: Finding a dynamical model of a social norm physical activity intervention. In: 2017 IEEE Second Ecuador Technical Chapters Meeting (ETCM), Salinas, pp. 1–6 (2017). https://doi.org/10.1109/ETCM.2017.8247450
8. Bandura, A.: Social Foundations of Thought and Action: A Social Cognitive Theory. Prentice-Hall series in social learning theory. Prentice-Hall, Upper Saddle River (1986)
9. Martín, C.A., Rivera, D.E., Riley, W.T., Hekler, E.B., Buman, M.P., Adams, M.A., King, A.C.: A dynamical systems model of social cognitive theory. In: Proceedings of the American Control Conference, pp. 2407–2412 (2014). https://doi.org/10.1109/ACC.2014.6859463
10. Nagarajan, K., Holland, B., George, A.D., Clint, K., Lam, H.: accelerating machine-learning algorithms on FPGAs using pattern-based decomposition. J. Sign. Process. Syst. **62**, 271–350 (2011). https://doi.org/10.1007/s11265-008-0337-9
11. Harries, T., Eslambolchilar, P., Rettie, R., et al.: Effectiveness of a smartphone app in increasing physical activity amongst male adults: a randomised controlled trial. BMC Public Health **16**(925), 1–10 (2016). https://doi.org/10.1186/s12889-016-3593-9
12. Vachhani, L., Sridharan, K.: Hardware-efficient prediction-correction-based generalized-Voronoi-diagram construction and FPGA implementation. IEEE Trans. Ind. Electron. **55**(4), 1558–1569 (2008). https://doi.org/10.1109/TIE.2008.917161

13. Rixen, M., Ferreira-Coelho, E., Signell, R.: Surface drift prediction in the Adriatic sea using hyper-ensemble statistics on atmospheric, ocean and wave models: uncertainties and probability distribution areas. J. Mar. Syst. **69**(1–2), 86–98 (2008). https://doi.org/10.1016/j.jmarsys.2007.02.015
14. Kanawaday, A., Sane, A.: Machine learning for predictive maintenance of industrial machines using IOT sensor data. In: 8th IEEE International Conference on Software Engineering and Service Science (ICSESS), pp. 87-90. IEEE (2017). https://doi.org/10.1109/ICSESS.2017.8342870
15. Azorín, J., Saval, M., Fuster, A., Oliver, A.: A predictive model for recognizing human behaviour based on trajectory representation. In: International Joint Conference on Neural Networks (IJCNN), Beijing, pp. 1494–1501 (2014). https://doi.org/10.1109/IJCNN.2014.6889883
16. Pentland, A., Liu, A.: Modeling and prediction of human behavior. Neural Comput. **11**(1), 229–242 (1999). https://doi.org/10.1162/089976699300016890
17. Khan, A., Lawo, M., Homer, P.: Wearable recognition system for physical activities. In: 9th International Conference on Intelligent Environments, Athens, pp. 245–249 (2013). https://doi.org/10.1109/IE.2013.50
18. Tirkaz, C., Bruckner, D., Yin, G.Q., Haase, J.: Activity recognition using a hierarchical model. In: IECON 2012—38th Annual Conference on IEEE Industrial Electronics Society, pp. 2814–2820 (2012)
19. Santhiranayagam, B.K., Lai, D.T., Jiang, C., Shilton, A., Begg, R.: Automatic detection of different walking conditions using inertial sensor data. In: the 2012 International Joint Conference on Neural Networks (IJCNN), pp. 1–6. IEEE (2012). https://doi.org/10.1109/IJCNN.2012.6252501
20. Upegui, A., Peña, C., Sanchez, E.: An FPGA platform for on-line topology exploration of spiking neural networks. Microprocess. Microsyst. **29**(5), 211–223 (2005). https://doi.org/10.1016/j.micpro.2004.08.012
21. Lacey, G., Taylor, G., Areibi, S.: Deep learning on FPGAs: Past, present, and future. arXiv preprint arXiv:1602.04283 (2016)

Complementary Admission Processes Implemented by Ecuadorian Public Universities Promote Equal Opportunities in Access: An Analysis Through Knowledge Discovery in Databases

Andrés Santiago Cisneros Barahona(✉), María Isabel Uvidia Fassler, Gonzalo Nicolay Samaniego Erazo, Gabriela Jimena Dumancela Nina, and Byron Andrés Casignia Vásconez

Universidad Nacional de Chimborazo, Riobamba, Ecuador
ascisneros@unach.edu.ec

Abstract. In Ecuador, since 2012, the admission process to public third level education has been regulated through an admission test called "Ser Bachiller". Starting on 2019, the National Leveling and Admission System (SNNA), through an update of its policies allowed public universities to create their own complementary admission processes. Therefore, the evaluation's grade stopped being the only resource to apply to an undergraduate program at a public Ecuadorian university. Under this frame, Universidad Nacional de Chimborazo (UNACH), implemented the "UNACH Complementary Admission Process", which exhaustively assesses the academic record of students' secondary education as well as the affirmative actions reported on their application. Through the use of knowledge discovery in databases (KDD), Unach generated an improved methodology that specifies the Educational Data Mining (EMD) in one of its sub-processes, establishing the Educational KDD methodology. As a result, once the student data was collected, a data warehouse (DW) was obtained. Additionally, through EDM and Bayesian networks, patterns and trends were identified in order to provide information and knowledge of the relationship between the admission process and the socio-economic and vulnerability conditions of the applicants, their ethnicity, disabilities and academic performance. In parallel, pertinent and valuable information was generated through business intelligence reports so that UNACH can make decisions that guarantee equal opportunities in the access to its undergraduate programs.

Keywords: Equal opportunities · Knowledge discovery in database · Educational data mining · Universidad Nacional de Chimborazo

1 Introduction

"Education is the most powerful weapon you can use to change the world" [1]. The fifth section of the Ecuadorian Constitution titled "Education", Article. 26, states that

M. Botto-Tobar et al. (Eds.): ICCIS 2020, AISC 1273, pp. 208–222, 2021.
https://doi.org/10.1007/978-3-030-59194-6_18

education is a right of all constituents throughout their lives; it is also an inexcusable duty of the State. It constitutes an area of priority for public policy in order to guarantee equality and social inclusion; it is an imperative condition to reach the "good living". Individuals, families and society as whole have the right and responsibility to participate in education processes [2].

The Ecuadorian Higher Education system through the Organic Law of Higher Education (LOES), in its article 63 "On Admission and Leveling" explains the admission process to third level education. The system is governed by the principles of merit, equal opportunities and freedom granted to aspirants to choose any academic program at the university of their interest. This admission process is implemented on a national basis by the Secretary of Higher Education, Science, Technology and Innovation (SENESCYT) through its National System of Leveling and Admission (SNNA). This government body also oversees that a system of quotas is implemented in each one of the public universities of the country.

The Ecuadorian state enforces the application of affirmative actions in order to guarantee that all public higher education institutions promote real equality to favor those who are in situations of inequality or vulnerability. Therefore, this admission system takes into account the capabilities and competences of the applicants as well as their academic background, socioeconomic status and aspects derived from the affirmative action policy. In addition, all applicants are required to take an admission test called "Ser Bachiller".

The score test in combination with the affirmative actions reported by the applicants are the instruments utilized by the Ecuadorian state to guarantee equal opportunities of access to third level education. However, those who obtain the highest scores will have greater opportunities of admission to the academic program of their choice, according to the offer at public higher education institutions of the country [3].

The implementation of information and communication technologies in administrative processes of Higher Education Institutions allows them to capture a high amount of data. This permits various analyzes to be carried out through Knowledge Discovery in Databases (KDD), and also to identify the patterns and trends that generate transcendent, valuable and useful knowledge, from which the institutions will be able to take significant and valuable decisions [4]. This analysis is applied within the core phase of Data Mining (DM), specifically related to Educational Data Mining (EDM). This methodology framework was used to study the processes and challenges related to the admission process at Universidad Nacional de Chimborazo (UNACH). This research was conducted with the purpose of guaranteeing equal opportunities in access to a third level degree at UNACH and also to replicate this methodology at any Ecuadorian public university.

2 Admission Process at Universities and Polytechnic Schools

Since 2012, the Ecuadorian Higher Education System, based on the Organic Law of Higher Education, in its article 63 titled "On Admission and Leveling", states that all senior students at public and private high schools around the country have to compulsory take an admission test, regardless of their decision to attend or not to a public university.

This evaluation is applied through SENESCYT and it is graded under 1.000 points. It measures the aptitudes and skills that applicants acquired upon the completion of their intermediate education. This knowledge is considered by the Ecuadorian government as necessary for student's successful development as citizens and also to thrive on the academic context. The test was initially called ENES (Exam for Higher Education); since 2016 it took the name of "Ser Bachiller" [5].

Since the second semester of 2019, SENESCYT modified its regulations in order to allow Higher Education Institutions (IES) to carry out their own complementary admission processes. The objective was that the Ser Bachiller test grade stops being the only mechanism by which students are admitted at an undergraduate program. The state also granted more autonomy to Ecuadorian public universities so they can design complementary processes that permits them to capture the greatest young talents but also to assure that the admission process remains democratic and meritocratic across the country. At Universidad Nacional de Chimborazo, only the students who complete the Ser Bachiller test and the complementary admission requirements are considered eligible for admission. This is further explained in the Admission and Leveling Regulations for Undergraduate students at Universidad Nacional de Chimborazo in its article 10.

> Complementary Admission Process.- All undergraduate applicants must register on the institution's online platform and then fill out the required information; it must be completed before applicants take the National Exam Ser Bachiller. This complementary process is taken in accordance to the Regulations of the National Leveling and Admission System and it will be adapted in case of policy changes by the state. It is advertised in order to promote a democratic admission through the different media that UNACH has in place for its communication efforts [6].

Any applicant who has the score from the Ser Bachiller test and completes the complementary admission process is eligible to apply to the academic program of his choice at Universidad Nacional de Chimborazo. The student can also use his score in any other Ecuadorian public university, "the applicant can choose up to five academic programs in order of preference. As a result of this selection process, the applicant will receive an admission offer and make it effective by enrolling at the academic program and institution of his choice [5] (Fig. 1).

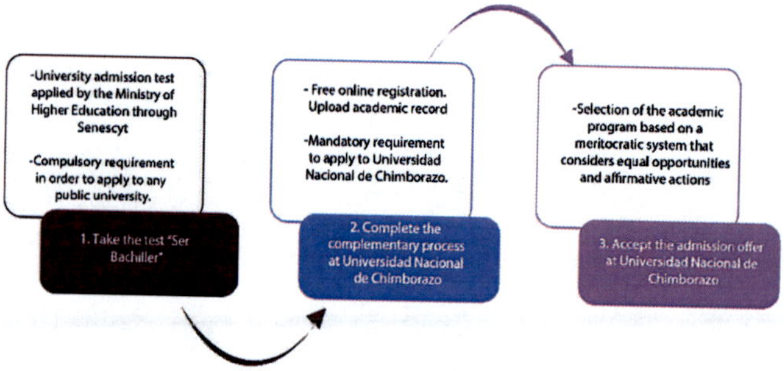

Fig. 1. Admission process at UNACH.

If a student receives an offer by UNACH and if he accepts it, he must enroll on the career leveling courses, offered by the same university as a compulsory requirement to start their first semester. These classes serve to improve student's knowledge and skills and also to facilitate their adaptation to the higher education system. The courses utilize various strategies to accomplish this purpose. The National Higher Education Academic Regulation, in the article 17 titled "Leveling strategies, explains that Higher Education Institutions can freely design curricular strategies that enable the leveling of student's knowledge as a mechanism to guarantee the successful completion of their studies" [7]. Once a student completes the leveling stage, he is able to enroll in the first semester of his academic program.

3 Socioeconomic Situation and Poverty Quintiles

There are several ways to stratify a population according to their consumption trends, as well as their income and working conditions. Based on the latest Survey of Employment, Unemployment and Underemployment (ENEMDU), applied on September 2019 on a sample of 59,963 individuals located throughout Ecuador, it was possible to obtain a quintile classification. This data is focused on the income received by families through their monthly salary. It is important to mention that the numbers shown in this article correspond to income per person, meaning that if the analysis is extended to the family nucleus, it should be multiplied by the number of people that is a part of the household [8] (Fig. 2).

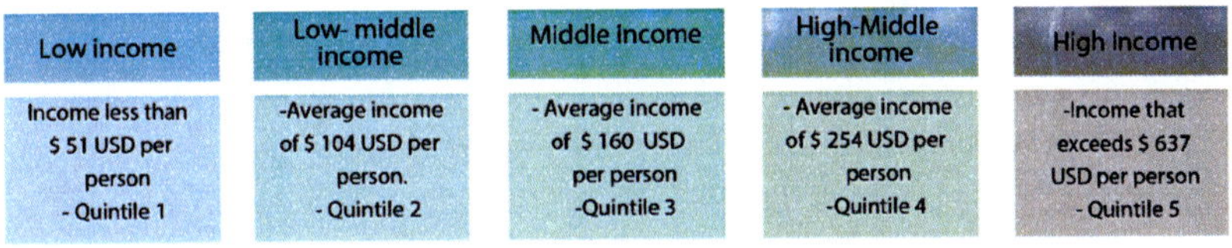

Fig. 2. Ecuador quintiles [8].

Income quintiles are used in economics to organize a population, according to their income. The first quintile (Q1) corresponds to those individuals who receive the lowest income, while the fifth quintile (Q5) represents the richest population. ENEMDU includes a question about the money people receive as a result of their work [8], therefore it was possible to create a classification by income quintiles in Ecuador. This study uses the official results published by the National Institute of Statistics and Census (INEC), based on the living conditions survey - 5th Round (poverty and extreme poverty line) as well as the income information from the employment, unemployment and underemployment survey. These results allow to see the trend of poverty measurements; it is important to mention that this is the only official data on poverty created by the Ecuadorian government [9].

The Ecuadorian government implemented important reforms to the higher education system in 2008 and 2012. These changes sought on one hand to eliminate the barriers of accessing to an undergraduate degree by offering free education, as it is stated on the

Ecuadorian National Constitution of 2008. On the other hand, they allowed to implement an admission test required by any public university that considers the merit and knowledge of applicants. As part of the reforms, the quality of public universities is constantly evaluated so that the ones that are not able to meet the minimum requirements proposed by the Ecuadorian Higher Education System can no longer operate. For instance, in 2012 the Ecuadorian government unprecedentedly closed 14 universities of poor quality across the country.

If the impact of the reforms is analyzed, it is possible to identify the positive effect of an admission process regulated by the Ser Bachiller test and the closure of poor quality Universities. The number of students enrolled at Ecuadorian public universities increased between 2006 and 2014; this growth also showed improvements in terms of equity. The enrollment concentration curve also became less regressive, this change was due to the greater access to higher education at the middle layers of the socioeconomic structure and a reduction in access to the richest decile. The econometric analysis of the factors associated with enrollment shows that the probability of being enrolled at a public university increased by 5% between 2006 and 2014 [10].

4 Affirmative Action Policies Related to the Access to Higher Education

The National Leveling and Admission System supports the country to guarantee equity in the access to higher education. Affirmative action policies guarantee equality in university admission. According to its regulation in the article 44 titled “Affirmative Action Policies” explains the way applicants with a poor socio-economic condition, rural background, geographic location, vulnerability conditions, disabilities and indigenous background receive additional consideration in their application process. For instance, a) Socio-economic status: fifteen additional points are given to the application of students that belong to the first poverty quintile; d) Conditions of vulnerability: five additional points and up to a maximum of ten additional points are given to people with disabilities. e) Indigenous ethnics and cultures: ten additional points are granted to people belonging to indigenous communities such as Afro-Ecuadorian and Montubio [11].

5 Knowledge Discovery in Databases

Knowledge Discovery in Databases (KDD) is an automatic, exploratory and modeling analysis of large data repositories. Fayyad mentions that “Knowledge discovery in databases is the non-trivial process of identifying patterns in data that are valid, novel, potentially useful, and ultimately understandable” [12]. Complementing this definition, authors Maimon & Rokach define KDD as an organized process of valid, novel, useful identification that generates understandable patterns of large and complex data sets; in addition, KDD process is iterative and interactive; it can be defined in nine steps [12] (Fig. 3).

For the KDD process of this study, a KDD - ESPOCH methodology was applied. This is the same one used on the Leveling and Admission Unit at the Higher Polytechnic

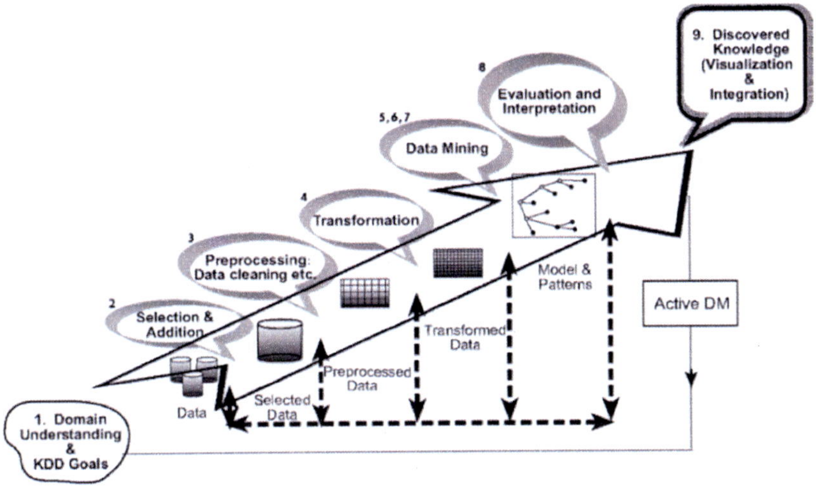

Fig. 3. The process of knowledge discovery in databases [12].

School of Chimborazo (ESPOCH), which was adapted to the information and knowledge needs of this institution as it is demonstrated in the article Moving Towards a Methodology Employing Knowledge Discovery in Databases to Assist in Decision Making Regarding Academic Placement and Student Admissions for Universities, published in Communications in Computer and Information Science, vol 798. Springer [13].

The methodology contemplates two subprocesses; the first one is Data Warehouse (DW), which refers to learning the application domain, selection and creation of a data set on which the discovery, pre-processing and cleaning, and the transformation of data along with the product DW was carried out. This is generated with validated and consistent data. The second subprocess is Data Mining (DM) and it includes choosing the appropriate DM task and selecting the DM algorithm; as an outcome, it was possible to generate patterns. Finally, in the evaluation and application steps of the discovered knowledge, the product obtained was the knowledge that allowed academic and administrative professionals to use this information in the context of Higher Education in order to inform their decisions. By demonstrating the usefulness and benefit of the proposed methodology, it became applicable to higher education in general. Specifically, it allowed the DM subprocess to become EDM. In addition, it was possible to improve the analysis methodology in relation to education by transforming the KDD into KDD- Educational (Fig. 4).

UNACH has gone through a significant transition of its admission process since the second semester of 2019. This was the first year that the institution implemented a complementary admission process that allowed it to have greater participation in the applicant's selection process. The university also had the opportunity to better allocate the quotas related to the affirmative actions through its Coordination of Admission and Leveling (CAN) of UNACH. This administrative unit integrated data that generates information and knowledge of inclusion and equal opportunities in the admission processes, based on the socio-economic and vulnerability conditions of the students

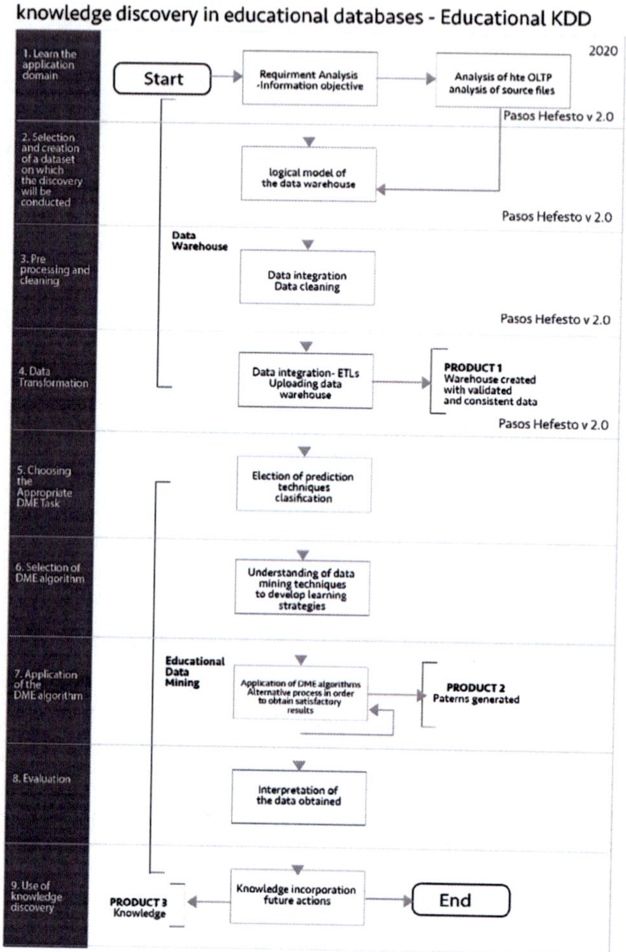

Fig. 4. KDD Educational Methodology.

through the KDD - Educational process. The EDM subprocess considered the Bayesian network prediction technique to obtain patterns and trends. Through Business Intelligence, important new knowledge was generated to demonstrate equal opportunities in access to third level education.

6 Educational Data Mining

"DM refers to the act of extracting patterns or models from data" [14]. Data mining is the core of the KDD (Knowledge Discovery in Databases) as this process allows to identify valid, useful and understandable patterns, as well as structures that convert data into useful information in order to organize the institution's work and implement pertinent actions accordingly [15].

The proposed KDD - Educational methodology, which in this case is specified as EDM, corresponds to the mining of educational data, that allows to conduct analysis and to take further decisions, according to the patterns created. It also helps to demonstrate the situation of the students in order to design the best actions on their benefit.

EDM offers numerous advantages compared to more traditional educational research paradigms, such as laboratory experiments, sociological studies or research design. In particular, the creation of public repositories of educational data allows the creation of an information base that makes possible the mining of educational data. The information from these repositories is valid, as it is gathered from the performance and learning of real students, in real educational environments, taken from different learning tasks. This information is increasingly easy to access to start any research. Data mining applied to the educational environment has the potential to extend a much broader set of tools to analyze important questions about individual differences [16]. The EMD approaches are explained below (Table 1).

Table 1. Approaches to Educational Data Mining [16].

Method category	Method objective	Key applications
Prediction	Development of a model that can infer a variable from the combination of the available data	Detection of student behavior (system tricks, distractions,'slipping'); Development of domain models; Prediction and understanding of a student's academic results
Grouping	Finding naturally grouped data sets, separating the entire set into a series of categories	Discovery of new patterns of student behavior; Investigation of similarities and differences between schools
Relationship Mining	Discovery of relationships between variables	Discovery of curricular associations in course sequences; Discovery of pedagogical strategies that guide a more effective learning process
Discovery through models	Modeling a phenomenon by prediction, clustering, or knowledge engineering is used as a component in future prediction or mining relationship	Discovery of relationships between student behavior and its characteristics or contextual variables; Analysis of research questions for a wide variety of contexts
Distillation from data	The data is distilled to allow a human to quickly identify or classify the data properties	Human identification of patterns in student learning, collaborative behavior; Data labeling for use in further development of predictive models

6.1 EDM Prediction Techniques - Bayesian Networks

The objective of the data mining prediction technique is to develop a model that can infer a variable from some combination of other variables included in the data [16].

Bayesian networks allow representing all the possible events of interest through a figure of conditional transition probabilities between events. It can be coded from the knowledge of an expert or it can be inferred from the data. It allows causal relationships to be established and predictions to be made [17].

7 Business Intelligence

Business Intelligence (BI) can be described as a concept that integrates storage on one hand and the processing of large amounts of data on the other, with the main objective of transforming it into knowledge and real-time decisions, through a simple analysis and exploration [18].

8 Case Study Results

In the application of the KDD - Educational methodology in the CAN at UNACH several results were obtained:

- In the DW subprocess, it was possible to consolidate the information from the students that obtained an admission offer in any of the university's undergraduate programs in 2019. UNACH's transition into a more autonomous admission process started on the second semester of this year (2019-2S). The first admission period (2019-1S) was conducted using only the score from the Ser Bachiller test and no other complimentary requirements. This research seeks to demonstrate if this change guaranteed equal opportunities based on the consideration of the following variables: students' socioeconomic and vulnerability conditions, performance ratio in relation to these conditions, relationship between applicants to certain academic programs and their social strata along with the Ser Bachiller score. It was also considered the information from geographic location, gender and disability. Various analyzes were conducted, as shown in the next subprocess of the KDD - Educational methodology.
- Several important analysis results were obtained in the EDM thread

a) Data mining technique

Bayesian networks was the projection technique used by Weka (platform for data mining) in the 2019 data at CAN - UNACH. Within the variables analysis, the data related to social information was the most important one because it represents the groups of socioeconomic conditions and vulnerability, the largest number of students and cases. Therefore, it is the one that best relates to the rest of the variables (Fig. 5).

To demonstrate the impact of the socioeconomic data, which includes the poverty levels of the country, according to the social registry, the following levels were identified according to the quintiles (Table 2):

Fig. 5. Bayesian Network - EDM

Table 2. Data levels and quintiles

Social register data	Description	Quintiles
Extreme poverty	Income under 51 per person	Q1
Poverty	Median income of \$ 104 per person	Q2
No poverty	Median income of \$ 160 per person	Q3
No registry	Median income of \$ 254 and \$637 per person	Q4 y Q5

In determining the relationship between social registry data and ethnicity, it was possible to establish that the indigenous groups are the most significant ones at UNACH. In addition, the highest concentration of Q1 is found in ethnic groups of mestizos and indigenous people in the same proportion of 48%, while in Q2 there was a notable difference of 76% in mestizos and 22% in indigenous population; in Q3 there is also 88% in mestizos and 10% in indigenous; finally, in Q4 and Q5 90% are mestizos and 8% are indigenous people. This implies that according to the ethnic identification of the students upon entering the university, there is a considerable number of indigenous people concentrated in Q1 of extreme poverty. This highlights a double vulnerability of the students due to their ethnic group and also to their economic situation (Fig. 6).

Probability Distribution Table By Etnia

Social Registry	Mestizo	Indigenous	White	Afrodescendent	Other	Mulato	Montubio	Black
Q4 and Q5	0,898	0,078	0,009	0,006	0,002	0,003	0,002	0,001
Q2	0,762	0,223	0,001	0,001	0,007	0,001	0,001	0,001
Q3	0,879	0,101	0,001	0,01	0,002	0,002	0,002	0,001
Q1	0,479	0,479	0,007	0,007	0,007	0,007	0,007	0,007

Fig. 6. Poverty distribution based on ethnicity

Within the equal opportunities framework, approximately 1% of the population have some type of disability and it is concentrated on the Q2 poverty group. This shows an additional level of vulnerability in a minimal percentage, but also as a priority (Fig. 7).

Currently, there are more women than men enrolled on third level education, for this reason it is also evident that there are more women in all quintiles. These numbers show at the same time the inclusion of the female gender within higher education (Fig. 8).

In regards to the place of origin of the students, Chimborazo is the province with the highest incidence of students in all quintiles with a 58%, also coinciding with the location of the university and its influence in the area. In addition, the province of Bolívar represents 10% of students in Q1, Tungurahua a 9% in Q2, 13% in Q3 and 9% in Q4 and

Probability Distribution Table By Disability

Social Registry	No	Yes
Q4 and Q5	0,993	0,007
Q2	0,981	0,019
Q3	0,999	0,001
Q1	0,993	0,007

Fig. 7. Poverty distribution based on disability

Probability Distribution Table By Gender

Social Registry	Female	Male
Q4 and Q5	0,561	0,439
Q2	0,6	0,4
Q3	0,584	0,416
Q1	0,545	0,455

Fig. 8. Poverty distribution based on gender

Q5. These are the provinces with the highest percentages of presence in the university at the country level (Fig. 9).

Students who are admitted at UNACH must compulsory take leveling courses which must be approved before they can start the first semester of their academic program. The approval rate is of 75%, demonstrating that in this academic cycle the economic situation does not directly influence academic performance (Fig. 10).

Finally, it is important to analyze the areas in which the most vulnerable students are enrolled at UNACH. People in Q1 chooses academic programs in the area of education with a 26%; 21% studies health sciences and 18% is enrolled in the engineering field. In Q2, 28% of the population selects academic programs related to education; 25% engineering and 20% health sciences. In Q3, 28% choses academic programs related to engineering; 27% studies social sciences and 23% education. In Q4 and Q5, 31% chooses a career in engineering; 25% health sciences, and 11% selects social sciences, such as journalism or law (Fig. 11).

b) Business Intelligence

In using the Microsoft Power BI application, several data analyses were additionally generated to demonstrate the importance that complimentary admission processes have had for UNACH since 2019 and the concentration of scores from the Ser Bachiller test.

The first analysis shows that despite of having a small number of students in the 2019-2S period in comparison to previous years, the number of students in Q2 and Q3 increased while it decreased in Q4 and Q5. This indicates that the complementary admission process is efficient to guarantee equal opportunities of accessing to an undergraduate degree at Universidad Nacional de Chimborazo (Fig. 12).

Probability Distribution Table By Pronvince

Social Registry	Q4 and Q5	Q2	Q3	Q1
Chimborazo	0,654	0,615	0,537	0,519
Cotopaxi	0,036	0,061	0,045	0,032
Napo	0,01	0,01	0,013	0,032
Pichincha	0,044	0,004	0,035	0,019
Orellana	0,006	0,015	0,009	0,032
Subumbios	0,007	0,007	0,015	0,006
Tungurahua	0,088	0,094	0,131	0,006
Azuay	0,002	0,004	0,003	0,006
Guayas	0,005	0,004	0,009	0,006
Cañar	0,011	0,044	0,018	0,032
Zamora	0,004	0,007	0,012	0,006
Loja	0,005	0,035	0,011	0,032
Morona	0,018	0,021	0,016	0,045
Bolivar	0,03	0,03	0,052	0,109
El Oro	0,014	0,01	0,006	0,006
Esmeraldas	0,009	0,001	0,015	0,006
Imbabura	0,009	0,001	0,015	0,006
Pastaza	0,01	0,001	0,003	0,006
Santo Domingo	0,024	0,01	0,017	0,045
Manabí	0,001	0,004	0,005	0,006
Santa Ana	0,001	0,001	0,003	0,005
Galápagos	0,001	0,001	0,001	0,006
Los Ríos	0,002	0,004	0,002	0,006
Carchi	0,003	0,001	0,003	0,006

Fig. 9. Poverty distribution based on the provinces

Probability Distribution Table By "Final Status"

Social Registry	Pass	Fail
Q4 and Q5	0,701	0,299
Q2	0,766	0,234
Q3	0,777	0,223
Q1	0,739	0,261

Fig. 10. Poverty distribution and the results of the leveling courses

In regards to the Ser Bachiller exam grade, in 2019, the highest concentration was between 778 and 871 which determines an acceptable trend of grades that allowed students to be admitted at the program of their choice (Fig. 13).

Probability Distribution Table By Area

Social Registry	Q4 and Q5	Q2	Q3	Q1
Business Administration	0,06	0,117	0,061	0,136
Engineering	0,31	0,252	0,275	0,179
Social Sciences	0,11	0,076	0,093	0,05
Arts and Humanities	0,02	0,007	0,024	0,021
Education	0,2	0,281	0,231	0,264
Health Sciences	0,245	0,197	0,272	0,207
Information Technologies	0,026	0,036	0,026	0,064
Services	0,029	0,033	0,018	0,079

Fig. 11. Poverty distribution according to academic areas

Quintil	2019-1S	2019- 2S	Total
Q1	37	29	66
Q2	167	176	343
Q3	469	493	962
Q4 and Q5	1229	1154	2383
Total	1902	1852	3754

Fig. 12. Distribution of the quintiles in 2019

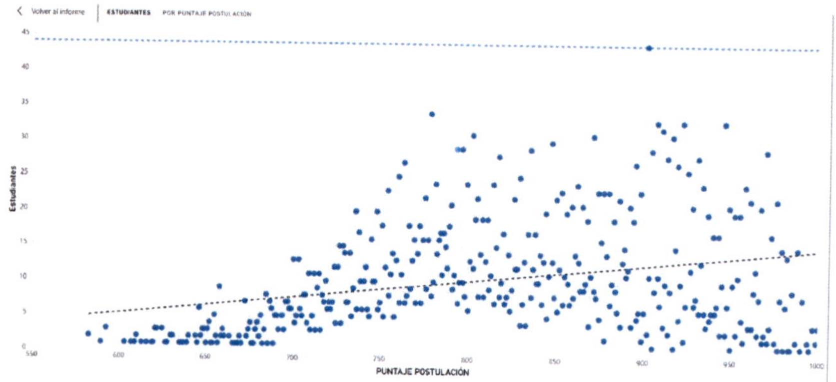

Fig. 13. Distribution of the Application Grade from the Ser Bachiller test - UNACH, 2019

9 Conclusions

Through experimentation in this research work, it was possible to demonstrate the importance of KDD methodology in the area of education. The KDD - Educational methodology was proposed in the search for an improvement process. Within the DW subprocess, it was possible to obtain a corporate database that integrated data from the students who applied to the 31 academic programs offered at UNACH and received classes in the Leveling Courses in 2019. This allowed the availability of data for the EDM subprocess that shows information and knowledge in the academic area, identifying the relevance of the complementary admission process implemented at the university. In this subanalysis it was possible to consider the socio-economic and vulnerability conditions of the students, and the relationship with their academic performance in order to guarantee equal opportunities in access to higher education.

The EDM prediction technique and the BI analysis, showed that the complementary admission process at UNACH guarantees equal opportunities of access as the inclusion of the poorest quintiles is evident. For instance, poverty quintiles 1, 2 and 3 are on average a 37% of the number of students enrolled each semester. In addition, indigenous people and multiple ethnics represent a participation of 10% within the university community; there is also an admission rate of 1% of people with disabilities. Definitely, equality of access opportunities is demonstrated. It is also evident that the largest number of students in the first three quintiles tend to choose academic programs related to the education field and envision a job as a teacher so they can alleviate the limitations of their peers in rural communities. In a second place, this population chooses careers in engineering and health. However, this decision also represents a risk to the completion of their academic program as there is a high demand of additional investment in certain classes, especially those hard subjects on the study of Health Sciences and Engineering.

Finally, it was possible to conclude that the university is attractive within its place of action. The majority of its students come from Chimborazo, Tungurahua, Bolívar and Cotopaxi which are closely located. Once the student scenarios were identified through the KDD - Educational process, it was possible to provide information and knowledge to make assertive decisions that guarantee equal opportunities not only in access, but also in the permanence of students, through the creation of financial aid supports and scholarships for the most vulnerable sectors of the Ecuadorian society.

References

1. Mandela, N.: Nelson Mandela por sí mismo. Plataforma, Catalonia (2011)
2. Ecuador, A.C.: Constitución del Ecuador (2008)
3. del Ecuador, P.: Ley organica de educacion superior. In: LOES, pp. 1–92 (2018)
4. Daza Vergaray, A.: Data Mining, Minería de Datos In: Primera. Lima: Editorial Macro (2016)
5. Uvidia, M.I., Cisneros, A., Dumancela, G., Samaniego, G., Villacrés, P.: Application of knowledge discovery in databases to predict the academic performance of university students based on their admission test process of application to universities and polytechnical schools in Ecuador, pp. 1–13 (2019)
6. de Unach, U.N.C.: Reglamento de Nivelación y Admisión para estudiantes de grado de la Universidad Nacional de Chimborazo, vol. 0052, pp. 1–12 (2019)
7. de Educación Superior: Reglamento de Régimen Académico (2019)
8. Gestión, R.: ¿Cuándo los ingresos de una persona son considerados altos?, no. Gráfico 1, pp. 0–2 (2019)
9. INEC, Medidas de Pobreza y Extrema Pobreza por ingresos (2008)
10. Ponce, I., Carrasco, F.: Acceso y equidad a la educación superior y posgrado en el Ecuador, un enfoque descriptivo. Access Equity High. Educ. Descriptive Approach 3(2), 9–22 (2016)
11. Senescyt: Reglamento del sistema nacional de nivelación y admisión snna (2019)
12. Maimon, O., Rokach, L.: Data mining and knowledge discovery handbook. In: Data Mining and Knowledge Discovery Handbook, pp. 22–38 (2015)

13. Uvidia Fassler, M.I., Cisneros Barahona, A.S., Ávila-Pesántez, T.F., Rodríguez Flores, I.E.: Moving towards a methodology employing knowledge discovery in databases to assist in decision making regarding academic placement and student admissions for universities. In: Technology Trends, pp. 215–229 (2018)
14. Fayyad, U., Piatetsky-Shapiro, G., Smyth, P.: From data mining to knowledge discovery in databases. AI Mag. **17**(3), 37 (1996)
15. Moine, M., Haedo, S., Gordillo, D.S.: Estudio comparativo de metodologías para minería de datos. In: CACIC 2011 - XVII CONGRESO ARGENTINO DE CIENCIAS DE LA COMPUTACIÓN (2011). http://sedici.unlp.edu.ar/bitstream/handle/10915/20034/Documento_completo.pdf?sequence=1. Accessed 03 Dec 2015
16. Jiménez, Á., Álvarez, H.: Minería de Datos en la Educación. In: Intel. en Redes Comun (2010)
17. Aluja, T.: La Minería de Datos, entre la Estadística y la Inteligencia Artificial. Qüestiió **25**(3), 479–498 (2001)
18. Bernabeu, R.: Hefesto, p. 146 (2010)

Cloud, IoT and Distributed Computing

LOLY 1.0: A Proposed Human-Robot-Game Platform Architecture for the Engagement of Children with Autism in the Learning Process

Dennys F. Paillacho Chiluiza[1,2](✉), Nayeth I. Solorzano Alcivar[1,3], and Jonathan S. Paillacho Corredores[1,2]

[1] ESPOL Polytechnic University, Escuela Superior Politécnica del Litoral, ESPOL, Guayaquil, Ecuador
{dpaillac,nsolorza,jspailla}@espol.edu.ec
[2] Facultad de Ingeniería en Electricidad y Computación – FIEC, CIDIS, Guayaquil, Ecuador
[3] Facultad de Arte Diseño y Comunicación Audiovisual - FADCOM, Guayaquil, Ecuador

Abstract. Existing studies indicate the use of the Socially Assistive Robotic (SAR) system has enormous potential to advance therapeutic interventions in the learning process of children with Autistic Spectrum Disorder (ASD). Interactive technology, including the use of educational video games, can motivate ASD children to engage by adopting the technology as part of their communications tools. The complexity in the integration of various components of SAR linked to educational/serious games hinder the modularity and scalability of these robotic systems. This work proposes the design of an architecture for a SAR platform through the integration and reproduction of robot prototype, "Loly 1.0". Loly 1.0 has mechanical movements, audio, and gestures connected to educational games in real-time. This will support the learning process of ASD children. The robotic platform architecture design is comprised of three main parts: robot, educational game, and cloud services. A proof of concept of the SAR prototype was per-formed with non-ASD and ASD children. Evaluation and analysis of "synchronized operational precision between robot and video game" and "degrees of attention and usability" will be conducted. The lessons learned during the testing process will help to improve the architectural designs of hardware/software integration and connectivity. The results of reproducing a SAR prototype will be to attract the attention of an ASD child through the use a video game to engage the process of "learn by playing". This will contribute to the design of engaging and personalized Human-Robot-Interaction (HRI) and Human-Computer-Interaction (HCI) technology beneficial for inclusive education.

Keywords: Robotic platform architecture · ASD · Socially Assistive Robot · SAR architecture · Learning process · Serious games · Educational videogames · Dashboard · Human-robot interaction

M. Botto-Tobar et al. (Eds.): ICCIS 2020, AISC 1273, pp. 225–238, 2021.
https://doi.org/10.1007/978-3-030-59194-6_19

1 Introduction

Research related to Human-Robot-Interaction (HRI) focused on the education and health area, dabbled in the use of Social Assistive Robot (SAR) systems; seek to contribute collectively to the improvement of people's lives [1]. In inclusive education, for example SAR studies are offering the possibility of helping people with Autism Spectrum Disorder syndrome - ASD who present social, communicational, and behavioral challenges [2]. In this sense, some authors explain that novel architecture based on HRI and supported in SAR platforms can benefit children with ASD by motivating their cognitive and interpersonal communication skills [1, 3, 4].

On the other hand, existing literature related to Human-Computer-Interaction (HCI), indicates that the use of interactive applications (apps) games for favoring ADS people. It is argued that Serious Games or educational videogames which apply audiovisual and animated resources used for mobile devices, can helps to motivate the apathy or distraction that children with ASD in the teaching-learning process [5–7]. Notably, the lack of interest they may have about some areas of knowledge, can be related to Language, Literature, or Natural Sciences, but there is no a defined pattern. In this sense, educational videogames apps or serious games include animated stories, seeking to support in an entertained way the learning process of a child about curricula content they should learns as part of their primary childhood education [8]. Besides, with the intention to monitor HCI, several of the serious game productions send out traces of a child playing to a repository monitored by a control panels (dashboard platform), such as the MIDI-AM game series applications, explained by Solorzano, Sornoza, and Carrera [9].

The use of interactive technology supported by HCI and HRI can offer roles that were probably not considered working together when they were created. However, scarcity of studies was found oriented to benefit children with ASD in social and cognitive interaction by adopting novel technology assisted by SAR joined to monitored Serious Game to make it part of their communication style. Robotic intervention linked to ludic interaction can act as a catalyst that removes social pressures and increases social interactions with others [10].

This research aims to explain the testing analysis of a robotic platform's architecture interconnected with an educational game for gather the attention of ASD children expecting to motivate their teaching-learning process.

1.1 State of Art

Research shows that there are several methods to treat the ASD in children. One of them is using technology as a therapy to improve ASD conditions. Some studies have identified an evident affinity for technology in children with ASD [7]. Specially computers applications with ludic entertainment purposes [8, 9, 11, 12] or new technologies such as SAR intervention [1, 3, 4].

Technology Intervention in ASD Using SOCIAL Assistive Robots. Assistive Robots (AR) or SAR platforms are defined as a broad class of robot architecture proposed under the concepts of HRI with the aims to help vulnerable people around the world [3].

Several studies indicate that children with ASD tend to be motivated using technological resources that appeal their attention. SAR platforms can bring technology-assisted instruction that support the development of ASD children's social skills by providing a combination physical and audiovisual factors to encourage communication and interaction skills [1, 3, 4]. Authors state that visual and auditory tools reinforce education by encouraging verbal responses, interaction, and social relationships in children with ASD [7]. Furthermore, these technologies as an innovative form of therapy using technology, include the use of audio devices, visual media, computers, and robots. Their require less human effort and they are more effective in therapies [6].

Technology Intervention in ASD Using Serious Games. The ASD is a human condition known as psychosocial development disability difficult to treat. It does not have an organic cause, but observable behaviors and manifestations. ASD is present throughout the life of the person who has it, but a person can be assisted from the first moment that the presence of ASD is evident in a child. Delaying early intervention may represent difficulties in the adult life of these subjects [2, 13].

According to López Reventós [8], in the educational field, researchers attempt to demonstrate the benefits of using educational video games or serious games in the classrooms and at home. Knowledge is developed through the academic contents of the game weft. On the other hand, cognitive abilities are expanded by the action of playing [14]. The use of interactive games to play can be a means of addressing the reluctance shown by children with ASD and an entertained way of learning academic content related to some specific topics such as Language, Literature, or Basic Sciences.

Solorzano [9] explaining the creation of a series of ludic apps MIDI-AM that monitor behavioral use and playability through a dashboard, suggest the inclusion of educational video games platforms as a means of learning. The learning is obtained because of tasks (level games challenges), stimulated by the content of animated stories that are visualized before play the game's levels. MIDI-AM apps for mobile devices use Android technology and an online dashboard platform architecture, for online data transmission in real time. The use of dashboard allows to store and present organized data for an orderly and detailed analysis of the action's children take when using a game [9, 12, 15].

HCI and HRI Intervention of Technology in ASD. Some research seeks to identify that through children's interest in technology, they can be approached in a different way of learning and awaken in them the motivations to learn other things. The combination of platforms using HCI and HRI by using mobile apps with AR or SAR intervention to address or encourage their correct use to play serious games can be a supporting alternative in the children's learning process supported in this research. With this technology assistance the learning process that comes from another person, such as the teacher or tutor it becomes not too invasive for a child with ASD. For these children, too much human intervention becomes invasive since their interaction in the social aspect is affected. The use of an educational AR platform joined to monitored serious game apps such as MIDI-AM, can benefit children's and children's guardians to have feedback about the human-robot-game intervention. For example, information about child's attention levels and exercises (games level) interest they have the most, during pedagogical session are relevant [9], particularly to encourage learning in children with ASD.

Next section presents the design of the hardware and software architecture, called the Human-Robot-Game - HRG architecture, implemented in the Loly 1.0 robotic platform. To validate the implementation of the Loly 1.0 architecture, pluralistic research methodology tools are applied to design the process and the applied. These technical tests as well as proofs of concept are carried out in Sect. 3. In Sect. 4, lessons learned on the implementation of the architecture are presented. Finally, the conclusions and future works are presented in Sect. 5.

2 Human-Robot-Game Architecture

The design of the Human-Robot-Game architecture involves a multidisciplinary team of researchers and students from the areas of computer science, electronic, audiovisual production, graphic design, psychology, and education. The goal is to achieve the intervention of HRI and HCI for pilot testing of a proposed Human-Robot-Game (HRG) architecture platform designed for educational purposes. This HRG platform using as a novel prototype of a SAR system linked to serious game apps, adapted for inclusive education, can contributes to the development of social and cognitive skills for ASD children through imitation using the senses of sight, touch, and hearing.

In order to have a modular and scalable hardware and software for the Loly 1.0 robotic platform, the system has been designed around 3 main parts: the robot, the educational game and the cloud services (as shown in Fig. 1).

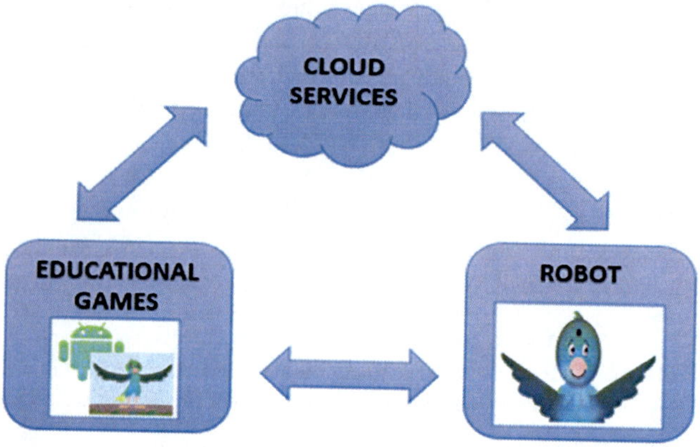

Fig. 1. General diagram of the Loly 1.0 Educational Robotic Platform Architecture

2.1 The Robot

The parameters that characterize a social robot are form, function, and context. Each robot is given an aesthetic shape depending on its function. Besides, each robot has different types of behavior, that is, variable functions. Finally, the social robot with its aesthetic form and its functions is found within various contexts, and it can also be assigned a character archetype and specific aesthetic design according to a wide range of types, each charged with satisfying different needs.

The main function of the head and torso designed as a robotic bust in this research is to interact with patients with ASD and to keep them focused on a single activity. In turn, it captures game data and real-time data to a dashboard. Expecting to give more expressiveness to the robot [16], an 8-inch LCD screen is used to display main facial expressions: happiness, sadness, anger, and enthusiasm. For verbal interaction, the robot uses a loudspeaker to emit previously established lines of dialogue obtained from the production of the game. The robotic head will produce pitching movements using servo motors controlled by a microcontroller.

For the adaptations of the games in functionality with the robot design, all the required components of programming and real-time linking previously used in apps programmed with Adobe Animate were considered. The file format for data transmission to the dashboard uses JSON records. The communication of the app with the robot uses Java for Android systems. This application communicates via socket directly to the NodeJS Server that receives the information to proceed through serial communication on the movement, audio, and gestures of the robot.

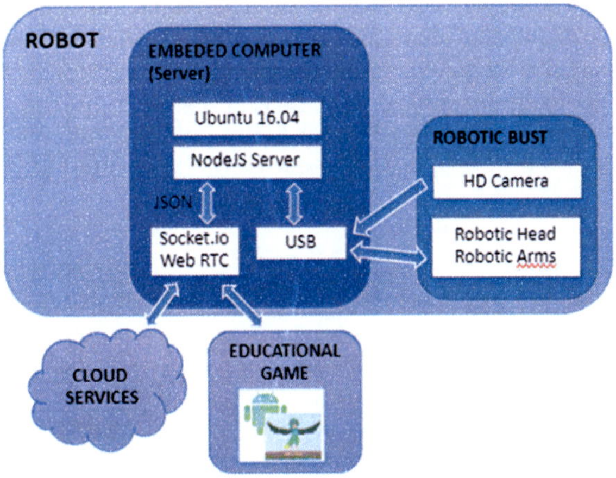

Fig. 2. Diagram of Hardware/Software Robot Architecture

As part of the design, it was implemented a real-time recording and transmission system of the interactive sessions. To accomplish this purpose, modularizing the mobile application with the robot and with the help of a camera installed on the robot were executed.

Communication is established between the mobile application and the robot so that each event produces the development of action in the robot that attracts the child's attention. For this purpose, adaptations are made in the game script to include interaction with the robot. Motor movements, audio, and video playback are included, transmitting usage data in real-time to a dashboard for operator management.

Fig. 3. Modularization of the robot-game-operator application

Aesthetic and Mechanical Design Aspects. As part of the game-robot implementation, the app used requires some design changes and adaptations in programing. For this purpose, as part of the design and production, to present a robot bust with a child-friendly image, it was decided to incorporate one of the main characters of the MIDI – AM project. The archetype of a friendly mealy parrot originally from the tropical forests of Ecuador named LOLY, who guides the use of the games and stories in this series, was then implemented. For the integration of the character's Loly in the app game, called Anibopi, with robotic accompaniment, the new version requires several adaptations as follows: design and illustration of the physical aspects of the character based on the characteristics and symptoms of the ASD children and specific physical characteristics of the mealy-ridden parrot. Subsequently, the physical structures and adaptations of the Loly character were made to the robot's schematic and mechanical structure, considering the anatomical limitations of the structure [17] (Fig. 4).

Fig. 4. Graphic representation and dimensions of robotic bust Loly1

The robotic head and bust design were performed in a 3D mechanical modeling program. To be built through 3D printing on most of its mechanical components to achieve head, beak, and wing movement. Finally, all the mechanical, electronic and control components are integrated to carry out the relevant tests (Fig. 5).

2.2 The Educational Games

For the adaptations of the games in functionality with the robot design, all the required components of programming and real-time linking previously used in apps programmed with Adobe Animate were considered. The file format for data transmission to the dash-board uses JSON records. The dashboard is developed in JavaScript, using the SailsJS

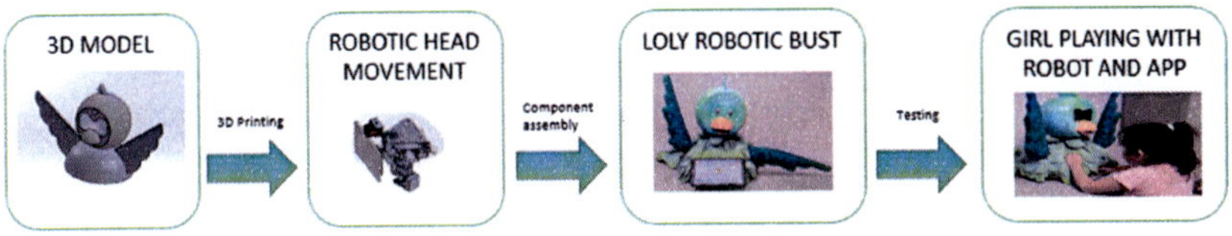

Fig. 5. Loly construction and assembly

framework. The database is implemented in PostgreSQL [15]. The communication of the app with the robot uses Java for Android systems. This application communicates via socket directly to the NodeJS server that receives the information to proceed through serial communication on the movement, audio, and gestures of the robot.

The game component chosen is a specific MIDI-AM game for inclusive education redesigned to be used with the robot prototype for testing purposes. The selected games, named Anibopi, helps the learning process of the Living and Non-Living Beings created mainly for children with high and medium performance ASD. These educational game take bases on active learning methods, considered as serious games. The apk of Anibopi Rb beta version can be upload for free from AppStore or from the MIDI[1] Website to be used for learning and feedback purposes. Anibopi as one of the MIDI-AM series games is monitored through a dashboard for gameplay evaluation, and other usage metrics [9, 15].

2.3 The Cloud Services

The cloud services that will be carried out on the robotic platform, as shown in Fig. 2, are the management of the information generated both in the educational game and in the robot through a dashboard, as well as monitoring and intervention of a specialist, such as a psychologist or an educator, as an operator.

Dashboard Service. Solorzano, Carrera [15] describe that the web architecture dashboard (Midiapi) is taking as reference the one used in the web application TeachTown. For the backend Node JavaScript programming language is used with Sails.js framework and for the frontend is used Angular. PostgreSQL is the database platform, and the file format for data transmission for the games is a JSON. The Midiapi dashboard shows detailed results with a statistical system, but only for its own game. However, the Midiapi modular division architecture seeks to be adapted to other educational gaming systems (Fig. 6).

Operator Service. As shown in Fig. 3, the design of the architecture includes the participation of an operator. The operator must be a specialist trained to carry out a therapeutic intervention, such as a psychologist. The specialist, who can be connected locally or remotely, can carry out a platform monitoring operation as well as control parameters of both the robot and the educational games that improve the HRG interaction. Expecting to have both video, audio and real-time data connectivity, the design includes the use of a web-based platform called WebRTC.

1 http://midi.espol.edu.ec/productos.

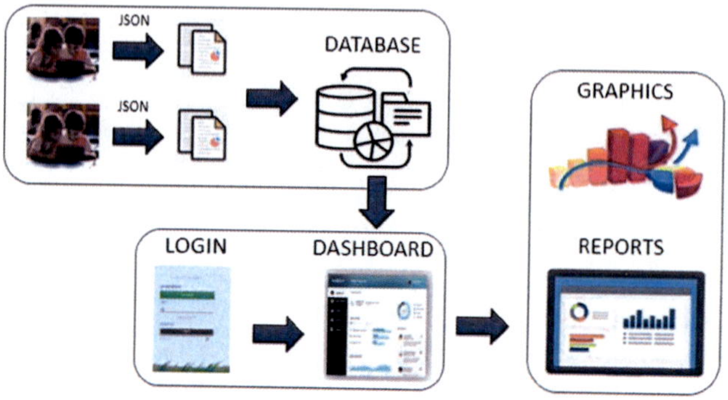

Fig. 6. Midiapi architecture. Illustration base on [15]

3 Human-Robot-Game Architecture Testing

In order to carry out a first evaluation of the Loly 1.0 architecture, this section presents both the development of technical tests related to the response times of several components, as well as several proofs of concept of human-robot interaction in real environments. Since the Loly 1.0 robotic platform is under development, these first tests only consider the integration of the Robot with the Educational Game.

3.1 Technical Tests

Communication tests with the robot, as well as the server response with the different robot devices, were carried out within the technical tests. The response times are measured since a signal is emitted from the mobile application when executing a button press event through the HTTP protocol through the X-Response-Time tool and the server response, receiving it. Also, the execution of audios and gestures of the robot from the server (see Table 1).

Table 1. Execution and response of times of the different servos.

Event	Http rsponse VS NodeJS server(sec)	Robot response VS server (sec)	Total (sec)
Script 1	0.95	0.36	1.31
Script 2	0.92	0.37	1.32
Script 3	0.80	0.50	1.30
Script 4	0.90	0.45	1.35

The server implemented with NodeJS technology communicates directly with the different components of the robot, and that includes the servos for movement, the display used for gesticulation, and the speakers that reproduce the audio. The response of the servos used in the four developed scripts is detailed below, which are executed depending on the different interactions or events according to the development of the story (see an example of one of the four scripts, Table 2).

Table 2. Execution and response of times of the different servos. Script 1

Servos	Head	To the left	To the right	Beak	Time	% Execution
Pitch	X				0.2 s	33
Yaw	X				0.3 s	36
Roll	X				0.3 s	42
Wing left		X			0.3 s	40
Wing right					–	0
Beak				X	0.3 s	72

3.2 Proofs of Concept

For testing the proposed HRG architecture platform of Loly 1.0, designed for educational purposes, technical and conceptual research tools for data analysis an experimental evaluation are applying from pragmatic point of view of researchers focusing on solving the problem instead of specific research [18].

First Proof of Concept The first proof of concept consisted of two HRG interaction sessions with Loly 1.0, one with a five-year-old regular boy and the second session with two girls with high and medium-performance ASD of nine years and five years, respectively (as shown in Fig. 7).

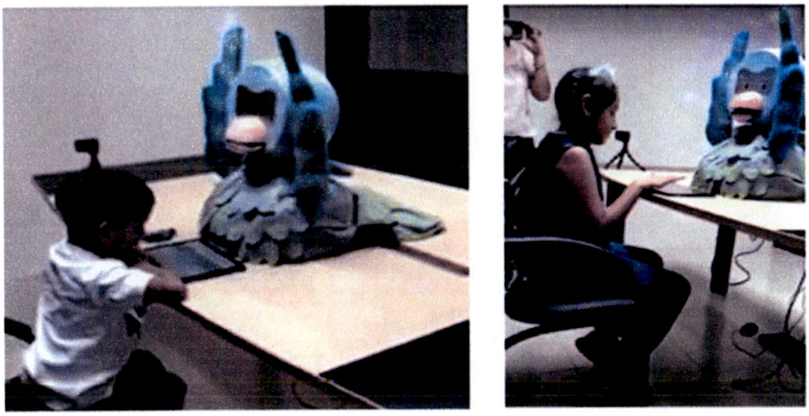

Fig. 7. Children playing with Loly

In the first session, we verified that the child mastered the application efficiently, attending to each of the instructions given by the same application, and performing each of the games. This game includes the teachings about plants, animals, rocks, and mountains, showing great skill in the execution and understanding of the game. Regarding physical interaction with the established elements, we obtain two different outcomes. From the session with regular children, they showed more affinity paying attention to the app rather than to the Loly robot. Besides, this was confirmed by asking the children, at the end of the session, what they liked the most? The answer to this question was "the tablet with the game" At the end of the session, the children were asked what they learn using the robot and the game? Then, they were asked to explain about living beings and non-living beings, which fulfills one of the objectives that are the educational play factor present in the game.

Second Proof of Concept. In this proof of concept, with an ASD children, it was observed, for example, that a nine-year-old girl mastered the application efficiently. She attended the instructions of the game and interacted with each of the sections and levels of the game. Also, she showed good mastery in the topics exposed, due to the general knowledge that the girl already had in understanding each of the topics. Referring to the physical interaction, the girl showed the same degree of affinity for both the mobile application (using a tablet) and the Loly's robot with a highly attraction to the Loly's robot gaze. In the session with the second five-year-old girl, we observed that she partially dominated the application with the support of her parents, with a great attraction to the actions of the robot, her eyes, voice, and gestures, as shown in Fig. 8. This attraction, together with pedagogical and psychological interpretations, gives a preliminary indication that for children with the autism condition, they offer a higher degree of attention to the physical interactions with elements present in the environment (movement, gestures, audio), in this case, on the robot and the Tablet.

Fig. 8. Girl playing with Robot and the App (Courtesy of (FEDEA: Federación Ecuatoriana de Espectro Autista - Ecuadorian Autism Spectrum Federation) FEDEA's President).

Third Proof of Concept. A work with 20 children between four and six years old from a primary school in Guayaquil city was carried out. It is worth mentioning that in the group, there were two children with Asperger syndrome. The child-robot interaction is recorded by a camera built into the top front of the robot's face. Using computer vision techniques allow knowing the moment when the child pays attention to the robot, that is, the moment he looks at the robot. The tests for the study involve three stages: pre-test, experiment, and post-test.

In the pre-test stage, the possible affinity of children with social robots is observed. The children are asked to select a card they like from among two possible ones expecting to identify their affinity with the robot. The tester shows the children four different cards, which contain a cartoon image of Loly's character, two cards with children's tv series cartoon characters, and another with Loly's social robot picture. for the test, we assume that those children who select the image of the robot were able to identify empathy with the Loly robot. While those who select the cards with other characters we assumed not to.

In the child-robot interaction experiment, the child followed Loly's instructions about what images to watch and drag inside the built-in Tablet. The moments when Loly spoke to the child, to give him the instructions, were the ones to determine if Loly's appearance likes them by observing their facial expressions. Besides, we seek to identify for how long the child looked at the robot or the app on Tablet, expecting to determine the focus of their attention while they are playing.

In Fig. 9, we can see the times that the children looked at the Loly robot during the interaction with the Tablet, obtaining a range between 3 and 7 times with their smiling gaze towards the robot, which indicates a high degree of attraction of the children with Loly.

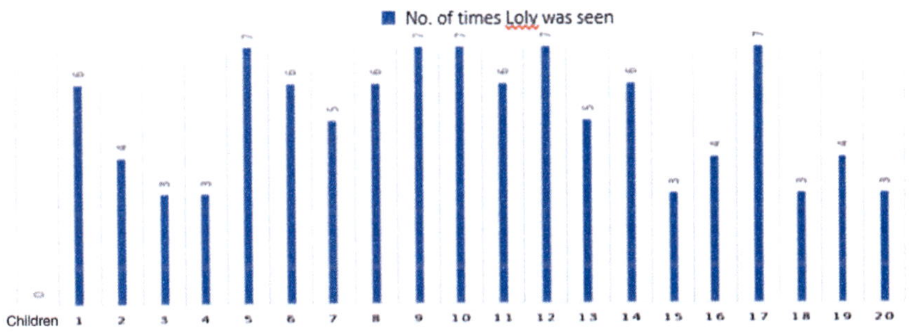

Fig. 9. Frequency of times the child watches the robot

The results of the post-test can be seen in Fig. 10, which indicates that 80% of children chose a card that represents Loly. For analyzing results, it was not possible to use only computer vision techniques. It was also necessary to analyze part of the videos manually. This happens because the quality of the camera located frontally in the robot was deficient, and in some cases, the height of the children did not allow them to capture the entire angle of vision. To sum up, the children mostly had a favorable reaction towards the robot, observing their enthusiasm and affinity for the game and the robot at the same time.

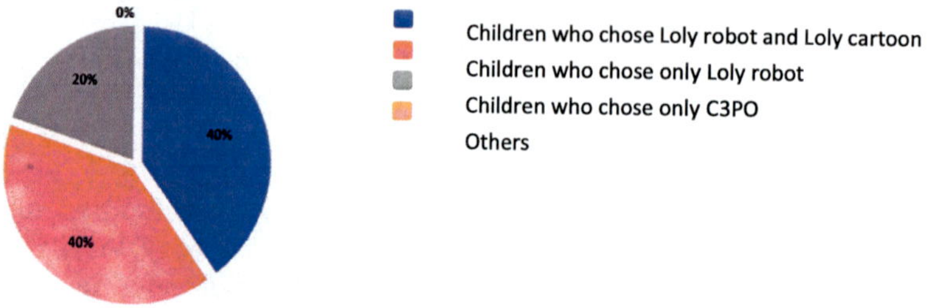

Fig. 10. Preference graph in character selection

In the case of children with Asperger's Syndrome through a social agent, which in this case is the robot, it was found they better retained the information that is being taught. This result is evident in the speed in which they carried out the interactive games that the application offered with a high level of success when playing the levels of the games of the app.

4 Lessons Learned

The lessons learned in the testing process helps to improve a proposed HRG architecture design in terms of hardware/software integration and connectivity, expecting to gather the attention of children with ASD helping them to use a game for succeeding in their teaching-learning process.

The design of engaging and personalized Human-Robot-Interaction (HRI) and Human-Computer-Interaction (HCI) is relevant for inclusive education proposing in this study a combination as Human-Robot-Game interaction explained as HRG.

In general, the testing outcomes helped to accomplish several tasks as follows: evaluate the overall performance platform and a new HRG architecture proposed; the accuracy of Platform of the SAR prototype (Loly 1.0) as the main component of the proposed HRG architecture; connectivity offline/online efficiency, accuracy to link the games between the interaction periods, aesthetic and mechanical issue with the Loly 1.0 prototype, the need of contingency plan or guiding instruction for technical support and pedagogy interventions. Furthermore, another stage of testing event is required to analyze data from the dashboard to evaluate behavioral use, playability and another adoption factors and technical matters such as online/offline, synchronic/asynchrony communications, security, and confidentiality of the entire platform.

On the other hand, focusing the lesson learned based on a first experience in the proof of concept on regards of testing the platform with children with ASD we realized a few matters that should be taken into consideration such as: in some way the child with ASD must be prepared with the human-robot interaction by introducing the robotic character in order to reduce the child's stress or anxiety and plan in advance the customized physical spaces to apply testing individually or to a focus group.

5 Conclusion and Future Work

This research seeks to promote the development, learning process, and social interaction of children with ASD syndrome. We identify many studies focused on helping to improve social skills in these children, but without much success so far. However, based on studies, these children tend to be motivated with the use of technological resources. We propose the design and implementation of a robotic platform with educational games to test the motivation of social interactions and learning in ASD children

The development of the platform and the prototype of a social robot linked to the creation of digital games apps that capture data and present results. With the developed platform, it is possible to make known and interpret information about the usability of these games. The bust of a social robot was built together with an application used in parallel with the robot, which yielded promising acceptance results for children with ASD. Despite not having conclusive results on acceptance, during the different experimentations, it was observed that, in the cases of ASD children, they respond positively to the interaction with the robot at a higher level than regular children, that is, they observe with more exceptional care and enthusiasm both Loly's face and the app.

In the same way, as a post-test through the selection of graphic cards, the positive effect on the degree of acceptance of the Loly robot in conjunction with the educational game could be corroborated.

As future work, this research does not only seek to record the robot-App gameplay results of regular children or children with ASD. It is expected to achieve a more detailed interpretation of how a child uses these educational apps, and the feedback that can be given to teachers, parents, and the developers of educational video games or digital games that interact with other human-machine apps. Technical improvements in the development of the robotic platform, the linked games, and metrics to evaluate the effectiveness of use factors were expected. Based on these premises, it is hoped to continue this research to complete the expected results.

The success of using future social robots as a type of innovative interactive toy to support inclusive education can contribute to evaluating ASD children's learning. Also, it is important to continue venturing into projects that seek to measure the level of children's acceptance, expecting to identify how much they can learn from a ludic game linked to the use of a social-robot interaction, instead of using regular teaching methods. Therefore, further studies for the evaluation of Children-Robot-Game interaction using the designed platform architecture need to be completed.

Acknowledgment. We extend a special thanks to the undergraduate students of the Faculty of Electrical and Computer Engineering - FIEC and the Faculty of Art, Design and Audiovisual Communication - FADCOM at ESPOL who participated in the Loly project: Kevin Alvarado, Joseph Cevallos, Keyla Toapanta, Sebastián Fuenzalida, Andrea León, Denisse Guizado, Joel Rodriguez and Fabricia Hurtado.

References

1. Jain, S., Thiagarajan, B., Shi, Z., Clabaugh, C., Matarić, M.J.: Modeling engagement in long-term, in-home socially assistive robot interventions for children with autism spectrum disorders. Sci. Robot. **5**(39) (2020). https://robotics.sciencemag.org/content/5/39/eaaz3791
2. Dugger, C.E.: The effects of early intervention on children with autism spectrum disorders. Research papers (2012)
3. Fridin, M., Yaakobi, Y.: Educational Robot for Children with ADHD/ADD, Architecture Design (2011)
4. Kozima, H., Nakagawa, C., Yasuda, Y.: Interactive robots for communication-care: a case-study in autism therapy. In: Paper presented at the ROMAN 2005. IEEE International Workshop on Robot and Human Interactive Communication (2005)
5. Contreras-Espinosa, R.S., Serra, A., Terrón, J.L.: Games and ADHD-ADD: a systematic mapping study. Acta Ludologica **2**(2), 4–26 (2019)
6. Herrera, G., et al.: Pictogram Room: Aplicación de tecnologías de interacción natural para el desarrollo del niño con autismo. Anuario de Psicología Clínica y de la Salud **8**, 41–46 (2012)
7. Moore, D., Cheng, Y., McGrath, P., Powell, N.J.: Collaborative virtual environment technology for people with Autism. Focus Autism Other Dev. Disabil. **20**(4), 231–243 (2005)
8. López Raventós, C.: El videojuego como herramienta educativa. Posibilidades y problemáticas acerca de los serious games. Apertura (Guadalajara, Jal.) **8**(1), 136–151 (2016)
9. Solorzano, N.I., Sornoza, L.I., Carrera, D.A.: Adoption of children's educational video games monitored with dashboards in the cloud. RISTI - Revista Iberica de Sistemas e Tecnologias de Informacao **2019**(19), 146–160 (2019)
10. Frauenberger, C.: Rethinking autism and technology interactions **22**(2), 57–59 (2015)
11. González, J.L., Cabrera, M., Gutiérrez, F.L.: Diseño de Videojuegos aplicados a la Educación Especial (2007). http://aipoes/articulos/1/12410.pdf
12. González, J.L., Gutiérrez, F.L.: Jugabilidad como medida de calidad en el desarrollo de videojuegos. Paper presented at the CoSECivi (2014)
13. Warren, Z., McPheeters, M.L., Sathe, N., Foss-Feig, J.H., Glasser, A., Veenstra-VanderWeele, J.: A systematic review of early intensive intervention for Autism spectrum disorders. Pediatrics **127**(5), e1303–e1311 (2011)
14. Knop, F., Domínguez, A.K.: El diseño y las características del videojuego en los juguetes lúdicos (2017)
15. Solorzano, N.I., Carrera, D.A., Sornoza, L.I., Mendoza, M.: Developing a dashboard for monitoring usability of educational games apps for children. Paper presented at the Proceedings of the 2019 2nd International Conference on Computers in Management and Business, Cambridge, United Kingdom (2019)
16. Danev, L., Hamann, M., Fricke, N., Hollarek, T., Paillacho, D.: Development of animated facial expressions to express emotions in a robot: RobotIcon. Paper presented at the 2017 IEEE Second Ecuador Technical Chapters Meeting (ETCM) (2017)
17. Tello, J.A., Zapata, M., Paillacho, D.: Kinematic optimization of the robot head movements for the evaluation of human-robot interaction in social robotics. In: Di Nicolantonio, M., Rossi, E., Alexander, T. (eds.) AHFE 2019. AISC, vol. 975, pp. 108–118. Springer, Cham (2020). https://doi.org/10.1007/978-3-030-20216-3_11
18. Creswell, J.W.: Qualitative Inquiry & Research Design: Choosing Among Five Approaches. Sage, Los Angeles (2013)

Alternative Psychological Treatment for Patients with a Diagnosis of Entomophobia with Virtual Reality Environments

Milton-Patricio Navas-Moya[1](✉), Paulina-Tatiana Mayorga-Soria[1], Germánico-Javier Navas-Moya[2], and Gonzalo Borja Almeida[1]

[1] Universidad de las Fuerzas Armadas ESPE, Sangolquí, Ecuador
{mpnavas,lgborja2}@espe.edu.ec, ptmayorga@espe.du.ec
[2] Academia de Guerra, Santiago, Chile
gjnavasm@ejercito.mil.ec

Abstract. Virtual reality is a term that emphasizes the establishment of virtual spaces that simulate the context. Thus, the use of this technology is an alternative for the treatment of phobias caused by irrational and unexplained fears, through the practice of some traumatic experience. Virtual reality helps to create places that simulate three-dimensional existence, in which objects or situations can be useful for the psychological treatment of patients with serious mental illnesses, such as phobias, which are considered an irrational fear of a stimulus that can cause dangerous consequences. Sometimes, fear is so aggressive that patients refuse to be treated, even knowing that they can free themselves from their fear. Nowadays, Entomophobia or insectophobia is the most common phobias that often do not justify the size or harmlessness of insects, the application of alternative treatments makes patients able to live with this type of fear or end the association of stimulus-response.

Keywords: Phobia · Virtual reality · Entomophobia · Environments

1 Introduction

Nowadays, many people suffer from phobias that are irrational fears before a certain stimulus and they cannot have a reasonable explanation since they are involuntary [1]. According to the DSM-5: Manual of Diagnosis and Statistics of Mental Disorders, it is stated that fears are classified into three types, from which entomophobia has been taken as part of the most common specific phobias that are developed from the age of thirteen in human beings, either because of a traumatic or genetic experience [2]. Every day around the world, people with anxiety caused by animals are added to this problem and hundreds of them are unable to cope with it [3].

In America during the last years of the 21st century, the treatment of phobias in patients has made great progress, including Information and Communication Technologies (ICT's) [3], since with the use of new sciences daily life is easier. As well, in the

M. Botto-Tobar et al. (Eds.): ICCIS 2020, AISC 1273, pp. 239–248, 2021.
https://doi.org/10.1007/978-3-030-59194-6_20

area of psychology in which these sciences have contributed in numerous ways to help healing mental conditions from medium to severe level [4, 5].

In Ecuador, one of the main problems is that live exposure is very expensive and complicated, because it is not easy to find the animals to which patients fear of [4], so it is always drawn to photos or treatments in museums or other places where a great variety of this species can be found.

This article is based primarily on an analysis carried out by the application of applied virtual environments using virtual reality technology for the psychological treatment of patients previously diagnosed with entomophobia [6]. The lack of studies using virtual reality in the country has caused an approach for the treatment of patients and its effects, especially because those affected, think that there are no treatment alternatives [7, 8]. The main feature of this type of application is to facilitate the overcoming of phobias, and in addition, people affected can interact with their main and greatest fears [5]. In this case, the research was based on the collection and analysis of a lot of literature based on scientific publications, psychology books, and it was complemented by a field research work in a psychiatric and psychological practice where it was obtained important results for the alternative treatment of entomophobia.

2 State of the Art

The generation of mobile applications for Virtual Reality can be used as a system in which the patient can interact and use it at any time. This program takes place through a technological device that performs a perfect combination between objects of physical and non-existent realities and shows the result to the user [3].

Phobias are a person's fears of feeling exposed to a situation that causes the fear, because of the possible consequences that may occur, whether death or pain [5].

2.1 Phobias

Phobias are a way to call fears, the same ones that are considered as fears of different stimuli which cause anxiety and have the ability to control people. They are usually presented by traumatic events that have occurred throughout their lives.

Phobias in children develop because of their obsession with a fact, object, animal or any type of situation that causes anxiety [1, 9]. But there are specific fears that are not attributed to any cause, one in twenty people, approximately suffers a phobia of any kind [3].

2.2 Phobias Treatment

Among the main treatments is medication, in which specialists prescribe patients whose fear is excessive and it does not allow them to lead a quiet life [10].

Hypnosis has the main objective of altering brain activity, in many cases, performing a regression over time and investigating when the fear factor was developed, in order to counteract with the necessary treatment imposed by the psychologist [11].

Exposure therapy, the therapist treats by desensitizing patients by means of phobic stimulus which are made through scheduled, progressive, and gradual exposures.

Cognitive therapy makes patients understand that their fears are not based on reality but rather on internal fears outside of it. Because it is assumed that phobias are acquired by some bad experiences.

For this research it was used virtual reality, in which context closely resembles systematic desensitization, which is a technique that uses imagination instead of exposure [12, 13].

2.3 Virtual Reality

Virtual Reality (VR) is a form of sensory immersion in a new world, which is basically real environments that were artificially created with animated or inanimate objects that look like they are real [14].

Virtual Reality is a computer-created environment that allows users to give a sense of immersion. The technology helps for a three-dimensional simulation and to achieve it there must have user perspective views and real size, the technology must be able to follow the user's eye movements so that it is reflected in the perspective [8, 15]. The sensations also called virtual environment successes can cause in users temporal immersion to other worlds away from the real one, interaction with virtual reality and telepresence that is to be in another world [13, 16].

3 Implementation

For the development of the application and the project, some tools were used. The same that were the medium to achieve the objective and are detailed below.

Blender: It is an open-source and cross-platform tool used to model objects in three dimensions and geometric features, which are mainly based on lighting, rendering, animations, and graphics creation. In addition, it serves to obtain photorealistic images of different elements [17, 18].

Unity: it was used to perform video games or simulations both in 2D and 3D. Simple to use for inexperienced people and very good with basic and advanced knowledge for cross-platform projects [19, 20]. Within desktop applications, Unity has the ability to create applications for both Linux and Windows, and it handles operating systems on mobile devices such as IOS, Android, Windows Phone and Blackberry, thus covering almost all current technology platforms [16].

Visual Studio: It is a tool that is composed of different languages and other development technologies, for example, C#, which is considered as a simple, modern and object-oriented programming language that combines high productivity and speed [21].

Figure 1 shows the structure of the application, which is included in an application on a mobile device, the virtual reality glasses that were responsible for interacting between virtual reality and the user of the application.

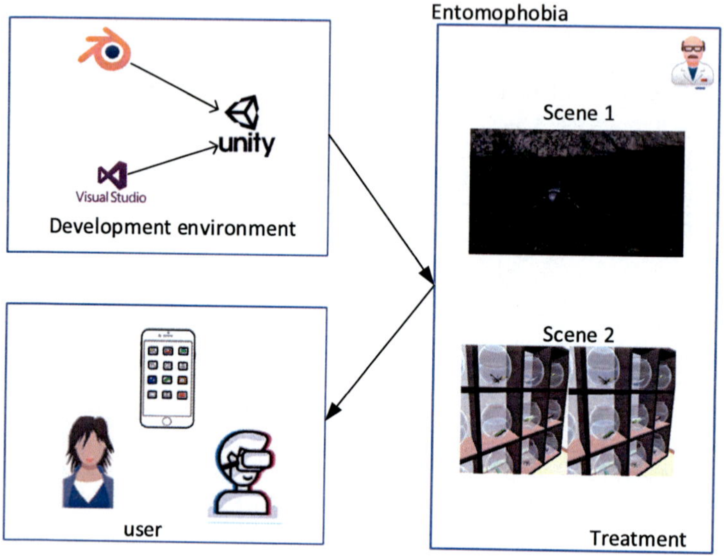

Fig. 1. Implementation Scheme

Capturing Requirements

Below there are the functional and non-functional requirements identified at the design and collection stage. These requirements facilitate that the application has functionalities that contribute to solving the fears of patients and help to overcome them.

Functional requirements of the mobile application:

- The mobile app will have the ability to display a menu, in which the customer chooses the first or second level depending on the state in which their treatment is and with it their fear.
- The mobile app should show on the first level a series of entomological animals in captivity.
- The mobile app will need to contain a gyroscope to work in Virtual Reality.

Non-functional mobile application requirements:

- The device on which the mobile app will be installed can have any operating system since the game is cross-platform.
- The mobile device must contain motion sensors.
- Virtual Reality goggles provide safety to the patient.

Application's Structure

The application structure comprises two phases also called modules. In the first one, the procedure consists of the diagnosis carried out by the specialist in which it is determined if the patient has irrational fear in the concerning area. And in the second module, the system is used to treat the patient's phobia. Making use of the information obtained, it is proceeded to enter the first scene of the application, requesting the patient to observe the animals until they no longer cause any fear. Once the first level has been passed, scene two continues and on it, the patient has to completely lose the fear. This may take as long as the psychologist deems appropriate.

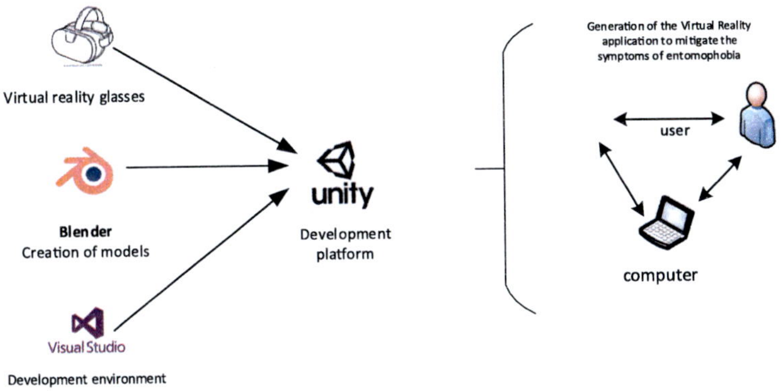

Fig. 2. Application scheme

In Fig. 2, the professional proposes the elaboration of a system that allows the application of a method for overcoming entomophobia, in which it is proposed a treatment by means of scenes created using virtual reality. These scenarios are increasing in complexity according to the needs perceived by the professional, needs that lead to the elaboration of two levels which are structured as follows:

Menu management level

The application permits the access to a screen called menu, in which there are two buttons with the options of entering the first or basic level and the second or advanced level.

First level: Administration Module

In this module, there is a scene which recreates a scientific laboratory in which the patient can enter and observe entomological animals. This level is used at the start of treatment because straightforward interaction with the animals is not necessary, therefore, patients will not feel excessive fear of seeing or being in contact with them. Specifically, it can be used by patients with entomophobia at the beginning of the treatment to make them able to observe entomological animals in bottles, generating an indirect exposition, which mitigates patients' fear since they are not directly exposed to danger.

Second level: Administration Module

In this module, an environment resembling the jungle is presented, in order to make patients walk and observe entomological animals in wild and movement. This level is used in patients during the advanced part of the treatment, due to it requires interaction with animals in their natural habitat through virtual reality (Fig. 3).

Fig. 3. Game app levels

Monitoring

The developed mobile application, adapted to virtual reality glasses, is a software that allows the entrée to a menu in which two buttons appear representing the options of entering a basic or advanced level of treatment for patients with an irrational fear of entomological animals.

In the second option of the menu, the system recreates a real environment in which animals are in a state of freedom, thus being able to move around the terrain, where the patient can interact with them. It should be noted that this level is considered advanced since the user must be close to insects and not feel irrationally threatened. From the point of view of the users, functionality tests on the system, and an evaluation of the graphical interfaces were carried out with the aim of rating the usability of the mobile application. This allowed the authors to observe the satisfaction or dissatisfaction with the interface and the functionality of such applications, attaining with it the fulfillment of stated goals.

Evaluation of results

This section describes the detailed solution, starting with the phase of determination of objectives. Below detailed are the established requirements that will be used within the design of the solution.

The following algebraic formula makes an approximate estimate of time to be used in the activities to be carried out.

It is calculated an average of the time it takes the specialist to get the animal that causes fear and the lapse of carrying out the treatment.

$$T_e = \frac{T_0 + 4T_n + T_p}{6}$$

In which:

Te: Estimated time

T0: Optimistic time is the value that is needed to perform the activity without having any problem.

Tn: Probable timeframe, the more accurate time required to fulfill an activity.

Tp: Pessimistic timeframe, the time requested to fulfill the activity in case one or more problems are generated.

The formula is applicable in a different time in order to make the most accurate estimate in performing treatment for patients with entomophobia every year, month and day (Table 1).

Table 1. Results to entomophobia treatment

				Weekly	Monthly	Annual
Patient	Optimistic timeframe (min)	Probable term (min)	Pessimistic term (min)	Estimated time (min)	Estimated time (min)	Estimated time (min)
Patient 1	168	180	240	51	188	2256
Patient 2	180	180	240	52,34	190	2280
Patient 3	152	180	240	49,38	185,33	2224
Patient 4	128	180	240	50,89	181,33	2176
Patient 5	172	180	240	49,54	188,67	2264
Total, monthly time (min)				253,15	933,33	11200

In the table above, it can be observed the variations of a population of 5 patients who contributed in the research with the time in the therapy and these will go within the term that was explained at the beginning. For the purposes of research, the results that were obtained through the application of patients in a month of considered treatment, are considered equally. The given periods according to the reality of the System were maintained between the most probable and the most pessimistic in the first week, while as the repetitions were applied, they were getting closer to the optimistic probable, which is the purpose in these types of treatments.

Being a project that involves many repetitions for a period of a year it could be obtained good results and these were in the most optimistic. Consistent with the psychology professional, interaction with the environment is important to be able to put into practice the results obtained in the applied experiments (Fig. 4).

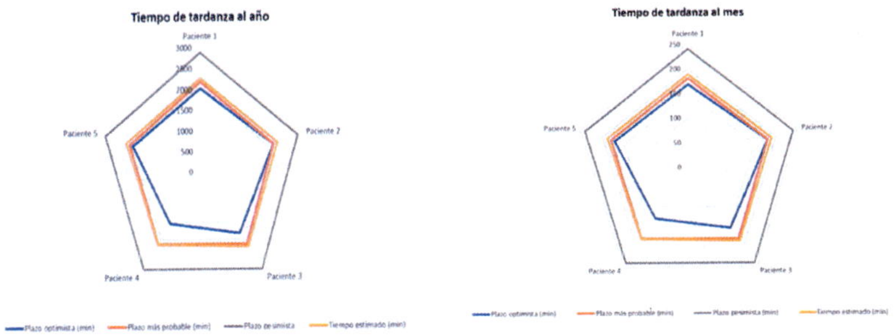

Figure 4 Graphic representation of entomophobia treatment

In the results, it can be seen that the graphs show that the values are within the proposed parameters between the pessimists and probable, however as more tests are performed the times continue towards the limit of the optimists, even though they do not go away from it even with a greater number of the tests done.

The majority of patients react in a different way because psychologically everyone does not have the same way to react to a treatment, but in general lines, it can be said that the treatment will always be the closest to the performance from psychologists to patients.

4 Discussion

Virtual reality is an alternative to the treatments for the mitigation of phobias and fears of animals, people, events. It improves the actions of those who suffer any disorder.

The application proposed was very well received despite its two levels, the same that served to make decisions about exposure to the environment by patients who are people between 15 and 45 years and have a certain phobia of insects. In all these cases, it was manifested that the treatment was effective by having both environments due to the interaction provided to overcome the fears of these animals.

Likewise, the use of virtual reality in a simple application that contains quiet environments, enabled the attainment of goal of increasing patients' confidence to interact with these kinds of animals, "insects", becoming them into something that they can handle and live with.

5 Future Research

For future research, it is proposed to improve the application with more levels, where it can be determined with precision which insects affect patients so that, psychology professionals could improve the treatment in a personalized way. In addition to improving socio-emotional and communicative skills through the application of artificial intelligence in terms of environments and insects, virtual environments can be improved by promoting animations and designs to be according to a level of realism.

6 Conclusions

The analysis performed on patients can determine the noticeable improvement in the phobias that they suffer, due to the correct use of the mobile application, taking into account the progress to show in simulated environments and the time in which the patient is treated, the phobia has been progressively dispelling in the patient, managing to control the irrational fear generated by insects.

The mobile application consists of two levels that allow the specialist to show the inadequate way to the patient, so that the specialist can control on a planned way the simulated environment and obtain better results, in comparison to real exposure which is direct and in which the level of exposure to the user cannot be controlled.

Concerning the above, it can infer that virtual reality contributes satisfactorily in the psychological field, specifically in the treatment of phobias to insects, gradually exposure to simulated environments allows the acquisition of confidence between the patient and the irrational fear of insects.

References

1. Fatharany, F., Hariadi, R.R., Herumurti, D., Yuniarti, A.: Augmented reality application for cockroach phobia therapy using everyday objects as marker substitute. In: Proceedings of the 2016 International Conference on Information and Communication Technology System. (ICTS 2016), pp. 49–52 (2017). https://doi.org/10.1109/icts.2016.7910271
2. Hnoohom, N.: Virtual reality-based smartphone application for animal exposure, pp. 0–5 (2017)
3. Shunnaq, S., Raeder, M.: VirtualPhobia: a model for virtual therapy of Phobias. In: Proceedings of the 18th Symposium on Virtual Augmented Reality. (SVR 2016), pp. 59–63 (2016). https://doi.org/10.1109/svr.2016.20
4. Wrzesien, M., et al.: The therapeutic lamp: treating small-animal Phobias. IEEE Comput. Graph. Appl. **33**(1), 80–86 (2013). https://doi.org/10.1109/mcg.2013.12
5. Horváthová, D., Siládi, V., Lackova, E.: Phobia treatment with the help of virtual reality. In: 2015 IEEE 13th International Scientific Conference Informatics. INFORMATICS 2015 - Proceeding, pp. 114–119 (2016). https://doi.org/10.1109/informatics.2015.7377818
6. Jensen, S.M., Foulds, R.: Adaptive videogame platform for interactive upper extremity rehabilitation. In: Proceedings of the 33rd Annual Northeast Bioengineering Conference, pp. 277–278 (2007). https://doi.org/10.1109/nebc.2007.4413385
7. Levis, D.: Diego Levis ¿Qué es la realidad virtual?
8. Multissensorial, E., Presença, D.: Criação de Ambientes Virtuais de Simulação Imersivos para Terapia de Exposição Exposure Therapy
9. De Oliveira, R.E.M., De Oliveira, J.C.: Virtual reality system for the treatment of acrophobia. In: Proceedings of the 19th Symposium on Virtual Augmented Reality. (SVR 2017), vol. 2017, pp. 74–77, November 2017. https://doi.org/10.1109/svr.2017.18
10. Cuellar-fierro, J.F., Orozco, A.A., Alvarez, M.A.: Processes for Enhancing Resolution over Diffusion Tensor Fields, vol. 1, pp. 371–381, August 2019 (2018). https://doi.org/10.1007/978-3-030-03801-4
11. Cavrag, M., Lariviere, G., Cretu, A.M., Bouchard, S.: Interaction with virtual spiders for eliciting disgust in the treatment of phobias. In: 2014 IEEE International Symposium on Haptic, Audio Visual Environment Games. HAVE 2014 - Proceedings, pp. 29–34 (2014). https://doi.org/10.1109/have.2014.6954327
12. Rosell, C., Jordana, G., Fargues, B., Sánchez-carbonell, X., Jordana, C.G., Fargues, B.: 77828306-3 (2007)
13. Jang, D.P., et al.: The development of virtual reality therapy (VRT) system for the treatment of acrophobia and therapeutic case. In: IEEE Transaction on Information Technology in Biomedicine, vol. 6, no. 3 SPEC, pp. 213–217 (2002). https://doi.org/10.1109/titb.2002.802374
14. Gigante, M.A.: Virtual Reality: Definitions. Academic Press Limited, History and Applications (2014)
15. Wille, D., et al.: Virtual reality-based paediatric interactive therapy system (PITS) for improvement of arm and hand function in children with motor impairment–a pilot study. Dev. Neurorehabil. **12**(1), 44–52 (2009). https://doi.org/10.1080/17518420902773117

16. Vajak, D., Livada, Č.: Combining photogrammetry, 3D modeling and real time information gathering for highly immersive VR experience. In: 2017 Zooming Innovation Consumer Electronics International Conference Galvaniz. Your Creat. (ZINC 2017), pp. 82–85 (2017). https://doi.org/10.1109/zinc.2017.7968669
17. Gadre, S., Rawalgaonkar, V., Warde, M., Balasubramanian, D., Gore, G., Joshi, J.: Implementation and design issues for augmented reality applications on mobile platforms. In: 2018 International Conference Advances Computing Communication and Informatics. ICACCI 2018, pp. 1341–1344 (2018). https://doi.org/10.1109/icacci.2018.8554456
18. Miloff, A., Lindner, P., Hamilton, W., Reuterskiöld, L., Andersson, G., Carlbring, P.: Single-session gamified virtual reality exposure therapy for spider phobia vs. traditional exposure therapy: study protocol for a randomized controlled non-inferiority trial. Trials **17**(1), 60 (2016). https://doi.org/10.1186/s13063-016-1171-1
19. Cruz, D.R.D., Mendoza, D.M.M.: Design and development of virtual laboratory: a solution to the problem of laboratory setup and management of pneumatic courses in bulacan state university college of engineering. In: 2018 IEEE Games, Entertainment and Media Conference. (GEM 2018), pp. 20–23 (2018). https://doi.org/10.1109/gem.2018.8516467
20. Peperkorn, H.M., Diemer, J.E., Alpers, G.W., Mühlberger, A.: Representation of patients' hand modulates fear reactions of patients with spider phobia in virtual reality. Front. Psychol. **7**, 268 (2016)
21. Baraka, K., Melo, F.S., Veloso, M.: Progress in Artificial Intelligence, vol. 8154, pp. 753–765 (2013). https://doi.org/10.1007/978-3-642-40669-0

Designing a Lighting System for Smart Homes

Marlon Moscoso-Martínez[1](✉), Nycolai Moscoso-Martínez[2], Jorge E. Luzuriaga[3], Pablo Flores-Sigüenza[4], and Willian Zamora[5]

[1] Universitat Politècnica de València, Valencia, Spain
marmosma@posgrado.upv.es
[2] Universidad Tecnológica Equinoccial, Quito, Ecuador
nmoscoso@ute.edu.ec
[3] ITACA Institute - Universitat Politècnica de València, Valencia, Spain
jorluqui@upvnet.upv.es
[4] Universidad Anáhuac, Juarez, Mexico
pablo.floressi@anahuac.mx
[5] Universidad Laica Eloy Alfaro de Manabí, Manta, Ecuador
willian.zamora@uleam.edu.ec

Abstract. The massive growth in the number of electronic circuits, their miniaturization, and the possibility of integrating them into integral systems has created many options for creating smart lighting. Smart lighting can save energy costs in homes, streets, and even public places. Thus, a general trend for the automation of homes called "smart homes" for builders and homeowners is noticed everywhere. In this article, we present the development of a lighting system as a prototype to offer the possibility of controlling the lighting of a whole residential building with a single button, a home of any kind up to approximately 1000 square metres to provide energy efficiency management, security, communication, and comfort services. Our prototype employs a control technology based on the creation of scenes and comfortable environments, which consists of integrating and managing all the elements of a residence to activate or deactivate them, discriminately, through a single instruction triggered from a lighting software with a friendly and intuitive interface. The user interface of the system allows reprogramming the central processing unit of the domotic system, according to the user's requirements, as is shown in its evaluation.

Keywords: Lighting systems · Smart homes · Automation · Control systems · Domotics · Energy efficiency

1 Introduction

In the residential sector, the luminous loads most commonly used to create lighting effects are incandescent and tungsten halogen (dichroic), whose electrical

M. Botto-Tobar et al. (Eds.): ICCIS 2020, AISC 1273, pp. 249–258, 2021.
https://doi.org/10.1007/978-3-030-59194-6_21

characteristic is resistive [1]; however, in recent years there has been a general trend towards the use of LED lamps and compact fluorescent bulbs (energy-saving bulbs) due to their low electrical consumption, compared to the former [2]. Incandescent loads and tungsten halogens are widely used for the quality of their light, for their characteristic of being dimmable and for being optimal for creating lighting scenes with light and shadow effects, by means of attenuation technologies [3]. A variant of the incandescent and tungsten halogen loads are those that work at low voltage (generally 12 V or 24V), which are known as magnetic low voltage loads (MLV), since they require a magnetic transformer -not electronic- (120V/12V or 120V/24V) for their connection to the electrical system [4]. In this case, the electrical characteristic of the assembly, transformer plus bulb, is inductive.

There are many types of LED lamps manufactured and available in the international market [5]; however, according to the construction characteristic of the driver that controls them, they may or may not be dimmable [6]. Similarly, fluorescent lamps may or may not be dimmable according to the construction characteristics of the ballast that controls them; however, compact fluorescent lamps are not dimmable because they have an embedded ballast that only allows on/off control [3]. The domotic system has been designed to handle LED, resistive and inductive luminous loads through dimmers, which implement a soft transition attenuation technology (STAT) that consists of creating on/off ramps, in very small discrete steps, imperceptible to the human eye, that give the sensation of continuity when increasing or decreasing the illumination level [7]; however, the compact fluorescent lamps can also be handled by the prototype through relays, using an on/off control.

The prototype shown in this article was made up of a network of devices designed to perform the control tasks required in a home, allowing from any point of the residence, using an input device, either a keypad or a CCI interface take control of the luminous loads connected to one or several lighting returns.

The main characteristic of the prototype is that it uses a control technology based on the creation of scenes and comfort environments, which consists of integrating and managing all the systems of the residence to activate or deactivate them, discriminated, through a single instruction, either manual or automatic. On the other hand, the prototype also has the capacity to integrate itself to the security and protection systems of a house, and to control the technical alarms, by means of the use of the CCI and CCO interfaces.

The rest of the article is organized as follows: Sect. 2 presents the architecture and operation of the home automation system; Sect. 3 presents the parts of the system; Sect. 4 provides the system evaluation and results obtained; and finally, Sect. 5 presents the conclusions of this work.

2 System Architecture

The prototype bases its architecture on the centralization of process control; that is, on the freedom to access any point of the residence through a central device or central processing unit (CPU), as shown in Fig. 1.

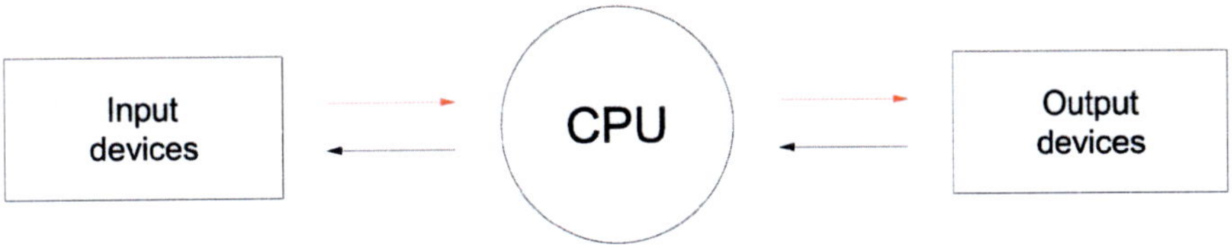

Fig. 1. System architecture.

To carry out a given task, there are three devices involved: an input device, a central device, and an output device. The role of the input device is to activate a system event. The role of the central device is to listen to the input devices, process the information received, and send the commands to the output devices. The role of the output device is to execute the event -specific task to be performed.

Figure 2 shows the sequence of event execution, for a keypad-dimmer communication, when a keypad button is pressed. The sequences labeled with 1 and 2 are related to the execution of an instruction: reading, processing and performing of an event; while, sequences labeled with 3 and 4 correspond to the monitoring of the executed instruction in order to show the event status on the LEDs of the keypad. In case communication is interrupted, the entire cycle is repeated several times to ensure the execution of an event.

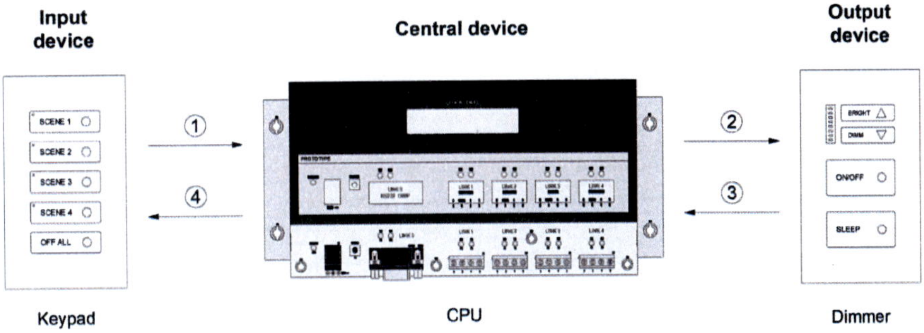

Fig. 2. Sequence of execution of an event.

3 System Description

Our prototype shown in this article was made up of a network of devices designed to execute the control tasks required in a home. The maximum connection capacity of the network consists of: 1 Central Processing Unit (CPU), 32 keyboards (keypads), 32 dimmers, 8 Wallbox Dimmer Modules (WDM), 4 Contact Closure Input Interface (CCI) and 4 Contact Closure Output Interface (CCO). The prototype can control up to 112 returns or lighting zones, 80 of them with dimmable control and 32 with on/off control. Even, other equipment, accessories and/or systems can be controlled through the integration interfaces.

Next, we describe each of the system elements, first the electronic parts as the hardware, then the software that have control over the electronic elements.

3.1 Central Processing Unit (CPU)

Fig. 3. Central Processing Unit (CPU).

The central processing unit or CPU (see Fig. 3) is the central device in charge of controlling the input and output devices, as well as the programmed time events. It is designed to control up to 80 devices and 100 hourly events.

It consists of five communication links. The first four links are used for communications with the system devices, such as: keypads, dimmers, WDMs, CCIs and CCOs; and the last link is reserved for communication with the computer.

The CPU contains the operating program of the prototype, which is elaborated and edited by the user in a lighting software discussed in Sect. 3.7.

3.2 Keypad

The keypad (see Fig. 4) is an input device designed to handle up to fifteen lighting scenes or comfort environments. With this accessory the user can interact with the system and take control of the devices connected to the CPU links.

It is integrated to the system through the link 2 (link of keypads) making use of its communications interface.

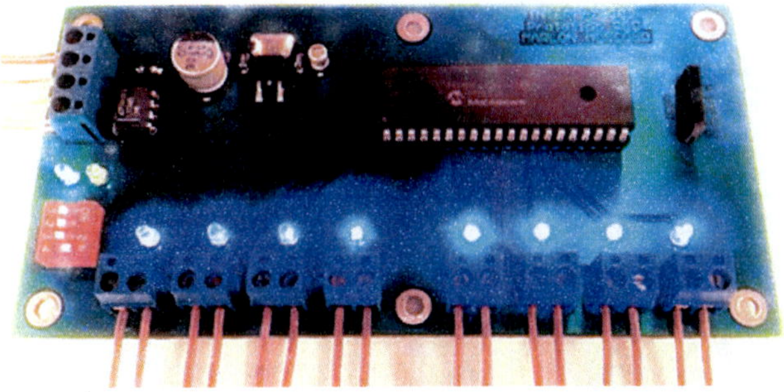

Fig. 7. Contact closure input interface (CCI).

3.6 Contact Closure Output Interface

The contact closure output interface or CCO (see Fig. 8) is a device designed to handle eight independent relay outputs. With this accessory the user can integrate up to eight external devices into the system that can be controlled by a contact closure.

Fig. 8. Contact closure output interface (CCO).

Each output of the interface can be programmed in the lighting software to be either pulsed or maintained. It is integrated into the system through link 4 (interface link) using its communications interface.

3.7 Lighting Software

The lighting software consists of a computer tool developed to program the prototype's CPU according to the lighting design of the house. It has a friendly and intuitive environment, as shown in Fig. 9, which facilitates the creation of a new project or the edition of an existing one.

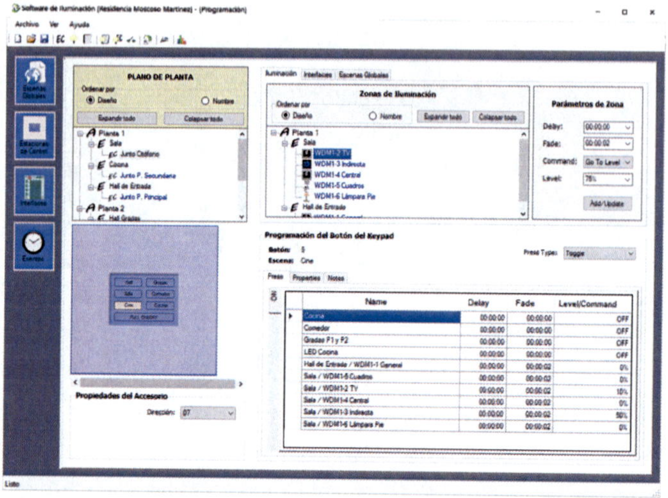

Fig. 9. Lighting software programming environment.

4 System Evaluation

In this section the results obtained and the discussion are presented. In accordance with the scope of the research work, the results are presented for a lighting actuator type: dimmer. The device network used is shown in Fig. 10, and has been tested in continuous operation.

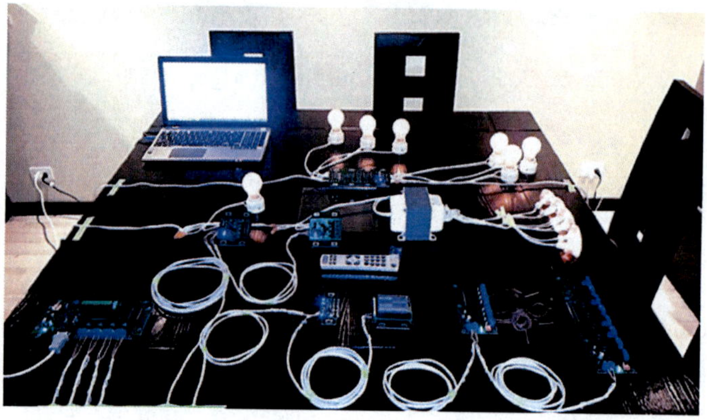

Fig. 10. Functional tests of the home automation system.

4.1 Functioning Tests of the Dimmer

The dimmer performance tests were performed using a dimmer for LED loads, under the following operating conditions: connection to link 1 (dimmer link) without terminal impedance; link connection length of 2.2m; connection of a 120V/7W LED load; use of an IR control. The measuring instruments used were: Tektronix TDS2012B digital oscilloscope, FLUKE 179 digital multimeter, FLUKE 80BK1 thermocouple, and standard chronometer. The results obtained are shown in Fig. 11, and in Table 1.

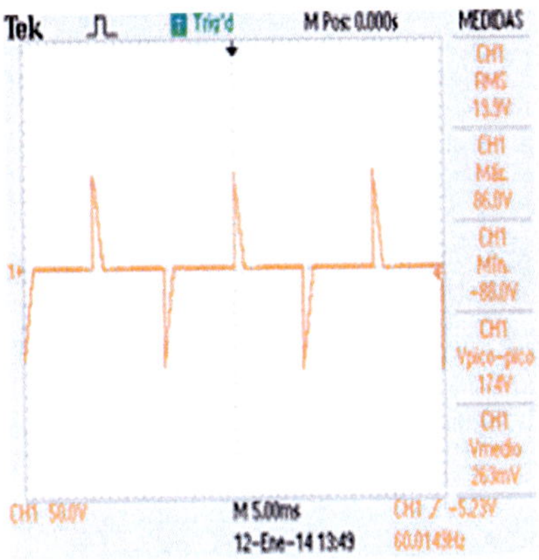

(a) 1% of illumination.

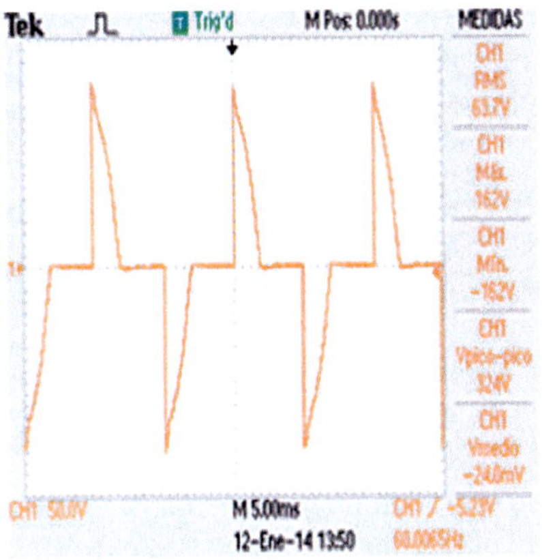

(b) 50% of illumination.

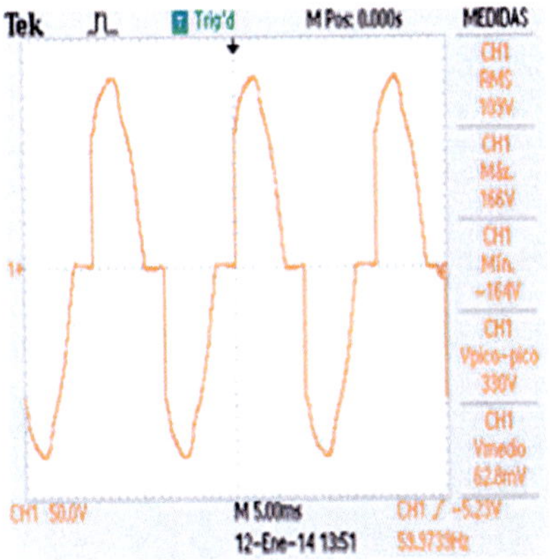

(c) 100% of illumination.

Fig. 11. Voltage waveform over a LED lighting load.

Table 1. Measurements made on the dimmer with LED lighting load.

Parameter			Symbol	Min	Max	Unit
Communication Interface	Input voltage		V_D	11.99	12.01	V
	Input current		I_D	6.8	15.1	mA
Control and power	Input voltage		V_{AC}	115.1	117.2	V
	Frequency		f	59.92	60.08	Hz
	Input Current	Without load	I_{AC}	39.2	40.2	mA
		With 7 W load	I_{AC}	1% → 16.14	100% → 50.56	mA
Ambient temperature			T_A	23.9	24.3	^oC
Measurement time			t	-	5.0	min

5 Conclusions

The general trend of homes towards process automation, together to the possibility of improving the safety of residents and turning their houses into "intelligent homes" is increasingly adopted in our societies. Intelligent lighting can save energy costs in these homes thanks to the appropriate design of home automation systems specialized in lighting.

In this work, the design of a centralized system for home automation was presented. This system consists of a network of 80 input and output devices, which handle light loads through continuous control based on the generation of on/off ramps to adjust the lighting levels to intermediate points between 0 and 100 % of brightness. The operation of this domotic system has given satisfactory results, without loss of data, link's drops or malfunctioning of the devices.

References

1. Lutron Electronics Co., I. Residential Lighting Controls Catalog (Coopersburg, PA, 2014) (cit. on p. 1)
2. Argentina, G. Iluminación Eficiente: primer paso hacia una revolución en- ergética (Coopersburg, PA, 2018) (cit. on p. 2)
3. Lutron Electronics Co., I. Dimmers
4. Lutron Electronics Co., I. Guide to Dimming Low-Voltage Lighting (Coopersburg, PA, 2012) (cit. on p. 2)
5. Electric, G. GE Lighting (Coopersburg, PA, 2017) (cit. on p. 2)
6. Fraile Vilarras, J. & Gago Calderón, A. Iluminación con Tecnología LED (Paraninfo, Madrid, 2012) (cit. on p. 2)
7. Moscoso Martínez, M., Zambrano Valverde, T. & Tigre Ortega, F. Tecnología de Atenuación de Transición Suave (TATS). Revista Científica de Investigación actualización del mundo de las Ciencias, 31–50 (2018) (cit. on pp. 2, 5, 6)

Communications

Green Energy for the Reception and Processing of Satellite and Microwave Signals

Daniel Icaza[1,2,3](✉), Diego Cordero Guzmán[2,3], and Santiago Pulla Galindo[1,3]

[1] GIRVyP Group, Cuenca, Ecuador
[2] Green IT Group, Toronto, Canada
[3] Universidad Católica de Cuenca, Cuenca, Ecuador
{dicazaa,dcordero,gpullag}@ucacue.edu.ec

Abstract. In recent years, the supply of energy by renewable systems has taken a lot of strength, particularly in this research is considered a strength in the application to telecommunications networks, specifically in the headend where the reception and processing of voice signals is carried out, video and data. In the headend it is transcendental that the supply of electrical energy exists, given that it is the most important area to be able to transmit and distribute the signals to the end customers. A field of interest is the use of renewable energy sources such as wind and solar energy, which even having access to the conventional distribution network, can be supported to guarantee continuity in the service to this sensitive area. In this document we examine existing works and present the fundamental principles and motivations necessary to a deep understanding of the use of renewable energy in the headend. In addition, we present a reference model for the renewable energy station. It provides a comprehensive overview and helps identify opportunities in this promising field of research related to telecommunications and renewable energy.

Keywords: Telecommunications · HFC networks · Signal reception · Data processing · Backup · Renewable energy · Smart grids

1 Introduction

Telecommunications services today have become a transcendental part in the lives of people, it does not distinguish race, sex or social level, we are all in a globalized network with prices that tend to decrease and quality of service every time of greater demand in favor of the consumer. The expansion of wire and wireless networks have no borders, their penetration levels were not foreseen.

A customer having alternatives with suppliers who are constantly reorganizing their processes, see the best options to contract their services, implying among the many demands the continuity and quality of the service. A customer who is not satisfied simply changes operator, which companies cannot afford to lose customers because in the end what translates is a reduction in their income. For this reason in this article our purpose is to guarantee sufficient energy to the headend where the signals of the different

M. Botto-Tobar et al. (Eds.): ICCIS 2020, AISC 1273, pp. 261–271, 2021.
https://doi.org/10.1007/978-3-030-59194-6_22

services are received and processed. The headend is the most sensitive part of the whole system, without the timely operation in this part it is not possible to reach any client.

For these reasons renewable energies are seen as an opportunity to guarantee electric power service, without relying on the public conventional network, a plant of this type either from a renewable energy source or the provision of a hybrid system will become a guarantee of supply, especially a source that does not pollute the environment, avoids having storage of fuels such as diesel or gasoline, which in some cases require quite special spaces due to the high risk of explosion that can be caused, comparatively, wind energy or energy. Solar do not present imminent risks and are rather friendly to the environment.

Energy generation processes using renewable sources to supply energy to telecommunications systems are systems that have been addressed in depth and have been very successful in different countries of the world, these analyzes have been addressed by Valerdi [1] and Zhou [2] to supply energy through renewable sources to base stations and cellular systems.

More recent studies were carried out by Cordiner S. [3] and Ahmed J. [4] referring to the power supply to the data routing centers and repetition stations for trunked systems. These investigations reinforce the methodologies of including renewable energies in the field of telecommunications with their strengths or weaknesses; however the results obtained in remote locations are still surprising, it is possible to take advantage of these options to generate electricity to supply base stations away from populated centers and thereby increase penetration rates in the rural sector [5–7].

In specific relation of energy supply to the headend, it is important to guarantee the service to this delicate part of the telecommunications system, generally people with a high knowledge are destined to this area, where there are a number of equipment operating and being refrigerated, the Leaving operation a team can generate demands by users and penalties by the control body. All precautions are taken to have the backup equipment and also have at least one additional source of energy to more than the public electricity grid. In this case our option is to implement hybrid systems of electrical energy provided by renewable sources to the headend. See Fig. 1.

2 Headend

The main mission of the headend is the reception and processing of television, telephony and Internet signals [8, 9]. See Fig. 2.

In the case of analog networks, signal processing consists mainly of the conversion of the original frequency and format of reception of the signal, for example from a satellite or a terrestrial broadcasting station, to a frequency and channel format established. for the cable network, mixing by means of frequency multiplexing with the rest of the signals received in the header for its "injection" into the system. The set of frequencies used in the cable network in which the different TV channels distributed by the network will be located is known as the frequency plan according to the standard of standardization established at the international level.

For this purpose, it must be taken into account in the transmission of data since the frequency chosen for the data transmission channel in the header-user direction must not interfere with the normal television service.

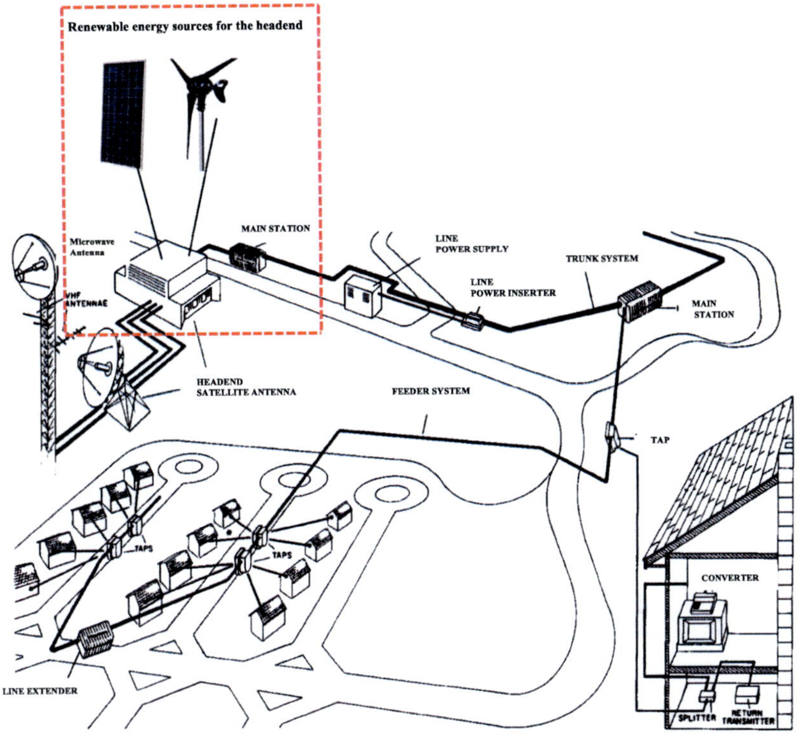

Fig. 1. General structure of the trunked network and power supply system to the headend.

In television, the signal processing is limited as we have said to the conversion in frequency and format (e.g. digital conversion, from satellite, to NTSC), and in the case of the cable network that uses digital transmission, we must add the digitization as a step in the processing of the signal. The channels digitized in MPEG2 are grouped in MPEG2 streams by means of digital multiplexers, giving rise to MPEG2 flows, these flows are modulated in 64 QAM, giving rise to the radiofrequency channel to be distributed in the network.

3 Headend Functions

1) The headend has the function of collecting digital video, data and RF signals [22, 27].
2) The collected signals include the delivery of analog and digital satellite signals, RF broadcast (VHF, UHF), pre-recorded video, AM and FM microwave, Video Baseband, Data and Telephony.
3) The collected signals are Multiplexed and/or Combined (Interleaved) for the input in the Distribution System via the distribution amplifiers or laser light amplification (Node).
4) The Headend also processes all return signals from the distribution system.

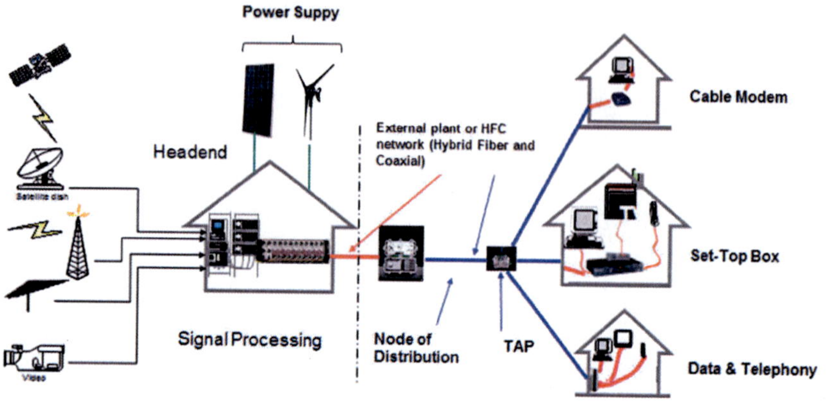

Fig. 2. Components of the telecommunications system with headend powered by renewable energy.

Next, in Fig. 3, the digital signal reception process verified in the spectrum analyzer is processed and then transmitted to the user with the appropriate transmission protocols.

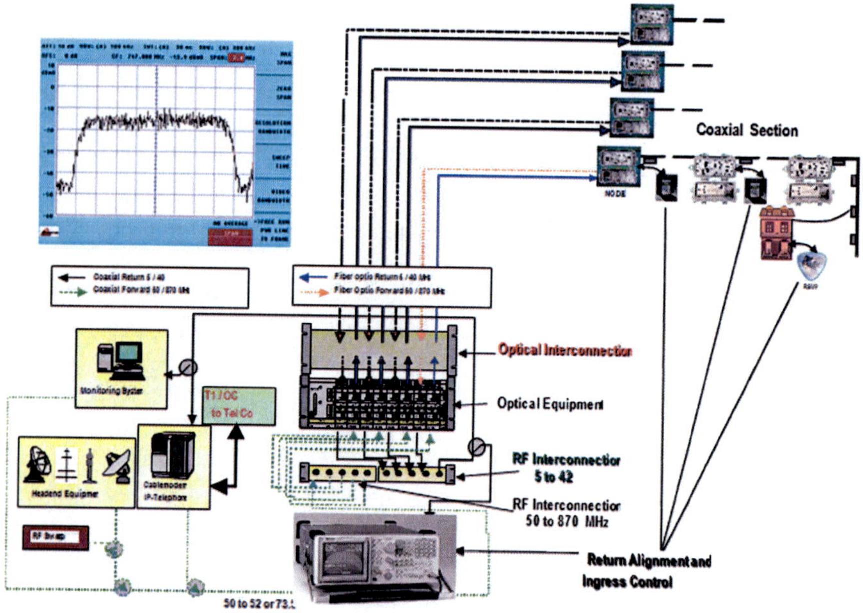

Fig. 3. Digital signal that is obtained in the headend

Figure 4 shows the number of problems generated in the headend in a telecommunications company in Ecuador with branches in Quito, Guayaquil and Cuenca. In these cities there is a backup system of energy by gasoline generator, implying that 30% of the problems occurred at headend level and the ups are attributed to start these generators, in many cases there is a blackout between 5 a 15 s until the backup goes to work. Headend equipment is very sensitive equipment and may be affected, as well as the continuity of the service.

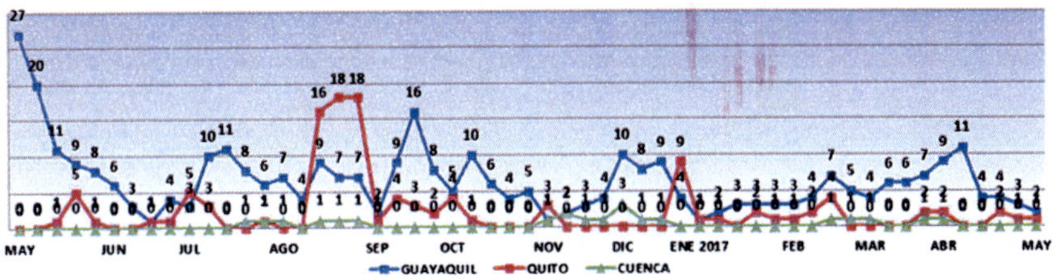

Fig. 4. Problem statistics May 2018–May 2019 at headend level.

According to the indicators shown in Fig. 4, the months of the year from May 2018 to May 2019 are observed on the x axis and the tickets generated week by week are represented on the y axis. We note that in the cities of Quito and Guayaquil there are different problems on a weekly basis, implying that one of the root problems is the lack of continuity in the electric power services inside the headend.

In Fig. 5, we can observe the total affectation accumulated in minutes by the different events aroused month by month in the three cities. It can also be observed that in several months there are out-of-target indicators, which makes the situation in the study area worrisome.

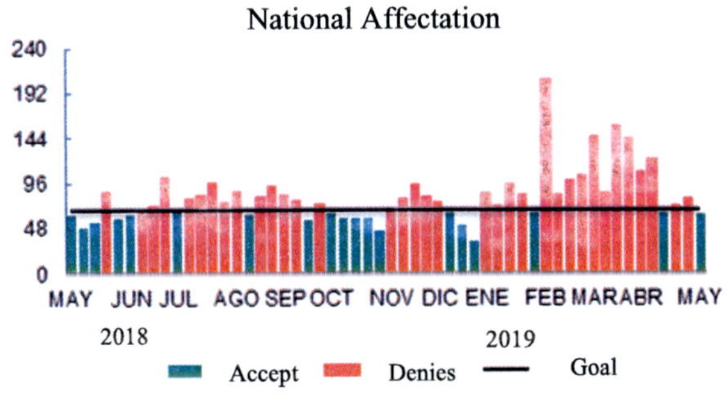

Fig. 5. National affectation May 2018–May 2019 at headend level.

Next, in Fig. 6, the basic design of the electric power generation system is shown, which consists mainly of supplying energy to the building where the headend works, identified as L1 [19–21].

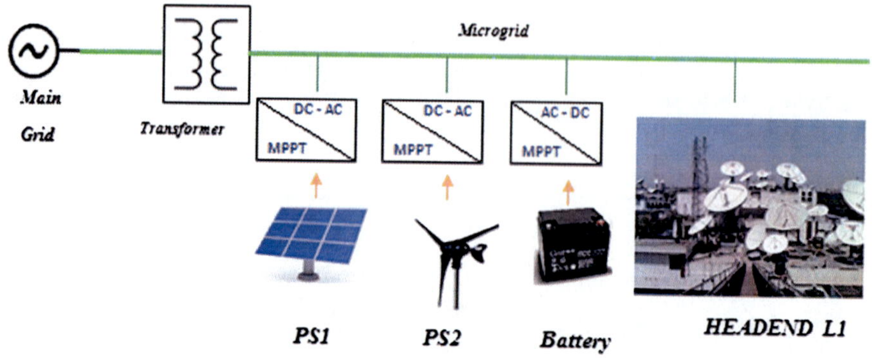

Fig. 6. Design of the backup system based on renewable energy.

4 Operating Scenarios

Below we have the following cases, in which a Headend is exposed to operate in correspondence to the electrical system by renewable energy, which is our purpose either through solar panels and wind turbines, is our proposal. However, in a new investigation it will be possible to analyze the feasibility of incorporating other sources of energy, for this occasion we consider only those mentioned [10–15].

a) **Solar panels and wind turbines supply power to the headend and store energy in the battery.**
When the total photovoltaic and wind power is higher than the demand power in the headend, the excess power will be used to store in the battery. It happens when there is high radiation from the sun during the day and also high wind speeds. The relationship of these power supplies and load in the headend can be written as:

$$P_{S1} + P_{S2} = P_{L1} + P_{battery} \tag{1}$$

b) **Photovoltaic panels and wind turbines supply power in the same proportion to the headend.**
The total power of the photovoltaic panels and the wind turbines is exactly the same to supply the power to the load of the headend. The battery in this case is not involved by the existing balance between generation and consumption.

$$P_{S1} + P_{S2} = P_{L1} \tag{2}$$

$$P_{battery} = 0$$

It can exclusively happen during the change between day and night.

c) **Batteries, photovoltaic panels and wind turbines supply power to the headend.**
During sunrise, sunset, or during cloudy days or times, also on days where there is no wind, the total power provided by these two renewable sources is less than the

power absorbed by the total load of the headend. In this case, the battery bank helps provide sufficient and necessary power to the headend.

$$P_{S1} + P_{S2} + P_{battery} = P_{L1} + P_{L2} \tag{3}$$

d) **The batteries provide sufficient energy to the headend.**
In this case, both the photovoltaic system and the wind turbines stop providing energy, for example, at night and without wind. The battery bank is controlled to discharge sufficient energy and directly to the load of the headend. It is represented as:

$$P_{S1} + P_{S2} = 0 \tag{4}$$

$$P_{battery} = P_{L1} \tag{5}$$

It should be noted that the reason for designing a hybrid system between wind turbines and solar panels, is to achieve greater efficiency in the supply of energy. It is not very common for the two sources of renewable energy to stop operating simultaneously, if that happens it is for short periods of time.

5 Simulations of the Hybrid System

The formulas related to each of the systems for energy production are described below and then simulated in matlab, taking into account the operating scenarios [16–18].

5.1 Wind Turbine

The output power of the wind turbine is expressed as: [23]

$$P_x = 1/2 * C_p * \rho_{air} * A * v^3 * \eta_x \tag{6}$$

Where P_x is the wind power sweep produced by the blades per unit area, C_p is the Betz power coefficient, ρ_{air} is the air density, A is the area swept by the blades, and v is the wind velocity [24, 25] (Fig. 7).

5.2 Photovoltaic PV System

The behavior of the I-V curve of the photovoltaic cell is described by (7)–(9), [26–28].

$$\mathrm{I} = \mathrm{I_L}(\mathrm{G_1}, \mathrm{T_1}) - \mathrm{N_p I_0}\left[\mathrm{e}^{\left(\frac{\mathrm{V+IR_s}}{\mathrm{V_t}} - 1\right)}\right] - \frac{\mathrm{V + IR_s}}{\mathrm{R_p}} \tag{7}$$

$$\mathrm{V}_t = \mathrm{mN_sK}\frac{(\mathrm{T_1}+273)}{\mathrm{qe}} \tag{8}$$

$$\mathrm{I_0} = \frac{\mathrm{I_{SC}} - \frac{\mathrm{V_{oc}}}{\mathrm{R_p}}}{\mathrm{e}^{\left(\frac{\mathrm{V_{oc}}}{\mathrm{V_t}}\right)} - 1} \tag{9}$$

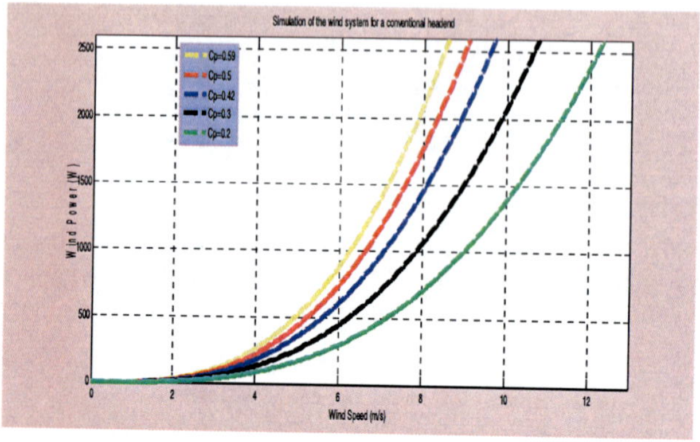

Fig. 7. Speed- power curves at different Cp values

Where;

N_s: Solar cells in series. N_p: Solar cells in parallel. K is the Boltzmann constant. qe: Charge of the electron. m: Diode ideality factor. T1: Working temperature of the solar panel in °C. R_s: Series resistor. R_p: Resistance in parallel. I_L: Photogenerated current = Isc. Io is the inverse saturation current of the diode. Voc: Open circuit voltage [23, 26].

The electric PV power output in DC taking into account the efficiency of conversion to electric energy is

$$P_{PV} = \eta_{c_2} I_{PV}.V_{PV} \tag{10}$$

Where η_{c_2} is the efficiency of conversion to DC and referred to V_{PV}, and I_{PV} (Fig. 8).

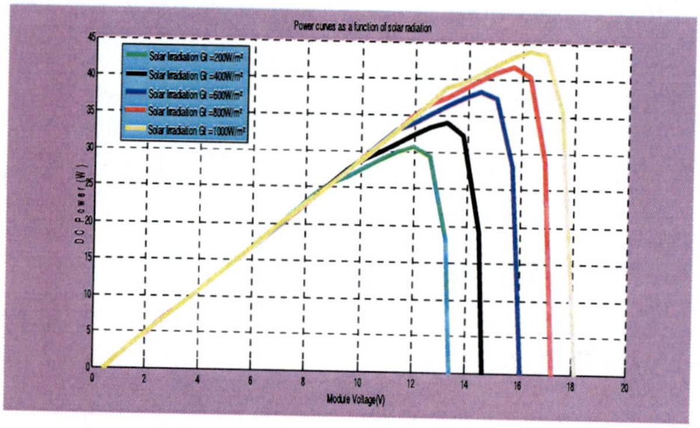

Fig. 8. Voltage-power at different radiation values.

6 Results and Discussions

We proceed to solve Eqs. (1)–(10) taking into account that the total power may not necessarily be simultaneous as already indicated in the analysis of the operation scenarios, and for data validation purposes later, this model numerical was used for the simulation in MATLAB. In addition, the intention in this type of project is in addition to guaranteeing the continuity of the service in such a sensitive area as the headend, it can be verified that the signals in both the waveforms, their magnitudes and frequencies are the same as obtained with the public electricity grid or with the proposed renewable energy system.

For validation purposes, the measurements were made with the spectrum analyzer shown in Fig. 9, it could be determined that the operation of the headend equipment has not been affected with the change of power supply. The registered waveforms are exactly the same. Figure 10 show the results of the measurements made, where it is confirmed that the frequency, amplitude and waveform are exactly identical.

Fig. 9. Signal comparisons between the two energy sources, both conventional and renewable energy [29].

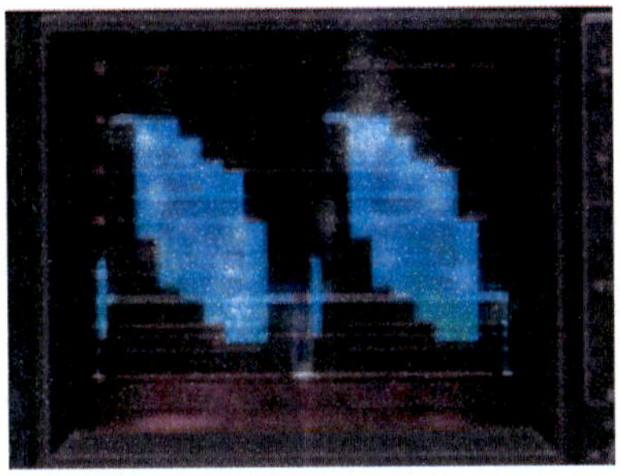

Fig. 10. Comparative analysis of the video signal spectrum to different energy sources. Derivative mode.

7 Conclusions

Although the production of electricity due to renewable energy sources in our environment does not generate confidence, it is no less true that in the near future they will be widely used. In the field of telecommunications we are going to have an important growth since in rural places far from the population centers it is lent so that it is a very interesting alternative.

In the specific headend where this study has been carried out, there is an important benefit such as avoiding the start-up of gasoline generators that are still being used as backup in the country's headend in practically all companies. It is important that trust in these types of energy sources be gained and insurance will see substantial improvements in the operation of the electrical system.

From the economic point of view it is also viable, because although it is necessary to make an initial investment with implementing these renewable energy systems, it is also important to consider that if breakdowns of telecommunications equipment are avoided, a good amount would be saved by the replacements that must be made. Note that telecommunications equipment has a high cost, here it is also intended to protect this type of investments, in many cases there is equipment that needs to be acquired abroad and also require important logistics.

These systems are friendly to the environment, are not polluting and do not obey too strict regulations, in comparison with the fuels that exist caution and state regulations in the purchase, where the amounts should not be greater due to the presumption of contraband. Permits for this fuel purchase must be justified from time to time.

It was found that the signals received in the headend are not altered at all by the transition from the power supply of the public network to the autonomous system of renewable energy. For the location of the solar panels and wind turbines, if the available space is to be considered, they are usually located on the roofs of the offices or an adjacent propitious space for the equipment.

A different proposal has been presented in the application of renewable energies. In Ecuador there are already some interesting applications in other fields, where renewable energy is supplied.

References

1. Valerdi, D., et al.: Intelligent energy managed service for green base stations. In: 2010 IEEE Globecom Workshops (GC Wkshps), pp. 1453–1457. IEEE (2010)
2. Zhou, J., et al.: Energy source aware target cell selection and coverage optimization for power saving in cellular networks. In: Proceedings of the 2010 IEEE/ACM Int'l Conference on Green Computing and Communications and Int'l Conference on Cyber, Physical and Social Computing, pp. 1–8. IEEE Computer Society (2010)
3. Cordiner, S., et al.: Fuel cell based hybrid renewable energy systems for off-grid telecom stations: data analysis from on field demonstration tests. Appl. Energy **192**, 508–518 (2017)
4. Ahmed, F., Naeem, M., Iqbal, M.: ICT and renewable energy: a way forward to the next generation telecom base stations. Telecommun. Syst. **64**(1), 43–56 (2017)
5. Zhang, H.L., Baeyens, J., Degreve, J., Caceres, G.: Concentrated solar power plants: review and design methodology. Renew. Sustain. Energy Rev. **22**, 466–481 (2013)
6. Saxena, N., Choi, B.J.: State of the art authentication, access control, and secure integration in smart grid. Energies **8**(10), 11883–11915 (2015)
7. Kawada, Y., et al.: Data access control for energy-related services in smart public infrastructures. Comput. Ind. **88**, 35–43 (2017)
8. Chiti, F., et al.: Performance evaluation of an efficient and reliable multicast power line communication system. IEEE J. Sel. Areas in Commun. **34**(7), 1953–1964 (2016)
9. Tsiropoulos, G.I., et al.: Radio resource allocation techniques for efficient spectrum access in cognitive radio networks. IEEE Commun. Surv. Tutorials **18**(1), 824–847 (2016)

10. Xia, J., et al.: Application of a new information priority accumulated grey model with time power to predict short-term wind turbine capacity. J. Cleaner Prod. **244**, 118573 (2020)
11. Ramazan, B., et al.: A comprehensive study on microgrid technology. Int. J. Renew. Energy Res. **4**(4), 1094–1107 (2014)
12. Bordons, C., Torres, F.G., Valverde, L.: Gestión óptima de la energía en microrredes con generación renovable. Revista Iberoamericana de Automática e Informática industrial **12**(2), 117–132 (2015)
13. Yoshimasa, A., Hiroyuki, O., Masahiro, M., Hirroki, I., Yasutaka, U., Takuo, I.: Demonstration of SOFC-micro gas turbine (MGT). Mitsubishi Heavy Ind. Tech. Rev. **52**(4), 47–52 (2015)
14. Turchi, C.S., Ma, Z.: Gas Turbine/Solar Parabolic Trough Hybrid Designs, To be presented at the ASME Turbo Expo 2011, 6–10 June, Vancouver, Canada (2011)
15. Browning, J.R., et al.: Calibration and validation of a spar-type floating offshore wind turbine model using the FAST dynamic simulation tool. J. Phy. Conf. Ser. **555**(1), 012015 (2014). IOP Publishing
16. Villena-Ruiz, R., et al.: Field validation of a standard type 3 wind turbine model implemented in DIgSILENT-PowerFactory following IEC 61400–27-1 guidelines. Int. J. Electr. Power Energy Syst. **116**, 105553 (2020)
17. Colak, I., Bayindir, R., Fulli, G., Tekin, I., Demirtas, K., Covrig, C.F.: Smart grid opportunities and applications in Turkey. Renew. Sustain. Energy Rev. **33**, 344–352 (2014)
18. Saha, N.C., Acharjee, S., Mollah, M.A.S., Rahman, K.T., Rafi, F.H.M.: Modeling and performance analysis of a hybrid power system. In: Proceeding of International Conference on Informatics Electronics and Vision (ICIEV), pp. 1–5 (2013)
19. Mehrjerdi, H.: Modeling, integration, and optimal selection of the turbine technology in the hybrid wind-photovoltaic renewable energy system design. Energy Convers. Manage. **205**, 112350 (2020)
20. Jones, G.G., Bouamane, L.: Historical trajectories and corporate competences in wind energy. Harvard Business School Entrepreneurial Management Working Paper, no. 11–112 (2011)
21. Cuviella-Suárez, C., et al.: Exergoeconomics in the sanitary-ware industry to reduce energy and water consumptions. In: AIP Conference Proceedings, p. 020061. AIP Publishing LLC (2019)
22. Elliott, I.K., et al.: Voice over data telecommunications network architecture. U.S. Patent No 6,614,781, 2 Sept 2003
23. Amutha, W.M., Rajini, V.: Techno-economic evaluation of various hybrid power systems for rural telecom. Renew. Sustain. Energy Rev. **43**, 553–561 (2015)
24. Cameron, P., et al.: International Energy Investment Law: the Pursuit of Stability. *OUP Catalogue*, Oxford (2017)
25. Hemeida, A.M., et al.: Optimum design of hybrid wind/PV energy system for remote area. Ain Shams Eng. J. **11**(1), 11–23 (2020)
26. Najafi-Shad, S., Barakati, S.M., Yazdani, A.: An effective hybrid wind-photovoltaic system including battery energy storage with reducing control loops and omitting PV converter. J. Energy Storage **27**, 101088 (2020)
27. Flores, V., Carlos, A., Marcos, S.D.C.: Modulación senoidal de ancho de pulso (SPWM) como etapa de control en módulos de pequeña y gran escala, para aplicaciones en variadores de frecuencia. BS thesis (2009)
28. Leloux, J., Narvarte, L., Trebosc, D.: Review of the performance of residential PV systems in Belgium. Renew. Sustain. Energy Rev. **16**(1), 178–184 (2012)
29. TVCable Group: Apuntes y presentaciones de capacitación recibida y otras. Ecuador (2012)

Network Design Defined by Software on a Hyper-converged Infrastructure. Case Study: Northern Technical University FICA Data Center

Santiago Meneses(✉), Edgar Maya, and Carlos Vasquez

Northern Technical University, Av. 17 de julio, 5-21, Ibarra, Ecuador
{sjmenesesn,eamaya,cavasquez}@utn.edu.ec

Abstract. The purpose of this document is to expose the integration of two technologies, Software Defined Networks (SDN) and Hyperconverged Architectures (HCI). SDN is a technology whose architecture works under two planes, the control plane and the data plane; On the other hand, HCIs are solutions that allow to virtualization of the network, storage, and computing environment to, allowing these three elements to work harmoniously. These two technologies mentioned above are currently presented as a solution to different problems. In the case of SDNs, they facilitate administration, flexibility, and programming, resulting in dynamic, programmable, centralized and flexible data networks; On the other hand, HCIs provide solutions to scalability, computing and availability problems in the Data Centers. The research works under an SDN together with an HCI deployed in a Data Center of North Technical University. For the deployment, we use Opendaylight as a controller in the control plane, OpenFlow as communication protocol, and mikroTik equipment in the data plane; the hyper-converged architecture with which it works is Proxmox VE. The results show the operability between these two technologies, SDN and HCI, underdetermined operating metrics.

Keywords: SDN · Openflow · Hiperconverged · HCI

1 Introduction

The innovation in digital services and platforms, as well as the ease of access to them, has allowed the generation of digital data and information, causing overload in hierarchical data networks. The generation of information by different digital platforms and applications have evidenced problems in this type of data networks [3], such is the case of the inflexibility of the network in the face of peaks in traffic demand, resulting in problems in the provision of services. On the other hand, the tendency to virtualize Data Centers forces us to look for solid options. Hyperconverged infrastructures satisfy this need, this type of solution works with three base elements; these elements are, computing, storage, and network, all defined by software working synchronously.

M. Botto-Tobar et al. (Eds.): ICCIS 2020, AISC 1273, pp. 272–280, 2021.
https://doi.org/10.1007/978-3-030-59194-6_23

Figure 1 shows a hierarchical network model operating under the concept of three physical layers [2], the access, distribution and core layer where each piece of equipment in the network is an individual entity that works together with the Other devices. This model has several drawbacks, such as flexibility, decentralized management, and automation [7].

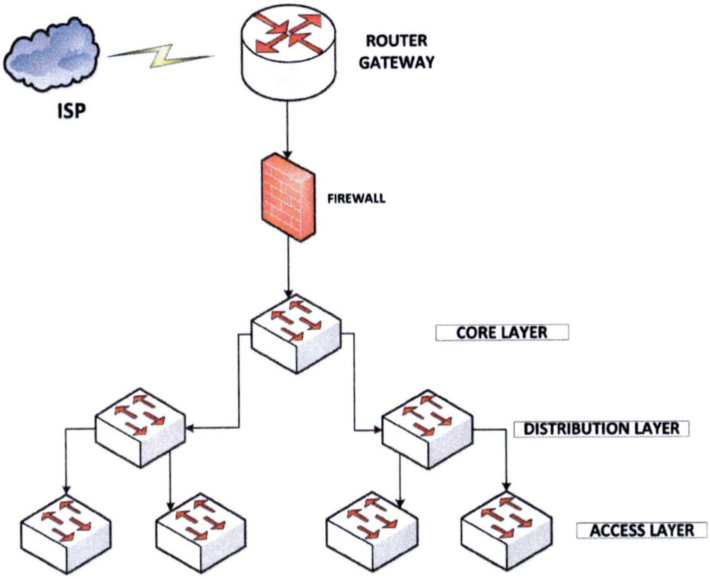

Fig. 1. Hierarchical network design.

SDN is s a solution to different problems in hierarchical networks, such as the administration, flexibility, and programming of data networks. SDN technology allows separating the data plane from the control plane (see Fig. 2). As a result of this solution we will have dynamic, programmable, centralized and flexible networks. They also allow an optimal harmony to work with hyperconverged infrastructures, having a complete virtualization in a Data Center [1].

In Software Defined Networks, the choice of controller is a critical point, because the performance and stability of the network depend on it. In the present work, OpenDayLight has been chosen as the network controller for its various advantages, it allows constant updates, it is open-source, and it has a graphic interface that allows basic configurations in traffic flows [11].

Proxmox VE is a platform that allows the management and virtualization of data centers, and it is open-source, whose implementation focuses on the hyperconvergence of three main components, computing, storage, and network, which allows working synchronously with an SDN environment [5].

The purpose of the research is to analyze performance by working together with these two solutions: software-defined networks and hyper-converged infrastructures in

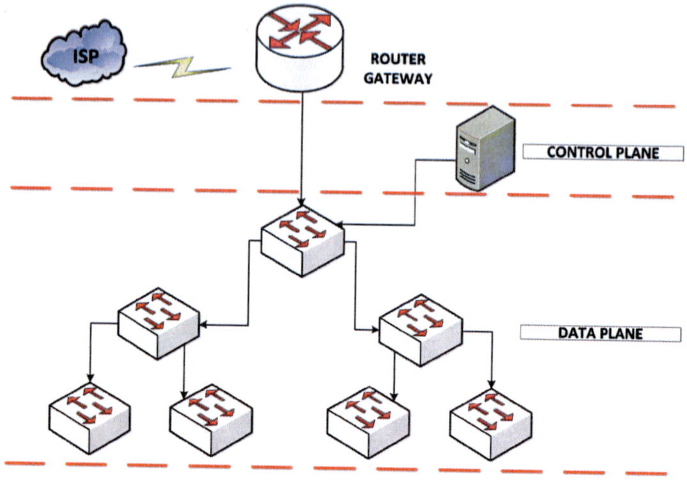

Fig. 2. SDN design.

Data Centers. In the hierarchical network and the SDN, performance tests are proposed, under a controlled and similar test environment for both cases.

2 Methodology

This project was developed in the FICA Data Center, in the North Technical University. The Data Center is a TIER I; on a physical level, the Data Center has twelve servers, the design of the data network is hierarchical. The administration is a non-centralized at the network level. In the virtualization environment, there is a hyperconverged infrastructure with Proxmox VE, on which there is an average of twenty-five virtual servers.

The test environment is the section of the Data Center framed in red, the switching equipment is Mikrotik (see Fig. 3).

For the implementation of the SDN we used three Mikrotik RB-750 devices, these devices have the OpenFlow protocol installed, the controller that works with the control plane is Opendaylightn is installed on a physical server. The hyperconverged infrastructure is made up of three physical HP Proliant G9 servers, which are in a cluster using Proxmox VE, Table 1 details the equipment used.

2.1 Test Scenarios

In the test scenarios proposed, we analyzed only the behavior of the network in a LAN environment using the services built on the hyperconverged infrastructure Proxmox VE. These services are at one end of the network; at the opposite end, you have test users and computers that allow you to analyze performance metrics on the network.

Scenario one: In this scenario there is three switching equipment, in one end of the topology there is a cluster with Proxmox VE, in the cluster, there are some services raised; connected users are at the other end (see Fig. 4).

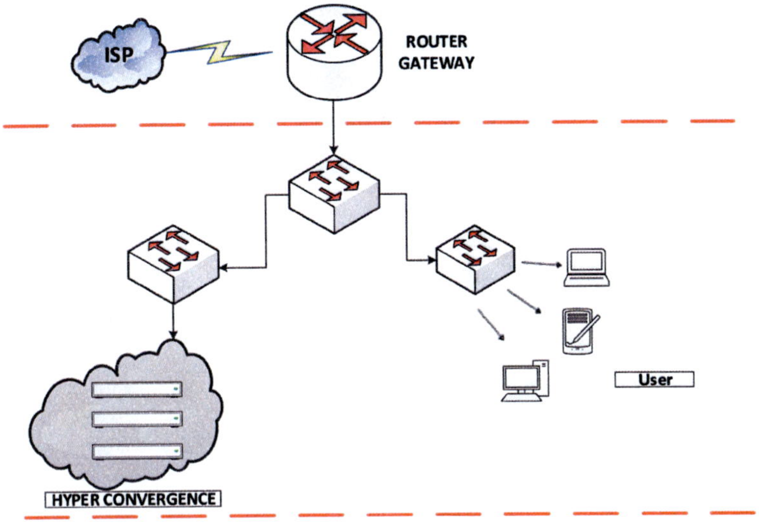

Fig. 3. Testing environment.

Table 1. Physical equipment used.

	Servers Proxmox VE			Mikrotik	OpenDayLight
Characteristic	Svr1-HP	Svr2-HP	Svr3-HP	RB 750	Toshiba
Ram	32 GB	32 GB	32 GB	256 MB	12 GB
HDD	2 × 600GB	2 × 600GB	2 × 600GB	–	900G
CPUs	16	16	16	MT7621A	4

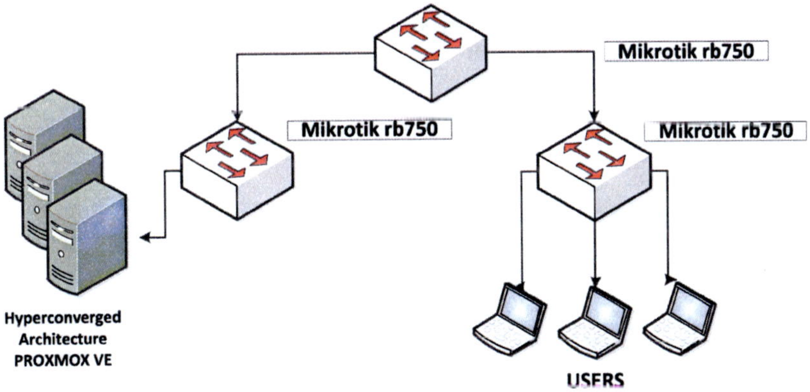

Fig. 4. Hierarchical network, test scenario.

Scenario two: In this scenario, the controller is connected to the central equipment; in this controller, the configuration of the traffic flows that allows communication between the users and the services is available (see Fig. 5).

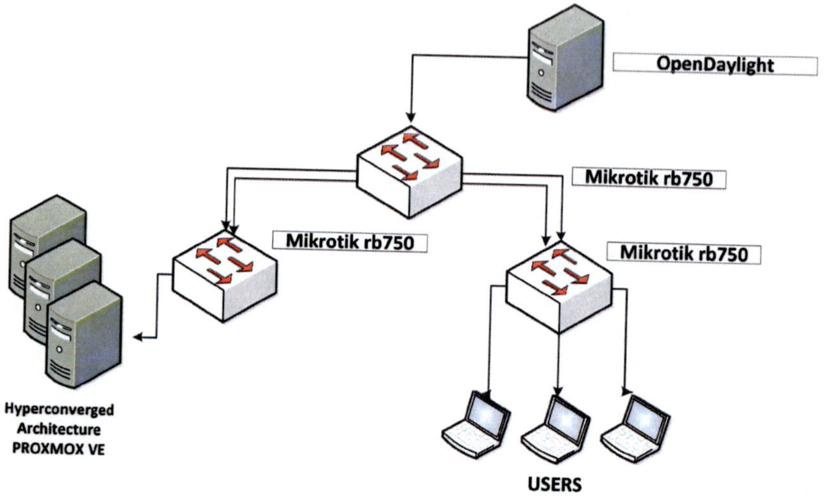

Fig. 5. SDN network, test scenario.

The SDN is made up of two main components, the control plane, and the data plane. The Open-daylight controller handles the control plane and the data plane are the Mikrotik teams that have the OpenFlow protocol enabled, Fig. 6 shows the flows raised in the controller for Mikrotik equipment, while Fig. 7 shows the image of the controller and flow enablement.

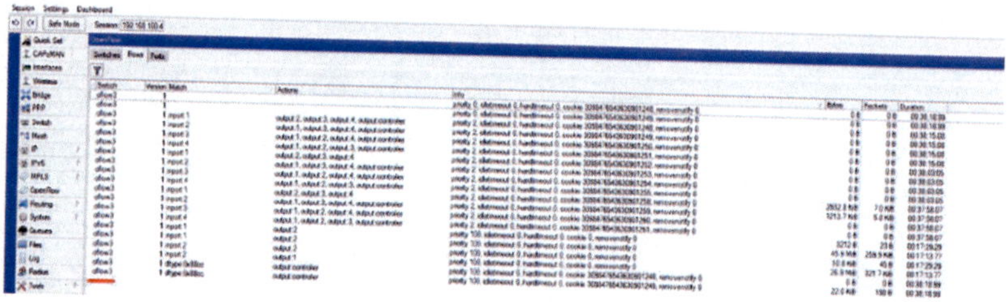

Fig. 6. Flows created for Mikrotik teams.

After enabling OpenFlow in the Mikrotik equipment and configuring the flows in the Opendaylight (ODL) controller, the topology of the SDN network can be observed with the network devices involved, in Fig. 8 the result obtained is observed.

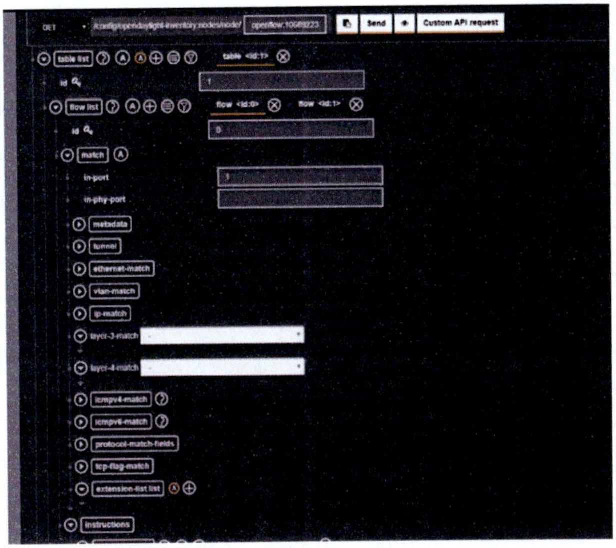

Fig. 7. ODL Streams Configuration Window

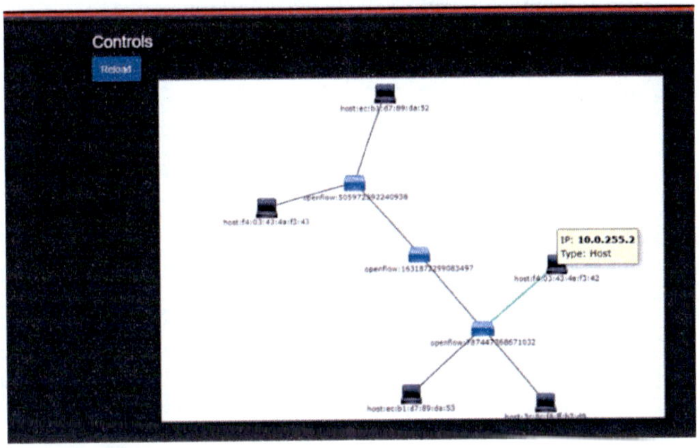

Fig. 8. SDN topology in the ODL window.

3 Results

Network load tests were performed for both the hierarchical network and the SDN, the two scenarios were evaluated under the same conditions and metrics such as latency and throughput, in the process we considered several tests from which we obtain ana average result.

One of the metrics evaluated is latency, through the JMeter software, JMeter measures latency from the moment before sending the request, until receiving the first response, the unit of measurement of latency is in ms [19].

The second metric considered is throughput; this parameter is calculated as requests/unit of time. The time is calculated from the start of the first sample to the end of the last sample. This includes any interval between samples, as it is assumed to represent the load on the server [19].

Table 2 shows the result of the metrics evaluated; to validate the results, we developed several evaluations related to the network metrics; in this case, the latency and throughput were measured. Based on the results of the metrics evaluated, we reached the following results: the SDN network was 20.71% more efficient in terms of latency measures, while in the throughput evaluation, the SDN network was 7.38% more efficient.

Table 2. Metrics evaluated.

Parameters	Hierarchical Network	SDN
Latency	7.82 ms	6.2 ms
Throughput	18,806.980/minuto	20,305.652/minuto

Table 2 summarizes the average values obtained in different tests performed; the metrics evaluated are latency and throughput for each case study.

In Fig. 9 we have one of the results evaluated in the load tests with JMeter, having a throughput above 20,000,000 requests per minute in the SDN. On the other hand, in Fig. 10, we have the result when evaluating this same metric in the hierarchical network whose value is much lower.

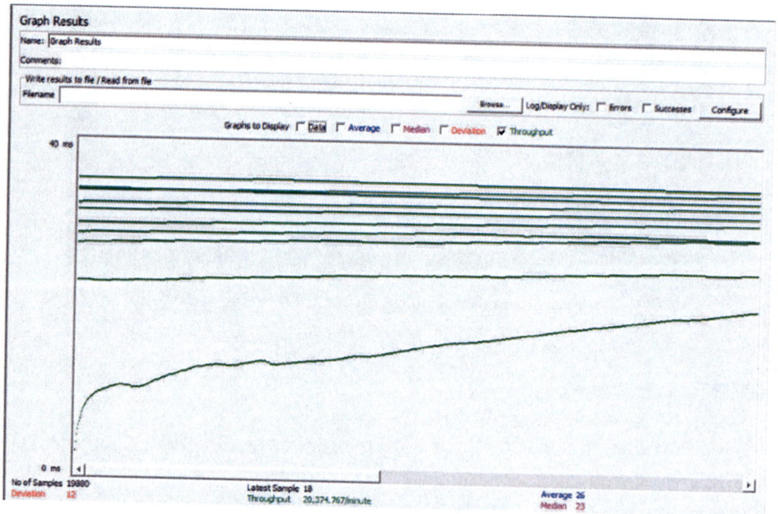

Fig. 9. Throughput SDN network.

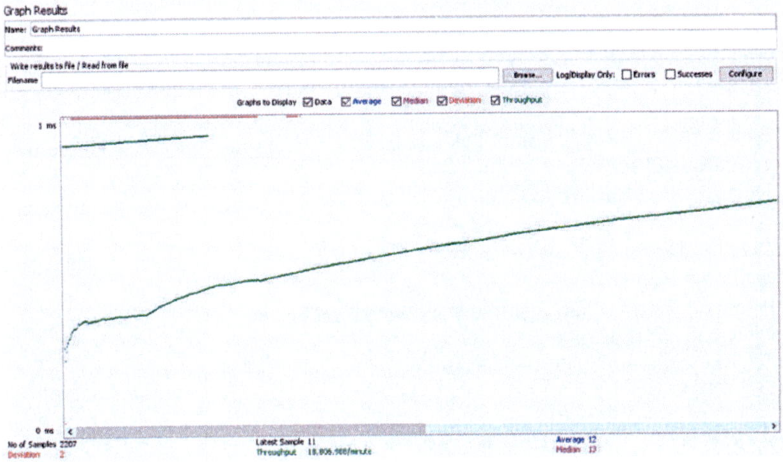

Fig. 10. Throughput hierarchical network.

4 Conclusions

Hyperconverged architectures allow data centers to be virtualized, solving various problems in non-virtualized environments, such as high availability of services, hot migration of servers, and data backups, allowing them to optimize their management and administration. On the other hand, SDN is a technology that aims to solve the multiple limitations of hierarchical data networks, such as administration, programming, and flexibility, innovating data networks, as we currently know. Software-Defined Networks allow us to get the most out of the data networks in a virtualized environment together with hyper-converged architectures.

This study shows the analysis of a Software-Defined Network vs. a Hierarchical network, where the SDN has worked with the OpenDayLIght controller, achieving better overall performance in the SDN; at the latency level, the SDN improves the result by 1.62 ms with respect to the hierarchical network, while at the throughput level the SDN responds to 1,498672 requests/minutes more than in the hierarchical network, in this sense, the SDN network is 20.71% more efficient in terms of latency measures, while in the evaluation of throughput the SDN network is 7.38% more efficient.

In the study carried out by Mohammes Jamal Salim, in his article An Analysis of SDN-OpenStack Integration, OpenDayLIght presents a high level of resistance, reliability and integration against other controllers, such as Ryu and Pox, having latency times of 10.48 ms; In the present study, under the planned test environments, a latency of 6.2 ms was observed.

We have many challenges in the area of network visualization, such as how to improve the centralized flow visibility, improve the discovery of the OpenFlow protocol, load balance, and to ease the programming through the controllers.

References

1. Yoon, B.Y.: Transport SDN architecture for distributed cloud services. IEEE (2014)
2. cisco.com. https://www.cisco.com/c/dam/r/es/la/internet-of-everything-ioe/assets/pdfs/en-05_campus-wireless_wp_cte_es-xl_42333.pdf. Accessed 2 Apr 2020
3. Hernandez, C.: OpenStackEmu - a cloud testbed combining network emulation with openstack and SDN. IEEE (2017)
4. Gedia, D.: Performance evaluation of SDN-VNF in virtual machine and container. IEEE (2018)
5. Maya, E.: Design and tests to implement hyperconvergence into a datacenter: preliminary results. Springer (2019)
6. Blenk, A.: Pairing SDN with network virtualization: the network hypervisor placement problem. IEEE (2015)
7. Erel, M.: Scalability analysis and flow admission control in mininet-based SDN environment. IEEE (2015)
8. Vora, J.: Performance evaluation of SDN based virtualization for data center networks. IEEE (2018)
9. Alcorn, J.: SDN On-The-Go (OTG) physical testbed. IEEE (2017)
10. Basu, K.: Performance comparison of a SDN network be-tween cloud-based and locally hosted SDN controllers. IEEE (2018)
11. Alsaeedi, M.: Toward adaptive and scalable OpenFlow-SDN. IEEE (2018)
12. Salim, M.J.: An analysis of SDN-OpenStack integration. IEEE (2015)
13. Beshley, M.: SDN/Cloud solutions for intent-based networking. IEEE (2015)
14. Singh, P.: Design and deployment of OpenStack-SDN based Testbed for EDoS. IEEE (2015)
15. Ventre, P.L.: Deploying SDN in G EANT production network. IEEE (2017)
16. Khorsandroo, S.: An experimental investigation of SDN controller live migration in virtual data. IEEE (2017)
17. He, T.: Performance evaluation of live virtual machine migration in SDN-enabled. Elsevier (2019)
18. Zhao, Y.: On the performance of SDN controllers: a reality check. IEEE (2015)
19. https://jmeter.apache.org. https://jmeter.apache.org/usermanual/glossary.html. Accessed 2 Apr 2020

Study of the Fronthaul WDM Applied to the Cloud RAN of 5G-CRAN

Stalin Ramírez[1](✉), Diana Martínez[2], and V. Marcelo Zambrano[1]

[1] Universidad Técnica del Norte, Ibarra Av. 17 Julio, Ibarra 5-21, Ecuador
smramirezr@utn.edu.ec
[2] Escuela Politécnica Nacional, Quito Av. Ladrón de Guevara, Quito 2-53, Ecuador

Abstract. In general, everyone involved with the telecommunications sector has reached a consensus that the Radio Access Network of existing 4G mobile networks will not be able to reach the volume of traffic and quality of service required for the 5G networks that are to come. Therefore, it is necessary to ensure that the migration process from the current scenario to 5G is carried out efficiently, eliminating operational limitations and guaranteeing interoperability between the reused infrastructure and the new technology to be implemented. This paper describes the process to be followed for an efficient migration from a 4G LTE to 5G CRAN cellular mobile network, utilizing the Public Telecommunications Company of Ecuador – CNT, as a case study. The proposal is based on the repurposing of the existing transport infrastructure, both for backhaul and fronthaul, and the incorporation of a virtualized and centralized Radio Access Network in the Cloud, which allows for optimal resource management, data processing and response times.

Keywords: 5G networks · Backhaul · Cloud RAN · Fronthaul · Mobil networks · Radio access networks

1 Introduction

The evolution of mobile networks from 4G to 5G is underpinned by the new requirements for bandwidth and access speed that new technologies and user applications demand today (eg IoT, video streaming, telemedicine, etc.). Mobile devices (smartphones, tablets, laptops, etc.) have become the main information and communication source for users; but to be able to take advantage of all its functionalities effectively, some radical changes will be required in the existing mobile communication networks. The Cloud Radio Access Network (CRAN) architecture applied to mobile networks is one of the alternatives for migration to 5G networks, as it would improve the performance of network communications and meet the quality of service that mobile users demand nowadays.

The main idea of this architecture is based on the implementation of a Baseband Units Array (BBU Pool), being centralized, software defined (SDN) [1], and interconnected with Remote Radio Headers (RRH) which are distributed throughout the entire

M. Botto-Tobar et al. (Eds.): ICCIS 2020, AISC 1273, pp. 281–294, 2021.
https://doi.org/10.1007/978-3-030-59194-6_24

geographical coverage area. It redefines the organization of mobile networks, accounting for the high density of cellular sites that 5G networks will possess, and improving their performance and operating costs [2]. This reorganization divides the mobile network into two sections: fronthaul and backhaul. The function of the fronthaul is to allow the BBU Pool to connect to the RRH in an optimal way, that is, without affecting the performance of radio communications. Separately, the backhaul functions as a connector of the BBU Pool and the core transport network (Fig. 1).

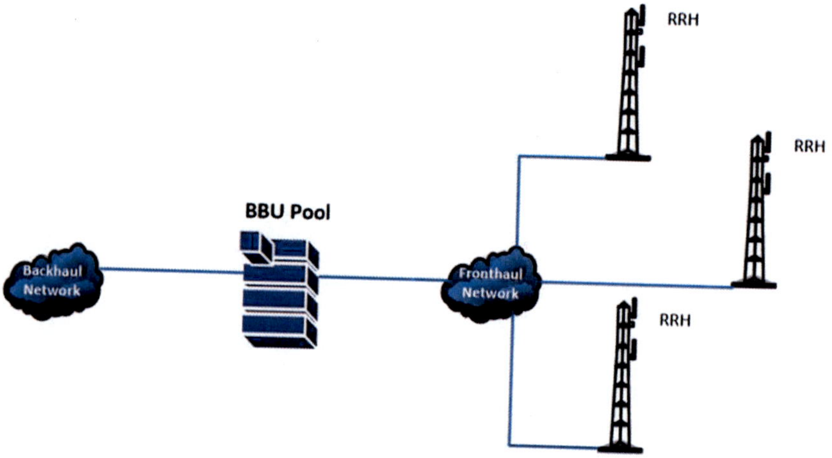

Fig. 1. General diagram of the CRAN architecture.

The BBU Pool, thanks to its software-defined architecture, can be virtualized either in a private or public data cloud, and facilitate the interoperability and integration of the mobile network with emerging technologies such as Big Data, Cloud Computing, and Artificial Intelligence, and others, in turn allowing for improved access times and information processing [2].

This paper describes the suggested process for the migration of the LTE 4G Mobile Network of the National Telecommunications Corporation (CNT) to a network with 5G CRAN technology. The document is divided into five sections: first, an introduction of the subject of study and the proposed objectives for the research; second, the current status and technology used in the CNT's mobile network already existing are described; third, the architecture, methodology and equipment which will be used for the migration are detailed; fourth, the conclusions and recommendations obtained after carrying out this work are presented; Fifth and finally, the references that have been used as the basis for this research are listed.

2 State of the Art

Currently, the CNT Mobile Network functions under the Huawei 4G LTE technology [3]. The architecture consists of three main parts: the Base band Unit (BBU 3900 Huawei)

[3], the Remote Radio Unit (RRU3832) [3] and the LTE AWS Transmission Antenna (A19451811) [3], (Fig. 2).

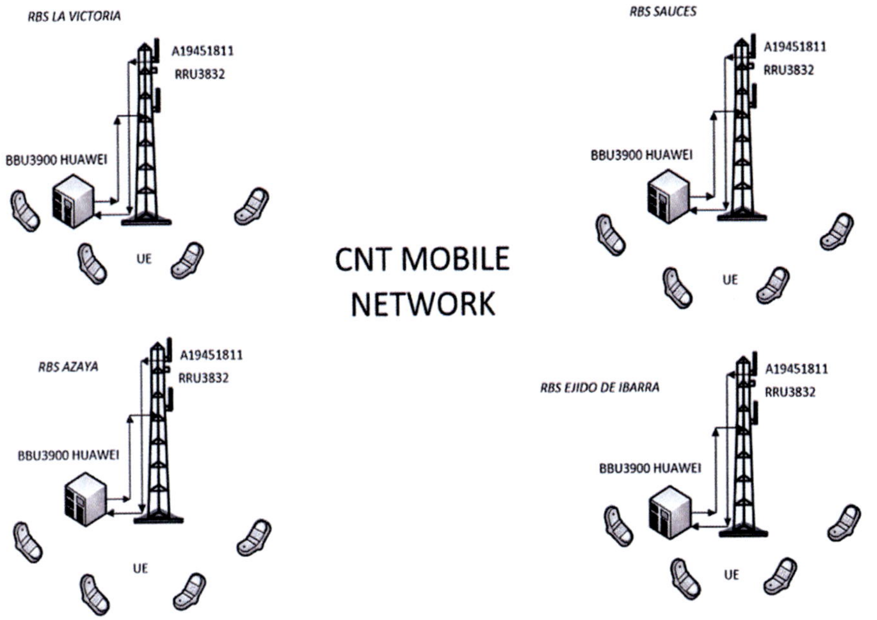

Fig. 2. CNT mobile network architecture.

BBU3900 Base Band Unit

The BBU3900 installed in the current CNT mobile network is based on a modular design granted by the manufacturer Huawei. In this structure, a base station can be configured as the company requires. This is done simply by adding new cards, to expand transmission capacity, expand the process capacity, or to implement a new technology in an existing node [4].

The BBU3900 is configured and consists of the following modules:

- Main Processing and Transmission Unit (UMPTb2): Manages the entire network, including O&M and signaling processing, and provides clock signals for the BBU.
- UMPTb2 or UMPTe2 Universal Master Transfer Unit
- Baseband Processing Unit (UBBPd4 or UBBPd6): Processes base band signals and CPRI signals.
- Fan: Controls the rotation of the blades, checks the temperature of the fan module, and dissipates heat for the BBU.
- UPEUd: converts –48 VDC power to +12 VDC and provides ports for the transmission of two RS485 signal inputs and eight dry node alarm signal inputs [4].

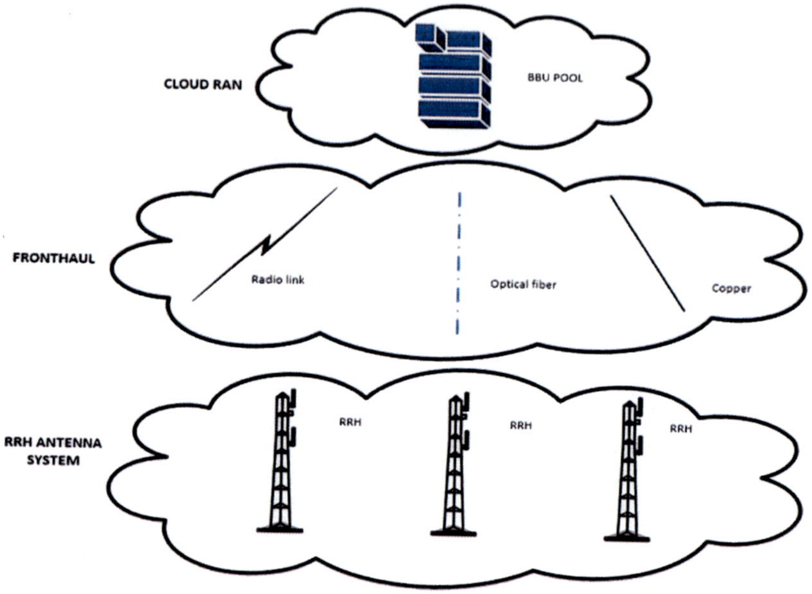

Fig. 3. CRAN architecture.

The BBU3900 is connected at one end to the RRU3832 using fiber optics through the CPRI ports that communicate with radio frequency (RF) modules and process uplink and downlink baseband signals. At the other end, it is connected to the transmission network through the FE/GE0 fiber optic port that reaches the IPRAN IBRCNTRA01 equipment located in the central structure [3].

Remote Radio Unit

The RRU is the remote outdoor radio unit, in charge of processing the baseband signal and the RF signal in GSM, and the RF signal in UMTS. It also integrates duplexers and transceiver units (TRX) [4].

The current CNT network uses the RRU3832, which works in the 1900 MHz band and is in charge of receiving, sending, modulating and demodulating the baseband and RF data, as well as multiplexing and amplifying the RX and TX signals on the RF channel and filters RX and TX signals [4].

Transmission Unit

The antenna is an essential part of the 4G LTE network. In CNT the transmission unit uses the A19451811 antenna which is connected to the RRU3832 and allows communication between the UEs and the BBU3900, and allows the transmission and reception of requested information [3].

The A19451811 works at the 1900 MHz frequency. It has optimal characteristics to work with the LTE network, since it has a gain of 18 dBi, a maximum input power of 300 W, an IM3 intermodulation, and presents an average impedance of 1.2 Ω [5].

RAN

The CNT EP manages its 4G LTE network using IPRAN technology. It has a team called IBRCNTRA01 [3] that is in charge of supervising and monitoring layer 2 links, and its function is to detect codes and erroneous frames with the BS. This occurs in order to identify and locate faults between and with BSs.

The connection of the IBRCNTRA01 with the BBU3900 located in each BS shelter is made by optical fibers towards the FE/GE0 port, that connects the transmission network to the CNT core network called the S1 service port [3]. The IP RAN has the characteristic of supporting and allowing mobility between multiple heterogeneous access networks, including legacy systems such as GPRS and UMTS (2G and 3G), and depending on their Wi-Fi, WiMAX or CDMA2000 configuration [4].

The main difference between 4G and 5G networks lies in the speeds and latency that each of them has. In the case of LTE Advanced (4G) networks [6], this means speeds greater than 1 Gbps with a latency of 10 ms. Likewise, LTE-Advanced Pro networks [6] are capable of offering speeds above 3 Gbps with latencies of less than 2ms. In contrast, 5G networks, according to the latest approved standard, suggest speeds above 10 Gbps and latency of less than 1 ms, with bandwidths of up to 500 MHz. For the short and medium term, 5G networks will not replace the 4G network, rather, they will coexist. Ericsson, one of the leading companies driving 5G, points out that only 20% of the planet's mobile users will have access to the fifth generation within the next five years [6].

In essence, 5G networks constitute the materialization of the digital transformation [6], with service-oriented architectures that allow improving and personalizing the service for all users, and where access speed will not be a limitation [6]. One of the proposals with the greatest acceptance to advance towards this digital paradigm is the one described by the CRAN architecture, and it is the one that we will take as the base for this work.

As its name implies, a CRAN is a "RAN developed on the cloud". In a 4G mobile network, radio units are mounted on top of the tower in conjunction with the antennas and the fiber optic channel connected to the BBU at the base of the tower. CRAN technology allows serving a large number of cell sites from a single central site (in the cloud), replacing traditional BBUs with a BBU Pool [6]. In accordance with the aforementioned information, the first step for migration to a 5G network is the virtualization and grouping of the BBUs, with a dedicated high-speed channel that takes advantage of this centralized architecture (Fig. 3).

2.1 Structure of the CRAN System

There are three possible ways to structure or centralize a CRAN: complete, partial or hybrid.

- **Complete centralization.** It is also called the Stacked BBU Structure. The functions of the BaseBand Layer (physical Layer or Layer 1), the Medium Access Control Layer (MAC Layer or Layer 2) and the Network Layer (Layer 3) of the conventional Base Stations (BS) move towards the BBU Pool. This option is the most common for the configuration of 5G networks, it is clear and simple, but carries a great load for the fronthaul. The BBU Pool contains all the processing and management functions of traditional BSs [7].
- **Partial centralization.** It is also known as the Independent Physical Layer Structure. The RRHs group the RF functions and some functions of the baseband processing, while the other functions of layer 1 and higher layers are housed in the BBU Pool. This option greatly reduces the overhead of the RRH-BBU link and introduces some special features, such as coordinated multi-point transmission and reception treatment (CoMP) [7] and cooperative processing. The interaction between layer 2 and layer 1 can be complex, increasing the difficulty of interconnection [7].
- **Hybrid centralization.** It is considered as a special case of complete centralization, in which some layer 1 functions, as well as cell signal processing functions, are removed from BBUs and assembled in a separate processing unit, which forms part of the BBU Pool. The benefit of this structure is its flexibility to support the redistribution of resources and its ability to reduce modifications and energy consumption in BBUs [7].

It is important to emphasize that in the case of a fully centralized structure, the performance of CRANs is clearly limited by the capacity of the fronthaul link; on the uplink, for example, RRHs must sample, quantize, and then send the received RF signals to the virtualized set of BBUs. With densely implemented RRHs (as in the case of 5G networks), fronthaul traffic generated from a single user equipment (UE), with multiple bandwidths in MHz, could easily scale to multiple gigabits per second (Gbps). In practice, a commercial fiber link with a capacity of tens of GHz could easily be saturated even with moderate mobile traffic [7].

2.2 Fronthaul Cran Transport Technologies

It is clear that one of the critical points of a 5G CRAN mobile network is located in the fronthaul, since, given the centralization of the BBUs, it happens to be much longer (several kilometers), and its capacity should be approximately 15 times that of backhaul due to new services such as eMBB [8], uRLLC [8], mMTC [8].

There are several technologies through which the Fronthaul of a CRAN architecture can be implemented; however, for this research, and taking into account meeting the demands of data intensive applications, the following options have been considered: dedicated fiber, Wavelength Division Multiplexing (WDM) and microwave.

- **Dedicated fiber.** In this option, each RRH is connected to the BBU by means of uni-directional or bidirectional point-to-point fibers; when the network has intermediate nodes, the connection is routed through optical distribution elements (ODF). This transport option is suitable for those scenarios where operators have a large amount of fiber optics installed, and it is categorized as a passive type solution.

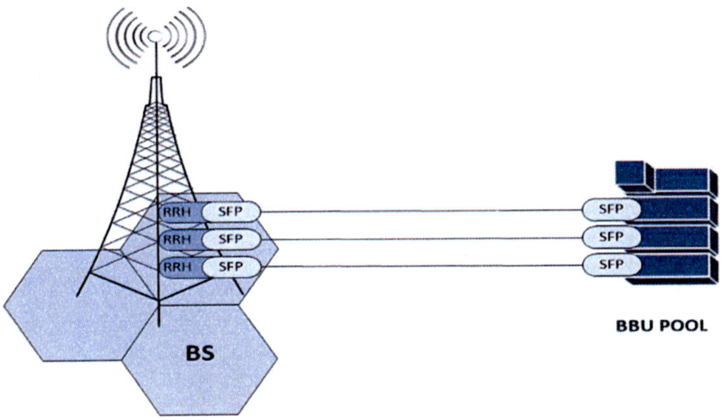

Fig. 4. Fronthaul through dedicated fiber optic.

- **Wavelength Division Multiplexing.** Fronthaul passive type. It allows transmitting on separate wavelength channels, using transceivers that operate according to WDM technology. This allows the multiplexing of channels with different wavelengths through CPRI ports [8]. Figure 4 shows the basic structure of a WDM network (Fig. 5).

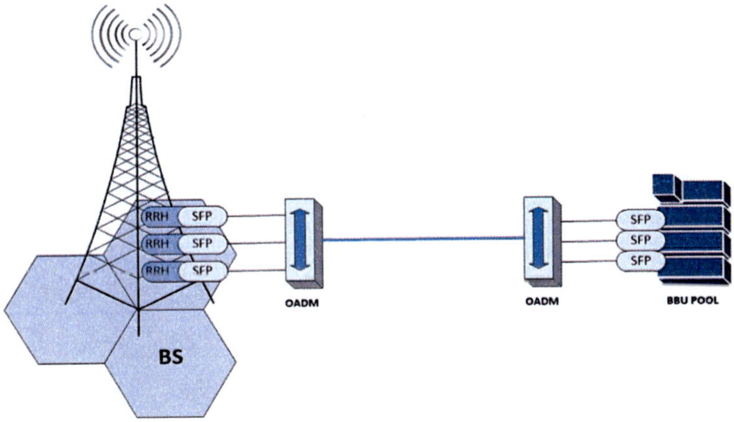

Fig. 5. Fronthaul through WDM

Dense Wavelength Division Multiplexing (DWDM) technology provides 16-channel wavelength multiplexing capability on a single fiber. In most cases, the maximum number of RRHs is less than this number of wavelengths. Compared to dedicated fiber, this solution is more efficient since it significantly reduces the amount of fiber optic to be installed.

- **Microwave.** It is an option for short distances between the RRH and BBU, where fiber optics is not available. It proposes a mobile fronthaul approach, for heterogeneous networks with small cells in the macro-sites. This transport of technology is only used when there are limitations for the interconnection of the CRAN by optical fiber; for this case study, it would not be the best option (Fig. 6).

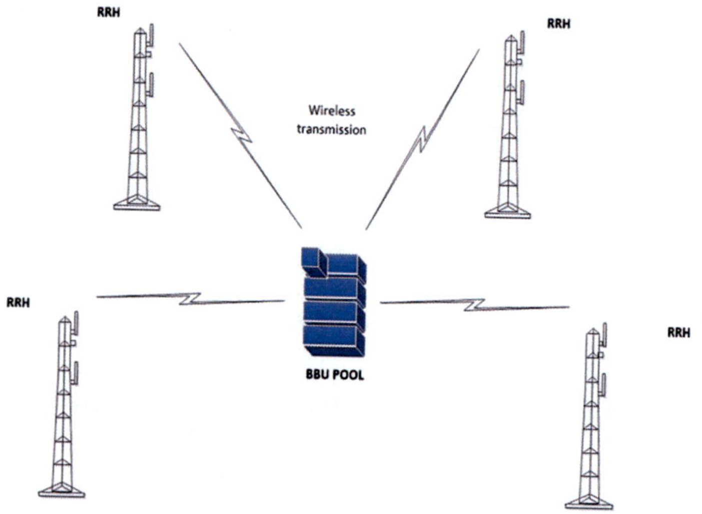

Fig. 6. Fronthaul via wireless access.

3 Methodology

This work describes the process to follow for the migration of the CNT's 4G LTE network, towards a new network with 5G CRAN architecture. The technical requirements, equipment, and procedures are detailed, emphasizing the criticality of the fronthaul and the interoperability between the previous infrastructure and the new infrastructure to be implemented.

3.1 Design Considerations

Based on all the above mentioned about the CRAN architecture, a passive solution is proposed for migration, with a complete centralization structure and WDM as technology for data transport on the fronthaul. This will give the network the flexibility to adapt to different scenarios and add new HR if necessary.

The elements of the passive network are made up of CPRI transceivers and optical multiplexers, complemented by supervision systems and protection against fiber breakage that guarantee the operability of the network at any time. Since OFDMA is used as a modulation scheme and multiple access strategy for LTE-Advanced (4G) networks, this same scheme has been used as a reference for the new 5G CRAN network [9]. Similarly, to optimize the transmission of data (with a considerably large bandwidth) on channels with limited capacity, the Intelligent Quality (IQ) compression scheme will be used [8–9].

The macro cells must be strategically distributed according to the demand of the mobile network, always complying with the service level agreements (SLAs) required by ARCOTEL [10].

In general, the new CNT 5G CRAN network will be made up of the following elements: Transport Core Network, BBU Pool, RRH, Fronthaul and Backhaul.

Transport Core Network

It is currently operational for the 4G LTE network; therefore, it should not be implemented. It is based on IP/MPLS [3] and DWDM [3] technology, and is responsible for providing connectivity to the BBU Pool and allowing connections with bandwidth capacities bordering the order of Terabyte (TB/s) and up to 192 Lambdas [3].

BBU Pool

It has a distributed architecture and a virtualized infrastructure (SDN), which allows separating the control plane from the user plane, grouping multiple BBUs in a single centralized arrangement, and will be in charge of the dynamic allocation of resources for HR, according to the requests and requirements of each of them. For this particular case study, it will be implemented on the CNT's private cloud, through a Huawei CBU5900 team [11] that will be part of the Core 5G All-Cloud network [11].

The BBU Pool is the main advantage of CRAN architectures, allowing the reduction of requirements and costs in terms of cooling and maintenance (decrease in operating costs) and information processing, as well as facilitating the synchronization of the satellite clock throughout the network.

RRH

They are the stations that allow communication between user equipment (UE) by sending and receiving radio frequency (RF) signals. For the proposed design, the RRH has to work in the 3.5 GHz frequency, which is the operating band assigned by the Telecommunications Regulation and Control Agency (ARCOTEL) for future 5G networks [10].

Since most of the signal processing functions in the BBU Pool, HRs can be relatively simple and can be cost-effectively deployed in a large-scale scenario [2]. In general, RRHs perform RF amplification, up/down conversion, filtering, analog to digital conversion, digital to analog conversion, and interface adaptation. A 32T32R Active Antenna Unit (AAU) will be implemented [12], which integrates both the radio and the antenna within the same hardware arrangement. The mobile broadband service (MBB) is performed in a decoupled manner, with the channel ascending at 1.8 GHz or 3.5 GHz and descending at 3.5 GHz, using scalable Orthogonal Frequency Division Multiplexing (OFDM) [12], which allows band widths up to 100 MHz band [12].

Fronthaul

It is the network segment that provides connectivity between HR and the BBU Pool, with high data transmission capacity and low time latency. The data link is made through different technologies such as fiber optics or radio links with protocols such as OBSAI, which use digitization of radio signals [1].

For the fronthaul of the CNT's 5G CRAN network, a transport network based on OTN/WDM technology is proposed [1]; multiple data signals share fiber optic resources based on Wavelength Division Multiplexing (WDM) [1], implementing functions such as port aggregation and transmission distance extension for IP/MPLS technology deployed on the CNT Core Network.

The suggested topology for the fronthaul segment is a highly concentrated ring, in which each station has a single input connection and an output connection. In order to obtain a backup of the entire fronthaul, it is proposed to implement a double ring. This way, the data is sent in both directions and redundancy is created with fault tolerance that will exist in the future.

Backhaul

The backhaul of the 5G C RAN network is the physical medium that allows the BBU5900 to be interconnected with the main base equipment for Cloud RAN (CB). This equipment will work as a data cloud (Cloud RAN) of the Core 5G network solution All-Cloud proposed by Huawei [11].

The CB works as a coordination server that receives problems, distributes them to the nodes, and adds and returns the results. This will be located in a centralized site allowing the control plane to be installed in a centralized data center and flexibly deploying the user map in accordance with service requirements [11].

The control plan of the proposed solution will be implemented in the Data Center that the CNT EP has in the city of Quito, which has high standards of quality, safety, availability, and redundancy, and has obtained Tier III certification and leadership in energy and environmental design (LEED). At this site the CB equipment will be installed, and will connect to the core equipment UIOMSCRB01 which is the one that will assign the necessary bandwidth capacity according to the number of BBU5900 and BS that we need to install. In turn, the CB is going to connect with the user plane to the BBU5900 through the backhaul network that has to be by optical fiber, because it is the only transmission medium that allows lower latencies and has higher speed and bandwidth capacity. The fiber optic backhaul allows the user plane to be installed on the edge of the CNT EP core network in order to minimize the traffic that has to travel through the network until reaching where the control plane is located in the CB [11] (Fig. 7).

3.2 Network Performance

To calculate the network performance, one must first obtain the probability distribution of the signal-to-noise and interference ratio (SINR) [13] for each cell. Then, the spectral efficiency for each possible SINR value is obtained based on the following formula:

$$\eta_{ISD} = \int \eta(SNIR) f(SINR) dSINR \quad (1)$$

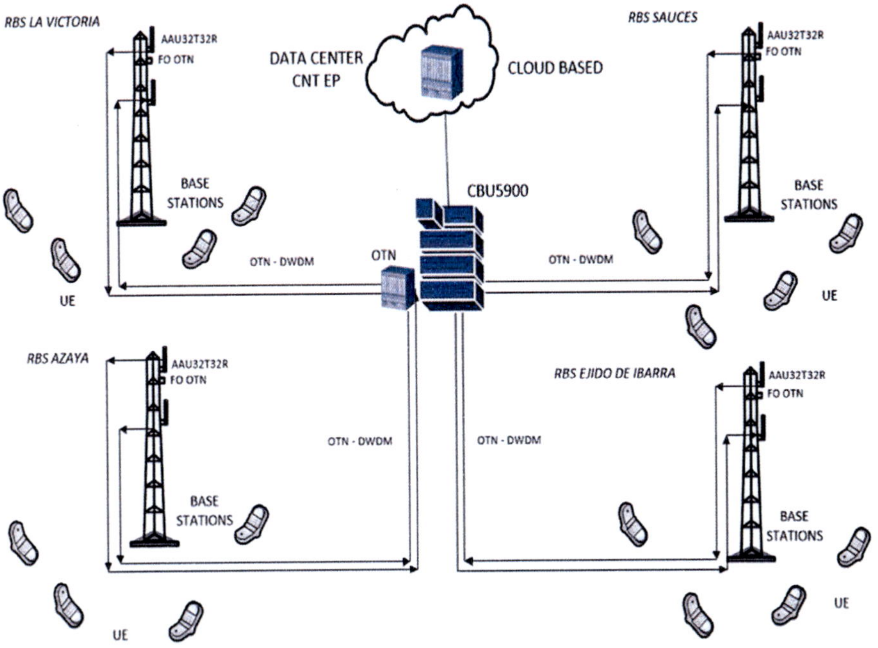

Fig. 7. Design of Huawei's new Core 5G All-Cloud network for the CNT EP.

where η_ISD is the average spectral efficiency (bps/Hz) for each cell.

The average throughput for each cell (in Mbps) is calculated according to the available bandwidth for each carrier frequency, as indicated in the following equation, and where factor 3 represents the sectoring of the cell (three sectors).

$$Throughput\frac{cell}{ISD} = 3\sum\nolimits_{f} \eta^{f}{}_{ISD}BW^{f} \tag{2}$$

Once all the possible spectrum has been integrated for each scenario, the unmet demand that is not satisfied through macro cells of three sectors, is routed through a layer of smaller cells that operates on the same spectrum. To estimate the number of small cells required, we consider a spectral efficiency similar to LTE (1.5 bps/Hz), an available bandwidth (100 MHz for non-sharing scenarios and 200 MHz for sharing scenarios), a download/75% upload, and a maximum coverage of 200m. The network parameters related to transmitted power, antenna height and propagation are used following the 3GPP technical recommendations issued for 5G [9–13].

3.3 Redundancy of the Network

The worst scenario proposed for the proposed design would be a fiber optic cut in the fronthaul of the network. This would cause the interruption in the mobile telephony and data services. For this reason, it is necessary that the macro cell is connected to the PBX by means of at least two alternative paths (double ring topology), if the channel or path by defect suffers a failure, the network configuration will automatically switch to the alternative path, all traffic between BBU and RRH [8] (Fig. 8).

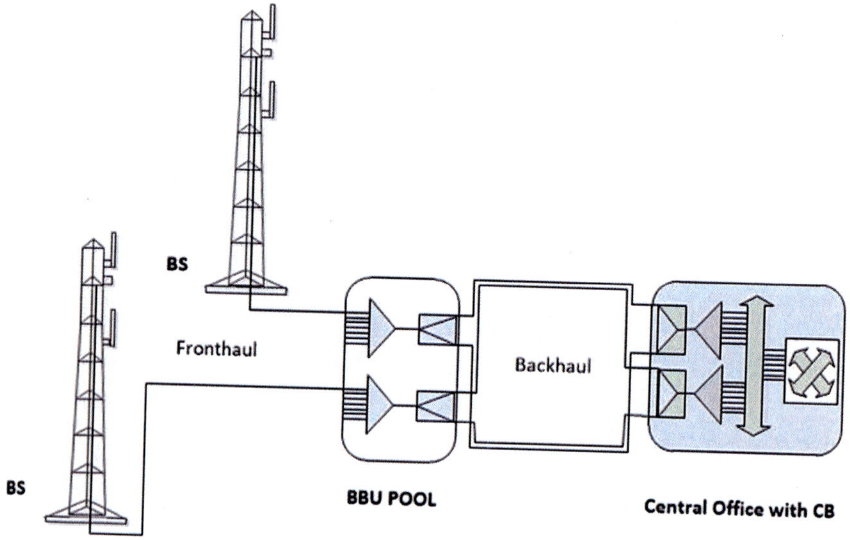

Fig. 8. Redundancy of the Fronthaul WDM network

3.4 Supervision and Control

From the operational point of view of the network, an important factor to consider is the monitoring of the quality of the fiber channel. To meet this objective, a passive solution is proposed, combined with optical multiplexing so that a pilot lambda is available within the optical channel, and a reflector at the end of each RRH that functions as its pair. In this way, through an Optical Time Domain Reflectometer (OTDR), from the control panel we can analyze the state of the fronthaul fiber and specifically observe if it has suffered any failure or attenuation [8] (Fig. 9).

3.5 Energy Efficiency

When considering a large-scale mobile network, energy consumption is an important factor, not only from an environmental point of view, but also from the perspective of network operating costs. Currently, more than 0.5% of the world's total energy is consumed by mobile networks [14].

A reduction in energy consumption is one of the main approaches of the 5G CRAN architecture. It is known that 5G networks will be much denser than 4G networks (larger number of RRHs), given the reduction in cell size, making it easier to foresee higher energy consumption. However, most of the energy consumption is related to the transmission powers, due to which, it is imperative to have virtualized and concentrated all

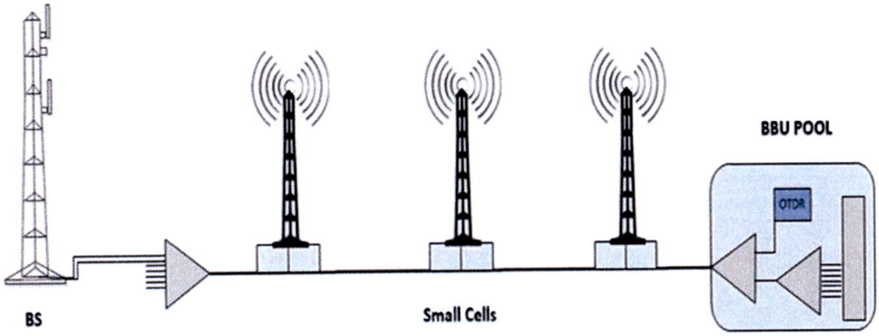

Fig. 9. Monitoring and control of the WDM Fronthaul network

the infrastructure of BBUs in the CRAN, reaching not only not to increase the energy consumption, but also improving the network efficiency [15].

A network saturated by reducing cell size has unavoidable network power requirements. As the network will have a larger number of smaller cells, the main component of energy consumption will transmit power, for it proposes a framework 5G candidate MAC software defined functional network virtualization with CRAN [16].

4 Conclusions

- The current requirements of mobile users make the evolution towards 5G networks imperative. 5G networks open up a new world of possibilities for most companies and organizations, where speed will not be a limitation and the communication needs of billions of devices can be met, with an equitable relationship between speed, latency, and cost.
- This paper describes the process to follow for the migration from a 4G LTE network to a 5G CRAN network, considering the reuse and optimization of the existing transport infrastructure.
- The CRAN or Cloud RAN is the core of the architecture described in this document. It groups the baseband resources in an infrastructure deployed on a communication cloud (private and/or public), maintaining the capabilities of a RAN implemented on a physical server, but obtaining the advantages of cloud applications, such as scalability, flexibility, ease of administration and maintenance, decrease in operating costs, among others.
- It is important to mention that the 5G technology implementation process for the National Telecommunications Corporation of Ecuador - CNT is already underway, and the proposal described in this paper is one of the possible options to follow.
- Finally, the importance of the timely deployment of 5G networks within the productive context of Ecuador is again emphasized, in order to meet the goals set by the State's industrial and development policy in its "Toda una Vida" plan, and to reduce problems related to the digital gap.

References

1. Asif, S.Z.: 5G Mobile Communications Concepts and Technologies. Taylor & Francis Group, Boca Raton, pp. 2742–33487 (2019)
2. Peng, M., Wang, C., Lau, V., Poor, V.: Fronthaul-constrained cloud radio access networks: insights and challenges. communications magazine. IEEE, 1536–1284 (2015)
3. CNT EP- Huawei - Site Survey – Ecuador - Imbabura – Ibarra. https://corporativo.cnt.gob.ec/intranet-4g/. Accessed 20 Apr 2020
4. HUAWEI TECHNOLOGIES eLTE DBS3900 V300R005C00 Product Overview (3GPP). https://e.huawei.com/es/material/wireless/mccs/3c65df00d7974b38be613c9281aae4f5. Accessed 20 Jan 2020
5. HUAWEI. https://www.huawei.com/ucmf/groups/public/documents/attachments/a19451811-pdf. Accessed 23Apr 2020
6. Tzanakaki, A., Anastasopoulos, M., Simeonidou, D.: Converged optical, wireless, and data center network infrastructures for 5G services. J. Opt. Commun, Networking **11**(2), A111 (2018)
7. Liu, X.: Initial study on the architecture of field observation in 5G era. In: IEEE 5G World Forum, 5GWF 2018 - Conference Proceedings, vol. 1, pp. 524–527 (2018).
8. Fiorani, M., Tombaz, S., Martensson, J., Skubic, B., Wosinska, L., Monti, P.: Energy performance of C-RAN with 5G-NX radio networks and optical transport. In: 2016 IEEE International Conference on Communications (ICC 2016), pp. 1–6 (2016)
9. Conectronica. https://www.conectronica.com/wireless/redes-wireless/fronthaulbackhaul-movil. Accessed 21 Dec 2019
10. Agencia de Regulación y Control de las Telecomunicaciones. Boletín Estadístico, vol. 1 (2019)
11. HUAWEI. https://www.huawei.com/es/press-events/news/es/2018/huawei-presenta-portfolio-completo-soluciones-5g-primer-estandar-3gpp. Accessed 01 Apr 2020
12. HUAWEI. https://www.huawei.com/ilink/en/download/HW_414150. Accessed 14 Mar 2020
13. Hadush Hailu, D.: Integrated hybrid optical networking for 5G access networks. Indonesian J. Electr. Eng. Comput. Science **17**(3), 47–54 (2020)
14. Checko, A., Christiansen, H.L., Yan, Y., Scolari, L., Kardaras, G., Berger, M.S., Dittmann, L.: Cloud RAN for mobile networks a technology overview. IEEE Commun. Surv. Tutorials **17**(1), 405–423 (2015)
15. Gomes, N., Chanclou, P., Turnbull, P., Magee, A., Jungnickel, V.: Fronthaul Evolution: from CPRI to Ethernet. Opt. Fiber Technol. **26**, 50–58 (2015)
16. 3rd Generation Partnership Project. https://www.3gpp.org/news-events/2117-ran_webinar. Accessed 19 Mar 2020

Evaluation of the Broadcast Storm Problem Based on Hybrid Retransmissions Algorithms in FANET Networks

Andrés Sánchez, Patricia Ludeña-González(✉), Francisco Sandoval, and Katty Rohoden

Departamento de Ciencias de la Computación y Electrónica, Universidad Técnica Particular de Loja, Loja, San Cayetano Alto s/n, 11-01-608, Ecuador
{adsanchez2,pjludena,fasandoval,karohoden}@utpl.edu.ec
http://www.utpl.edu.ec

Abstract. A flying ad hoc network (FANET) is a mobile ad hoc network (MANET) which comprises a group of flying nodes, unmanned aerial vehicle (UAVs), that communicate with each other. These networks are mainly used in emergency scenarios due to easy deployment and operation. In these networks, the method used to transmit packets is known as flooding. This method makes the nodes retransmit the received packet to all neighbors' nodes, causing three issues in the network: redundant retransmissions, contentions, and collisions. These issues constitute the broadcast storm problem (BSP) which causes increased latency, packet loss, and unnecessary power consumption by the nodes. Some basic packet retransmission schemes have been proposed to mitigate these drawbacks such as applying probabilistic retransmissions, packet counter and distance between the sending and receiving node. Based on these techniques, hybrid algorithms have been analyzed to improve the network efficiency. The present work studies the performance of hybrid packet retransmission algorithms to reduce the BSP problem in FANET networks by comparing with the traditional operation. The results show the improvement of reachability, the reduction of retransmissions, and the lowering of power consumption.

Keywords: Broadcast storm problem · FANET · Flooding · MANET · UAV communications · Wireless ad hoc communication.

1 Introduction

The technological advance within the sensor networks, electronics and wireless communications has made it possible to have unmanned aerial vehicles (UAVs) systems that incorporate sensors and telecommunications systems according to their application. A FANET (flying ad hoc network) is a subtype of MANET

M. Botto-Tobar et al. (Eds.): ICCIS 2020, AISC 1273, pp. 295–306, 2021.
https://doi.org/10.1007/978-3-030-59194-6_25

network (mobile ad hoc network) which is made up of UAV systems that communicate with each other and can operate autonomously or remotely [1]. Given the easy deployment and installation, versatility and flexibility of these networks, the application scenarios can go from civil to military, carrying out missions such as: environmental variables monitoring, natural disasters monitoring, detection and tracking of objects on the ground, search and pursuit of objectives, etc. [2,6,7,9,10,13].

These networks do not use packet router devices, e.g., access points or base stations. The network nodes are responsible for carrying the packets from the sending node to the receiver [4]. When the packets must pass through the intermediate nodes that are between the sending and receiving node, the method of packet transmission used is known as flooding [8]. The flooding method forces the nodes to relay the received packet to all their neighbors' nodes causing three main problems in the network. First, a node can be found within the coverage area of several nodes, therefore, many of the retransmissions that reach the node will be redundant and there would be unnecessary waste of energy. Second, due to the nodes proximity, intense contentions can occur between the nodes when trying to retransmit the packet. Third, as a result of the contentions, packet collisions can occur causing loss of information in the network. As a result, these problems may lead to the so-called Broadcast Storm Problem [15].

In this paper, some of the basic schemes and hybrid packet retransmission algorithms proposed to reduce the effects of BSP in MANET networks are adapted to FANET networks in order to analyze their performance in terms of reachability, saved rebroadcast, and saved power consumption metrics. The techniques are evaluated in low and high density scenarios.

The rest of the paper is organized as follows. Section 2 presents the BSP concept and the basic schemes that have been proposed to reduce its effects. In Sect. 3, the hybrid packet retransmission algorithms are evaluated. The proposed scenarios for evaluating the performance of the algorithms and the metrics used for the analysis are described in Sect. 4. The results of the simulations are presented in Sect. 5 while conclusions are set out in Sect. 6.

2 Background

2.1 Broadcast Storm Problem

As stated before, FANET networks do not use packet router devices, thus, when a node requires to send a packet to another node, two cases arise. In the first case, the receiver node is located inside the coverage area of the sender node, thus, communication is realized in a direct manner. The second case considers a multi-hop scenario where the receiver node is not located within the coverage area of the sender node. In this case, the packet generated by the sender node is broadcast by all the nodes that are between the receiver and the sender [8].

The flooding method is used to relay packets in FANET networks. This implies a number of retransmissions equal to the number of nodes in the network. If the nodes are able to identify duplicate packets (every packet is associated with an ID, a counter, etc.), they can prevent infinite floods and thus avoid all network failure [15]. The flooding method causes the appearance of three phenomena in the network:

- **Redundant retransmissions:** It is produced when multiple nodes share the same coverage area causing to replicate several times the retransmitted packet. Therefore, the nodes will receive the packet more than once.
- **Contention:** It occurs after that neighboring nodes receive the packet from the sender node. The nearest nodes will generate interference among them when transmitting.
- **Collisions:** They are the result of the contentions caused in the packets retransmission between nearby nodes. When a collision is produced, the sender node is forced to retransmit the packet again. As a result, more redundant retransmissions are generated.

The three phenomena are known as Broadcast Storm Problem (BSP) [15]. BSP generates congestion due to the excessive amount of packets circulating in the network, increases the latency and the packet loss, makes inefficient use of the bandwidth. Furthermore, it causes an unnecessary waste of energy when retransmitting redundant packets which limit the availability of one of the most important resources in FANET networks.

2.2 Basic Packet Retransmissions Schemes

One way to mitigate the BSP is to prohibit that some nodes retransmit, in order to reduce the amount of redundant retransmissions, contentions and collisions that may be generated. Some techniques that limit the retransmission of the nodes are presented in [15]. We have selected three of them for implementation in this work. These schemes require selecting a parameter of the network and defining threshold values of the network to decide if the node should retransmit.

Probabilistic Scheme. The easiest way to reduce retransmissions is to add a probability value to the packet that must be retransmitted. When the node receives a retransmission packet for the first time, the node will retransmit the packet with a probability P, e.g. $P = 1$ meaning that flooding is taken place. Additionally, a threshold probability value is selected which will serve as a reference to determine the retransmission. After receiving a retransmission packet, the node assigns the value of P to the packet and it is compared with the threshold value. If P is less than the threshold probability, the node retransmits the message. Otherwise, the node is prohibited from retransmitting the message [15].

Counter-Based Scheme. Conter-based technique [15] implements a C packet counter in order to keep track of the number of times the broadcast packet was received. Besides, a threshold counter value is selected as the basis for determining the retransmission of the packet. When the node receives a broadcast packet for the first time, it is retransmitted and the counter is initialized at $C = 1$. If the message has already been received before, the value of counter C is increased by one. If C is less than the value of the threshold counter, the packet is retransmitted, otherwise the node is prohibited from doing so.

Distance-Based Scheme. In this scheme, the distance between the nodes is used as a metric to determine if the packet should be retransmitted. Once a broadcast packet is received for the first time, the node initializes the d variable with the distance between the sending and receiving node, and retransmits the packet. If the message has been received previously, d is updated with the new distance value between the nodes. If d is less than the threshold distance the packet is retransmitted, otherwise the node is prohibited from retransmitting it [15].

3 Hybrid Algorithms to Rebroadcast

In the reviewed literature, hybrid algorithms based on basic schemes to broadcast have been proposed. These algorithms use the best characteristics of individual schemes in order to improve the performance.

3.1 Probability Plus Counter

Fixed Probability-Based Algorithm. The Fixed Probability-based scheme [11] aims to improve the performance of the counter-based scheme by adding a fixed probability value (probabilistic scheme) to the retransmission decision process. The process of this algorithm is similar to the counter-based scheme when receiving a package for the first time. If the message has been previously received, the value of C is increased by one. If C is greater than the threshold counter value, the node is prohibited from retransmitting the packet. Otherwise, the node assigns the probability value P to the packet. If P is less than the threshold probability value the packet is retransmitted, otherwise the node is prohibited from doing so. The algorithm is described in Fig. 1(a).

Dynamic Probability-Based Algorithm. Defining arbitrarily the value of threshold probability for packet retransmission may not be optimal globally. The algorithm performance can be affected by other network parameters such as topology, simulation area, nodes transmission range, nodes quantity, etc. In this work, we propose to use such parameters to calculate the optimal probability value for any type of scenario. The threshold probability value (P_{th}) is calculated as in [11]

$$P_{\text{th}} = \exp\left(-\frac{C}{\lambda}\right), \tag{1}$$

where C is the current value of the packet counter and λ is equal to

$$\lambda = (N - 1) \times \frac{\pi R^2}{A}. \tag{2}$$

Here, N is the total quantity nodes, R is the nodes transmission range, and A is the simulation area. Upon receiving a packet for the first time, the algorithm operates as the Fixed Probability-based scheme, however it uses the new threshold probability, P_{th}. The algorithm is detailed in Fig. 1(a) according to [11].

3.2 Probability Plus Distance

It improves the distance-based scheme, adding the concept of the retransmission probabilistic scheme. The rebroadcast scheme is based on distributed vehicular broadcast (DV-cast) [14]. This algorithm has two variants:

DV-cast-fixed. Tonguz et al. [14] describe that in fixed probability DV-cast scheme, when a packet is received for the first time, d is initialized with the value of the distance between the sending and receiving node. If the package has been previously received, the value of d is updated and compared with the threshold distance value. If d is less than the threshold distance value, the node is prohibited from retransmitting the packet. Otherwise, the probability value P is assigned to the package. If P is less than the threshold probability value, the packet is retransmitted. Otherwise the node is prohibited from doing so. The algorithm is detailed in Fig. 1(b).

DV-cast-dynamic. As mentioned before, defining a value for threshold probability can be not optimal for multiple scenarios. Thus, we propose to calculate the value of threshold probability based on the network parameters such as in [14]:

$$P_{\text{th}} = \frac{D_{ij}}{R} \tag{3}$$

where D_{ij} is the distance between the sending node and the receiving node, and R is the nodes transmission range. Thus, the algorithm operates as the DV-cast-fixed, but with dynamic probability. The algorithm is detailed in Fig. 1(b) according to [14].

4 Scenarios and Methods

4.1 Scenarios

In order to evaluate the performance in different scenarios, this work considers high density scenarios (HDS) with areas 1×1, 3×3 and 5×5 units, and low density scenarios (LDS) with areas 7×7, 9×9 and 11×11 units. Where a unit has a length equal to the transmission range of the nodes (500 m).

To simulate the nodes mobility, the random walk algorithm is implemented. With this algorithm the nodes can choose their speed and direction autonomously and based on predefined ranges [3]. The velocity for each scenario is chosen randomly from the ranges showed in the Table 1. In [5] all mobility algorithms applicable to FANET network scenarios are detailed. Furthermore, the simulation parameters are shown in the Table 2.

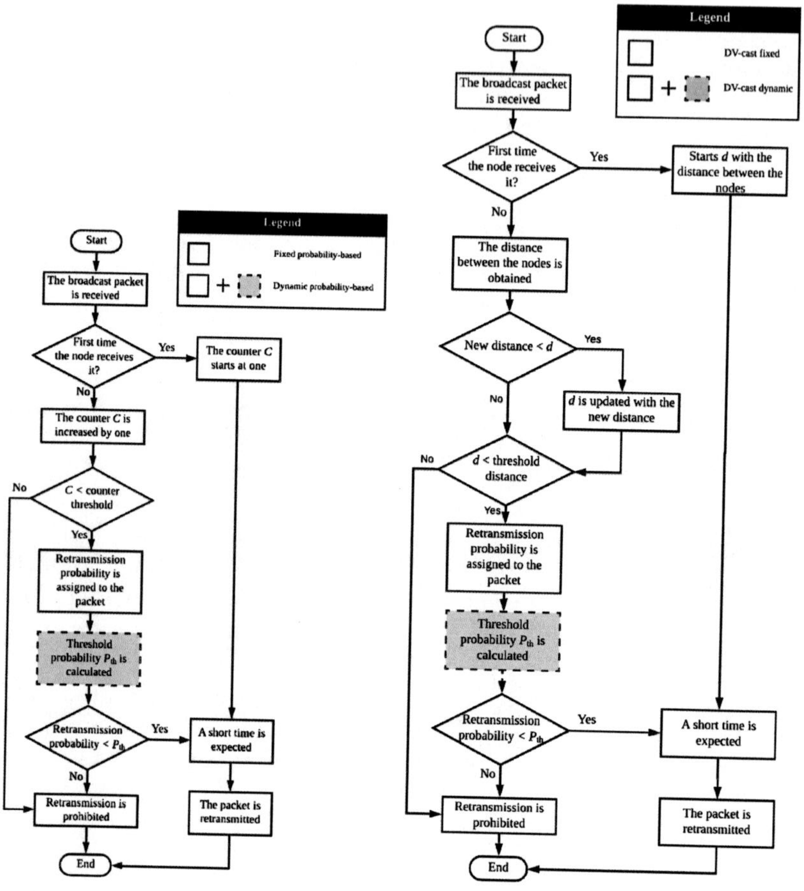

(a) Probability-based algorithm (b) DV-cast algorithm

Fig. 1. Flowchart of the fixed and dynamic algorithms.

4.2 Metrics

In order to analyze the performance of rebroadcast schemes, three metrics are used: reachability, saved rebroadcast and saved power consumption.

Reachability (RE). It is the amount of nodes that received the packet, at least once, divided by the total number of nodes that are reachable. If n represents the nodes that received the packet at least once and N is the total quantity nodes, the reachability is computed by

$$\mathrm{RE} = 100 \times \frac{n}{N}. \tag{4}$$

Table 1. Scenarios and speed range

Type	Scenario	Speed range
HDS	1×1	0 km/h to 10 km/h
	3×3	0 km/h to 30 km/h
	5×5	0 km/h to 50 km/h
LDS	7×7	0 km/h to 70 km/h
	9×9	0 km/h to 90 km/h
	11×11	0 km/h to 110 km/h

Saved Rebroadcast (SRB). SRB is the percentage of nodes that have received the message but have not retransmitted it. Based on [15], it is given by

$$\text{SRB} = 100 \times \frac{n - T}{n}. \tag{5}$$

where n is the nodes that received the packet and T is the nodes that actually retransmitted the packet.

Saved Power Consumption. It is the amount of save power (dB mW) in the network. It is computed through the path propagation evaluation [12]. First, it computes reception power (P_{rx}) with link margen ($M = 10\,\text{dB}$) and reception threshold (U_{rx}) (Eq. 6).

$$P_{\text{rx}} = M + U_{\text{rx}}. \tag{6}$$

Then the transmitted power (P_{tx}) can be obtained with the link balance equation (Eq. 7).

$$P_{\text{tx}} = P_{\text{rx}} + L_{\text{tt}} + L_{\text{tr}} - G_{\text{tx}} - G_{\text{rx}} + L_b. \tag{7}$$

where L_{tt} and L_{tr} are line/connector losses in transmission and reception, respectively. G_{tx} and G_{rx} are antenna gain in transmission and reception, respectively. The link losses L_b for specific frequency (f) is computed by

$$L_b = 32.45 + 20\log(f) + 20\log(d). \tag{8}$$

When a node starts transmitting a broadcast packet, the area around it is divided into three sections. All nodes that are in the first section are called nearby nodes. The nodes that are between the second and the third section are called intermediate nodes. The rest of nodes are called distant nodes. Then, the energy consumption in each section is calculated.

5 Results

In this section, we analyze the performance of basic and hybrid schemes in terms of reachability (RE), saved rebroadcast (SRB), and saved power consumption.

Table 2. Simulation parameters

Parameters	Values
Number of nodes	100
Transmission range	500 m
Initial distribution	Random
Mobility algorithm	Random walk
Threshold probability	0.2, 0.4, 0.6, 0.8, 1
Threshold counter	2, 3, 4, 5, 6
Threshold distance	11 m, 20 m, 37 m, 72 m, 147 m
Reception threshold	–110 dB[a]
Antenna gain	2 dBi[b]
Operating frequency	2.45 MHz[b]

[a]DJI Drone Inspire 2 (https://www.dji.com/inspire-2/info#specs)
[b]DJI Lightbridge 2 (https://www.dji.com/lightbridge-2/info#specs)

5.1 Basic Schemes

Probabilistic Scheme. The results of the probabilistic scheme are depicted in Fig. 2(a). This scheme can achieve between 15% and 75% retransmission savings for HDS, and between 15% and 55% for LDS. In addition, for HDS, the reachability and the saved rebroadcast for small p values (0.2 to 0.6) are greater than 90% and 75%, respectively. LDS needs higher probability values (close to 1), in order to achieve reachability greater than 50%.

Counter-Based Scheme. In this section, the counter-based scheme results are shown in Fig. 2(b). This scheme keeps constant reachability levels. However, for HDS scenarios, the percentages of saved rebroadcast are the highest (between 70% and 95%). Since the nodes are dispersed to occupy the entire simulation area and the number of neighbor nodes is reduced, the saved retranmissions levels decreases dramatically.

Distance-Based Scheme. The percentages of saved retransmissions obtained are the lowest of all the basic schemes. If the distance is very small it acts as flooding, thus, only higher threshold distance values are analyzed in Fig. 2(c). For higher distance values (400 m to 500 m) the reachability decreases dramatically. Since, only the farthest nodes are allowed to transmit the packets, the amount of retransmissions is reduced. For HDS scenarios, the saved retransmissions rates are between 10% and 70%, while for LDS scenarios are between 5% and 40%.

5.2 Hybrid Schemes

Probability Plus Counter-Based Scheme. The results of fixed probability-based algorithm are presented in Fig. 3(a). It shows that this algorithm achieves saved retransmissions between 50% and 95% for HDS scenarios and between

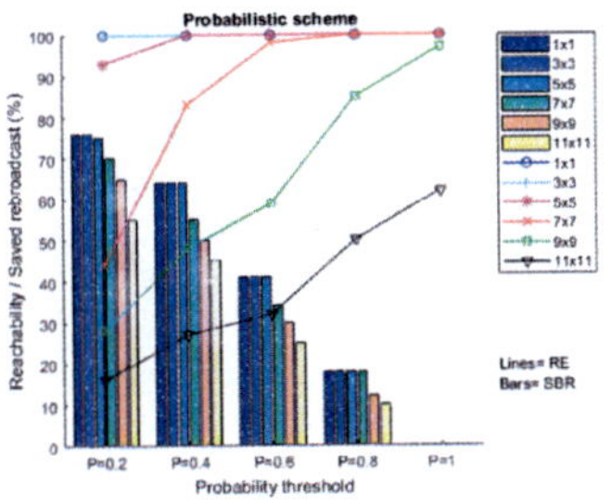

(a) Probabilistic scheme

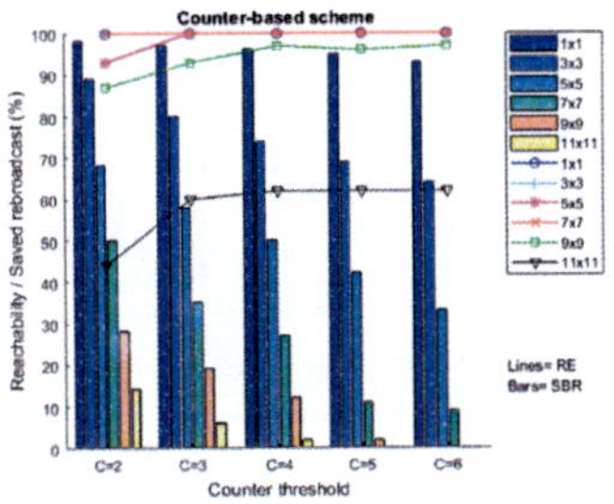

(b) Counter-based scheme

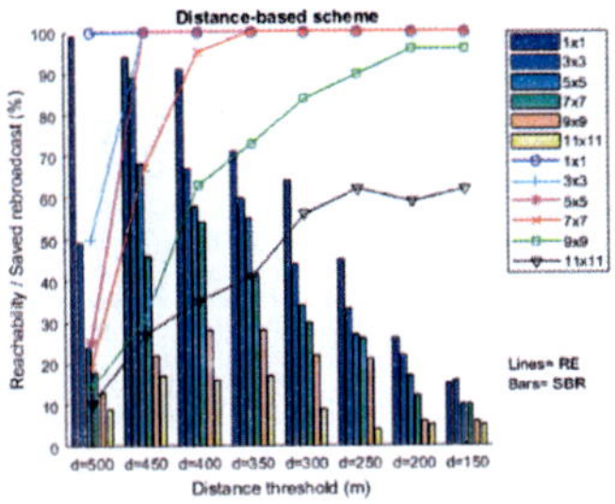

(c) Distance-based scheme

Fig. 2. Reachability (RE) and save rebroadcast (SBR) results with basic schemes.

15% and 50% for LDS scenarios. The results of the dynamic probability-based algorithm are presented in Fig. 3(b). It achieves reachability between 95% and 100% and percentages of saved retransmissions between 70% and 97% in HDS scenarios. In LDS scenarios, the reachability is smaller.

Probability Plus Distance-Based Scheme. With this algorithm, percentages of saved retransmissions between 50% and 85% are obtained for HDS scenarios, while for LDS scenarios between 20% and 50%, as can be seen in Fig. 3(c). The results obtained from the implementation of dynamic DV-cast algorithm are presented in Fig. 3(d). Constant reachability is achieved up to a threshold distance of 400 m. The percentages of saved retransmissions for HDS scenarios are between 50% and 90%, and for LDS scenarios between 20% and 50%. Thus, this algorithm has a better performance compared to the fixed DV-cast algorithm and the distance-based scheme. The Fig. 4 shows the relationship between reachability and saved rebroadcast for different threshold values. For HDS scenarios, the dynamic probability-based algorithm allows to obtain the highest reachability and saved retransmissions, as shown in Fig. 4(a), 4(b), and 4(c). For LDS scenarios, the DV-cast algorithm allows to obtain high reachability but low saved retransmission, according to the Fig. 4(d), 4(e), and 4(f).

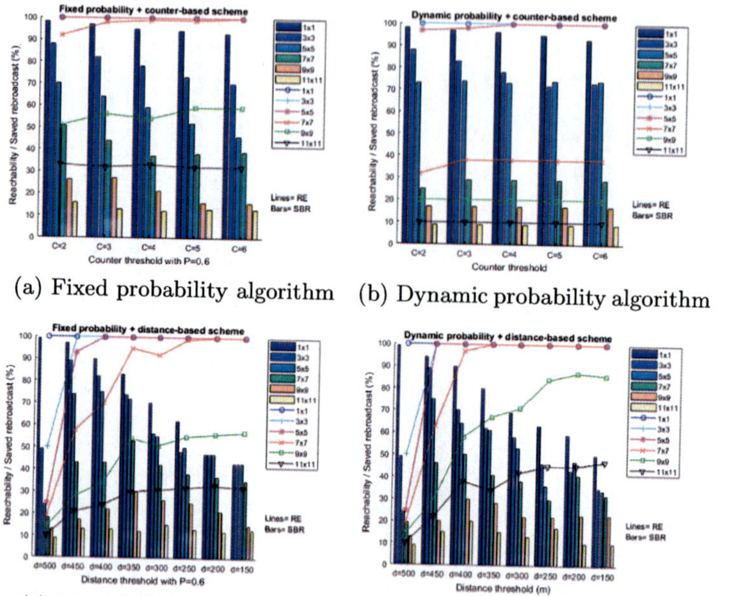

(a) Fixed probability algorithm (b) Dynamic probability algorithm

(c) Fixed DV-Cast algorithm (d) Dynamic DV-cast algorithm

Fig. 3. Reachability (RE) and save rebroadcast (SBR) results with hybrid schemes.

5.3 Power Consumption

The Fig. 5 shows the power consumption for each broadcast packet sent. In the case of flooding, the network consumes about 50 mW for each packet that is transmitted. With the implementation of the dynamic DV-cast algorithm, the energy consumption is reduced by 30%. On the other hand, when implementing the dynamic probability-based algorithm, the power consumption is reduced by 70%. Thus, the algorithm with better performs in reachability, saved retransmissions, and energy saving is the dynamic probability-based algorithm.

6 Conclusion

In this work, rebroadcast schemes to reduce the BSP effects are applied to FANET networks. On MATLAB framework a several scenarios were deployed testing basic and hybrid algorithms. The results show that, with basic schemes, redundant retransmission are reduced between 10% and 90% by implementing the counter-based scheme. The dynamic probability-based algorithm has a higher savings retransmissions, with constant reachability for the different threshold values analyzed, and the power consumption of the network is reduced in 70%.

Finally, the results show that the best scheme to avoid BSP is the probability dynamic scheme due to its stable reachability. In addition, the saved retransmissions are between 70% and 90% while reducing power consumption by 70%.

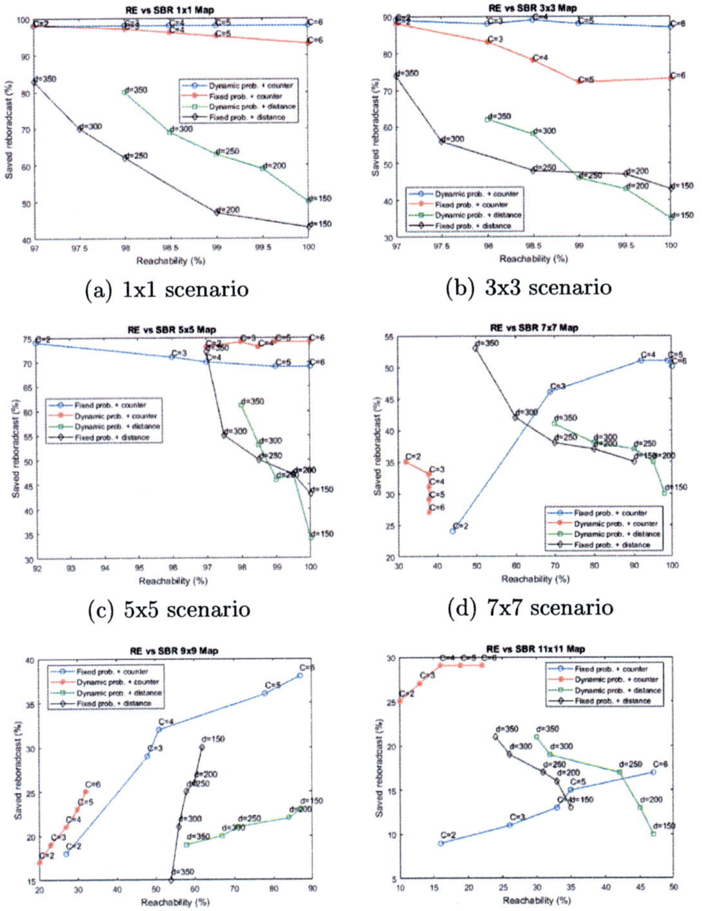

(a) 1x1 scenario (b) 3x3 scenario

(c) 5x5 scenario (d) 7x7 scenario

(e) 9x9 scenario (f) 11x11 scenario

Fig. 4. Hybrid algorithms performance comparison, reachability (RE) versus saved rebroadcast (SBR).

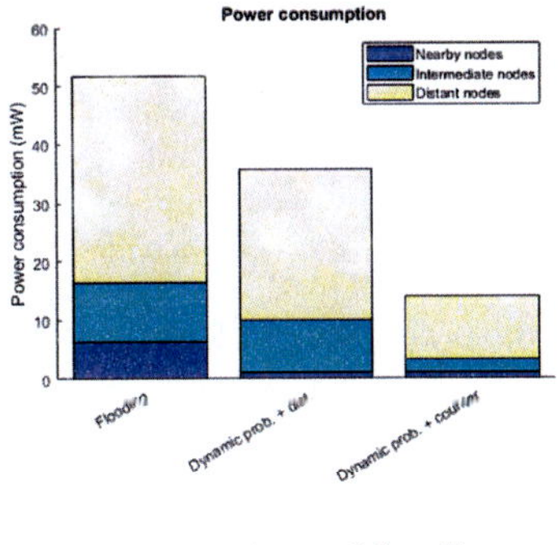

Fig. 5. Power consumption comparison of flooding and hybrid algorithms.

This work can provide the basis for future research on the evaluation of other rebroadcast schemes such as algorithms based on topology and based on genetic algorithms.

References

1. Bekmezci, I., Sahingoz, O.K., Temel, Ş.: Flying ad-hoc networks (fanets): a survey. Ad Hoc Netw. **11**(3), 1254–1270 (2013)
2. Bethke, B., How, J., Vian, J.: Multi-uav persistent surveillance with communication constraints and health mangement. In: AIAA Guidance, Navigation, and Control Conference, p. 5654 (2009)
3. Bujari, A., Calafate, C.T., Cano, J.C., Manzoni, P., Palazzi, C.E., Ronzani, D.: Flying ad-hoc network application scenarios and mobility models. Int. J. Distrib. Sens. Netw. **13**(10), 1550147717738192 (2017)
4. Guillen-Perez, A., Cano, M.D.: Flying ad hoc networks: a new domain for network communications. Sensors **18**(10), 3571 (2018)
5. Kumari, K., Sah, B., Maakar, S.: A survey: different mobility model for fanet. Int. J. Adv. Res. Comput. Sci. Softw. Eng. **5**(6) (2015)
6. Manathara, J.G., Sujit, P., Beard, R.W.: Multiple uav coalitions for a search and prosecute mission. J. Intell. Robot. Syst. **62**(1), 125–158 (2011)
7. Maza, I., Caballero, F., Capitán, J., Martínez-de Dios, J.R., Ollero, A.: Experimental results in multi-uav coordination for disaster management and civil security applications. J. Intell. Robot. Syst. **61**(1–4), 563–585 (2011)
8. de Melo Pires, R., Arnosti, S.Z., Pinto, A.S.R., Branco, K.R.: Experimenting broadcast storm mitigation techniques in fanets. In: 2016 49th Hawaii International Conference on System Sciences (HICSS), pp. 5868–5877. IEEE (2016)
9. Merino, L., Caballero, F., Martínez-De-Dios, J.R., Maza, I., Ollero, A.: An unmanned aircraft system for automatic forest fire monitoring and measurement. J. Intell. Robot. Syst. **65**(1–4), 533–548 (2012)
10. Olsson, P.M., Kvarnström, J., Doherty, P., Burdakov, O., Holmberg, K.: Generating uav communication networks for monitoring and surveillance. In: 2010 11th International Conference on Control Automation Robotics & Vision, pp. 1070–1077. IEEE (2010)
11. Pires, R.D.M., Pinto, A.S.R., Branco, K.R.L.J.C.: The broadcast storm problem in fanets and the dynamic neighborhood-based algorithm as a countermeasure. IEEE Access **7**, 59,737–59,757 (2019)
12. Rábanos, J.M.H., Tomas, L.M., Salis, J.M.R.: Transmisión por radio. Editorial Universitaria Ramón Areces (2013)
13. Scherer, J., Yahyanejad, S., Hayat, S., Yanmaz, E., Andre, T., Khan, A., Vukadinovic, V., Bettstetter, C., Hellwagner, H., Rinner, B.: An autonomous multi-uav system for search and rescue. In: Proceedings of the First Workshop on Micro Aerial Vehicle Networks, Systems, and Applications for Civilian Use, pp. 33–38. ACM (2015)
14. Tonguz, O.K., Wisitpongphan, N., Bai, F.: Dv-cast: A distributed vehicular broadcast protocol for vehicular ad hoc networks. IEEE Wireless Communications **17**(2), 47–57 (2010)
15. Tseng, Y.C., Ni, S.Y., Chen, Y.S., Sheu, J.P.: The broadcast storm problem in a mobile ad hoc network. Wireless Netw. **8**(2–3), 153–167 (2002)

Database System and Application

Proposal of a Framework for Information Migration from Legacy Applications in Solidarity Financial Sector Entities

Marcos Guerrero(✉), Marco Segura, and José Lucio

Facultad de Ingeniería de Sistemas, Escuela Politécnica Nacional, Ladrón de Guevara E11-25 y Andalucía, Quito, Ecuador
{marcos.guerrero,marco.segura,jose.lucio}@epn.edu.ec

Abstract. Software development tools evolve rapidly, which allows the design and implementation of new software applications that optimize every time the automation of processes required in any type of business sectors. New applications replace legacy systems including also the upgrade of operating systems and infrastructure. However, information is a resource that cannot be replaced and requires the cleaning, improvement and movement from the legacy system to the new target application; this process is called information migration. In literature, migration focused on infrastructure and information is limited only to the creation of tools for Extraction-Transform-Load, which in most of the cases is enough. In developing countries, Solidarity Financial Sector Entities are entities that grow quickly and often require the migration of software systems and therefore their corresponding information. In this context, we propose a framework called Financial Migration Framework that exposes guidelines for the process of information migration that protects the legacy information and ensures the quality of information for correct operation of the target application. For validation purposes, the framework was applied in two Entities, allowing the evaluation of the financial information clients' privacy and the process duration. To visualize these results, surveys were applied to the stakeholders who participated in each activity, and the results indicate for both Entities the quality of the migrated information allows the correct operation of the new target system, even though the process duration did not meet their expectations.

Keywords: Financial data migration · Secure legacy data migration · Migration Framework

1 Introduction

The evolution in the use of software products in all industries is inevitable and especially in the financial environment where all processes are related to the economic information of their customers [1]. The new software products enable administration of this sensitive information and consequently the timely fulfillment of the requirements established by

M. Botto-Tobar et al. (Eds.): ICCIS 2020, AISC 1273, pp. 309–320, 2021.
https://doi.org/10.1007/978-3-030-59194-6_26

the control entities [2]. The updating of tools requires the transfer of information from the legacy system to the new information manager, this process is called information migration [3]. Additional challenges facing data migration (collection, lack integrity, absence, etc.), is added the generation of new information fulfills entities requirement [4].

Financial information possessed by Solidarity Financial Sector Entities (SFSE), evidently, requires more rigorous management. The data migration process is not a business process for financial institutions. Therefore, each time it is improvised according to the circumstances of each project, which can cause inconsistencies during and in the result of the migration [1]. The development of automatic ETL tools (Extraction-Transformation-Load) requires manipulating sensitive information such as account numbers, values, emails, movements, transfers, etc. [5]. The confidentiality of this information is more critical than in other types because an uncontrolled exposition of it would allow theft or any other type of misintended acts that affect the economy of customers, partners and financial institutions [3].

There are limited migration methodologies that suggest a general guideline for the process; however, these are limited to technological activities and neglect the management characteristics required for this type of information [6]. The "Chicken Little" methodology [7] consists of 11 steps that focus on progressive migration to the target system while both environments are operating simultaneously in production. On the other hand, the "Butterfly" methodology [8] suggests remove the legacy system immediately after the end of the migration. Butterfly consists of five steps that allow a complete transition between the legacy and target system.

1. Determine the semantics of the legacy system and develop the scheme in the objective.
2. Create a data repository on the target system with a sample of records.
3. Migrate all components, except data to the target system.
4. Gradually migrate data and train staff in the target system of migrated data.
5. Remove the legacy system and use only the target system.

The two mentioned methodologies were considered as a reference by other researchers who contributed with variations in the number of target systems (Emmrich, 2001) [9] and the use of resources for their execution (Ratib, 2003) [9].

On the other hand, the ISO/IEC 27001: 2013 standard is the current framework for information security management, applicable to organizations of any size and line of business [10]. Security aims to manage and maintain three fundamental characteristics of information, confidentiality, integrity and availability.

The main objective of this standard ISO/IEC 27001: 2013 is to provide the best information security practices in processes that involve people, processes and technology. Annex A of the standard presents the applicability matrix, which defines the control objectives and controls established for the certification of an entire institution or a single process [11].

Financial Migration Framework (FMF) tries to perform a migration process which result generates information required by control entities and protects their integrity and privacy used in this transition. Therefore, the contribution of the research is to create a framework for data migration between software systems of regulated financial institutions, so that these specific problems can be solved:

- Limitation for mass data extraction.
- Control of access to data by the actors involved in the process.
- Inconsistency or absence of financial or accounting information at legacy source.

FMF is generated by a combination of the best parts of available methodologies, information security standards and good practices acquired in the participation of research migration processes. Its validation was carried out with two SFSE in different situations. The different scenarios allow validating that the framework meets the purposes described in the problems to be solved. The validation is carried out with these guidelines:

- Staff Surveys measure the level of satisfaction of the process performance.
- Error rate between financial reports of the legacy application and information loaded in the target application.
- Number of load attempts in each structure in the first monthly period closing after the start-up of the data in the production environment.

2 Methodology

To evaluate the research hypothesis, FMF was applied in two financial institutions in similar situations. Despite the similarity of the evaluated SFSE, certain variables were identified during the execution of the process. Adjustments were made to FMF so that it is possible to standardize the resources and results of the processes.

2.1 Detail of Experiment Environment

In Ecuador, the SFSEs are categorized by the Monetary Board of Financial Regulation according to the value of the assets that are reported in their monthly balances. There are five levels called segments, segment 5 has the lowest asset value and segment 1 has an amount greater than $10 million as assets. When an SFSE improves its segment, its information to be delivered is recurrent and control regulations are more rigorous.

A problem has been found in the transition from segment 4 to 3 because the quality of information generated by promoted SFSE does not meet the requirements of control entities. The experiment was performed in two SFSE that are in segment 3. One was recently promoted from segment 4 to segment 3 and has not previously performed migration processes. The other one has its second migration process and its staff that carried it out are the same that will carry it out with FMF. Both scenarios are important for comparing FMF with another migration process and measuring their expectations of the process in both different cases. It is important to specify that the staff of both institutions is limited and can carry out more than one activity in the institution. For this reason, on several occasions the same person participates in FMF but with different roles. Table 1 shows the difference in the quantity and frequency in which the institutions deliver reports according to their segment.

Table 1. Structure of control for each segment

Structure of Control	Segment 1	2	3	4	5
B11 – Financial Statement (monthly or quarterly)	X	X	X	X	X
B13 – Financial Statement (diary)	X				
D01 – Deposit	X	X	X	X	X
C01 – Loans granted	X	X	X	X	X
C02 – Loans Balances	X	X	X	X	X
C03 – Loans Guarantees	X	X	X		
C04 – Titles of value	X	X	X		
C05 – Credit Cards granted	X				
C06 – Credit Cards consumes	X				
C07 – Credit Cards Provision	X				
I01– Investments and available funds granted	X	X	X		
I02 - Investments and available funds granted balance	X	X	X		
F01 – Financial services	X	X	X		
S01 – Customers	X	X	X	X	X

2.2 Framework Definition

Figure 1 shows the three typical environments that are identified in SFSE in a transition of software systems. Each environment requires actors and resources that must be used to maintain privacy and ensure the quality of data from source to destination (Fig. 2).

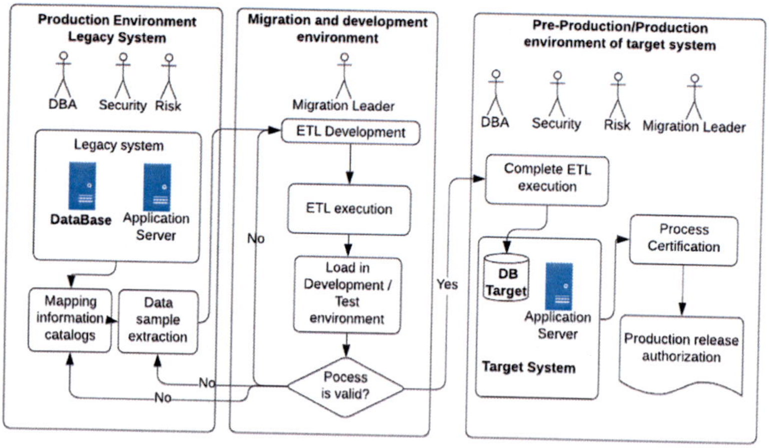

Fig. 1. Environments required for migration process

Depending on the circumstances of the SFSE, the production environment of the new application is also used as a certification or testing environment for a replacement software application. The framework is divided into 6 phases. They must be performed in chronological order, each defining details of the actors and conditions required for their proper accomplishment.

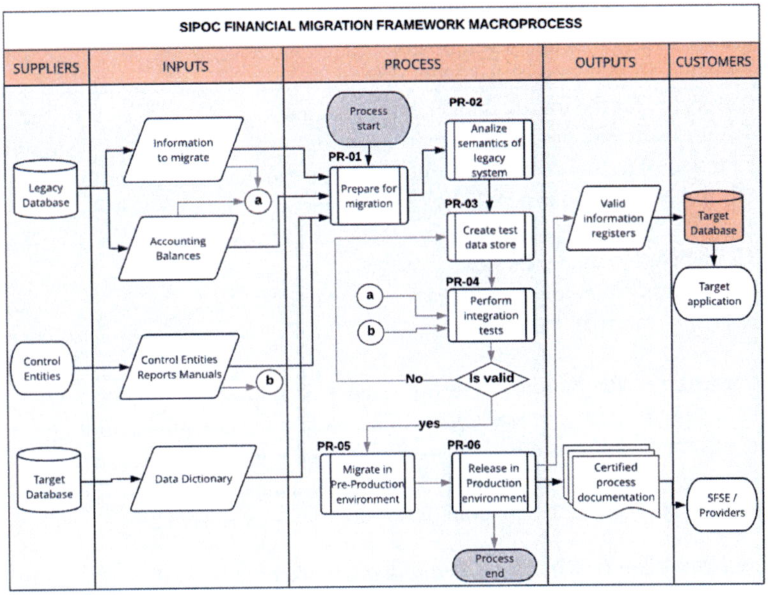

Fig. 2. FMF model process

Phase 1: Prepare for Migration. *Require database documentation from the legacy system.* The information owner, in this case SFSE, should demands the legacy database structure to the source provider. This is achieved through a data dictionary; the migration team needs to identify the table and the field where the data is stored to migrate in the target system.

Define the Way of Data Delivery. First, legacy data must load in neutral database, where the legacy data is controlled while it is transferred from the source. The second option is provided of templates, which requires a structuration based on the target system, the selection of the method depends of migration manager. The sample of information provided should be minimal so as not to expose the information privacy, but it should cover all the conditions of the information that could be found at the source [12]. It is a complicated task that must be carried out considering the business experience of department heads, risk management department and database administrators.

Define Data Extraction Managers. The SFSE must assign a manager to control the technical part of the process. For the information security, the administrator must provide information using the SFSE's ISMS manual.

Define Schedule for Loading Test Information, Parallels and Production Release. It is important to coordinate testing with the software application development and implementation team. The developers of the target application must test a customization of the software with the final structure of the database and a sample of migrated information. Parallels are a good practice in a migration process. It consists in simultaneous operation of the legacy system and the target system. Its objective is to evaluate the new system and find incidents related to the development or customization of the new application. It is recommended to do this on a day when the operational or transactional load is minimal so that users reflect the transactions in both systems.

The release for the production environment must necessarily take place on the first day of a month. This is because the control reports use information for the entire month. In conclusion, to define the dates, the following aspects should be considered:

- SFSE inoperative time (ATMs).
- Days with less operational load and transactionality.
- Delimit and prioritize the use of processes.

Phase 2: Analyze the Semantics of the Legacy System.
Analyze available and required information. Determine the information that is missing, the information that needs to be mapped, and the information that must be created to incorporate in the new database [13]. The new database already has the necessary schema to comply with the deliverables required by the control entities.

Develop Extraction-Transform-Load application. According to the structure of the target application and availability of data, the developers must create a tool that automates the transfer of information from the source to the destination database. This stage is the most critical in the entire migration process. Automating the transfer of information will allow the transition between the test and production environment not to stop the activities of the financial institution. In a migration project, customization of the target application is typically required. If customization involves changes in the database structure, it is very important to coordinate both development processes, so the migration considers the necessary adjustments in the ETL.

Phase 3: Create a Test Data Store. *Create an isolated and secure channel for information exchange.* The selected channel depends on the agreement between SFSE and the provider established in the service contracts. It is essential to follow the instructions of information security manual. It is suggested that a single person receives the information so that the control is focused on just one person. In the same way, it is recommended that the information be delivered in database tables in a financial institution environment. This option allows use software tools like firewalls or VPN to control access. In case of using the template option to avoid doing it by emails, use a private FTP server to control access to information.

Make Sure that the Information Management Agreement is Followed. The information security manual must consider all conditions for manipulating it. These manuals must be continuously updated depending on the circumstances of each migration project. There must always be a document, physical or digital, that supports each activity carried out with the information, including the names of those responsible for the SFSE and the providers.

Phase 4: Perform Integration Tests. *Determine the legacy system data sample.* The person responsible for the source information must present a sample of information to be used in the ETL. It should be considered that the less information used at this phase, the better its security is guaranteed. For this reason, it is important to coordinate with the leaders of each area to obtain a sample that contains the greatest number of variations in the information to be migrated.

Run ETL. The information delivered from the SFSE is used as input for the ETL application. In this step, ETL developers and application developers perform unit tests. This cyclical task depends on the result of the unit tests. Tests are done only with the technical staff of the migration team.

Review Migrated Data. After the ETL is confirmed to run correctly, the DBA should perform an inspection of the data inserted into the destination database. It is recommended that the DBA have a set of queries to confirm that the tables were correctly populated.

The development team also performs unit tests of the application with the migrated data. Test results could return the framework development cycle to phase 2 or later depending on the type of incident found.

ETL Delivery-Reception Agreement. When the tests reach the expected percentage of success, the providers must deliver the ETL to the SFSE. The ETL can be an application or a script depending on the source information to be migrated. The providers must deliver the ETL with all the necessary documentation so that the SFSE's staff can use the tool as many times as necessary in their internal tests.

Phase 5: Migrate in Pre-Production Environment. *Migrate all modules with ETL.* In this phase, the SFSE's staff has the custodian of the ETL. Security is more controllable when there are no third parties involved. The privacy of the information and its quality only correspond to staff of the institution. Therefore, at this point the necessary migration tests must be done with the complete information to be migrated. The information upload must be done in an environment like the production environment so that all the SFSE's staff can access the environment and validate the migrated data and its interaction in the target system.

Reviews with Users and Entity Control Validators. In this phase, the review is carried out with operatives from the institution. This review is the last one performed by the operatives to certify the requirements requested in the target application and consequently the migrated information.

Documentation of Completion of Migration Contract. In this last activity, the provider still has access to the information. Although the SFSE controls the ETL, the provider maintains its support for any incident from the time of delivery of the ETL. In this phase, it is considered that the SFSE has reached an expected percentage of compliance with migration tests. Besides, the areas involved are aware that the information complies with integrity from the legacy system. For this reason, at this phase, documents must be signed, guaranteeing that the SFSE is only responsible for its security and quality of the information that will be used in the production environment. It is important that the highest authority of the SFSE is aware of this agreement and for this reason, the provider may not be accused of any inappropriate handling of information or lack of quality.

Phase 6: Release in Production. *Stop Legacy System.* The legacy system and the satellite applications that depend on it are totally suspended. The information of the legacy system must remain enabled only for consultation. This process must be carried out on the first day of the month for the collection of information to send to the control entities.

Migrate all Modules with ETL to Production Environment. At this point of the process, the ETL and the target application must run correctly, and this must be documented. In the pre-production environment, the complete information to be migrated was validated. This allowed us to analyze the approximate time that it takes to perform. The staff in charge of the Accounting area are essential to certify that the migration process has been successfully completed. To determine that the migration was successful, the corresponding areas must compare the accounting and financial reports between the legacy system and the new target system. This comparison is the support for the areas involved in the process indicating that the migration process has been successfully completed.

Process Monitoring. All staff of the SFSE must act as if the migrated information was not correct. Each process performed on the target application must be carefully examined at least the first week of the system operation. Most incidents do not occur due to errors in the migration, but rather due to user or system parameterization errors. The error must be distinguished and assigned to the corresponding area and, if necessary, go to the migration staff to confirm the incident and find the solution.

3 Results

The questions in the surveys have been classified into quantitative and qualitative according to the type of response, the tabulation of the results was carried out independently for

Table 2. Quantitative results

Question	SFSE-1		SFSE-2	
	FMF	Others	FMF	Others
# internal participants	3	n/a	5	2
# external participants	1	n/a	1	1
# month migration duration	6	n/a	3	1
How many times was the data migrated in a test environment?	4	n/a	2	0
How many hours of downtime does the release take to the production environment?	1	n/a	2	92

each case. Table 2 shows the quantitative results, these questions were applied to the IT team who participated directly or indirectly in all the activities of the process. Questions find out about staff or technology resources used in FMF and other migration process. Table 3 shows the qualitative results that were applied to operational staff of the SFSE, who is involved in the reviews and documentation of the activities carried out by FMF. These questions compare the activities carried out in the migration processes. To measure the quality of migrated information, the number of loading attempts was tabulated before receiving correct confirmation from the control entities; this information is found in Table 4. In the result tables, the acronym SFSE-1 represents the institution that carried out a migration process for the first time and SFSE-2 corresponds to the institution whose staff previously participated in a migration process. Finally, Table 5 shows the combined result of two SFSEs about execution or results of FMF. The questions were applied to 15 persons in each institution including administrators. When the response has a percentage greater or equal to 75%, it is considered as an answer in favor of the use of FMF.

Table 3. Qualitative results

Question	SFSE-1		SFSE-2	
	FMF	Others	FMF	Others
Test environment location	SFSE	n/a	SFSE	SFSE/External
Migration type	Automatic	n/a	Automatic	Semiautomatic
Format used for share information	Templates	n/a	Controlled DB	Data Base
Channel used for share information	One Drive controlled by SFSE	n/a	Direct access	Direct access
Security standard applied	ISO/IEC 27001	n/a	ISO/IEC 27001	None
Do you have documentation of the process?	yes	n/a	yes	No
Was the Risk staff involved?	yes (TI)	n/a	Yes (CEO)	Yes (TI)
Was the target application tested with migrated information?	yes	n/a	yes	No
Did the provider suggest information security policies?	yes	n/a	yes	No
Does the migration process require ISMS for information management?	Yes (SFSE don't have)	n/a	Yes (SFSE don't have)	No
Did control entities monitor the process?	no	n/a	no	n/a

Table 4. Upload reports efficient

Report type	SFSE-1		SFSE-2	
	Attempts	%	Attempts	%
B11	1	100	1	100
D01	3	33	2	50
C01	1	100	1	100
C02	4	25	2	50
I01	1	100	1	100
I02	1	100	2	50

Table 5. Acceptance of FMF execution

Questions	Yes	No
Do you consider that the application of the framework brings benefit in the result of the migration?	100%	0%
Would you recommend the use of FMF for future migration processes?	85%	15%
Do you know other formal migration methodologies?	0%	0%
Do you consider that the information keep privacy after the process?	75%	25%
Do you think the FMF runtime is offset by the results?	90%	10%
Do you consider that this process requires staff from the SFSE?	80%	20%
Do you consider that control entities should monitor the migration process?	50%	50%
Do you consider that the security and quality of the information are ensured with FMF?	100%	0%
Do you consider that it is necessary to create the documentation of each activity in FMF?	100%	0%
Do you consider that only IT staff should be only responsible of process?	60%	40%
Do you consider that FMF requires a lot of time from personnel?	80%	20%

4 Discussion

The methodologies that until now have been applied to migration processes focus on the transfer of information. However, they accept the loss of information while executing it [7]. Most of the literature related to migration focuses only on technical activities [5, 14]. For FMF technical and operational activities are equally important to achieve a safe migration and with integrity of information. This leaves many gaps in a process where the replication of activities is important, as are the people who accessed this information. FMF uses the combination of migration methodologies and information security standards that are summarized in six clearly defined phases. The sequential execution of the FMF phases allowed filling the deficiencies that these methodologies did not consider in their execution. The privacy of a considerable amount of sensitive information was preserved and the service suspension time was reduced to a minimum. There is enough documentation to reproduce the activities carried out and determine the people responsible for their execution. With this framework, it was possible act to solve any incident presented during or when migration finished. Once the process was completed, the migrated information preserved its integrity and its quality was improved with the insertion of new fields necessary for the delivery of reports to the control entities.

5 Conclusions

With this it was, was possible act to solve any incident presented during or when migration finished. Once the process was completed, the migrated information preserved its integrity and its quality was improved with the insertion of new fields necessary for the delivery of reports to the control entities.

However, the participants suggest that the documentation to support the privacy of information is too extensive. In addition, the survey indicates that the number of staff reviews should be reduced. It is important to highlight that both SFSE consider for the case studies are of medium size (Segment 3), for which, the staff is scarce, and the time dedicated to the migration process represents a high cost for the SFSE.

Finally, researchers are suggested to carry out the same experiment with a Segment 1-2 SFSE. In these segments, the control entities oblige the SFSE to have specialized staff in security and risks areas, in these conditions, the response time of the framework application could be reduced considerably. Finally, in a new investigation, the proposed framework could be used in other types of institutions where the information requires the same treatment in terms of privacy.

References

1. Kantawala, A.: Case Study: Data Migration for a Global Semiconductors Manufacturer. In: Rightshore! Successfully industrialize SAP® Projects Offshore (2008)
2. Patricio, A.G.: Política y regulación de la aplicación de nuevas tecnologías para el avance de la inclusión financiera por las cooperativas de ahorro y crédito del Ecuador, in Dinero y operaciones bancarias. Universidad de Salamanca, Salamanca (2017)
3. Mulagala, D.: Data migration. Int. J. Adv. Res. **6**, 431–444 (2018)
4. Zahari, N.M., et al.: Data quality issues in data migration, pp. 33–42 (2015)
5. Biswas, N., et al.: A new approach for conceptual extraction-transformation-loading process modeling. Int. J. Ambient Comput. Intell. **10**, 30–45 (2019)
6. Oladipo, F.O., Raiyetumbi, J.O.: Re-engineering legacy data migration methodologies in critical sensitive systems. J. Global Res. Comput. Sci. **6**, 5–6 (2015)
7. Stonebraker, M., Brodie, M.L.: Migrating legacy systems: gateways, interfaces & the incremental approach. In: Kaufmann, M. (ed.) Morgan Kaufmann Series in Data Management Systems. Morgan Kaufmann Publishers, Burlington (1995)
8. Wu, B., et al.: The butterfly methodology: a gateway-free approach for migrating legacy information systems, pp. 200–205 (1997)
9. Ratib, O., et al.: Multigeneration data migration from legacy systems. In: Proceedings of SPIE - The International Society for Optical Engineering (2003)

10. Standardization, I.O.f., ISO/ICE 27001: 2013 Information technology - Security techniques - Information security management systems - Requirements. International Organization for Standardization (2013)
11. Shojaie, B., Federrath, H., Saberi, I.: Evaluating the effectiveness of ISO 27001: 2013 based on Annex A, pp. 259–264 (2014)
12. Morales, F., Toapanta, S., Toasa, R.M.: Implementation of a perimeter security system as an information security strategy. RISTI – Rev. Iberica de Sistemas e Tecnologias de Informacao **2020**(E27), 553–565 (2020)
13. Muheet, E., et al.: Migrated legacy data issues: recovery, concurrency & security (2019)
14. Hwang, K.S., Cui, J.F., Chae, H.S.: An automated approach to componentization of java source code, vol. 2, pp. 205–210 (2009)

Financial Technologies (FinTech), Economics and Business Engineering

Path Planning for Mobile Robots Applied in the Distribution of Materials in an Industrial Environment

Sylvia Mercedes Novillo Villegas[1], Allison Rodríguez[2], Danilo Chavez[2](✉), and Oscar Camacho[2]

[1] Universidad de Las Américas, Quito, Ecuador
Sylvia.novillo@udla.edu.ec
[2] Escuela Politécnica Nacional, Quito, Ecuador
{allison.rodriguez,danilo.chavez,oscar.camacho}@epn.edu.ec

Abstract. This paper presents a new approach to solve the problem of logistics in a factory by automating the distribution of materials. Thus, a general algorithm of path planning is applied for the physical distribution of the production area, in the case of PowerTech Transformer Factory and Man Busses Factory. A modification to the basic RRT algorithm is used to attain the minimum distance to be followed when exploring the entire work area. Accordingly, simulated results demonstrate how the mobile robots distribute the materials from the warehouse to each workstation and return to the warehouse.

Keywords: Mobile robot · RRT algorithm · Path planning · Coordinate system · Lyapunov control

1 Introduction

Large factories have achieved a high percentage of automation of their processes. When correctly implemented, the delivery of materials might be 100% automatic. As a result, these factories register a reduction in the amount of time need to distribute the materials from the warehouse to each work station, increasing their productivity by decreasing product handling. It also prompts the possibility of a continuous supply in the production lines, improves efficiency in the use of a transport fleet, enhancing speed and stability when stowing over other products as well as increases the safety of the personnel involved in the handling of goods (Sun et al. 2010). However, there are many challenges still to be properly solved, one of which is the use of mobile robots for internal logistics.

Mobile robots are generally used in several applications such as exploration, research, trajectory tracking, among others. here exists a wide array of uses for these robots to load and offload materials (Nascimento Martins et al. 2008). For instance, (Grijalva et al. 2018) proposed a path planning for mobile robots based on the physical disposition of the factory where the robots distributed raw material, or as (Sun et al. 2010) presented a decentralized controller of autonomous robots for intralogistics. A prime example is

M. Botto-Tobar et al. (Eds.): ICCIS 2020, AISC 1273, pp. 323–337, 2021.
https://doi.org/10.1007/978-3-030-59194-6_27

factories at the forefront of the distribution of materials such as KUKA ("KMP 1500," 2017), MIR ("The future in motion," 2018) or Amazon Robotics.

The control of several mobile robots becomes highly complex due to the required coordination among them, the creation, the following, and the navigation of paths in the area (Andaluz and Andaluz 2012). Hence, an analysis of a dynamic model is necessary to perform the coordination, the validation of constants, and the nonlinear control for trajectory tracking of mobile robots (Nascimento Martins et al. 2008). (Sabattini et al. 2015) introduced an analysis of the dynamic model of a unicycle-type mobile robot with nonholonomic restriction for positioning control, trajectory tracking and path following, where the dynamic and kinematic model was presented for control.

There are some algorithms capable of creating paths in a known environment. For instance, (Barraquand 2014) used artificial potential fields distributed through the work area to attain the paths free of collisions. These fields generate forces that attract the robots to the goal and others which move them away from the obstacles. Thus, the robots are capable of carrying out a point-to-point follow-up between the obstacles. However, when considering the robot as a particle, it has certain disadvantages. (Garrido et al. 2006) presented the planning of paths for mobile robots using Voronoi Diagrams, that are responsible for dividing the plane into regions (Voronoi regions). They defined the locus of the points of the plane equidistance to the obstacles which are responsible for moving the robot away from the nearest obstacles. To solve the disadvantages of a mobile robot as a particle, (LaValle 1998) first introduced the algorithm RRT as a new tool for tracking paths without problems. This algorithm integrates nonholonomic restrictions and high degrees of freedom. Hence, the RRT algorithm is used by scholars in several change processes. For instance, (López et al. 2016) explored the bidirectional RRT algorithm by extending the branches of the tree both from the initial configuration as well as from the final configuration. By doing this change, the authors obtained greater scan speed and the desired route in the shortest time. (LaValle 2001) explored different environments using the basic RRT algorithm. In this article, we present a modification to the RRT algorithm based on (LaValle 1998) exploring the entire work area to plan the paths to be followed by the mobile robot regarding the minimum distance due to the radius analysis.

This paper is structured as follows: Sect. 2 presents the system for coordinating the mobile robots to deliver materials from the warehouse to each workstation and vice versa. Section 3 describes the kinematic and dynamic model with improved holonomic restriction as well as the control used for paths following. Section 4 introduces the algorithm used to create paths in the respective factories related to this research. Section 5 presents the results in industrial environments. Finally, states the conclusions.

2 Coordinate System

2.1 Description

In this section, we will analyze how the path planning is put together with the control of the mobile robot to coordinate the system. To organize several mobile robots that follow a referential path generated by an algorithm RRT completely independent of the mobile robots, it is necessary to join these two elements under a unified system (Fig. 1).

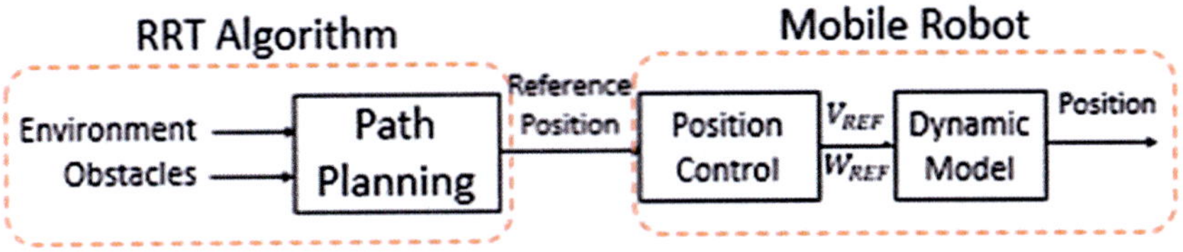

Fig. 1. Coordinate system

Moreover, it is important to mention that each mobile robot works independently, i.e., each mobile robot performs the tasks of transporting material to a specific work station and returning to an independent warehouse, although they are part of a coordinated system of multiple robots that work together for a common goal.

The coordination system requires data such as paths generated (X_{ref}, Y_{ref}) by the RRT algorithm, work station's coordinates, number of robots, and the work area (factory's layout). It verifies the order in which the materials must be delivered to each work station and the number of robots. The warehouse delivers material to each work station by each mobile robot. Then, it is verified if the robot has reached its destination to return to the warehouse. When the mobile robot has reached the warehouse, it goes into an inactive state ready to be used again when another delivery is necessary.

3 Model and Control of Mobile Robots

3.1 Description

In this section, we describe the kinematic and dynamic model of the unicycle-type mobile robot and the realization of the control to be able to follow paths. Mobile robots use a nonlinear cascade controller based on Lyapunov's theory (Nascimento Martins et al. 2008). It is highly important to analyze the dynamic model of the mobile robot due to the applications where heavy loads at high speed are transported. A detailed analysis of the dynamic model of the mobile robot is presented in (Nascimento Martins et al. Nascimento Martins et al. 2008) and in (Andaluz et al. 2012).

3.2 Dinámic Model

Figure 2 shows the model of the nonholonomic unicycle mobile robot, used in (Andaluz et al. 2012) where (x, y) defines the position of the robot, a is the distance from the center of gravity of the robot to the center of the axis of the wheels, v is the linear speed, w is the angular speed, ψ is the orientation, G is the center of gravity, B is the center of the axis of the robot's wheels, E is the point of location of some work tool, C is the point of the castor wheel, F_rlx and F_rly are the lateral and longitudinal forces respectively in the tire of the left wheel, F_rrx and F_rry are the lateral and longitudinal forces in the tire of the right wheel.

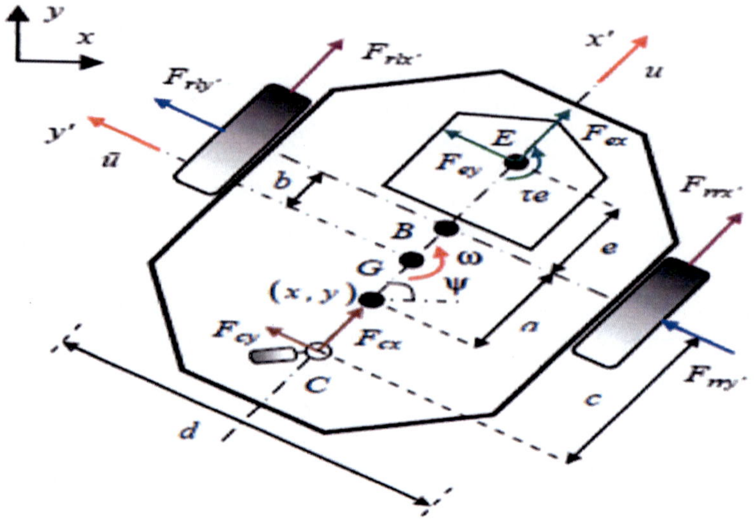

Fig. 2. Unicycle type mobile robot model

Therefore, from Fig. 2 is the mathematical model of the nonholonomic unicycle mobile robot.

$$\begin{bmatrix} \dot{x} \\ \dot{y} \\ \dot{\psi} \\ v \\ \dot{w} \end{bmatrix} = \begin{bmatrix} \cos(\psi) - a * \sin(\psi) \\ \sin(\psi) + a * \cos(\psi) \\ w \\ \frac{\theta_3}{\theta_1} w^2 - \frac{\theta_4}{\theta_1} v \\ -\frac{\theta_5}{\theta_2} vw - \frac{\theta_6}{\theta_2} w \end{bmatrix} + \begin{bmatrix} 0 & 0 \\ 0 & 0 \\ 0 & 0 \\ \frac{1}{\theta_1} & 0 \\ 0 & \frac{1}{\theta_2} \end{bmatrix} \begin{bmatrix} v_{ref} \\ w_{ref} \end{bmatrix} + \begin{bmatrix} \delta_x \\ \delta_y \\ 0 \\ \delta_v \\ \delta_w \end{bmatrix} \quad (1)$$

Where v_{ref} and w_{ref} represent the reference inputs of the linear velocity and angular velocity respectively, the parameter vector of the model is represented by $\theta = [\theta_1\, \theta_2\, \theta_3\, \theta_4\, \theta_5\, \theta_6]$, the uncertainty vector is represented by $\delta = \left[\delta_x\, \delta_y\, 0\, \delta_v\, \delta_w\right]^T$, the parameters of the vector θ are validated in (Nascimento Martins et al. 2008) and exhibited in Table 1.

3.3 Controller

From control of the mobile robot, the kinematic and dynamic parts of (1) are divided. In (2) the kinematic part of the robot is illustrated.

$$\begin{bmatrix} \dot{x} \\ \dot{y} \\ \dot{\psi} \end{bmatrix} = \begin{bmatrix} \cos(\psi) - a * \sin(\psi) \\ \sin(\psi) a * \cos(\psi) \\ 01 \end{bmatrix} \begin{bmatrix} v \\ w \end{bmatrix} + \begin{bmatrix} \delta_x \\ \delta_y \\ 0 \end{bmatrix} \quad (2)$$

Table 1. Parameters of the mobile robot unicycle type

Parameters	Value
θ_1	$0.2604\,s$
θ_2	$0.2509\,s$
θ_3	$-0.000499\,\frac{sm}{rad^2}$
θ_4	0.9965
θ_5	$0.00263\,\frac{s}{m}$
θ_6	1.0768

Regarding, δ_x and δ_y, (3) presents the law of nonlinear control based on system errors.

$$\begin{bmatrix} v_{cref} \\ w_{cref} \end{bmatrix} = \begin{bmatrix} \cos(\psi) & \sin(\psi) \\ -\frac{1}{a}\sin(\psi) & \frac{1}{a}\cos(\psi) \end{bmatrix} \begin{bmatrix} \dot{x}_d + l_x \tanh\left(\frac{k_x}{l_x}\tilde{x}\right) \\ \dot{y}_d + l_y \tanh\left(\frac{k_y}{l_y}\tilde{y}\right) \end{bmatrix} \tag{3}$$

The dynamic part of the robot exhibited below by Eq. (4).

$$\begin{bmatrix} \dot{v} \\ \dot{w} \end{bmatrix} = \begin{bmatrix} \frac{\theta_3}{\theta_1}w^2 - \frac{\theta_4}{\theta_1}v \\ -\frac{\theta_5}{\theta_2}vw - \frac{\theta_6}{\theta_2}w \end{bmatrix} + \begin{bmatrix} \frac{1}{\theta_1} & 0 \\ 0 & \frac{1}{\theta_2} \end{bmatrix} \begin{bmatrix} v_{cref} \\ w_{cref} \end{bmatrix} + \begin{bmatrix} \delta_v \\ \delta_w \end{bmatrix} \tag{4}$$

With respect to δ_v and δ_w, control law for the dynamic part is defined by Eq. (5)

$$\begin{bmatrix} v_{ref} \\ w_{ref} \end{bmatrix} = \begin{bmatrix} \theta_4 & -\theta_3 w \\ \theta_5 w & \theta_6 \end{bmatrix} \begin{bmatrix} v \\ w \end{bmatrix} + \begin{bmatrix} \theta_1 & 0 \\ 0 & \theta_2 \end{bmatrix} \begin{bmatrix} \dot{v} \\ \dot{w} \end{bmatrix} \tag{5}$$

4 Path Planning

In this section, we present the RRT basic algorithm and its modification to find the necessary paths to track unicycle type robots. Additionally, we analyze the distribution of the factories used as proof of the path planning.

4.1 RRT Algorithm

(LaValle 1998) introduced the RERT algorithm (Rapidly Exploring Random Tree) as a new perspective that aims to solve the existing problems for generating paths or trajectories in certain environments under nonholonomic restrictions and degrees of freedom. To understand the RRT algorithm Table 2 defines several concepts necessary for the RRT.

Table 2. RRT variable algorithm

(x, y)	Coordinate points
q_{ini}	Initial configuration
q_{goal}	Final configuration
q_{rand}	Random configuration
q_{near}	Configuration closed to q_{rand}
q_{new}	New configuration
ε	Tree growth length
K_{max}	Maximum number of interactions for tree generation

The purpose of the RRT algorithm is to extend a scan tree through the entire collision free space, selecting an arbitrary point out q_{rand} of the obstacles.

Figure 3 details the growth of the tree through it generates a random state q_rand in the direction of the configuration obtained previously.

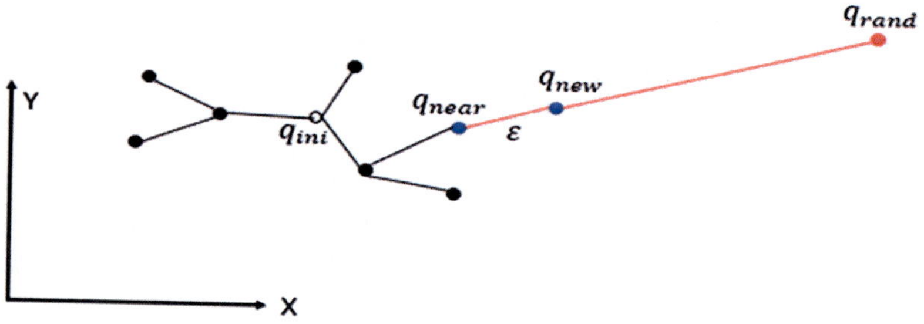

Fig. 3. Tree growth

4.2 RRT Algorithm Modified

As observed in Fig. 9 and Fig. 14, all the robots start from the same configuration (warehouse) q_ini to the different configurations (workstations) q_end. The modified RRT algorithm consists in generating several paths from q_ini to each q_end Fig. 4 so that every mobile robot can complete the assigned task. In order to obtain shorter distances and fewer inflections points, the modified algorithm allows the analysis of the configurations, q_rand, q_near, q_new, q_min are analyzed outside an obstacle. Considering the shortest distance D from q_new to q_neighbor it is necessary to analize the valid configurations in a predetermined radius r, where the value C_min = min(D) is the minimum distance. Thus the configuration q_min belongs to C_min.

```
RRT Algorithm
Inputs: q_ini, q_end, K_max, # obs, obs, X_max, Y_max
T [0] = q_ini;
For K = 1 to K_max
 q_rand = random_configuratio(X_max , Y_max);
   D = [ ];
   For j = 1 to length(T)
     n = T(j)
     D = [ D, distance(n, q_rand)];
   end
   Dmin = min(D) ;
   q_near = n(Dmin);
   nc = nocollision(q_rand, q_near, obs);
   if nc=1
     q_new = steer(q_rand, q_near, Dmin, ε);
     r=search radius minimum distance
     q_neighbor = [ ];
     Neighbor_count=1;
     For i = 1 to length(T)
       T = [T q_new];
       n = T(i);
       nc1 = nocollision(n, q_new, obs);
       d = distance(n, q_new);
       If nc1=1&&d<r
         q_neighbor = T(i);
         Neighbor_count=Neighbor_count+1;
       end
     end
       q_min = q_near;
      D = [ ];
       For k=1 to length(q_neighbor)
         q_min = q_neighbor(k);
         D = [D, distance(q_neighbor(k), q_new)];
       end
       C_min = min(D) ;
      q_min = q_min (C_min)
   end
   T = [T   q_min ];
 end
```

Fig. 4. RRT algorithm implemented

The function random configuration charges on to generate the random state. Figure 5 is valid if and only if q_{rand} is in the production area.

```
random_configuration(X_max, Y_max);
q_rand = [q_rand(1) * X_max , q_rand(1) * Y_max ];
While q_rand (1) > X_max &&q_rand (2) > Y_max
    q_rand = [q_rand(1) * X_max, q_rand(1) * Y_max];
end
return q_rand
```

Fig. 5. Random configuration

The function steer serves to find. It depends on the minimum distance between q_{rand} and q_{near} as detailed in Fig. 6.

```
steer(q_rand, q_near, Dmin, ε);
If Dmin ≥ ε
    q_new = q_near + (q_rand − q_near) * ε / distance(q_rand, q_near);
else
    q_new = q_rand;
end
return q_new;
```

Fig. 6. Steer function

The function *nocollision* Fig. 7 analyzes each of the configurations generated *nc* which take the last *T* value in order to avoid the generation of points on the obstacle.

```
nocollision(q1, q2, obs);
If q_1 && q_2 ∈ obs area?
    nc = 0;
else
    nc = 1;
end
return nc;
```

Fig. 7. No collision function

4.3 Creation of Paths

Figure 8 shows the process of creating paths. First, the entire work area is explored to obtain the points for the vector T. Then, the minimum distance is analyzed to create the paths. The vector nodesval stores all the valid points to create of paths. Finally, the algorithm generates the paths followed by the mobile robot.

```
Inputs: q_ini, q_end, # q_end, T
For h = 1 to # q_end
  D = [];
  For j= 1 to length(T)
    D = [ D, distance(n, q_rand)]
  end
  [Dmin, idx] = min(D);
  q_goal = q_end(h);
  nodesval(h) = [];
  While idx ≠ 0
    nodesval(h) = [nodesval(h), q_goal];
    q_goal = T(idx);
  end
  [Path_go(h).X_ref. Path_go(h).Y_ref. ] =
  Path_go_gen(nodesval(h));
  [Path_ret(h).X_ref. Path_ret(h).Y_ref. ] =
  Path_ret_gen(nodesval(h));
end
```

Fig. 8. Paths generate

5 Simulated Results

This section presents the experimental results of the RRT algorithm simulation, path planning, and path following of mobile robots in the production area of PowerTech Transformers and the assembly area of Man Busses. The K_max used in RRT algorithm, was implemented 2000 interactions to explore the entire production area of PowerTech Transformers and the entire area assembly Man Busses.

It is necessary to explain how the algorithm works in each factory. For this purpose, Fig. 9 depicts the plant configurations at PowerTech Transformers.

Figure 10 depicts the tree expansion after the exploration of the entire area of the plant by the RRT algorithm.

The value of the radius chosen is r = 8.5 o find the minimum distance due to the distribution of the obstacles in the work area. It is important to bear in mind that it is not feasible to choose a very large radius. Then, the distance is analyzed by considering all the configurations within the radius $q_{neighbor}$ and the configuration of the minimum distance q_{min} (Fig. 11).

Finally, the generated paths appear in Fig. 12

Figure 13 presents the results of the coordination of the robots. Bear in mind that mobile robots have dimension and they do not pose the risk of collision with obstacles due to a security area that surrounds each one of the obstacles.

Likewise, Fig. 14 depicts the configurations at the Man Busses' plant.

Then the RRT algorithm explores the entire area of the plant (Fig. 15).

The chosen radius is r = 15 to find the minimum distance. In this case, the radius is greater since the length of the area and the distribution of the obstacles are wider at Man Busses than PowerTech Transformers. The distance is analyzed for all the configurations within the radius $q_{neighbor}$ and the configuration of the minimum distance q_{min} (Fig. 16).

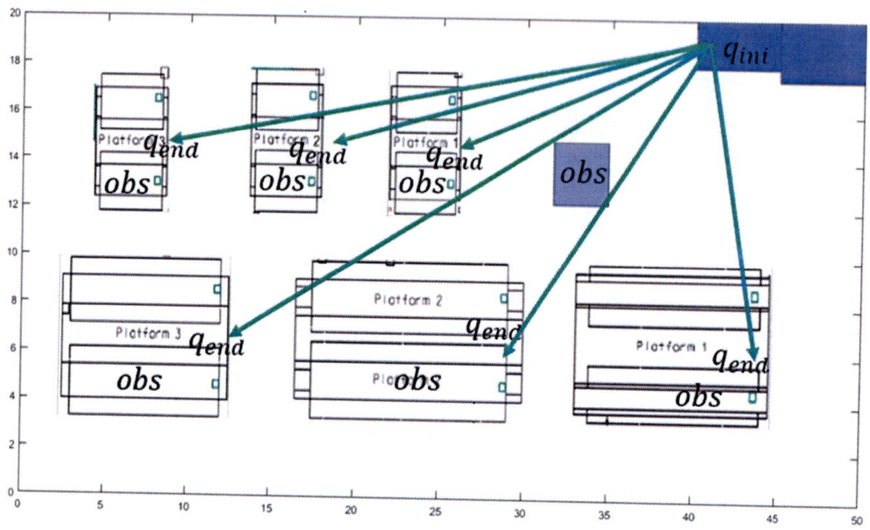

Fig. 9. PowerTech transformers' configurations

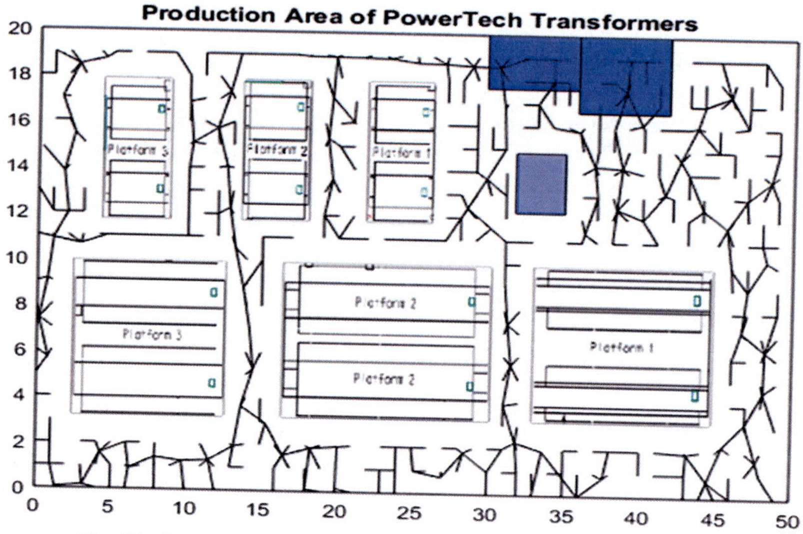

Fig. 10. Tree expansion throughout the production area of PowerTech

Finally, Fig. 17 presents the generated paths.

Figure 18 depicts the results of the coordination. In this case, the mobile robots have dimension and they do not pose the risk of collision with obstacles due to a security area that surrounds each one of the obstacles.

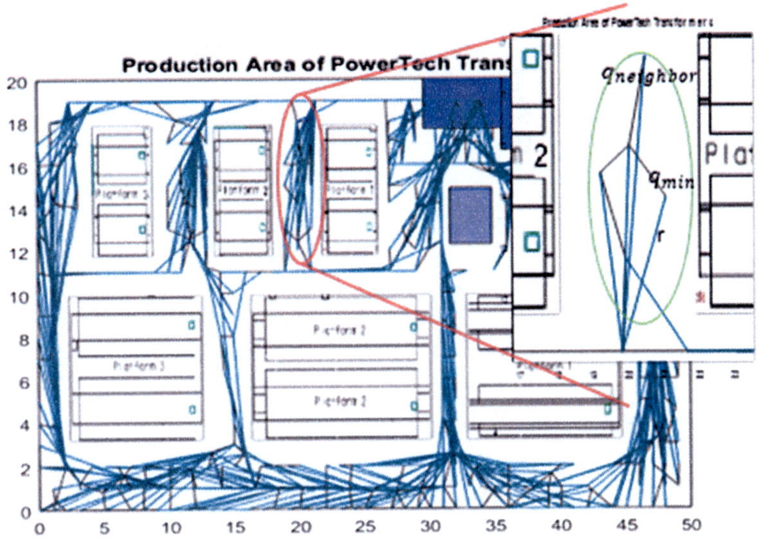

Fig. 11. Radius for minimum distance in the production area of PowerTech

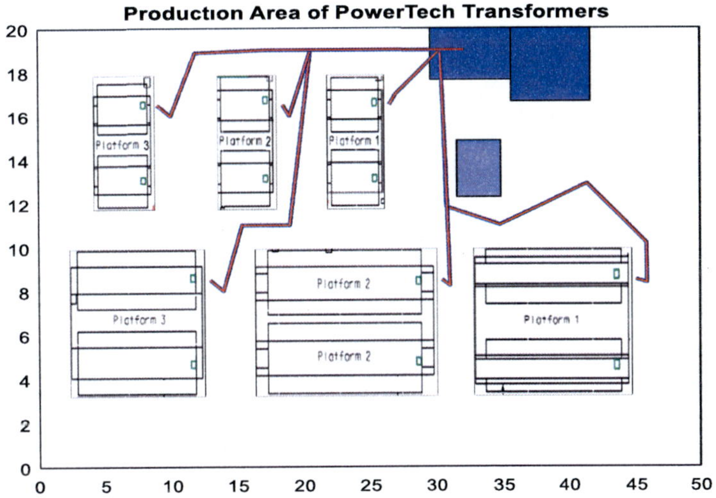

Fig. 12. Paths planning in the production area of PowerTech

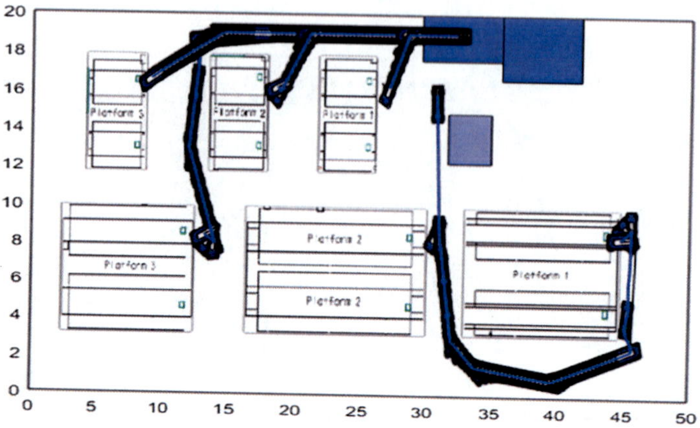

Fig. 13. Path Following in the production area of PowerTech

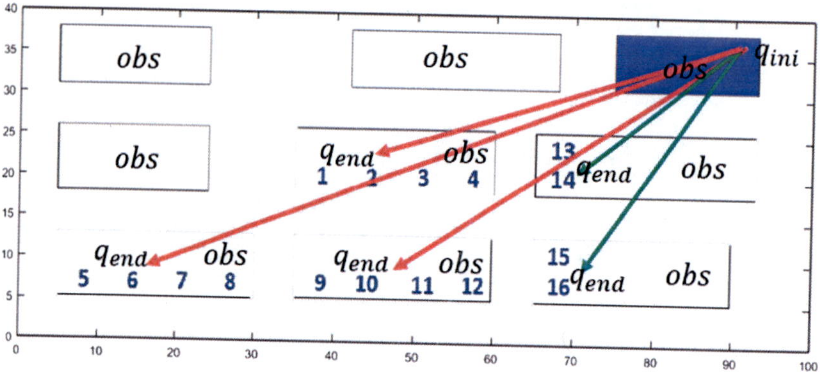

Fig. 14. Man Busses configurations

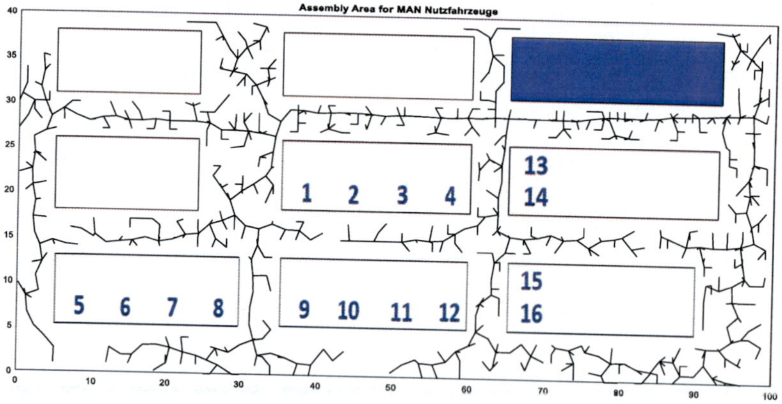

Fig. 15. Tree expansion throughout the assembly area of Man Busses

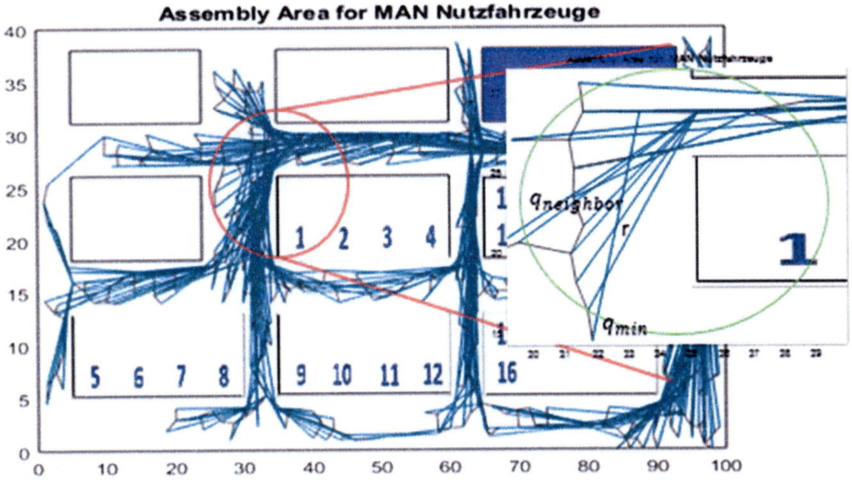

Fig. 16. Radius for minimum distance in the assembly area of Man Busses

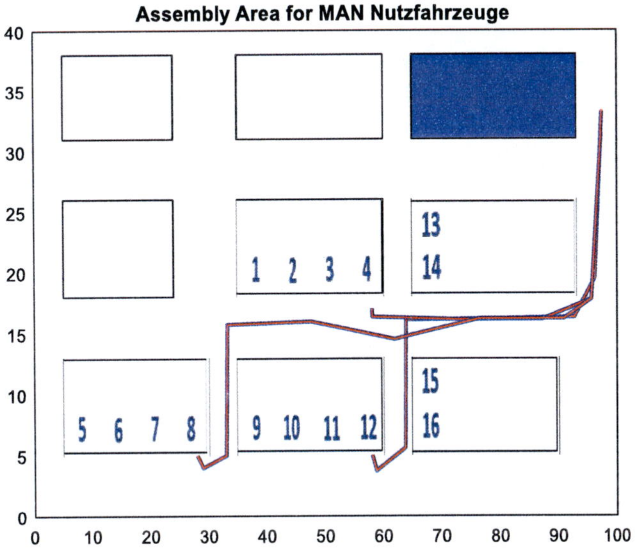

Fig. 17. Paths for tracking robots in the assembly area of Man Busses

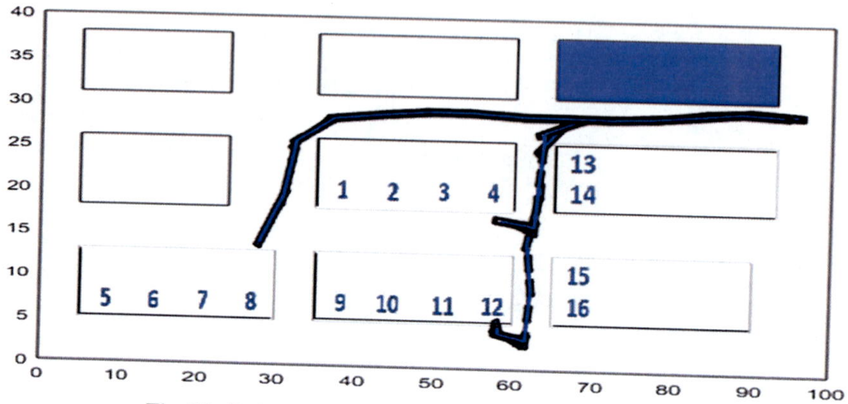

Fig. 18. Path following in the assembly area of Man Busses

6 Conclusions

The RRT algorithm is analyzed in an environment containing several obstacles and through the selection and search of valid configurations through a determined radius. To get the shorter path for mobile robots, we reduced the computational cost and the inflection points for creating the paths. The implemented modification to the RRT determines the best radius to find the minimum distance to create the paths for mobile robots.

The use of the RRT algorithm in this research has advantages such as nonholonomic restrictions and the mobile robots are not considered as particles.

Universidad de Las Americas and the PIJ 17-10 project from Escuela Politécnica Nacional are thanked for providing us with full availability of equipment and technical assurance in order to carry out the validation tests of the project and to implement it properly.

References

Andaluz, G., Andaluz, V., Rosales, A.: Modelación, Identificación y Control de Robots Móviles. Escuela Politécnica Nacional (2012)

Barraquand, J., Latombe, J.-C.: Robot Motion Planning: A Distributed Representation Approach, vol. 10. Sage Publication (2014). https://doi.org/10.1177/027836499101000604

Garrido, S., Abderrahim, M., Moreno, L.: Path planning and navigation using voronoi diagram and fast marching. Carlos III University **39**(15), 346–351 (2006)

Grijalva, G., Chávez, D., Camacho, O.: Material Distribution with Mobile Robots in an Industrial Environment. KMP 1500 (2017). Retrieved from KUKA Mobile robotics website. www.kuka.com

LaValle, S.: Rapidly-Exploring Random Trees: A New Tool for Path Planning. Ames, IA 50011 USA (1998)

LaValle, S., Kuffner, J.: Randomized Kinodynamic Planning (2001)

López, D., Gómez-Bravo, F., Cuesta, F., Ollero, A.: Planificacion de Trayectorias con el Algoritmo RRT. Aplicacion a Robots No Holonómicos. Revista Iberoamericana de Automática e Informatica Industria, **3**(3), 56–67, July 2016

Nascimento Martins, F., Carelli, R., Sarcinelli-Filho, M., Freire Bastos, T.: Modelo Dinámico y Controlador de Seguimiento para Robots Móviles Tipo Uniciclo. Universidad Nacional de San Juan (2008)

Sabattini, L., Digani, V., Lucchi, M., Secchi, C., Fantuzzi, C.: Mission Assignment for Multi-Vehicle Systems in Industrial Environments. Department of Sciences and Methods for Engineering, pp. 268–273 (2015)

Sun, D., Kleiner, A., Schindelhauer, C.: Decentralized hash tables for mobile robot teams solving intra-logistics tasks. In: International Foundation for Autonomous Agents and Multiagent Systems, pp. 923–930 (2010)

The future in motion: Retrieved from MIR (Mobile Industrial Robots), August 2018. https://www.mobile-industrial-robots.com/en/

General Track

Towards Industry Improvement in Manufacturing with DMAIC

Patricia Acosta-Vargas[1(✉)], Edison Chicaiza-Salgado[1], Irene Acosta-Vargas[2], Luis Salvador-Ullauri[3], and Mario Gonzalez[1]

[1] Intelligent and Interactive Systems Lab, Universidad de Las Américas, 170125 Quito, Ecuador
{patricia.acosta,edison.chicaiza.salgado, mario.gonzalez.rodriguez}@udla.edu.ec

[2] Universidad Regional Amazónica Ikiam, Kilómetro 7, Tena, Ecuador
irene.acosta@ikiam.edu.ec

[3] Department of Software, University of Alicante, 03690 Alicante, Spain
lasu1@alu.ua.es

Abstract. Today, the accelerated development of industry 4.0 and the immersion of technologies with the DMAIC methodology, Lean Manufacturing allows higher productivity and continuous improvement of industrial processes. There are countless manufacturing companies, of which very few have invested in technology and methodologies, so they have not been able to prepare themselves to advance at the same pace as the industrial revolution. This research proposes to include improvements in the manufacture of furniture by applying software tools and methodologies such as DMAIC, which consists of combining the application of some tools such as Lean, visual management, and standardized work. As a case study, we applied to the manufacture of furniture; the results indicated that by applying DMAIC and simulation tools, a significant increase in production was achieved, with a decrease in delivery time, in addition to the optimization and reduction of waste. This research can serve as a reference for future studies related to manufacturing processes when applying DMAIC, Lean Manufacturing, and simulation software.

Keywords: DMAIC methodology · Industry 4.0 · Lean manufacturing · Process · Productivity · Simulation · Software

1 Introduction

Currently, small and medium-sized enterprises (SMEs) have become an essential source of employment in the economies of countries, and their contribution is fundamental to generating jobs in a more dynamic and competitive environment [1].

The problems in SMEs have related to the lack of 1) organization in the processes; 2) continuous training; 3) implementation of new technologies; and 4) knowledge of those who lead the organization [2]. A high number of companies vary according to the tasks they perform.

M. Botto-Tobar et al. (Eds.): ICCIS 2020, AISC 1273, pp. 341–352, 2021.
https://doi.org/10.1007/978-3-030-59194-6_28

The term Lean Manufacturing means fat-free [3, 4], so in the Lean industry and service field, it is agility and flexibility. Lean Manufacturing aims to eliminate all activities that do not add value [5], focusing on process improvements and optimization. In this research, we take as a case study, this methodology to apply in the manufacture of metal furniture. This article proposes, through the DMAIC [6] to identify the process to which the study will be focused, to measure the time of the furniture manufacturing cycle, to analyze the problems employing technological tools such as simulators to 1) identify the problems, 2) propose improvements based on the application of reading tools and 3) include control by establishing management indicators.

At each stage of the DMAIC [7], different tools were used to enable the organization to 1) to increase productivity and profitability, 2) to understand the current situation of the organization, 3) to look for the root of the problems, and 4) to propose alternatives for change.

This work proposes a change in the production process of metal furniture, with the immersion of digital technologies within the factories. The development of this field implies a considerable digital transformation in the current production processes. In the industry, all the information concerning the production process is available in real-time and can be simulated with technological tools to 1) reduce production time, 2) optimize quality levels, 3) obtain more considerable cost savings, 4) generate greater security in the processes, 5) achieve more flexible production, 6) achieve a more efficient data flow and 7) achieve greater business competitiveness.

This research can serve as a point of reference for future studies related to the improvement of the production process through the application of DMAIC, Lean Manufacturing, and embedded technology to achieve new ways of organizing the means of production and improving Industry 4.0.

Currently, social distancing [8] and connectivity are vital parameters for reducing COVID-19 infections [9]. Promoting it implies redesigning our abilities, accelerating the future through disruptive technologies, and understanding the strategy of change within Industry 4.0, which implies effective automation.

As future work, we suggest applying this methodology and simulations before the manufacture of any product. Besides, for post coronavirus era offices we suggest 1) applying data science to predict trends and preferences in furniture design, 2) design furniture that optimizes spaces and at the same time apply materials and structures that revolve around health security, 3) create furniture that applies the concept of social distancing.

The rest of the document is structured as follows: Sect. 2 presents studies related to DMAIC, Lean, and Industry 4.0. Section 3 explains the method, case study, and materials used to improve the process in furniture manufacturing. Section 4 discusses the evidence for the results. Finally, Sect. 5 presents the conclusions and possible future work that may emerge from this study.

2 Background

Today, the evolution of consumer areas, the growing demand, and the continuous search for innovation are essential elements in the industrial business model called 4.0 [10], guided by technological and digital tools. The Industrial Revolution 4.0 [11] has been the central axis to achieve a technological implementation in this sector. In these times of pandemic, the technology, especially artificial intelligence with support in mobile applications helped to control and combat the expansion of COVID-19, as well as [12] the Internet of things, the big data, the robotics, among others, allow processes to be more optimal and efficient to drive higher productivity.

Cost management can drive profitability; because of this, most industries can find some waste to cut. It is important to remember not to cut costs at the expense of the quality of your products and services [13] since it is essential to consider suppliers, finances, and production. It is also vital to check if production processes can be adapted to be more agile, using fewer hours of work or resources to reduce labor costs.

The competition is an essential factor in fostering innovation, promoting best practices, and strategic management decisions. Besides, it enables performance improvements to be driven in various industrial activities, including cost reductions, increases in productivity levels, which involves promoting business efforts [14]. For the development of this research, we propose the following definitions that are part of this article.

Process Management: In the current context of globalization and competitiveness, companies must be more efficient every day and seek to increase profitability. In this sense, process management allows companies to identify how activities are carried out and establish the indicators that are part of a set interrelated with each other [15, 16].

Working Measure: It is the quantitative part of the study of work that indicates the result of the physical effort developed as a function of the time allowed for an operator to complete a specific task, following a regular rhythm, or a predetermined method. The measurement of work is a technique that has the objective of establishing a time standard to control a specific job, conserve resources, and minimize the costs present in it [17].

Lean Manufacturing: It is a management and administration philosophy that was born in Japan with Toyota, the objective of this philosophy is the identification of Mudas (waste), Muri (overload) and Mura (variability), that is, of all those activities that do not add value to production or service processes, and which generate high costs and long response times. Lean is based on four principles 1) long-term philosophy, 2) correct processes, 3) providers and developed personnel, and 4) problem solving [18]. For the application of Lean, tools that fulfill different functions are used, the essential tool is 5's that promote the order and cleanliness of jobs to be more productive [19]. Value Stream Mapping (VSM) [20] helps to map the entire process from the moment the customer places his orders and how the company should organize with suppliers and plan production. This tool helps identify opportunities for improvement. The tools allow us to achieve continuous improvement or known as Kaizen, that seeks to discover opportunities for improvement based on the identification of the root cause of the problems. The Overall equipment effectiveness (OEE) [20] is an indicator that aims to improve productivity by reducing costs by determining the waste present in the process.

DMAIC: It consists of five stages: 1) Define, we limit the objectives to be achieved, and the definition of the object of study. 2) Measure, we collect project performance information. 3) Analyze, we determine the root of the problems once the variables have been identified. 4) Improve, we propose all activities to improve. 5) Control, we support the changes so that the previous stage is maintained over time [7]. We applied the DMAIC

[21] to reduce waste in furniture manufacturing; this methodology can be used to find such dimensional variations. This study has been carried out successfully in a medium-scale industry that has quality management in practice. Six Sigma DMAIC was necessary for the identification and reduction of defects that arose in furniture manufacturing, and we applied to solve the problem of waste and to increase quality.

Industry 4.0: The concept of Industry 4.0 was mentioned for the first time at the 2011 Hannover Fair [22], to put into practice the concept and development of the intelligent factory associated with the fourth industrial revolution, a vision of manufacturing that includes digital technologies with all their processes interconnected through the use of the Internet of Things (IoT), now called the Industrial Internet of Things (IIoT).

Industry 4.0 [10] consists of the digitalization of industrial processes through the interaction of artificial intelligence with machines and the optimization of resources focused on the creation of effective business methodologies.

One of the objectives of this revolution is the union between digital technologies that allow the creation of new markets and the interaction between economic actors.

Industry 4.0 [22] through digitalization and the use of connected platforms allows 1) a capacity of constant adaptation to the demand, 2) serve the client in a more personalized way, 3) provide after-sales service, 4) design, produce and sell products in less time; 5) add services to the physical products, 5) create shorter and more profitable production series, and 6) use the information for analysis from multiple channels and exploit it in real-time.

3 Case Study and Methodology

3.1 Case Study

The case study we apply to an SME in the city of Quito is a small company dedicated to the manufacture of metal furniture, stainless steel showcases, and metal shelves.

4 Methodology

For the development of DMAIC, we carry out an analysis in each phase; we use different tools that allow us to define the process under study, obtain data with the measurement of times and movements and the OEE, analyze the problem, propose improvements to the process and control Using key management indicators, Fig. 1, describes the stages of DMAIC.

Stage 1 - Define: In this stage, we determine the processes to be improved; it begins with the voice of the client, who indicates the expectations, preferences, and comments on a product or service on which the improvement project be defined. In this stage, we use tools such as the Kano model that allowed us to determine the critical points of the client.

Once the client's voice has been captured, we transfer the critical quality points for the selection and establishment of a project charter that mark the beginning of the improvement project.

After determining the product that is the object of this study, we proceed to survey the production process in each of the stages of sub-processes to determine the activities that do not add value. To do this, we use tools such as the SIPOC (Supplier, Input, Process, Output, and Customer) diagram and the use of flow charts that allow us to model the process.

The Pareto diagram is another tool that we use for the selection of the process under study. In this case, it was done based on the units manufactured and the invoicing data, in a period of analysis of six months, this analysis showed that the manufacture of metallic furniture represents 54% of the company's production.

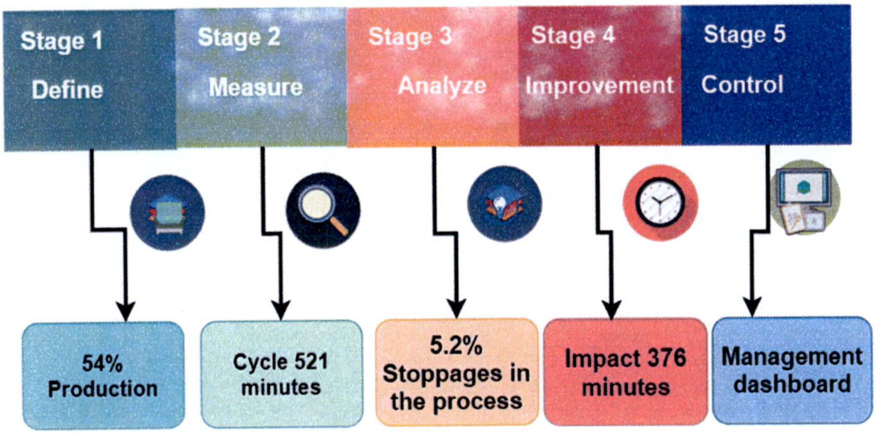

Fig. 1. DMAIC methodology.

Stage 2 - Measure: In this stage, we apply sampling techniques and take the data to analyze it and see the behavior through control charts. Then we make the control chart to understand the stability of the data, in which we identify the causes of the variation.

Besides, the OEE of the process was calculated, the value obtained was 76%, a value that represents an excellent opportunity for improvement, especially in the efficiency, we determined that the production capacity was 20 pieces of furniture based on the set times, considering that the company only makes 16 pieces of furniture on demand.

Table 1 shows include the operation, the description, the minimum time, and the takt. Table 1 presents the times and movements; we determine that the cycle time of a piece of furniture is 521.75 min; in the same way with the thread and flow diagram, we identify that 20% of this time represents the waste and 168 meters of travel.

Stage 3: analysis: At this stage, the EEO analysis, related to availability, indicates that 5.2% represents a percentage of downtime in the process, due to waiting and the nature of manufacturing. At this stage, we determine that the company can manufacture more furniture than the current demand. If demand increases by more than 20 pieces of furniture, the company has delivery problems. Based on the study of times and movements, we carry out the mapping of the value chain (VSM), where improvement opportunities can be identified.

Table 1. Current cycle time furniture production.

Operation	Description	Minimum time	Takt
1	Cut pieces	61.09	512
2	Tracing	15.36	512
3	Blunt	7.87	512
4	Bent	75.96	512
5	Armed	145.67	512
6	Prepared	78.55	512
7	Painted	87.95	512
8	Final enlistment	49.3	512
Total cycle time		521.75	

The generation of waste is the result of the inadequate design of the production plant; this cause can be analyzed with the threading diagram, see Fig. 2.

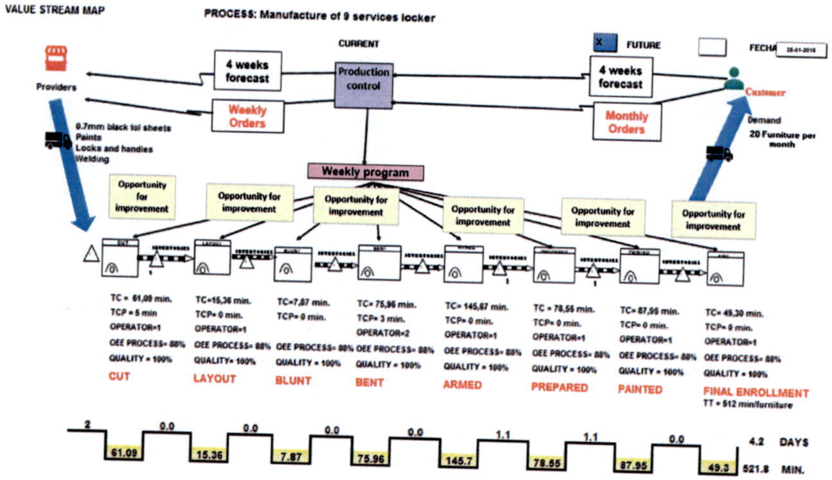

Fig. 2. VSM process

Stage 4 - Improvement: In this stage, we take as a base the root cause found in the previous stage; in this phase, some tools and methodologies can be applied, such as Lean Manufacturing, optimization, process standardization, and software simulations.

In this case study, we apply the tools of Lean Manufacturing and the software FlexSim [23]. Then we start with an implementation approach for the 5'Ss program, said program includes a training plan for the personnel, to subsequently carry out the implementation of each "S" with the assistance of red cards. To control the OEE, we propose the generation of formats that help control the start of activities and the stoppages generated in the production process.

To reduce waste and manufacturing times, we propose a change in the physical distribution of the machines; this change allows a higher flow to the process. This change added to the improvements of each thread has an impact on the new cycle time, which is now 376.31 min with which the production capacity can be increased to 28 units, the waste times in the proposal are 36.65 min, and the distance traveled is 69 meters, the times are shown in Table 2.

Table 2. Cycle time proposed furniture manufacturing.

Operation	Description	Minimum time	Takt
1	Cut pieces	22.39	512
2	Tracing	13.53	512
3	Blunt	7.46	512
4	Bent	68.70	512
5	Armed	126.17	512
6	Prepared	54.45	512
7	Painted	57.15	512
8	Final enlistment	26.46	512
Total cycle time		**376.31**	

For quality improvement, we apply Poka-Yoke [24] that allows preventing errors at the source before the defect occurs.

With the data obtained, we applied simulations with FlexSim [23], a 3D modeling and simulation analysis software. The FlexSim tool was selected in this case study because it allows: 1) 3D visualization, so each phase of the simulation can be observed. 2) Greater ease of use compared to other simulation software is a friendly and very intuitive software. 3) It is not necessary to be an expert in programming to use the tool. 4) It offers several user levels. 5) All option list-objects make modeling more accessible. 6) Eliminates waste, by analyzing a system with a well-constructed simulation model, you get the relevant data to discover and eliminate waste. 7) Increases usability, allows modification and testing to find the changes that optimize the process, and generate more excellent usability and productivity. 8) Customization in report generation and analysis. One of its limitations is that it is a paid software.

With FlexSim, we transformed the existing data into predictions and obtained the results presented in Fig. 3, we observed that with the new production times, a piece of furniture could be made in the eight working hours, and the second piece of furniture is in process.

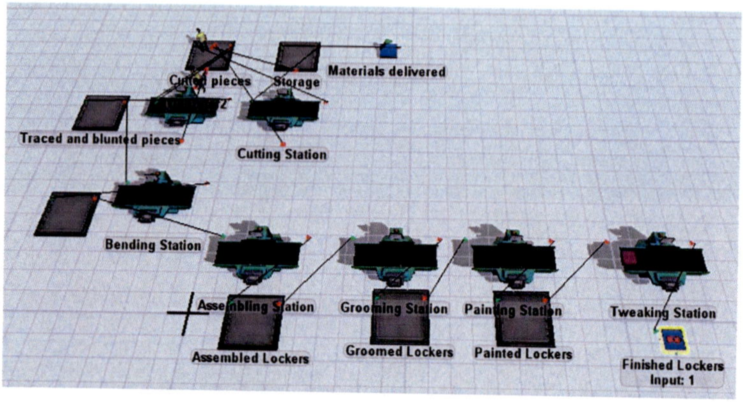

Fig. 3. Simulation of production times with FlexSim

Figure 4 shows that at the end of the working day, the second piece of furniture is in the final preparation process.

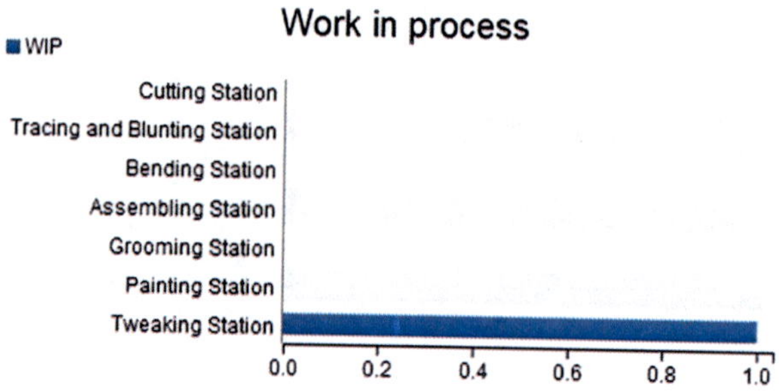

Fig. 4. Final enlistment process

Figure 5 shows that the waste represented in Setup has decreased concerning the current process.

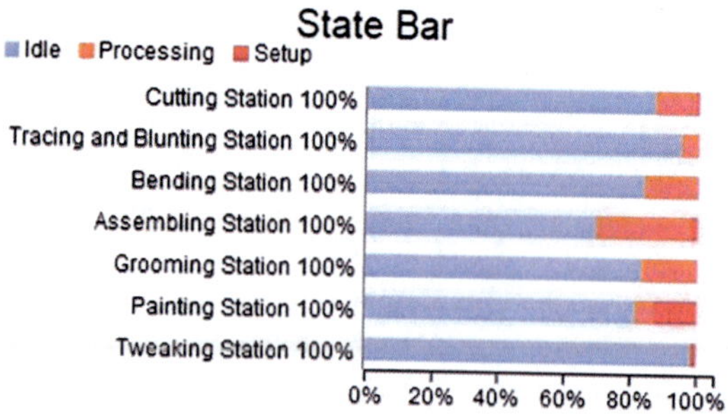

Fig. 5. Current process

Stage 5 - Control: In this stage, we propose management indicators that allow the organization to control the operation; these indicators are based on the activities performed in each previous stage.

To manage the indicators, we suggest the creation of a visual management and administration scorecard that is fed with data collected monthly to allow for continuous improvement in the process cycle. To standardize the new process, we developed standardized worksheets that help in the implementation of the new method.

To generate a culture of order and cleanliness, we designed a checklist that allows the organization to perform periodic audits of the work in the production area.

5 Results and Discussion

With the implementation of Lean tools, the results can be shown in Table 3, indicating the improvements concerning the cycle time of the manufacture of a piece of furniture and the impact on production capacity and productivity. More detail on experimentation and data analysis is available in the Mendeley repository[1].

Table 3. Evaluators' Comments.

Indicator	Current situation	Proposal	% improvement
Partial productivity in relation to the time spent	0.11 furniture/hour	0.16 furniture/hour	45%
Partial labor productivity	5 furniture/operator	7 furniture/operator	57%
Production process time	417.65 min	339.66 min	77.99 min optimization
Waste time	104.11 min	36.36 min	67.76 min optimization
Total standard manufacturing time	521.76 min	376.31 min	145.45 min optimization
Distance traveled in the manufacturing process	168 meters	69 meters	99-meter optimization
Production capacity	20 furniture/month	28 furniture/month	8 pieces of furniture/month representing a 40% improvement

Figure 6 presents a summary of the development, availability, efficiency, and quality of EEO compared to real-time and available. Figure 6 shows the unscheduled stops present in the production process that affect availability. During the analysis phase, some

[1] 10.17632/69wsz4p8n9.1.

interruptions related to the absence of operators and delays in time and movements were identified. As a result, we obtained an availability value of 94.8%.

The efficiency is related to the real production over the theoretical production; that is derived from the capacity that has the company; as a result, we obtained 80%.

The quality is related to defective products or problems that have been identified during the manufacturing process, in this case, no defects were found, so the quality is 100%.

The calculation of the OEE is made by multiplying Availability by Efficiency and Quality, which results in 76%, this result indicates that the process must be improved.

The generation of waste is the result of the inadequate design of the production plant; this problem can be identified by applying an analysis with the yarn diagram.

In the analysis, it is fundamental to determine the problem presented by the company, identifying the root cause of the problem. At this stage, we apply the cause-effect diagram or Ishikawa diagram [25] that shows us the root causes based on the 6 M's (Machinery, Manpower, Management, Money, Materials, Milieu).

In the study applied with the cause-effect diagram, we found that the root causes of the problem are found in the poor physical distribution of the operating space and the lack of technology in machinery used in the manufacturing process.

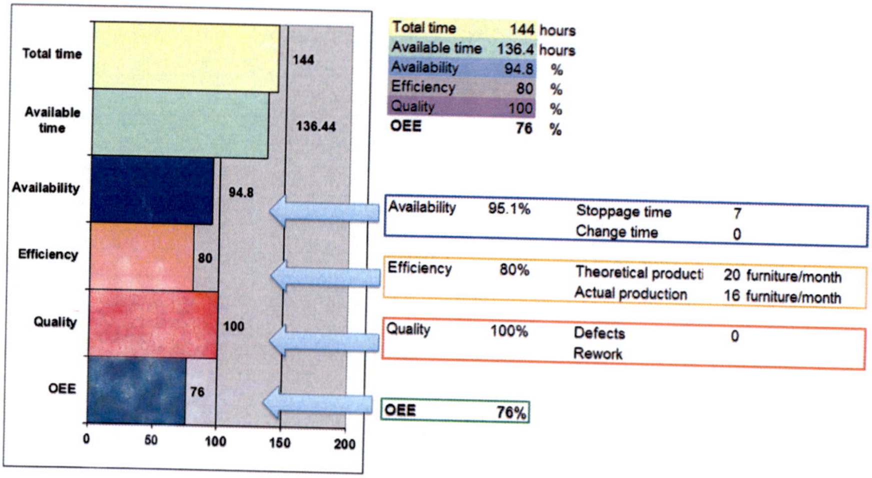

Fig. 6. Preparation of OEE.

6 Conclusions and Future Work

The organizational culture must be global with the inclusion of technological tools that generate a capacity for constant adaptation to the demand and the environment.

This work proposes a change in the furniture manufacturing process by including Lean Manufacturing tools, together with DMAIC methodology and Industry 4.0, through the incorporation of technological tools that promote the development of processes.

We suggest the use of production line simulation software such as FlexSim that allows the simulation of manufacturing processes and the identification of phases in which processes can be optimized and waste reduced.

Simulation tools are innovative tools with predictive technology used to know in advance the behavior of a system under different configurations or operational policies. The methodologies and tools applied can be applied to any type of organization and product, regardless of size, location, or market segmentation. The limitation would be in not committing to change, so it is essential to break the paradigms that organizations have that resist change.

As future work, we suggest: 1) incorporating technology based on Lean Manufacturing in each phase of the process to design, produce, and distribute the products. 2) Store the information generated in databases to analyze and make timely decisions. 3) Apply DMAIC, Lean Six Sigma, and Industry 4.0 in furniture design with the concept of social distancing to avoid COVID-19 infections.

Acknowledgment. The authors thank Universidad de Las Américas, Ecuador, for funding this research through the project FGE.PAV.19.11.

References

1. Carvajal, C.Y., Granda, L.S., Villamar, I.B., Hermida, L.H.: La importancia de las PYMES en el Ecuador. Rev. Obs. la Econ. Latinoam. (232) (2017)
2. Choudhary, A.K., Harding, J.A., Tiwari, M.K., Shankar, R.: Knowledge management based collaboration moderator services to support SMEs in virtual organizations. Prod. Plan. Control. **30**, 951–970 (2019)
3. Negrão, L., de Sousa Jabbour, A.L., Latan, H., Godinho Filho, M., Jabbour, C.C., Ganga, G.: Lean manufacturing and business performance: testing the S-curve theory. Prod. Plan. Control. **31**(10), 771–785 (2020)
4. Henao, R., Sarache, W., Gómez, I.: Lean manufacturing and sustainable performance: trends and future challenges. J. Clean. Prod. **208**, 99–116 (2019)
5. Kęsek, M., Bogacz, P., Migza, M.: The application of lean management and six sigma tools in global mining enterprises. In: Conference Series: Earth and Environmental Science, p. 012090 (2019)
6. Abualsaud, A.S., Alhosani, A.A., Mohamad, A.Y., Al Eid, F.N., Alsyouf, I.: Using six sigma DMAIC methodology to develop a facility layout for a new production line. In: International Conference on Modeling Simulation and Applied Optimization, pp. 1–5 (2019)
7. Kiatcharoenpol, T., Seeluang, R.: Six sigma methodology for reducing warpage defect of sanitary ware production. In: International Conference on Mechanical and Intelligent Manufacturing Technologies, pp. 104–108 (2019)
8. Pan American Health Organization (PAHO): WHO characterizes COVID-19 as a pandemic. https://www.paho.org/
9. Gallo, G., Trompetto, M.: The effects of COVID-19 on academic activities and surgical education in Italy. J. Investig. Surg. **33**(7), 687–689 (2020)
10. Fu, S., Li, Z., Liu, K., Din, S., Imran, M., Yang, X.: Model compression for IoT applications in industry 4.0 via multi-scale knowledge transfer. IEEE Trans. Ind. Inform. **16**(9), 6013–6022 (2019)

11. Dewi, M.V.K., Darma, G.S.: The role of marketing and competitive intelligence in industrial revolution 4.0. J. Manaj. Bisnis. **16**, 1–12 (2019)
12. Cakmakci, M.: Interaction in project management approach within industry 4.0. In: Advances in Manufacturing II, pp. 176–189. Springer, Cham (2019). https://doi.org/10.1007/978-3-030-18715-6_15
13. Annarelli, A., Battistella, C., Nonino, F.: Product Service Systems' Competitive Markets. Springer, Cham (2019)
14. Agarwal, R., Brown, P.J., Bajada, C., Stevens, P., Green, R.: The effects of competition on management practices in New Zealand–a study of manufacturing firms. Int. J. Prod. Res. 1–18 (2019). https://doi.org/10.1080/00207543.2019.1672901
15. Farfan-Quintanilla, Z., Caira-Jimenez, M., Sotelo-Raffo, F., Raymundo-Ibañez, C., Perez, M.: Applying SLP in a lean manufacturing model to improve productivity of furniture SME. In: International Conference on Intelligent Human Systems Integration, pp. 690–696. Springer, Cham (2019). https://doi.org/10.1007/978-3-030-39512-4_106
16. Rentes, V.C., de Pádua, S.I.D., Coelho, E.B., Cintra, M.A.D.C.T., Ilana, G.G.F., Rozenfeld, H.: Implementation of a strategic planning process oriented towards promoting business process management (BPM) at a clinical research centre (CRC). Bus. Process Manage. J. **25**(4), 707–737 (2019)
17. Filbeck, G., Krueger, T.M.: An analysis of working capital management results across industries. Am. J. Bus. **20**(2), 11–18 (2005)
18. Socconini, L.: Lean Company. Más allá de la manufactura. Marge Books (2019)
19. Sangani, R., Kottur, V.K.N.: Enhancement in productivity by integration of 5S methodology and time and motion study. In: International Conference on Intelligent Manufacturing and Automation, pp. 541–550. Springer, Singapore (2019). https://doi.org/10.1007/978-981-13-2490-1_50
20. Shakil, S.I., Parvez, M.: Application of value stream mapping (VSM) in a sewing line for improving overall equipment effectiveness (OEE). In: Intelligent Manufacturing and Energy Sustainability. p. 249 (2019). https://doi.org/10.1007/978-981-15-1616-0_24
21. Abhilash, C.R., Thakkar, J.J.: Application of six sigma DMAIC methodology to reduce the defects in a telecommunication cabinet door manufacturing process. Int. J. Qual. Reliab. Manag. **36**(9), 1540–1555 (2019)
22. Santos, M.Y., e Sá, J.O., Andrade, C., Lima, F.V., Costa, E., Costa, C., Galvão, J.: A big data system supporting Bosch braga industry 4.0 strategy. Int. J. Inf. Manage. **37**, 750–760 (2017)
23. FlexSim: 3D simulation modeling and analysis software. https://www.flexsim.com/es/
24. Abu, F., Gholami, H., Saman, M.Z.M., Zakuan, N., Streimikiene, D.: The implementation of lean manufacturing in the furniture industry: a review and analysis on the motives, barriers, challenges, and the applications. J. Clean. Prod. **234**, 660–680 (2019)
25. Liliana, L.: A new model of Ishikawa diagram for quality assessment. Conf. Ser. Mater. Sci. Eng. **161**, 012099 (2016)

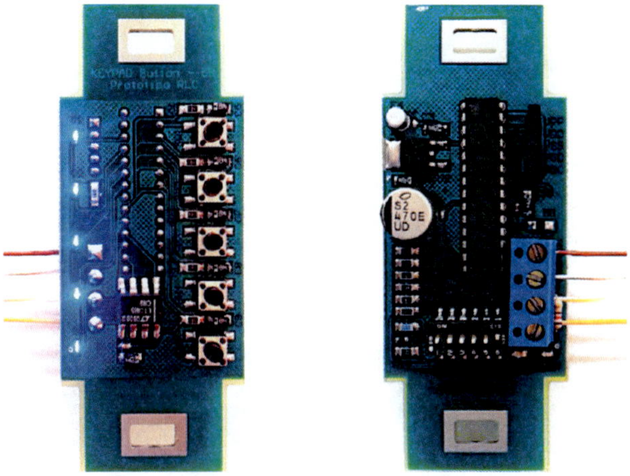

Fig. 4. Keypad button.

3.3 Dimmer

The dimmer (see Fig. 5) is an output device designed to drive a lighting zone using soft transition dimming technology [7].

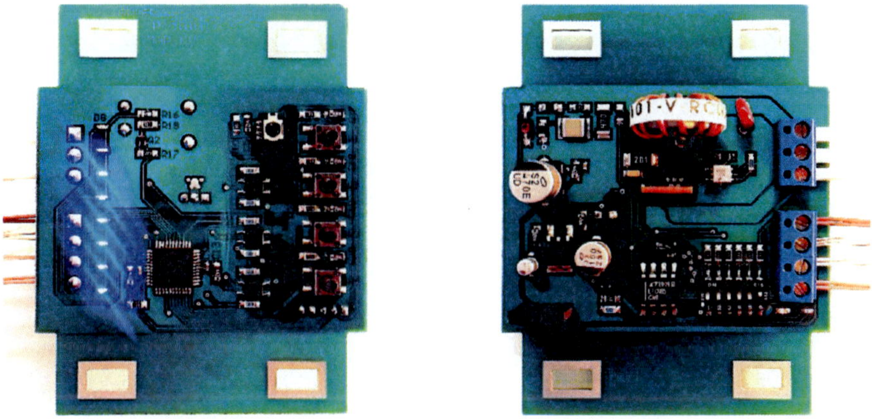

Fig. 5. Dimmer for LED and resistive light loads.

It incorporates a four-button local keypad: bright, dimm, on/off and sleep, used to raise the light level, lower the light level, turn on/off, or timer off; respectively. In addition to the local keypad, the dimmer has an infrared (IR) receiver, through which it can be remotely controlled with an IR remote control. In this mode, additional functions are implemented, such as: favorite scene, four-scene stage, and emergency scene.

It is integrated to the system through the link 1 (link of dimmers) using their communications interface.

3.4 Wallbox Dimmer Module

The wallbox dimmer module or WDM (see Fig. 6) is an output device designed to handle six independent lighting zones, using soft transition dimming technology [7].

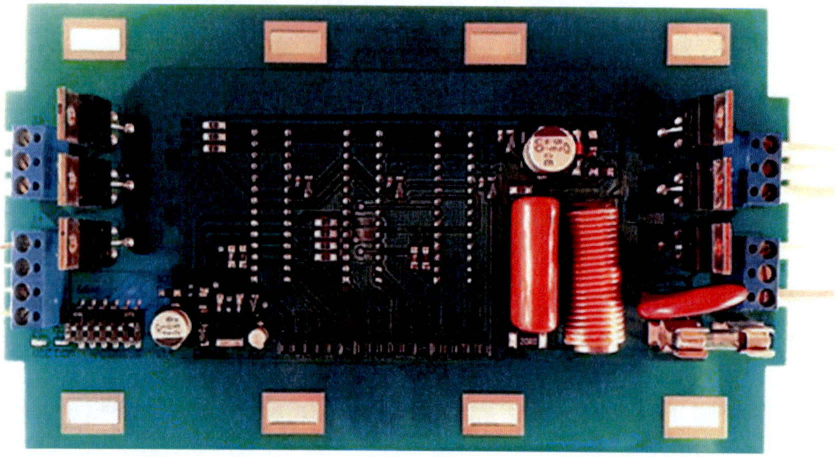

Fig. 6. Wallbox dimmer module (WDM) for LED and resistive light loads.

It incorporates a local push button to turn on or off the six lighting zones at the same time. It is integrated into the system through link 3 (WDM link) using its communications interface.

3.5 Contact Closure Input Interface

The contact closure input interface or CCI (see Fig. 7) is a device designed to handle eight independent relay inputs. With this accessory the user can integrate up to eight external devices to the system that have a contact closure as output.

Similar to the keypads, each input of the interface is programmed in the lighting software to execute a scene or a comfort environment. It is integrated into the system through link 4 (interface link) using its communications interface.

An Application of MVMO Based Adaptive PID Controller for Process with Variable Delay

Estefania Salazar, Marco Herrera, and Oscar Camacho(✉)

Departamento de Automatización y Control Industrial, Escuela Politécnica Nacional, Quito, Ecuador
{estefania.salazar,marco.herrera,oscar.camacho}@epn.edu.ec

Abstract. In this work, a Mean–Variance Mapping Optimization (MVMO) based adaptive PID controller is developed for a chemical process with variable time delay. For this, an adaptive tuning equation is obtained based on delay time variation. In order to verify the performance of the proposed controller, a comparison against an adaptive Smith Predictor, an adaptive control tuned by Dahlin equations and a Gain Scheduling controller through the ISE, TVu indexes is made. The simulations results show that the response of the system with the proposal approach improves the performance of the process with variable time delay.

Keywords: Adaptive control · MVMO · Smith Predictor · Optimization problems

1 Introduction

Most industrial processes are inherently nonlinear, and this feature makes the control problem challenging because nonlinearities could make the system not always kept at operating conditions. Another problem in industrial processes is the time-delays, which could vary due to the dynamic behavior of the system, strong disturbances or the delay time that comes from a material that moves from one point to another [1,2]. When time-delay varies, the controller could not be effective for new scenarios and this implies possible plant stoppages and economic losses [3], in these situations, the process needs a new tuning or adjustment of the control parameters.

The PID controller is widely used in the industry due to its simple structure and ease of implementation. Studies have shown that 90% of all controllers in the process industry have a PID structure [4]. However, the performance of this controller is limited for complex systems, such as those with time-varying parameters, and more than 60% of them require re-tuning after only 6 months [5].

Thus, PID control schemes that can adapt their parameters to the new scenarios have been developed, the main applications of adaptive PID control

M. Botto-Tobar et al. (Eds.): ICCIS 2020, AISC 1273, pp. 353–365, 2021.
https://doi.org/10.1007/978-3-030-59194-6_29

(APID) has been made on non-linear systems, where the analysis and control implementation are difficult [6]. However, most of the APID parameters are tuned based on classical methods such as Ziegler-Nichols, consequently, their performance could not be the best [7].

In order to tune PID controllers, several authors had studied different optimization algorithms. For instance, Verma S. and Mudi R. developed Particle Swarm Optimization (PSO) based adaptive PID controller for a chemical process [8]. Another optimization algorithm used in order to tune a PID controller for a tank coupled system is based on Genetic Algorithm (GA) [9].

In the present work, an adaptive control scheme based on Mean-Variance Mapping Optimization (MVMO) for tuning the PI controller parameters for a process with variable time-delay is developed. MVMO is the optimization algorithms proposed in [10], has been widely applied in Electrical Power Systems, where the MVMO has shown a better performance than some basic evolutionary algorithms, specifically in relation of convergence speed [11].

This work is organized as follows: Sect. 2 shows the basic concepts of PID controller and adaptive control. Section 3 presents the optimization algorithms that are applied for the tuning of classic PID controllers. Section 4 the proposal for the adaptive control scheme is presented. Section 5 describes the results that were obtained in this work for the mixing tank with variable time-delay. Finally, the conclusions are presented.

2 Basic Concept

In this section a brief introduction of the classical PID and adaptive controller concepts are presented.

2.1 PID Controller

A PID is a feedback controller that improves dynamic response and reduces steady the state error by combining the control actions: proportional, integral and derivative [12]. The transfer function of a PID controller is described by:

$$Gc(s) = Kc\left(1 + \frac{1}{T_i s} + T_D s\right) \tag{1}$$

Where:

Kc : proportional gain
T_i : integral time constant
T_D : differential time constant

These parameters adjust or correct a process variable in relation to a given reference by means of a control element [13].

2.2 Adaptive Control

Adaptation is to change a behavior in front of new conditions. Thus, in order to provide greater robustness it is convenient that controllers can modify its behavior to conform to changes in the dynamics of the process or disturbances [6]. In Fig. 1 is presented the general diagram of an adaptive scheme. Some kind of adaptive control schemes are described below:

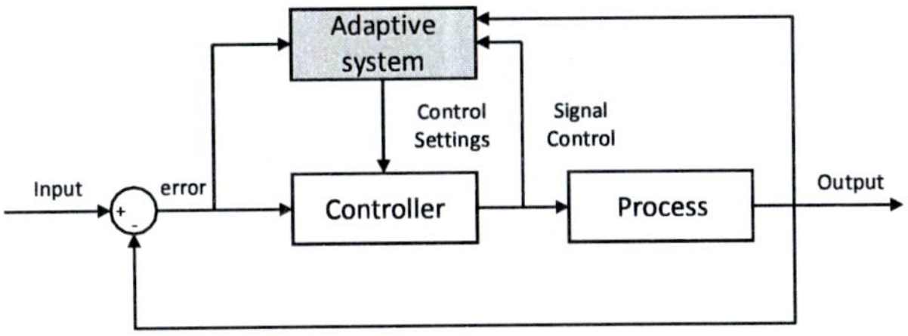

Fig. 1. General scheme of adaptive control

Switching Controller Gains. In this adaptive control scheme, when the dynamics of the system are previously known according to the operating conditions. Therefore, by continuous measurement of strategic process variables, the current operating point of the system is determined, and the controller parameter values are switched to the values that provide better performance under the given conditions [14].

Continuous Adjustment of Controller Gains. In this adaptive control, the controller parameters are a function of one or more process variables, which can be continuously measured; therefore, the controller is continuously updated. This feature makes it suitable for applications involving noisy measurements, such as flow or level [14].

3 Optimization Algorithms

The optimization algorithms are used to tune controller parameters in order to improves adaptive control performance to deal with time-variant dynamics [14]. The key differences among the algorithms are the mechanism to generate the new populations and the number of total iterations, but all of them seek to minimize a fitness function, for this work Integral Square Error (ISE) index, where error value corresponds to the difference between output and reference associated with differences or variations between the measured output and the reference input resulting in an error signal.

Therefore, this section is focused on two aspects. The first one consists on describing four optimization techniques in order to solve the problem of PID controllers tuning and secondly the methodology to perform the comparison to choose the better performance optimization method. The better optimization algorithm is selected to tune the parameters of the proposal approach.

3.1 Optimization Techniques

Genetic Algorithm (GA): is a search algorithm based on genetics and the concept of natural selection, where the fittest individuals survive [15].
Monte Carlo Method (MC): is a subset of computational algorithm that use the process of repeated random sampling to make numerical estimations of unknown parameters [16].
Particle Swarm Optimization (PSO): an optimization technique inspired by social behavior of bird flocking and fish schooling [8].
Mean–Variance Mapping Optimization (MVMO): belongs to the family of evolutionary algorithms, its operating principle is based on a special mapping function, which is applied to the product (child) of the processes of mutation and reproduction of individuals [17].

3.2 Methodology and Optimization Techniques for First Order Plus Dead Time Systems

In order to select the better optimization algorithm among GA, MC, PSO and MVMO, each one is used for tuning the PID controller parameters of each of four processes. These processes are described by a First Order Plus Dead Time (FOPDT) model as is indicated in Eq. (2); where, K_p is a static gain, τ is a time constant and t_o is a time delay. These models are obtained by varying the parameters K_p, t_o and τ for an interval of 0.5 into 1,3 and we have chosen the models for which all algorithms are able to obtain a maximum overshoot in control signal less than 10% (Table 1).

$$Gp(s) = \frac{K_p}{\tau s + 1} e^{t_o s} \tag{2}$$

Table 1. Parameters of the process models.

	Kp	τ	t_o
Model 1	2.5	2	1.5
Model 2	2	1.5	1
Model 3	2	2	1.5
Model 4	1	2	2. 5

Table 2. Parameters of the algorithms.

	MVMO	PSO	MC	AG
Number of generations	200	50	1019	50
Population size	1	20	1	20
Total iterations	200	1000	1019	1000

In Fig. 2 the ISE performance indexes (fitness function) for all models using optimization algorithms are presented, thus, in Model 4 the time delay is longer than the others. The convergence time in Fig. 3 indicates that the MVMO algorithm obtains a valid solution to the problem in less time than the other algorithms, because the number of total iterations of MVMO is approximately five times less than the other algorithms, as detailed in Table 2. By considering this analysis, MVMO algorithm was chosen to be applied in tuning the proposed adaptive control scheme.

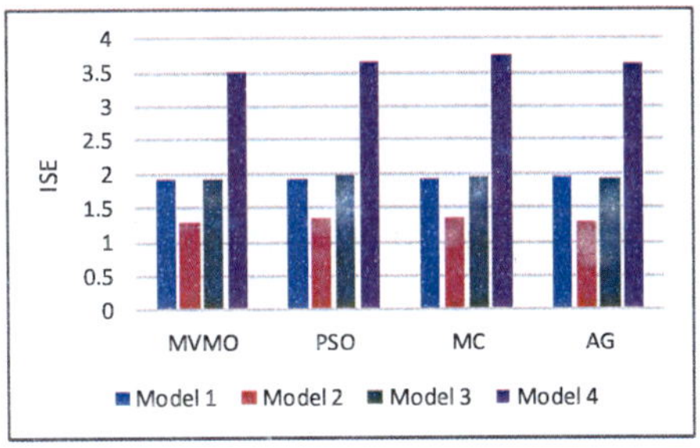

Fig. 2. Performance ISE by each algorithm in four models.

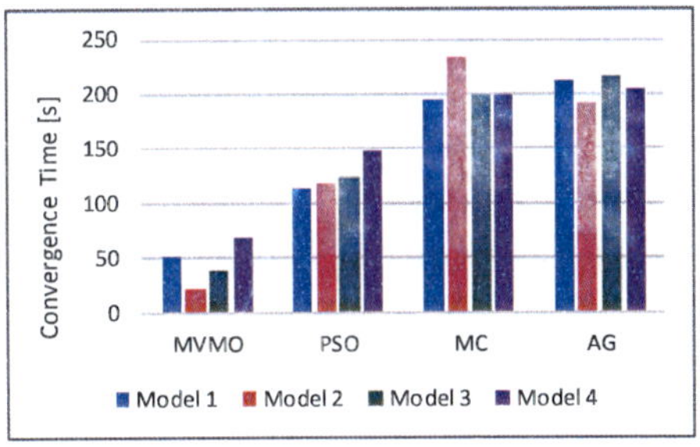

Fig. 3. Convergence time of the algorithms in four models.

4 Adaptive Control Scheme Proposed Approach

In this section a procedure for tuning of PI controller parameters and the control adaptive scheme based on MVMO is presented.

4.1 PI Tuning by Using MVMO

MVMO special mapping function is based on the average and the variance of the set comprised of the n-best solutions obtained up to that moment. MVMO saves the best solutions in a memory file that is continuously updated between iterations and the algorithm selects the first ranked solution in the file as the parent solution, so adopts an elitism criterion [17].

A notable feature of the MVMO implementation is its focus on single-particle evolution within a search space, and optimization variables are normalized in a range of 0 to 1, [18]. The procedure of MVMO for the optimal tuning of the PI parameters is as follows and the classical MVMO flowchart is schematically shown in Fig. 4. Where, the MVMO algorithm is based on the following steps:

Initialization. A total number of iterations is defined N and the possible initial solutions of the Kc and T_i parameters are generated randomly within their bounds by using Eq. (3), where $rand$ is a random number between $[0,1]$ and D is the number of variables to optimize. This array of possible solutions is normalized within the range $[0,1]$.

$$x_i^{ini} = x_i^{min} + rand(x_i^{max} - x_i^{min}), i = 1...D \tag{3}$$

Fitness Evaluation and Termination Criterion. The fitness of possible solutions was evaluated by ISE index. The termination criterion is governed by the value of N.

Solution Archive. The n-best solutions that MVMO has found so far, are stored in the solution archive. The solutions are sorted in a descending order of fitness, and the archive is updated in each iteration. Furthermore, the mean and shape values for each optimization variable are calculated after each update of the archive.

Parent Selection, Mutation, and Crossover. The first ranked position in the solution archive is assigned as the parent. One of the two dimensions (K_c, T_i) of the optimization problem is randomly selected to mutate based on the mean and shape values. For the next generation, a child vector is created by combining the parent vector and the mutated dimensions vector.

4.2 Adaptive Scheme

The proposal corresponds to an adaptive control by continuous adjustment of controller parameters, whose control scheme is shown in Fig. 5. These parameters vary according to the variation of time delay by means of tuning equations obtained by MVMO.

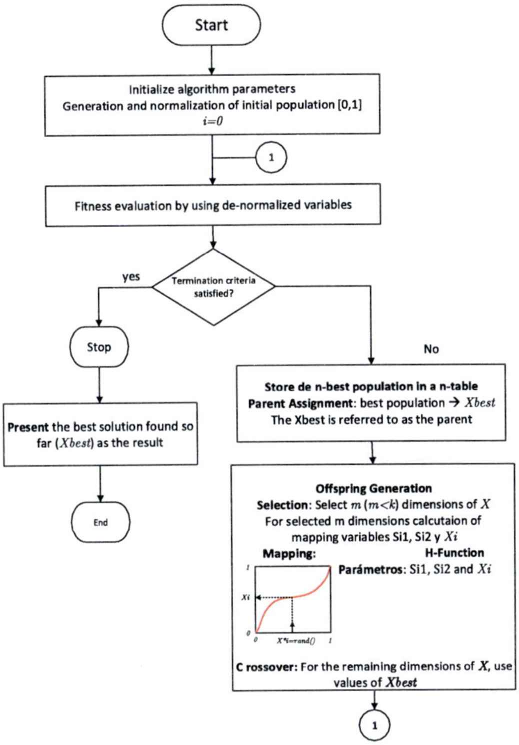

Fig. 4. MVMO flowchart for optimal tuning of PID

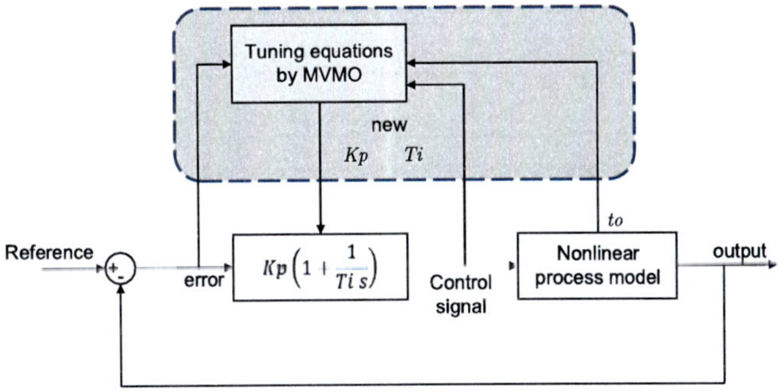

Fig. 5. Control scheme by continuous adjustment of controller parameters

5 Simulation Results

The algorithms were implemented using Simulink-Matlab® R2018a on a Dell computer Intel® CoreTM i5-7200U CPU @2.50 Ghz, 8 GB RAM and Windows 10 enterprise, 64-bit OS.

5.1 Proposed Case Study

The case of study is a variable time delay process presented in [19], which consists of a tank that mixes two input streams that have different temperatures, a hot one $W1(t)$ and a cold one $W2(t)$ as is shown in Fig. 6.

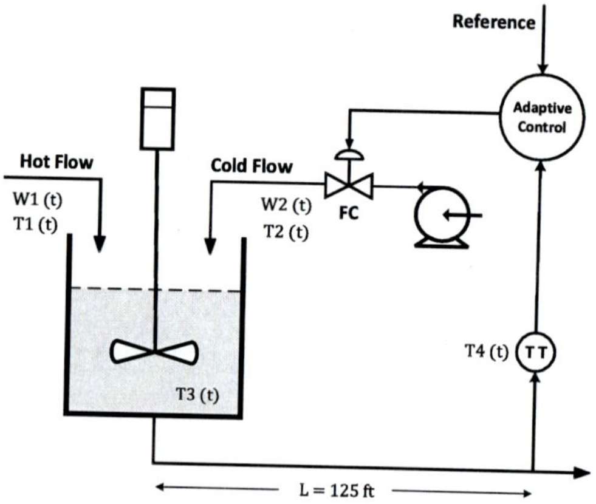

Fig. 6. Mixing tank with variable delay

The differential equations model of this process is presented in [19]. The amount of cold stream input to the tank is handled by opening or closing of a valve which the temperature of the product inside the tank is controlled. To measure this temperature a transmitter is located 125 ft downstream from the product outlet of the tank (see Fig. 6), this transmitter operates in a range of 100 to $200^o F$.

5.2 Proposed Control Scheme

In order to find the tuning equations (4) and (5), seventeen changes in the hot water flow $W1(t)$ are made and using MVMO algorithm a PI controller is tuned for each change in the flow. These disturbances are carried out within the permitted range of variation in the plant, from 250 $[lb/min]$ to 120 $[lb/min]$, and with the data obtained from the tuning and time delay changes the linear equation of the PI controller adjustment parameter are found and presented in Fig. 7 for Kc and Fig. 8 for Ti.

Where, the linear equation is given by:

$$Kc(t_o) = 0.0938t_o - 1.0378 \quad (4)$$

$$T_i(t_o) = 0.5923t_o - 1.9536 \quad (5)$$

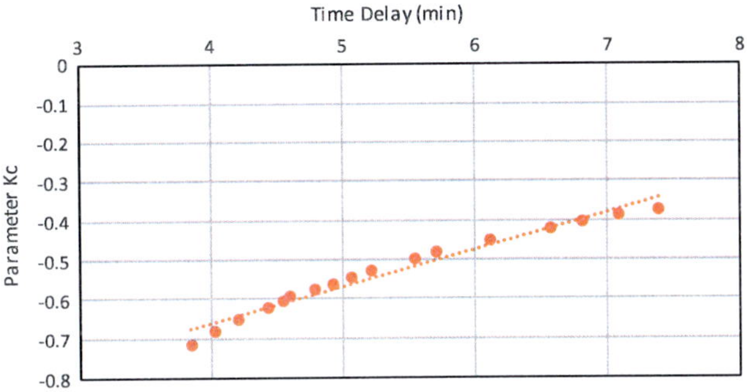

Fig. 7. Trend line of Kc parameter.

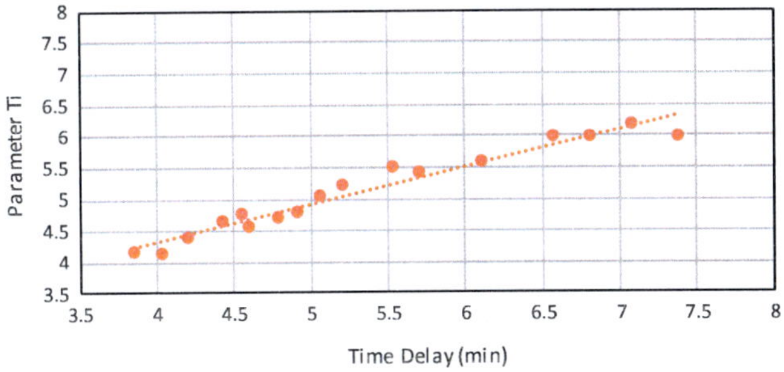

Fig. 8. Trend line of T_i parameter.

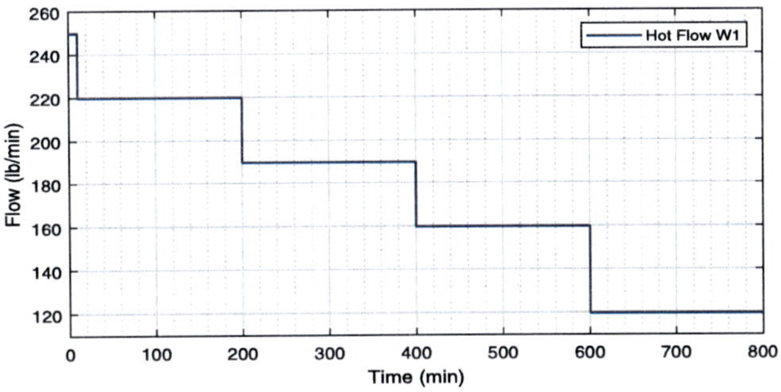

Fig. 9. Inlet flow change.

5.3 System Response for Disturbances Test

The input hot water flow varies, from initial value 250 $[lb/min]$, then this value decreases to 220 $[lb/min]$, 190 $[lb/min]$, 160 $[lb/min]$ to finally stabilize at 120 $[lb/min]$. These variations are observed in Fig. 9.

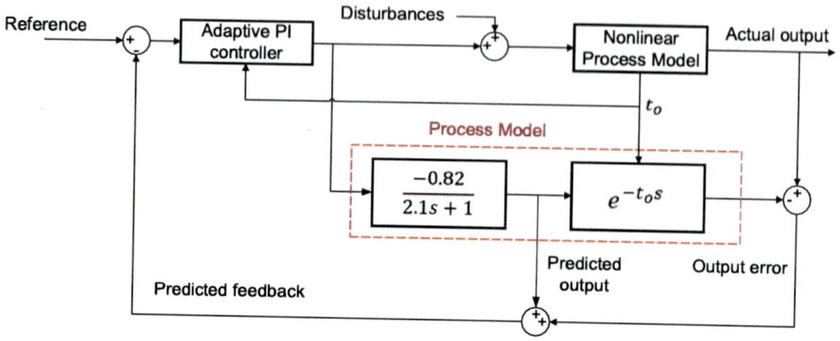

Fig. 10. Adaptive smith predictor scheme.

Table 3. PI controller parameters table for AC 2.

Hot flow $W1(t)$	Kc	T_i
220 [lb/min]	−0.8974	5.931
190 [lb/min]	−0.8448	7.6622
160 [lb/min]	−0.7272	8.2197
120 [lb/min]	−0.5749	8.9649

The proposal control is compared against the following schemes: Adaptive Control 1 (AC1) is an Adaptive Smith Predictor. Kc and T_i parameters are adjusted as a function on Time-delay, in Fig. 10 the corresponding scheme is shown, where the reduced order model was taken from [20]. Adaptive Control 2 (AC2) represents a Gain Scheduling Controller where Kc and T_i parameters change depending on the disturbance variation to certain predefined values tuned by MVMO, this parameters are presented in Table 3. Adaptive Control 3 (AC3) denotes a PID controller using Dahlin equations. Kc parameter is adjusted as a function of the continuously measured Time-delay of the nonlinear process model (Eq. 6). Finally, the proposed control scheme, is called Proposal.

Figure 11 shows the increase of time-delay when the flow of hot water decreases. In addition, Fig. 13 shows how the controller output is affected by increasing the delay for all adaptive controllers.

$$Kc = \frac{1}{-0.82}\left(\frac{t_o}{2.1}\right)^{-1}, T_i(s) = \left(1 + \frac{1}{2.1s}\right) \tag{6}$$

Figure 12 shows the evolution of output for disturbances test, and the adaptive controller schemes 1, 2 and 3 have stronger oscillations and more settling-time than the proposed control scheme when the last disturbance occurs, and the hot water flow drops to 120 $[lb/min]$.

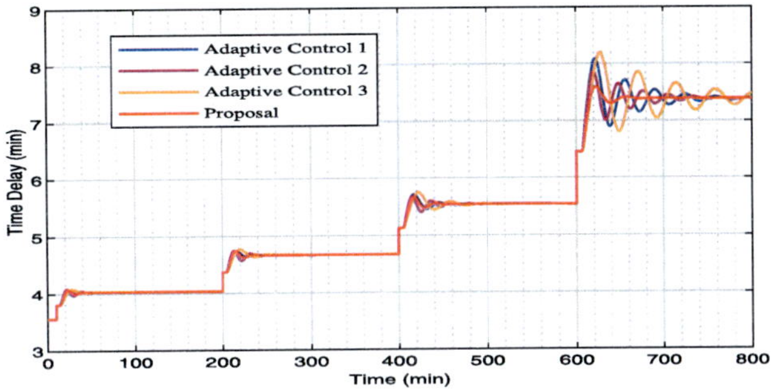

Fig. 11. Time delay variation.

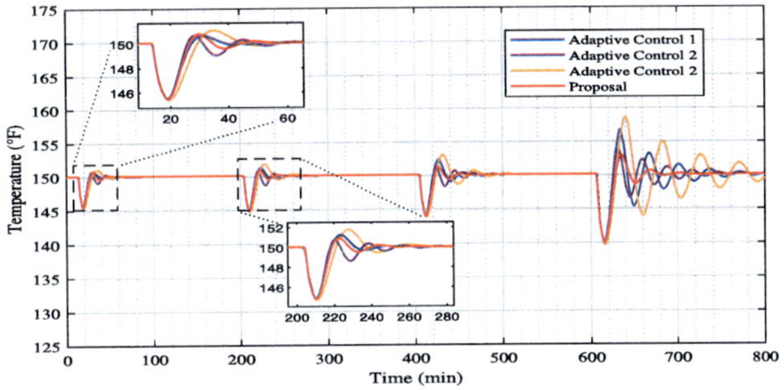

Fig. 12. Output process.

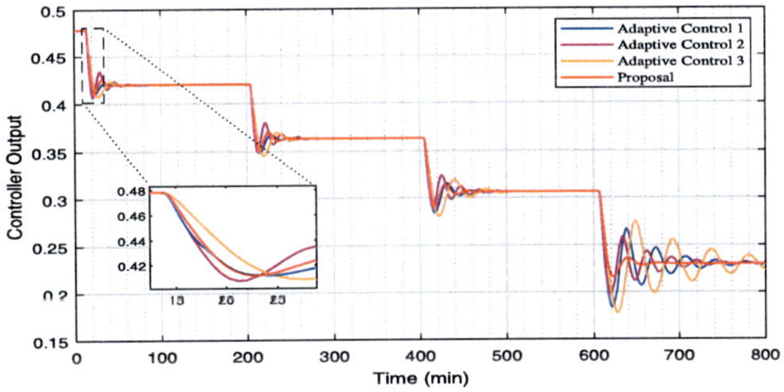

Fig. 13. Control signal.

According to Table 4 the proposed adaptive control scheme presents the lowest ISE and TVu compared to the other schemes when a disturbance occurs. In Fig. 13 can be seen the signal control, where the proposed scheme has smoother control signal than the other adaptive controllers.

5.4 Analysis of Results

The values of ISE and TVu indexes obtained for each adaptive control scheme when changes in the input flow occur are shown in Table 4. The ISE index is associated with the output reference stabilization.

The proposed scheme presents an ISE index reduction of 13.39% compared to the Adaptive Smith Predictor, 55.17% less compared to the Gain Scheduling, and 61.37% less compared to Dahlin based adaptive control. On the other hand, the proposed scheme presents a TVu index reduction of 17.54% compared to the Adaptive Smith Predictor, 51.07% less compared to the Gain Scheduling, and 37.6% less compared to Dahlin based adaptive control.

Table 4. ISE and TVu indexes.

	AC1	AC2	AC3	Proposal
ISE	0.2017	0.3897	0.4523	0.1747
TVu	0.5371	0.9051	0.7098	0.4429

6 Conclusions

An Adaptive PID controller adjusted using a MVMO algorithm was designed for a process with variable time delay, where the proposal adaptive controller presents the lowest ISE and TVu performance indexes for disturbance test. According simulation results, the MVMO algorithm has a lower convergence time than the other algorithms studied due to the directly proportional relationship of the number of total iterations and the convergence time.

Acknowledgements. Authors thank to PII-19-03 and PIGR-19-17 Projects of the Escuela Politécnica Nacional, for its sponsorship for the realization of this work.

References

1. Camacho, O., Martinez, J.: Procesos con Retardo de Tiempo Dominante: Diseño, Análisis y Comparación de Estrategias de Control. Editorial Académica Española. Saarbrücken, Alemania (2017)
2. Camacho, O., Leiva, H.: Impulsive semilinear heat equation with delay in control and in state. Asian J. Control **22**(3), 1075–1089 (2020)
3. Baquero, D., Ochoa, C., Chávez, D., Camacho, O., Colina E.: Adaptive sliding mode control based on fuzzy logic for variable dead time processes. In: 4th International Conference on Control, Decision and Information Technologies (2017)
4. Samad, T.: A survey on industry impact and challenges thereof. IEEE Control Syst. **37**, 17–18 (2017)
5. Copot, D., Ghita, M., Ionescu, C.M.: Simple alternatives to PID-type control for processes with variable time-delay. Processes **7**, 146 (2019)

6. Astrom, K.J., Wittenmark, B.: Adaptive Control. Dover Publications Inc., New York (2008)
7. Dey, C., Mudi, R.: Performance improvement of PI controllers through dynamic set-point weighting. ISA Trans. **50**, 220–230 (2011)
8. Verma, S., Mudi, R.: Particle swarm optimization based adaptive PID controller for pH-neutralization process. Adv. Intell. Syst. Comput. **327**, 159–166 (2015)
9. Colorado, O., Hernández, N., Touh, J.S., Medina, J.: Algoritmo genético aplicado a la sintonización de un controlador PID para un sistema acoplado de tanques. Pädi **10**, 50–55 (2018)
10. Erlich, I., Venayagamoorthy, G.K., Worawat, N.: A mean-variance optimization algorithm. In: 2010 IEEE Congress on Evolutionary Computation, pp. 1–6 (2010)
11. Universität Duisburg-Essen, MVMO Homepage. https://www.uni-due.de/mvmo/. Accessed 26 Apr 2020
12. Van Rensburg, P., Shaw, I., Van Wyk, J.: Adaptive PID control using a genetic algorithm. In: 1998 Second International Conference on Knowledge-Based Intelligent Electronic Systems, Australia, pp. 133–138 (1998)
13. Dos Santos, A., Victhoff, M.: A tunning strategy for multivariable PI and PID controllers using differential evolution combined with chaotic Zaslavskii map. Expert Syst. Appl. **38**, 13694–13701 (2011)
14. Lipták, B.G.: Instrument Engineers' Handbook: Process Control and Optimization, pp. 265–273. CRC/Taylor and Francis, Connecticut (2018)
15. Goldberg, D.E.: Genetic Algorithms in Search, Optimization and Machine Learning. Addison-Wesley Publishing Co. Inc., Massachusetts (1989)
16. Tempo, R., Ishii, H.: Monte Carlo and Las Vegas randomized algorithms for systems and control: an introduction. Eur. J. Control **13**, 189–203 (2007)
17. Rueda, J., Loor, R., Erlich, I.: MVMO for optimal reconfiguration in smart distribution systems. IFAC-PapersOnLine **48**, 276–281 (2015)
18. Loor, R.X.: Reconfiguración de Redes de Distribución de Energía Eléctrica basada en Optimización de Mapeo Media-Varianza. Escuela Politécnica Nacional, Quito (2014)
19. Camacho, O., Smith, C.: Sliding mode control: an approach to regulate nonlinear chemical processes. ISA Trans. **39**, 205–218 (2000)
20. Mejía, C., Camacho, O., Chavez, D., Herrera, M.: A modified smith predictor for processes with variable time delay. In: 4th Colombian Conference on Automatic Control, pp.1–6 (2019)

PID and Sliding Mode Control for a Reactor-Separator-Recycler System: A Regulatory Controllers' Comparison

Roberto Arroba, Karina Rocha, Marco Herrera, Paulo Leica, and Oscar Camacho(✉)

Departamento de Automatización y Control Industrial, Escuela Politécnica Nacional, Quito, Ecuador
{jorge.arroba,karina.rocha,marco.herrera,paulo.leica, oscar.camacho}@epn.edu.ec

Abstract. One of the most common methods to obtain propylene glycol (PG) is using a Re-actor-Separator-Recycle (RSR) system. The control of temperature and reagent concentration during the process operation are essentials for the optimal performance of this chemical process. Those variables are regulated using control techniques to maintain the process in the operating points in the presence of disturbance. However, the non-linear behavior of the process limits the range of operating point using classical controls, being robust control an alternative for nonlinear processes. Here, the performance of classical and robust control techniques in the presence of disturbance in the mixing point is simulated in an RSR system to obtain PG. The results show through of the Integral Square Error (ISE) index that robust control techniques have better efficiency and soft control action than classical techniques when a stronger disturbance is applied.

Keywords: Regulatory control · Propylene glycol · Chemical process · Recycling system · SMC

1 Introduction

In chemical processes, the increasing demand for the reuse of raw materials has resulted in increasingly integrated processes. Most chemical processes in the industry contain recycle streams; this type of stream, also called re-circulation, connects the output to the input of a process, equipment, or group of equipment [1]. The benefits of recycling systems are recovery and reuse of unconsumed reagents, recovery of catalysts, especially in reactors, because they use substances that accelerate or retard a chemical reaction without participating in its [2]. One of the most frequent recycling systems in chemical processes consists of a reactor coupled to a separation system that allows the regain components from the product stream.

M. Botto-Tobar et al. (Eds.): ICCIS 2020, AISC 1273, pp. 366–377, 2021.
https://doi.org/10.1007/978-3-030-59194-6_30

The components of a recycling system combine heat exchangers, reactors, distillation columns, for example., which must remain at a defined point of operation so that the system does not destabilize when a disturbance appears [3]. This occurs because by increasing the integration of both matter and energy in the plant, when the disturbance appears input of a process, it is feedback to the process itself through the recycling streams, increasing the gains and the time constants of the process [4].

Due to the dynamic integration of these processes, it is necessary to keep variables such as temperatures, concentrations and flows within the reference points that allow the process to work under safe operating conditions [5]. The type of control allows a certain number of variables to be maintained in the set-points is called regulatory, whether they are defined by an operator or by higher layers of control. In the regulatory control, there are single loop controllers among these the classic Proportional-Integral-Derivative PID [6].

PID controllers are the most widely used in industrial applications because they are the easiest to implement in digital systems. Besides, for these controllers have different tuning methods, among the most used methods are Ziegler-Nichols and Dahlin. The controller tuning is based on first approximating the process by a first- or second-order plus delay model, thus the values of K, t_0 and τ are obtained [7]. In the case of systems that have a single input and multiple outputs, there is the tuning method proposed by Austin [7]. Although the classic PID controllers continue to be one of the most widely used, it has a significant disadvantage, it only allows control of systems that operate in a restricted area, and due to the chemical processes are highly nonlinear, robust controllers are required.

Sliding mode control (SMC) is a well-known technique, based on a control law designed to guarantee the state trajectories reach a switching surface previously chosen, according to the desired dynamical behavior of the closed-loop system. The fundamental advantages of this method are robustness against a large class of perturbations or model uncertainties and a reduced amount of information needed, compared with other classical control techniques. Nevertheless, the use of the sign function in the control law to guarantee the passage of the approach phase to the sliding mode produces the chattering phenomenon, which consists of a rapid variation of the control signal, exciting the high frequency of the process, and leading to damages. SMC has been known as an appropriate method for handling nonlinear systems with uncertain dynamics and disturbances. Because of these properties, the diverse applications of the sliding mode control methodology can be found in the areas of electric motors, manipulators, power systems, mobile robots, spacecraft, automotive control, and chemical processes [8].

This work presents the comparison of regulatory control, applying classic control techniques, and robust control. Finally, the performance of these controllers is compared using the ISE performance index to determine which of the controller should be recommended for this plant. The paper is organized as follows: section two describes some fundamentals about control, then in section three the process is shown, in section four is presented the regulatory control approaches, section five contains the results and finally some conclusions are presented.

2 Control Fundamentals

2.1 Approach to Reduced Order Systems

Models developed with this method provide the dynamic relationship between the selected input and output variables. The model approximation is obtained by making small changes in the input variables on a nominal operating condition [9]. The resulting dynamic response is used to determine the model. The method used in this work uses the graphical calculations shown in Fig. 1. The intermediate values determined from the graph are the magnitude of the input change, which is called δ, the magnitude of the steady-state change in the output denoted $\varDelta$, and the times in which the output variable reaches 28 and 63% of its final value.

$$Y(t_0 + \tau) = 0.632\varDelta \tag{1}$$

$$Y(t_0 + \frac{\tau}{3}) = 0.283\varDelta \tag{2}$$

$$t_{28\%} = t_0 + \frac{\tau}{3} \tag{3}$$

$$t_{63\%} = t_0 + \tau \tag{4}$$

$$\tau = 1.5(t_{63\%} - t_{28\%}) \tag{5}$$

$$t_0 = t_{63\%} - \tau \tag{6}$$

$$K = \frac{\varDelta}{\delta} \tag{7}$$

The expressions taken from [9], are shown in Eq. 1, and Eq. 2 are those that represent the values of the output variable at 28 and 63% of its final value. The two-time values at 28% (3) and 63% (4), are selected to determine parameters such as the delay time t_o (6) and the time constant of the model τ (5) is obtained, and with (7) the gain of the system model is determined. With this procedure we obtain the first-order-plus-dead time (FOPDT) model.

2.2 Proportional-Integral-Derivative Controllers

It combines three control components that are proportional, integral, and derivative, and each one has a specific function on the process output. The classical law of control expressed as a function of time is expressed in (8). The proportional control action provides a faster system speed in case of reference value changes or disturbances. The integral control action eliminates the error in steady-state, and the derivative action means that changes in the desired value can be anticipated, and therefore an appropriate correction can be added before the actual change [10].

$$u_c(t) = K_p e(t) + K_i \int_o^t e(t)dt + K_d \dot{e}(t) \tag{8}$$

Where K_p, K_i and K_d are the proportional, integral and derivative constants of the controller, respectively, being tunable. $e(t)$ is the error between the desired

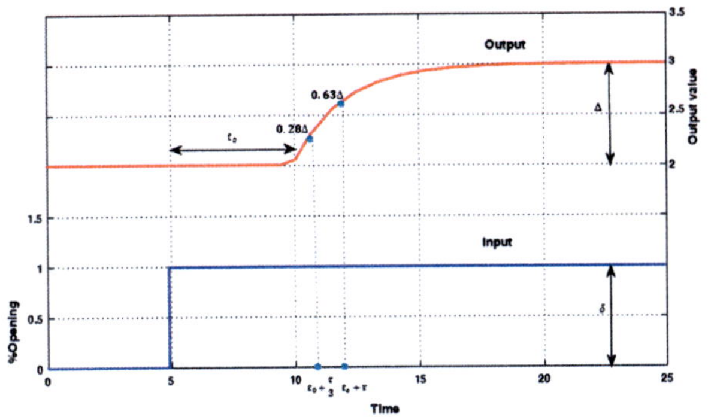

Fig. 1. Process reaction curve method

value and the measured value. The control law expressed in Laplace's domain is expressed:

$$U_c(s) = K_c\left[1 + T_d(s) + \frac{1}{T_i(s)}\right] \tag{9}$$

2.3 Controller Tuning

To tune a PID type controller from a First-Order-Plus-Dead Time (FOPDT), multiple methods can be found in the literature, in this work we will talk about the chosen tuning methods, which are from Skogestad, due to the advantages over other methods as explained in [10] among them showing a tuning method for systems without delay, and Dahlin for its performance when having plants with delays as explained in [7].

2.4 Sliding Mode Control

Sliding mode control (SMC) is a robust nonlinear control technique that has been developed in the last decades. This type of control is especially attractive due to its ability to work in nonlinear systems, variable time systems, as well as uncertainties and disturbances directly [11]. The SMC-based design consists of defining a surface along which the process should slide to the desired final value, as shown in (10), where n is the degree of the system.

$$\sigma(t) = \left(\frac{d}{dt} + \lambda\right)^n \int_o^t e(t)dt \tag{10}$$

The control objective is to ensure that the controlled variable is equal to its reference value at all times, which means that $e(t)$ and its derivatives must be zero.

$$\dot{\sigma}(t) = 0 \tag{11}$$

The control law, $U(t)$ in an SMC controller defined in (12), consists of two parts; a continuous part $U_c(t)$ and a discontinuous part, $U_D(t)$.

$$U(t) = U_c(t) + U_d(t) \tag{12}$$

In Camacho and Smith [11], the SMC is developed based on a FOPDT model of the process, which can be seen in (13) and (14), the sliding surface, this with the aim to have a function of parameters that can be tuned.

$$U(t) = \frac{t_{0T}}{K_d}\left[\frac{X(t)}{t_{0T}} + \lambda_0 e(t)\right] + K_d \frac{\sigma(t)}{|\sigma(t)| + \delta} \tag{13}$$

$$\sigma(t) = sign(K)\left[-\dot{X}(t) + \lambda_1 e(t) + \lambda_0 \int_o^t e(t)dt\right] \tag{14}$$

2.5 Regulatory Control

The central control system is the regulatory control level; this consists mainly of single-loop and ratio controllers. Multivariable controllers are rare at this level. The regulatory level maintains many controlled variables near their set points. This mainly serves to stabilize the plant and maintain safe operating conditions [12].

3 Reactor-Separator- Recycler (RSR) System

In this paper is taken the model proposed by [13] including linear valves and linear transmitters in each operating unit, as shown in Fig. 2. The RSR system has a heat exchanger, reactor CSTR, flash distillation column, and a mixing point where the recirculating fluid is mixed with propylene oxide that is at temperature, flow, and initial concentration. In Table 1 shows the operation point of the system RSR. All operating values and all the differential equations describing the process can be found in [13].

4 Regulatory Control Design

In this work are developed PI type, cascade and SMC controllers at the regulatory level, because at the point of mixing the propylene oxide can possibly enter with an excess of initial concentration or at a high temperature. The objective of the control is that each operating unit of the RSR system remains at the operating points when concentration or temperature disturbances appear.

4.1 Classic Controller (PID)

PI-type controllers were developed to cause a variable to return to its reference point when a disturbance occurs, causing the steady-state error to be zero. For this reason, in the heat exchanger and flash separation column PI controllers were developed, controlling the temperature and concentration of propylene oxide, respectively. Cascade control is developed in the CSTR reactor because there is only one variable to control and two outputs (temperature inside the reactor and propylene glycol concentration) that need to be controlled (Fig. 3).

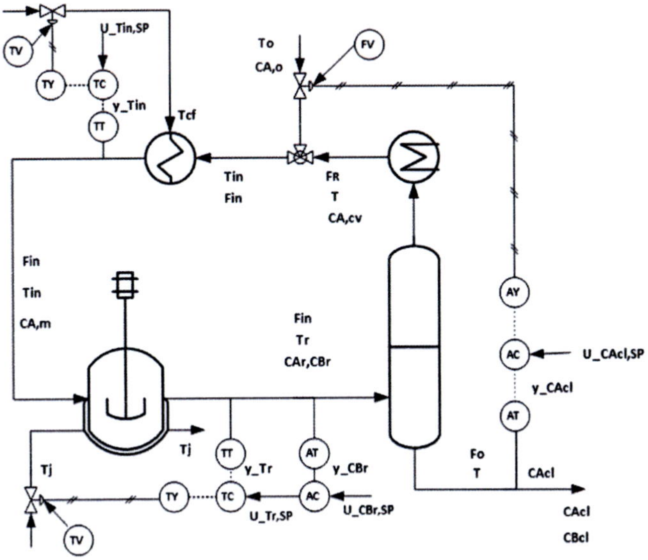

Fig. 2. Reactor-Separator-Recycler (RSR) system

Table 1. Systems approach and tuning methods.

Controlled variable	Approximate model	Tuning
$y_{C_B,r}$	$\frac{289.9e^{-10.833s}}{227.979s+1}$	External loop-Austin
y_{T_r}	$\frac{28.87e^{-0.437s}}{27.981s+1}$	Internal loop-Dahlin
y_{T_in}	$\frac{15.7}{0.957s+1}$	Skogestad
$y_{C_A,cl}$	$\frac{81.8}{23.573s+1}$	Skogestad

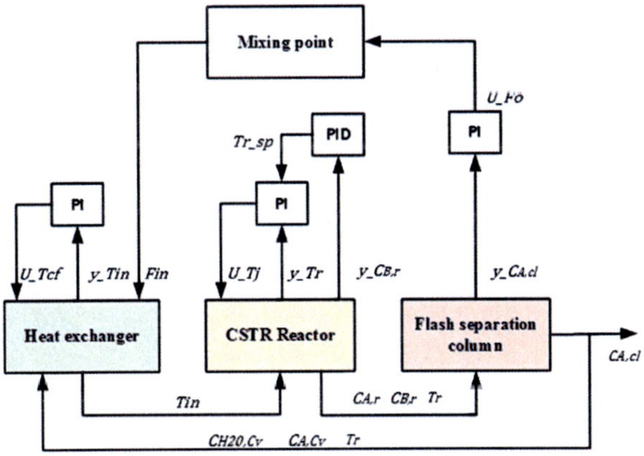

Fig. 3. Regulatory control with classic controllers

4.2 Robust Controller (SMC)

The design of the SMC controller was based on the FOPDT model previously obtained, then with the method proposed by [11] the tuning parameters were also obtained in [11]. The variables to which this control was applied are shown in Fig. 4.

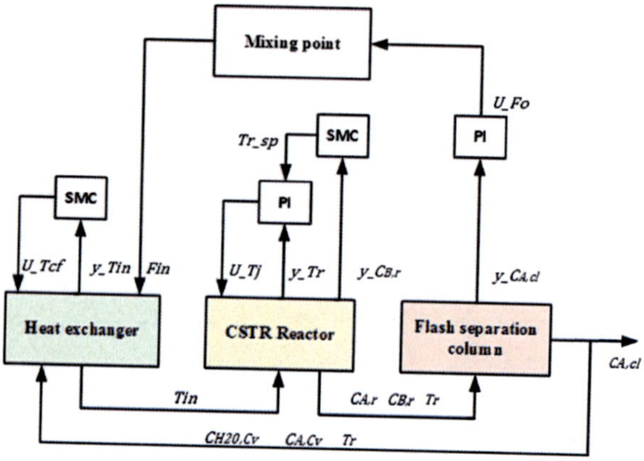

Fig. 4. Regulatory control with robust controllers

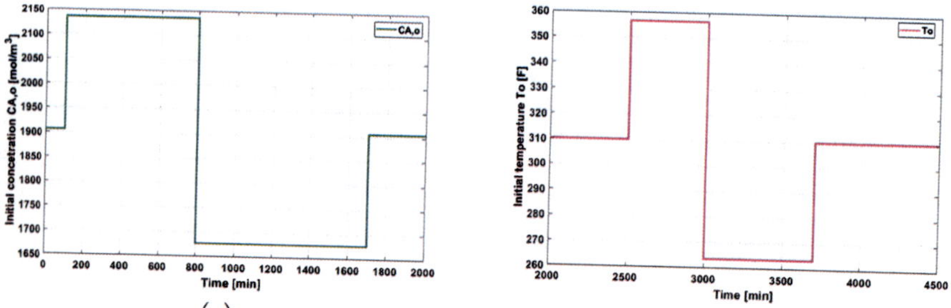

(a) (b)

Fig. 5. Disturbances at the mixing point (a) Concentration (b) Temperature

5 Results

To test the performance of the controllers, the system was tested with two disturbances that appear at the mixing point of the process. The first disturbance is the initial concentration, at 100 min it is increased to 12% of its nominal value, at 800 min it is reduced on 24% and at 1700 min it is increased on 12% until reaching the nominal value as shown in Fig. 5a. The second disturbance is the initial temperature, at 2500 min it is increased to 15% of its nominal value, at 3000 min it is decreased on 30% and at 3700 min it is decreased 15% until reaching the nominal value as shown in Fig. 5b.

In heat exchange, the mixture is preconditioned to a necessary temperature before entering the CSTR reactor. For this reason, when the initial temperature disturbance appears at the mixing point, it is observed at 2500, 3000 and 3700 min the overshoot is 4% with the PID controller is greater compared to the SMC that is 1%, as shown in Fig. 6. Additionally at 100, 800 and 1700 min when the initial connection disturbance appears, there is a minimal interaction its doesn't exceed 0.04% of its nominal value. In addition, Fig. 7 shows the control action with the PID controller has less over-shoot with respect to the SMC controller.

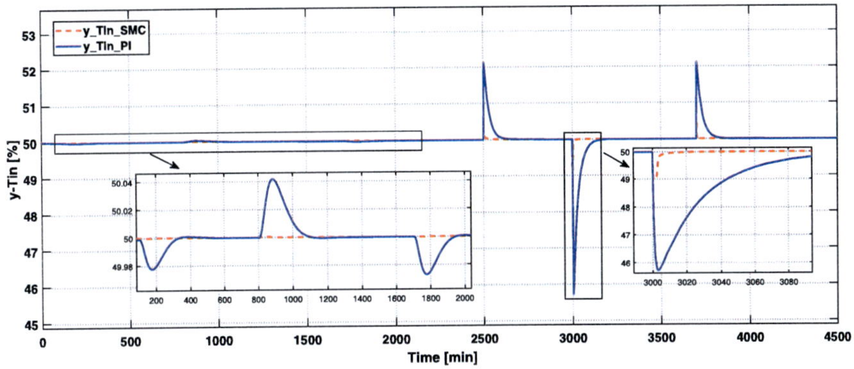

Fig. 6. Temperature inside the heat exchanger

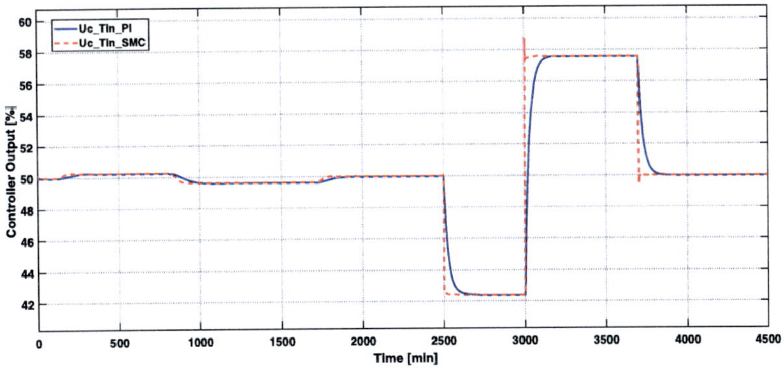

Fig. 7. Controller output in the heat exchanger

The CSTR reactor of the RSR system has only one control valve, the variables to be controlled are the concentration and the jacket temperature, for this reason, a cascade controller was designed. The external loop variable is the concentration of propylene glycol and the internal loop variable is the temperature. In this case, when the initial concentration disturbance appears at the mixing point at 100, 800 and 1700 min, the overshoot with the PID-PI controllers are greater respect to the SMC-PI controllers as seen in Figs. 8 and 9.

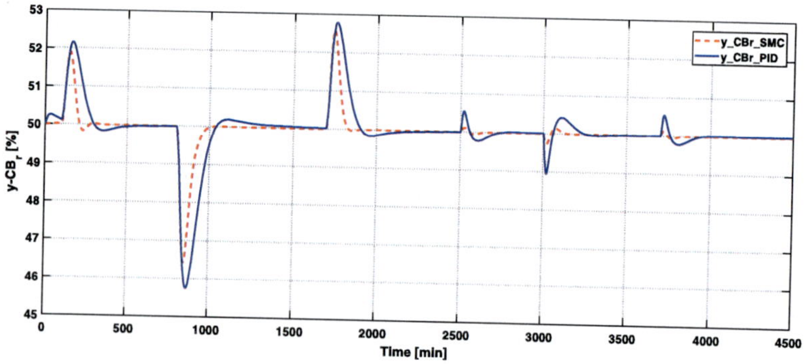

Fig. 8. Propylene glycol concentration in the CSTR reactor

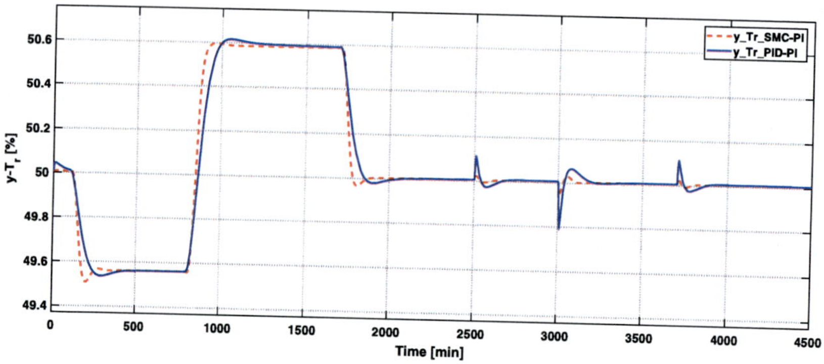

Fig. 9. Temperature inside the CSTR reactor

At 2500, 3000 and 3700 min the initial temperature disturbance appears at the mixing point, in this case in the internal loop it presents the dynamics observed in Fig. 8 that allows the concentration of propylene glycol to remain at the operating point. Furthermore, the effect of the SMC controller on the output loop reduces over-shoots on the internal loop as shown in the Fig. 9. The control effort shows over-shoots of the 2% with the PID-PI controllers, while the overshoots with the SMC-PI controller are 0.01% as shown in Fig. 10.

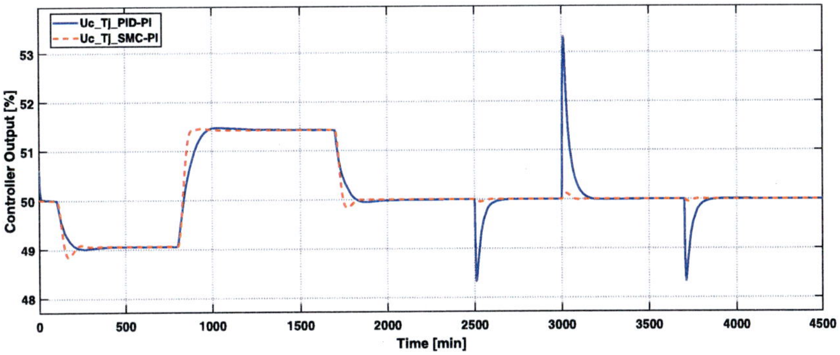

Fig. 10. Controller output in the CSTR reactor

In the flash distillation column, as in the CSTR reactor, it is observed that the concentration disturbance at the mixing point produces overshoots greater than 1%, while the temperature disturbance generates minimal interactions. Since it is an interconnected system, it can be seen that making the system robust with SMC controllers reduces the effect of disturbances in this unit, as shown in Fig. 11. The settling times in the operating units of the RSR system are shorter with the SMC controllers, in addition to the overshoots generated with the SMC controllers are higher than with the PID controllers, despite this, the control actions are smooth, see Table 2 (Fig. 12).

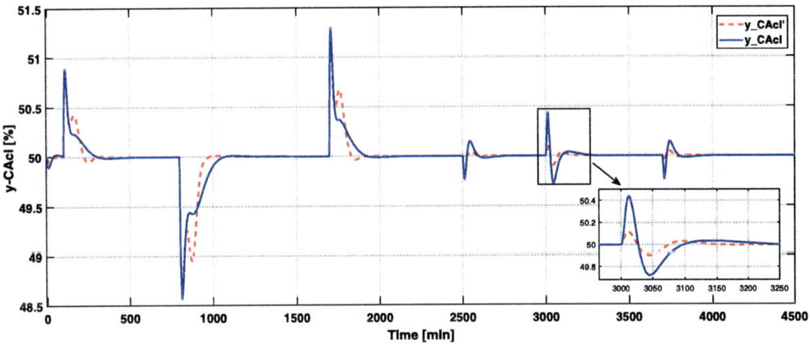

Fig. 11. Propylene oxide in the flash separation column

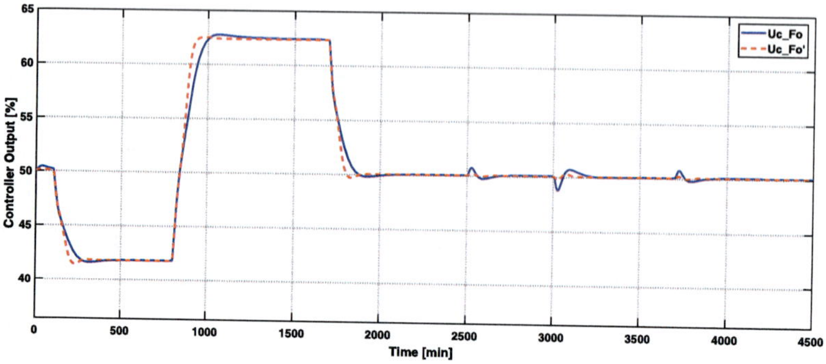

Fig. 12. Controller output in the flash separation column

Table 2. Comparison of classic and robust controllers

Variable	Index	SMC	PID type
y_$C_{B,r}$	ISE	1073	2497
	TVU	4.2	6.3
yT_r	ISE	6.12	8.02
	TVU	1.2	1.7
yT_{in}	ISE	49.97	457.3
	TVU	17.3	16.9
y$C_{A,cl}$	ISE	142.4	145.9
	TVU	8.1	9.2

6 Conclusions

Two controllers were designed and compared. The PID continues being of great help in control systems, and it is the reason to be the most widely used in the industry, but despite this, it still has problems when there are disturbances that can cause to stop working the system at a certain point of operation and therefore decrease the performance of the process. SMC controllers are an alternative to be used at an industrial level since the approach based on a FOPDT model becomes less complex to implement in electronic devices.

Acknowledgements. Authors thank to PII-19-03 and PIGR-19-17 Projects of the Escuela Politécnica Nacional, for its sponsorship for the realization of this work.

References

1. Yin, X., Lin, J.: Subsystem decomposition of process networks for simultaneous distributed state estimation and control. Am. Inst. Chem. Eng. J. **65**(3), 904–914 (2019)

2. Patle, D.S., Gadhamsetti, A.P., Sharma, S., Agrawal, V., Rangaiah, G.P.: Plantwide control of the formic acid production process using an integrated framework of simulation and heuristics. Ind. Eng. Chem. Res. **57**(40), 13478–13489 (2018)
3. Wu, K.L., Yu, C.C.: Reactor/separator processes with recycle - 1: candidate control structure for operability. Comput. Chem. Eng. **20**(11), 1291–1316 (1996). Elsevier, Great Britain
4. Zheng, R.A., Mahajanam, V., Douglas, J.M.: Hierarchical procedure for plantwide control system synthesis. Am. Inst. Chem. Eng. J. **45**(6), 1255–1265 (1999)
5. Brogan, W.L.: Modern Control Theory, 3rd edn. Prentice-Hall, New Jersey (1991)
6. Tipett, M.J., Bao, J.: Distributed control of chemical process networks. Int. J. Autom. Comput. **12**(4), 368–381 (2015). Springer, Heidelberg
7. Zhi-gang, Z., Ben-guo, Z., Zhen-fu, B.: Dahlin algorithm design and simulation for time-delay system. In: Proceedings of the 21st Annual International Conference on Chinese Control and Decision Conference, pp. 5849–5852. IEEE Press, Guilin (2009)
8. Musmade, B., Patre, B.: Sliding mode control design for robust regulation of time-delay processes. Trans. Inst. Meas. Control **37**(6), 699–707 (2015)
9. Marlin, T.: Designing Processes and Control Systems for Dynamic Performance, 2nd edn. McGraw Hill, Hamilton (2008)
10. Skogestad, S.: Simple analytic rules for model reduction and PID controller tuning. Model. Ident. Control **25**(2), 85–120 (2004)
11. Camacho, O., Smith, C.: Sliding mode control: an approach to regulate nonlinear chemical processes. ISA Trans. **39**(2), 205–218 (2000)
12. Alvarez, T.R.R., Codorniú, R., Pina, I. F. B.: PI and PID Temperature and Product Concentration Control in a Chemical Plant producing propylene glycol. J. Eng. Technol. Ind. Appl. **4**(14) (2018)
13. Moscoso, H., Bravo, G.M., Alvarez, H.: Model-based supervisory control structure for plantwide control of a reactor-separator-recycle plant. Ind. Eng. Chem. Res. **53**(52), 20177–20190 (2014)

YaniWawa: An Innovative Tool for Teaching Using Programmable Models over Augmented Reality Sandbox

Sonia Cárdenas-Delgado[1,2](✉), Oswaldo Padilla Almeida[1,3], Mauricio Loachamín-Valencia[1,2], Henry Rivera[1,4], and Luis Escobar[1,4]

[1] Universidad de las Fuerzas Armadas - ESPE, Sangolquí, Ecuador
{secardenas,ovpadilla,mrloachamin,herivera,lfescobar}@espe.edu.ec
[2] Departamento de Ciencias de la Computación - Grupo Geoespacial, Sangolquí, Ecuador
[3] Departamento de Ciencias de la Tierra y la Construcción - Grupo Geoespacial, Sangolquí, Ecuador
[4] Departamento de Ciencias de la Energía y Mecánica - Laboratorio de Mecatrónica y Sistemas Dinámicos, Sangolquí, Ecuador

Abstract. In the last decade, new technologies have emerged and proved being useful in the teaching-learning process. In this paper, a playful interaction system for learning about Andean Mountain range in Ecuador is presented. The interaction was achieved using natural gestures, images visualization, frontal and vertical projection over an augmented reality sandbox. Additionally, a flexible learning mode has been developed, where children can learn by exploring. An initial study was carried out to compare whether the learning outcomes were achieved by a free exploratory experience or traditional learning methods. Forty children from 10 to 12 years old participated. The analysis of the pre-tests and the post-tests indicated children increased their knowledge about Ecuador's mountains after using YaniWawa. Our study reveals the potential of exploratory learning games. The questions about satisfaction and usability showed that 80% of children had fun while learning; 60% would recommend it to their friends; and 80% of the participants liked YaniWawa, moreover YaniWawa was scored with a mean of 4.70 over 5 .

Keywords: Augmented reality · Elementary and secondary education · Topographic maps · Graphics systems and interfaces · Experiential learning modes · Interactive learning games.

1 Introduction

The teaching-learning process cannot be restricted to traditional classrooms, children require new styles of learning and new ways of teaching [11]. They require methods that capture and hold their attention, engaging them in the learning process when they are in a state of flow [5]. Children and adults like to

M. Botto-Tobar et al. (Eds.): ICCIS 2020, AISC 1273, pp. 378–388, 2021.
https://doi.org/10.1007/978-3-030-59194-6_31

"play" with games, puzzles and perform activities which provide an opportunity to experience, interact and learn. They prefer learning through a well-designed and effective concrete experience, tools easy of use and most natural [14]. Game interactive based learning are considered appropriate approach for teaching and engaging children in a more successful way than traditional learning methods [4].

Exploration is crucial when the development of innovation and creativity is needed. High levels of exploration imply variance seeking rather than mean seeking learning processes. Learning is more effective when the tasks operate with autonomy with respect to goals and supervision [12]. In this context, educational theorists Kolb and Fry developed the Experiential Learning Model (ELM) [9]. In this model of learning, individuals are stimulated to reflect on their actions and consequences, in order to foster understanding and applying these skills in future actions. The model helps teachers design lessons experiential, interactive in order to develop critical and creative thinkers.The model consists of four stages: concrete experience, reflective observation, abstract conceptualization, and active experimentation. ELM is an integrated process, each stage being mutually supportive of and feeding into the next [6,9,15]. It is possible to start the learning at any stage, the important is to follow the sequence, as seen in the Fig. 1 [9].

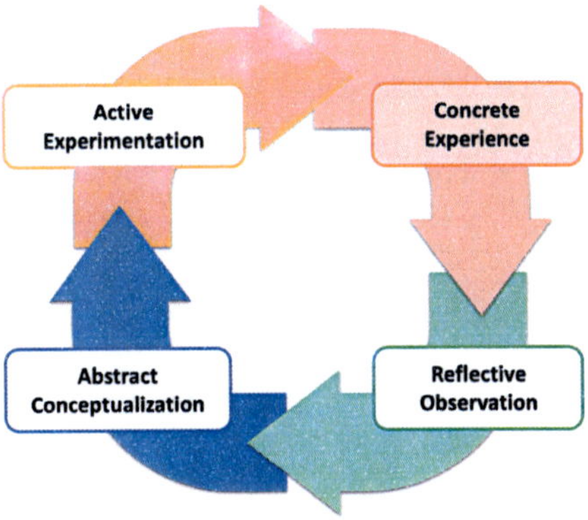

Fig. 1. Learning model [9].

On the other hand, design-based-learning focuses on using the game as a tool for learning. In this line, new and interesting opportunities have presented to develop game-based learning (GBL). Within game-based learning, a particular sub-genre is learning through sandboxes. The sandbox-style game play, suggests more-or-less indirect free-play, less-structured, something pure and free, promotes creativity, suggests imaginative capability, nurtures decision making, critical-thinking skills, and self-awareness.

Nowadays, the teaching-learning processes demand new challenges to provide more and better interactive tools. In this paper, a novel playful interaction system for learning about Andean Mountain Range in Ecuador is presented. It incorporates Natural Human-Computer Interaction based on video game technology sandbox-style, and exploratory learning method with natural gesture interaction. The game was called YaniWawa.The name of our game is in a native language of Ecuador, Quechua. Yani means thought and Wawa means children, therefore, YaniWawa can be understood as children's thoughts. The objective of YaniWawa is to reinforce the learning of topics related to Ecuadorian Geography, such as the Andean Mountain Range, the relief, the highest volcanoes and snow falls in Ecuador.

The first objective of this work was to develop a game that included playful interaction with sandbox-style and exploratory learning. The second objective was to carry out a study to find out if children had a higher increase in knowledge playing in pair with the exploratory learning method or traditional learning method. Our main hypothesis is that children will learn more by playing the game in the exploratory mode than in the traditional mode.

There are plenty amount of research about interactive games and systems used to teach like: storytelling interfaces [16] for preschool children that is a two-page book, it uses picture cards to trigger animations; CheMO [17] is a system where users can perform virtual chemistry experiments and learn about all the procedures that involve experimentation; FlowBlocks [8] used to teach children how interconnection works; AR Sandbox [10] is a system used to learning about Geoscience Education by the use of a 3D physic model that shows with layers and colors the height of landscapes; even in Ecuador there are some research about this topic, like NAR [7] that is a robot developed to help children to learn numbers and colors, but there is no research about the use of Augmented Reality (AR) to teach children about our geography.

The paper is organized as follows. Section 2 mentions Materials and Methods. Section 3 presents the results obtained after pre and post test, and Sect. 4 is a brief discussion of results about this research. Section 5 presents conclusions and future work.

2 Materials and Methods

This sections contains the materials and method that were used to perform the research. As it was said in the previous section in Ecuador there is no study about this kind of systems, however at countries like Australia, Tailand, USA and Spain among others there are much rich information in this matter. In [10] is said that there are more than 150 AR sandboxes around the world and in the web page [2] can be seen their distribution. It is important to denote tree aspect involved to implement the systems:

2.1 Learning Process

There are works developed to motivate in training and learning, of this technically speaking [1], who have developed a work in which they show how the Kinect sensor can be used, together with Unity, and Gamification techniques to develop virtual 3D Game-Like training environments, where users are highly motivated during training, maximizing at the same time the learning and retention of contents during the process. This case study is the prevention training environment and fire escape aimed at children that was developed for the Cochabamba Fire Department explaining the advantages over traditional methods.

Ayala [3], mentions that learning mathematics using technology offering a kinesthetic learning results in better understanding of mathematical concepts, graphics and formulas. the AR provides a big boost in the learning curve but is limited to the use of markers and the amount of movement that is why the complement with the Kinect becomes a fundamental equipment for kinesthetic learning, the main objective of this work is the development and the graphic application position versus time that allows the student to replicate with their hands several different types of movements and the software will detect, calculate and generate graphs based on the positions in the selected period of time.

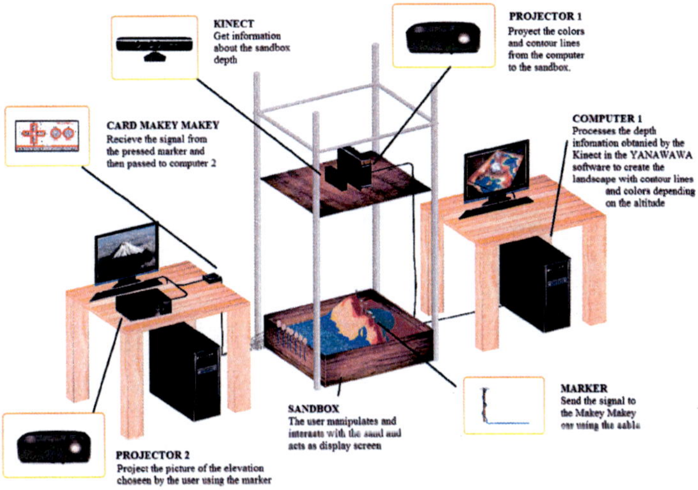

Fig. 2. YaniWawa physical structure.

2.2 Structure

The structure proposed in the present research, is a generic way uses two systems that interact with each other. A steel support was built to hold the two devices. The first one is the basic ARSandbox as showed in [10,13] and the second one consist in a computer used to project the elevation chosen by the user thought a marker connected to a Makey Makey card as can be seen in Fig. 2.

2.3 Software

YaniWawa application was programmed in C# language. The game combines NUI and Frontal Projection. It simulates a tactile screen. The required elements are a sort throw projector and a Kinect. The screen area is a box filled with sand. To achieve the user interaction, the Kinect device was used. It was located next to the top projector. The top projector and the Kinect devices were located above the sandbox. A interactive interface were connect with the Sandbox by means of a Makey Makey and alligator clips. Buttons were built in the model for each mountain. This card was configured to perform the interaction among the sandbox, the the buttons, and the application hosted on a laptop. The main interface includes a map of Ecuador that has been adapted to the model and presented eight of the most important mountains learned in school. The interfaces of each mountain show its image, a video of of the local wildlife, in addition, an audio describing the most important characteristics of it. YaniWawa architecture is selfexplained in Fig. 3.

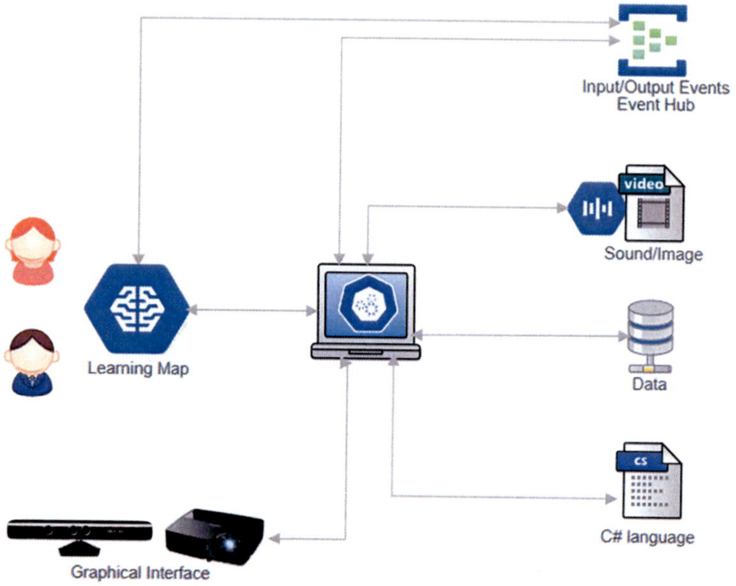

Fig. 3. YaniWawa architecture.

3 Results

This section presents the results obtained in the research, for a better explanation there are three subsections:

3.1 Game Description

There are two game modes, the first is to teach the Ecuadorian Andean System, here the children have to complete a test to know how many mountains and snows they know, then with YaniWawa the teacher show them how many and where the mountains are. The markers in the system help the children to get information about a specific mountain or snow by touching them. At the end of the lesson the children have to answer another test remembering as much mountains as they can. The second mode uses the Ecuadorian coastal profile to teach children about peninsulas, inlets, gulfs, etc. The first mode can be seen in the Fig. 4.

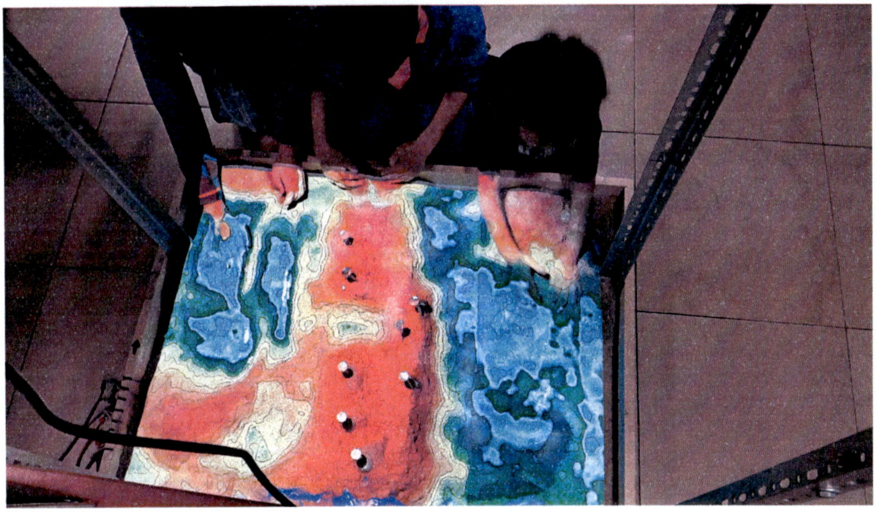

Fig. 4. The main game interface.

3.2 Study Description

The study was carried out taking four groups of 10 participants each, with ages of 11, 12 and 13 years, with 23, 45 and 32% respectively.

Before playing, each child answered a questionnaire writing how many mountains and snowy mountains they know. After playing with the platform, a new questionnaire was conducted asking how many mountains and snowy mountains they remember. The phases of the study are shown in the Fig. 5.

After the experiment all the data were stored and organized and showed that the children were able to learn whit this method as can be seen in the Fig. 6.

The mean values for each series in the Fig. 6 are 3.3 correct answers in the pre-test and 5.8 correct answers in the post-test.

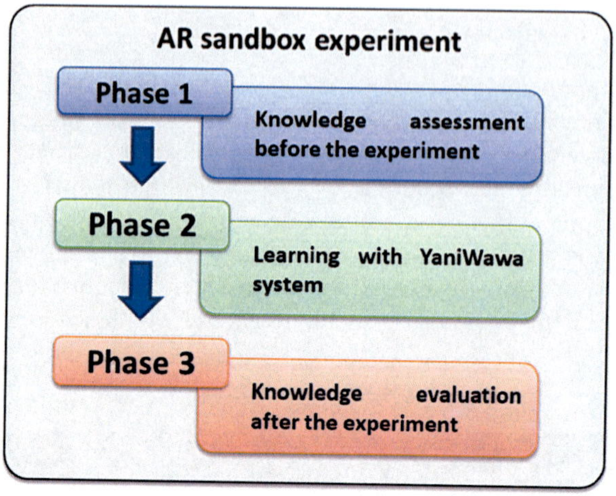

Fig. 5. Study phases of research.

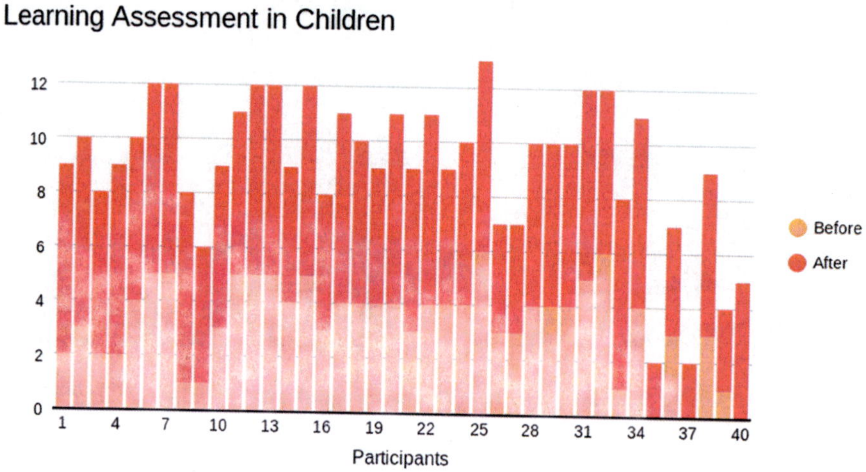

Fig. 6. Results of learning assessment before and after playing YaniWawa.

3.3 Measurements

In the current research, the impact that the system generated on children was also evaluated thought a short questionnaire. It has seven items that evaluate ease of use, understand and learn, also a part of the questionnaire ask the children if they had fun and recommend this game to a friend. All the questions were scored from 1 to 5, being 5 (very satisfied) the greater value.

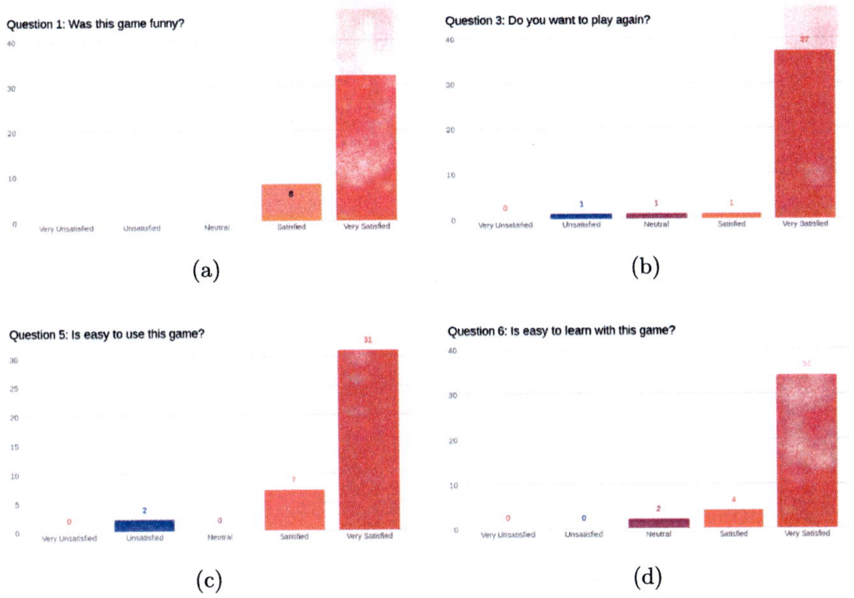

Fig. 7. a) Question 1 b) Question 3 c) Question 5 d) Question 6.

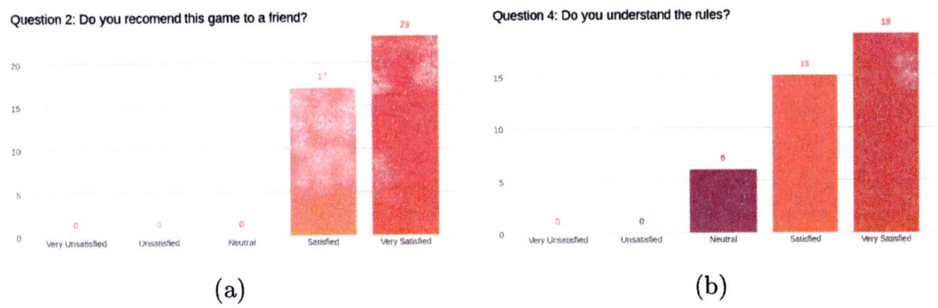

Fig. 8. a) Question 2 b) Question 4.

The first, third, fifth and sixth questions show that more the 77% of the participants give the greatest score as can be seen in the Fig. 7. being this questions related to how fun the game was, if the user want to play again and how easy was to use and to learn with tit.

In the questions two and four the largest number of participant rated the game with a value of 4 or more as it is show in the Fig. 8. Question 2 is related to recommending the game to a friend and the question 4 is used to know if the rules of the game were easy to understand.

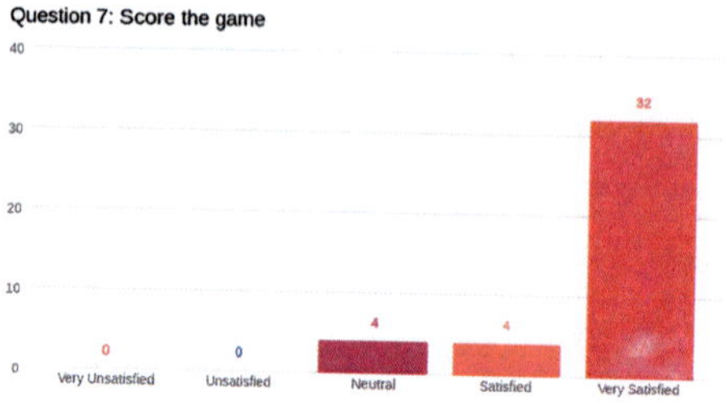

Fig. 9. Results obtained on the game score.

Finally, with the last question evaluated all the system as a game. As it is showed in the Fig. 9 the 80% of children said the game scored 5 points (very satisfied).

4 Discussion

The results obtained in the test about mountains and snowy mountains revels important information about how the system helps children remember twice the names of the mountains taught. Moreover, almost all the children believe that the game is fun, easy to use, and learn whit it as can be seen in the Fig. 7. The lower score in questions 2 and 4, was mainly due to the scarce experience of the staff in charge of giving the instructions to deal with children, and finally question 5 showed that children found easy to play with it.

Unlike other works [10,13], an extra device help to specify information about landscape, and it help children to get more interested in the game, because it allows to get a deeper interaction with the game.

5 Conclusions and Future Work

A game was created that allow children to have a playful interaction with the ARSandbox system called YaniWawa. It uses a exploratory learning method to enhance the memory capacities of children, with an increase to twice of information retained, as it can be seen in the Fig. 6. The research showed that tools as Makey Makey in conjunction with other technologies and methods led children to had a meaningful experience in a short period of time, increasing the knowledge, enriching and complementing the traditional method of teaching. As a future work, it is intend to use printed 3D models to show how different physical phenomenon may affect a landscape, and what kind of consequences it can have.

Acknowledgement. This paper and the research behind it would not have been possible without the support and colaboration of Universidad de las Fuerzas Armadas ESPE, specially the Center for Military Applications (CICTE), the Mechatronis and Dinamyc Systems laboratory and IEEE-ESPE student branch, whose efforts and colaboration has been key to develop and test our prototype.

References

1. Arispe, R.M.S., Collarana, V.D.: Virtual training platforms using Kinect sensor, Unity and Gamification techniques. Acta Nova **8**, 109 (2017)
2. Augmented Reality Sandbox. https://arsandbox.ucdavis.edu/. Accessed 20 Apr 2020
3. Ayala, N.A.R., Mendívil, E.G., Salinas, P., Rios, H.: Kinesthetic learning applied to mathematics using kinect. Procedia Comput. Sci. **25**, 131–135 (2013). https://doi.org/10.1016/j.procs.2013.11.016
4. Blair, K., Murphy, R.M., Almjeld, J.: Cross Currents: Cultures, Communities, Technologies. Cengage Learning, Boston (2013). google-Books-ID: T8aiUYoQvBAC
5. Craig, S., Graesser, A., Sullins, J., Gholson, B.: Affect and learning: an exploratory look into the role of affect in learning with AutoTutor. J. Educ. Media **29**(3), 241–250 (2004). https://doi.org/10.1080/1358165042000283101
6. Dalvi, B., Cohen, W.W., Callan, J.: Exploratory Learning. In: Blockeel, H., Kersting, K., Nijssen, S., Železný, F. (eds.) ECML PKDD 2013. LNCS (LNAI), vol. 8190, pp. 128–143. Springer, Heidelberg (2013). https://doi.org/10.1007/978-3-642-40994-3_9
7. Espinoza, E.N.A., Almeida, G.R.P., Escobar, L., Loza, D.: Development of a social robot NAR for children's education. In: Advances in Emerging Trends and Technologies. Advances in Intelligent Systems and Computing, pp. 357–368. Springer International Publishing, Cham (2020). https://doi.org/10.1007/978-3-030-32033-133
8. Grouse, P.: "Flowblocks": a technique for structured programming. ACM SIGPLAN Not. **13**(2), 46–56 (1978). https://doi.org/10.1145/953422.953424
9. Kolb, D.A.: Experiential Learning: Experience as the Source of Learning and Development. FT Press, Upper Saddle River (2014). google-Books-ID: jpbeBQAAQBAJ
10. Kundu, S.N., Muhammad, N., Sattar, F.: Using the augmented reality sandbox for advanced learning in geoscience education. In: 2017 IEEE 6th International Conference on Teaching, Assessment, and Learning for Engineering (TALE), pp. 13–17 (2017). https://doi.org/10.1109/TALE.2017.8252296. ISSN: 2470-6698
11. Leung, M.Y., Wang, Y., Chan, D.K.K.: Structural surface-achieving model in the teaching and learning process for construction engineering students. J. Prof. Issues Eng. Educ. Pract. **133**(4), 327–339 (2007). https://doi.org/10.1061/(ASCE)1052-3928(2007)133:4(327). http://ascelibrary.org/doi/10.1061/%28ASCE%291052-3928%282007%29133%3A4%28327%29
12. McGrath, R.G.: Exploratory learning, innovative capacity, and managerial oversight. Acad. Manage. J. **44**(1), 118–131 (2001). https://doi.org/10.5465/3069340
13. Pantuwong, N., Chutchomchuen, N., Wacharawisoot, P.: Interactive topography simulation sandbox for geography learning course. In: 2016 8th International Conference on Information Technology and Electrical Engineering (ICITEE), pp. 1–4 (2016). https://doi.org/10.1109/ICITEED.2016.7863262

14. Risner, R., Ward, T.E.: Concrete experiences and practical exercises: interventions to create a context for a synergistic learning environment. Technical report, Working paper (2007)
15. Sadler-Smith, E.: The relationship between learning style and cognitive style. Pers. Individ. Differ. **30**(4), 609–616 (2001). https://doi.org/10.1016/S0191-8869(00)00059-3
16. Solenthaler, B., Bucher, P., Chentanez, N., Müller, M., Gross, M.: SPH based shallow water simulation. Eur. Assoc. (2011). https://doi.org/10.2312/PE/vriphys/vriphys11/039-046
17. Song, K., Kim, G., Han, I., Lee, J., Park, J.H., Ha, S.: CheMO: mixed object instruments and interactions for tangible chemistry experiments. In: CHI 2011 Extended Abstracts on Human Factors in Computing Systems, pp. 2305–2310. CHI EA 2011, Association for Computing Machinery, Vancouver, BC, Canada (2011). https://doi.org/10.1145/1979742.1979907

m-Learning and e-Learning

Google Classroom as a Blended Learning and M-learning Strategy for Training Representatives of the Student Federation of the Salesian Polytechnic University (Guayaquil, Ecuador)

Joe Llerena-Izquierdo(✉) and Alejandra Valverde-Macias

Salesian Polytechnic University, Guayaquil, Ecuador
jllerena@ups.edu.ec, avalverdem1@est.ups.edu.ec

Abstract. This paper presents a new way of training youth aspiring to participate in the Salesian Polytechnic University (Guayaquil, Ecuador) (UPS-Guayaquil) Student Federation (FEUPS), proposing specific content for each federation position using blended learning and M-learning strategies with the help of Google Classroom. The methodology uses technologies that combine both classroom and distance learning activities, supported by the infrastructure of Google's online platform, and allows the Salesian Polytechnic University's current Student Federation to participate in efficient, accessible, timely and responsible learning; the use of technologies increases participants comfort and mobility as well as improving their time management. Collaborative applications permit real time access to information. Course planning is conducted according to the objectives of the FEUPS and the effectiveness of course implementation is documented through the verification of obtained knowledge. As the study shows, participants are highly satisfied with the innovative methodology used to implement the course. This study opens the possibility of uniting learning strategies for training through a combination of methodologies suitable for each intended purpose.

Keywords: Blended learning · Computer-supported collaborative learning · M-learning

1 Introduction

With the arrival of new mobile applications for collaborative work activities, users must be increasingly adept in their use, using this new technology in a motivating way that enriches the users' knowledge and learning process, in collaboration with others [1]. Blended Learning first emerged in 2004 and today there are various studies examining the benefits of combining traditional learning with electronical learning through the application of technology, technological services, and the Internet [2, 3, 4, 5]. Experiences of immersive technologies as well as new technological services applied in different fields of study [6, 7, 8] highlight the usefulness of these technologies in education.

M. Botto-Tobar et al. (Eds.): ICCIS 2020, AISC 1273, pp. 391–401, 2021.
https://doi.org/10.1007/978-3-030-59194-6_32

Combined methodologies generate positive results based on the pedagogical criteria of user-friendliness, relevant content and leaning improvement [4, 9, 10, 11].

With Google Classroom, Google creates an integrated learning space where Google technologies (Docs, Sheets, Slides and Forms) can be used to perform integrated learning activities. In a traditional educational institution or organization these activities would require a physical platform or infrastructure. By using Google Classroom an instructor can provide a mobile experience in which participants use their own personal devices to experience content with greater autonomy as compared to traditional learning management systems (LMS). Users can share their activities and tasks with other participants (students and teachers) who are using the same online platform [12, 13].

M-learning generates an improved and more efficient teaching-learning methodology that is no longer limited by lack of access to education structures. Technology allows teaching and learning to be integrated into course participants daily life [14].

Young people are familiar with and interested in technology; mobile applications and digital platforms can be easily understood by this new generation of students and innovative technology helps to maintain students' attention. The learning experience is only possible when a correct method is applied, and must go hand in hand with motivation, ranging from wanting to pass a course to wanting to acquire more knowledge [14, 15].

Through mobile technologies, courses are produced and are accessed by students to initiate a self-learning process. This process generates learning in an unlimited educational field no longer bound to the physical limitations of a traditional classroom. Values such as discipline and responsibility are required to support a successful learning experience in this learning model [15].

Participants who have constant access to their courses' learning environment and broad access to technology adapt better to these new methodologies. Students are able to learn and practice anywhere provided they have a smart device linked to the internet [16]. Over time, educational platforms bring new teaching-learning resources to improve knowledge and skill acquisition. This teaching-learning process can be measured at the end of each course through planned evaluations.

The Salesian Polytechnic University (Ecuador) (UPS) is a higher education institution founded in 1994. It promotes student organization and representation through a Student Federation (FEUPS) and its respective executive board so that students can represent their peers through boards and/or official activities. The Salesian Polytechnic University's FEUPS is active in each of the three cities where it maintains campuses: Quito, Guayaquil and Cuenca (main campus).

The executive board positions include that of president, vice president, secretary and treasurer; board members are known as student representatives. Each student government proposes activities and reforms that generate change; each successive government evaluates the work of previous governments to continue to improve upon them or to propose new ideas for the benefit of all. Board positions are made available according to regulations and curious students can freely choose to participate. Students predisposed to participate in this governing body must self-educate on their own merits; Salesian Polytechnic University students are by and large studying engineering, and have not studied law or politics, careers generally associated with governing positions.

Each semester the youth who participate in the FEUPS must be trained for the specific role and responsibilities they aspire to assume within the University. The training for new participants is not mandatory; it is carried out voluntarily by the outgoing directives and the new board members freely participate. Up until now it was not possible for all entering members to participate in specific training and hand over activities for their positions as they could not always coordinate with the outgoing members due scheduling difficulties.

This research paper plans to address how the training needs of future Executive Board Officers can be met through blended learning and M-learning strategies.

2 Materials and Methods

This research uses an empirical-analytical methodology with a quantitative approach. It is quasi-experimental and uses the survey technique to sample a group of young people who are being trained to assume responsibilities within UPS's FEUPS.

Fig. 1. "Introduction to FEUPS- UPS" opened through the Google Classroom App in a smartphone, a) courses registered in the user's account; b) display of available courses; c) collaborative work teams; d) registry of teachers.

The Google Classroom application is proposed to host course information (see Fig. 1a), and a course denominated "Introduction to FEUPS- UPS" (see Fig. 1b) is developed to generate a learning space for new FEUPS members.

Course content includes the following: What is the FEUPS- UPS?; FEUPS- UPS bylaws; FEUPS- UPS events; FEUPS- UPS election regulations; evaluation tools to measure participants autonomous learning (see Fig. 1c). The course methodology and content was designed by members of the FEUPS- UPS Executive Board interested in training future generations, researchers from the GIEACI[1] group (Research Group for Teaching-Learning of Engineering Sciences) and researchers from GIE-TICEA group (Educational Innovation Group – Information Technologies and Communications Teaching – Learning Process) (see Fig. 1d).

The target group is comprised of 15 students postulating for positions in FEUPS- UPS elections. Participants are qualified to participate in the elections according to current FEUPS- UPS election regulations.

The following Google Classroom tools are used to design the learning space: creation and management of virtual classroom (see Fig. 2a); subdivisions per study chapter (see Fig. 2b); file hosting to review documents; notifications of uploaded activities or lists for evaluation; incorporation of participants through an associated Gmail account (see Fig. 2c); assignments; creation of multiple classes and activation of all settings designed by the Google Classroom team (see Fig. 2d) to carry out meaningful activities formatted for computers and mobile devices.

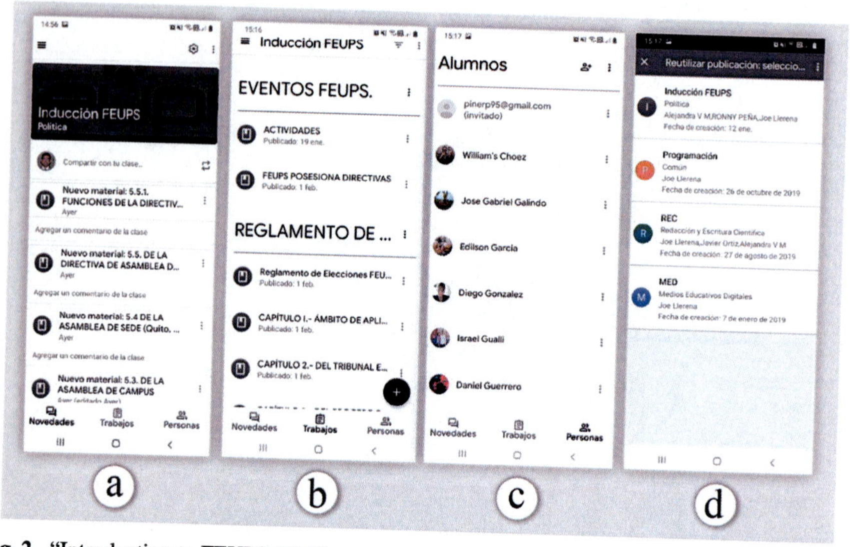

Fig. 2. "Introduction to FEUPS-UPS" screens a) notifications b) assignments organized by unit c) course registry d) registry of posts to reuse or share with other courses.

Study participants have varied activity level: the youngest participants are generally solely dedicated to their studies whereas older participants often work and study. The training platforms allow participants to combine classroom and distance learning activities according to each participants' specific academic interests.

Assignments and reading activities are conducted according to a course syllabus. The course is designed to inform participants about the goals, functions, and responsibilities of the FEUPS. Course material is prepared in accordance with the policies established

[1] GIEACI website available at: https://gieaci.blog.ups.edu.ec/.

by the academic directors of the university and in accordance with FEUPS statutes that allow its current Executive Board to form leaders who will continue ongoing FEUPS activities for the good of the student community.

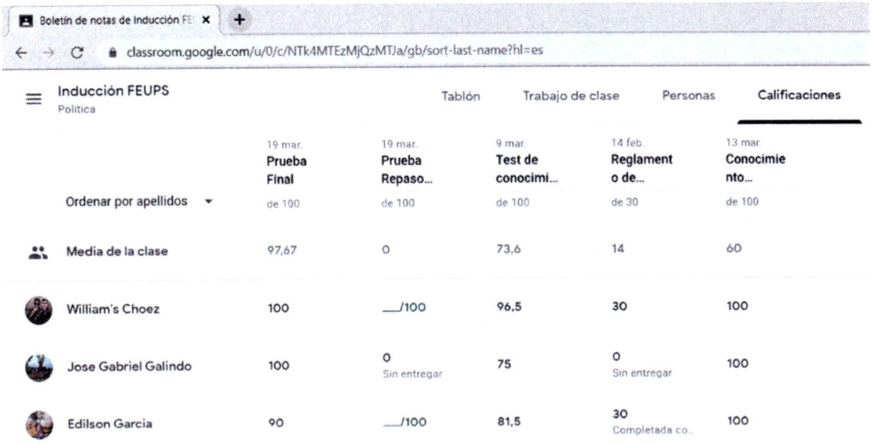

Fig. 3. A list of grades and evaluations per activity.

This research is conducted in three phases: the first phase consists of the development of the FEUPS- UPS course, including the identification of interactive tools to plan the course, Blended Learning and M-Learning support strategies for classroom and distance learning activities, and the preparation of programmatic content according to current regulations and statutes.

The second phase is comprised of the invitation to voluntarily participate in the study: participation in the study is considered as participation in an extracurricular activity (see Fig. 3). The third phase is the application of a course satisfaction survey.

The training space (virtual classroom) is available to participants for one month, the length of the course (defined according with course content). When the course is completed participants are invited to evaluate course content. The online course is seen as a complement to the face-to-face training required to become a member of the FEUPS's Executive Board.

3 Results and Discussion

To measure the effectiveness of Google Classroom participants began the course by completing an initial assessment to evaluate participants' knowledge of FEUPS functions and activities. Results were compared with the final evaluation completed at the end of the blended learning course. The average grade of the initial assessment was 73.6% whereas the average grade of the final evaluation, after a month of learning, was 97.67% (Fig. 4).

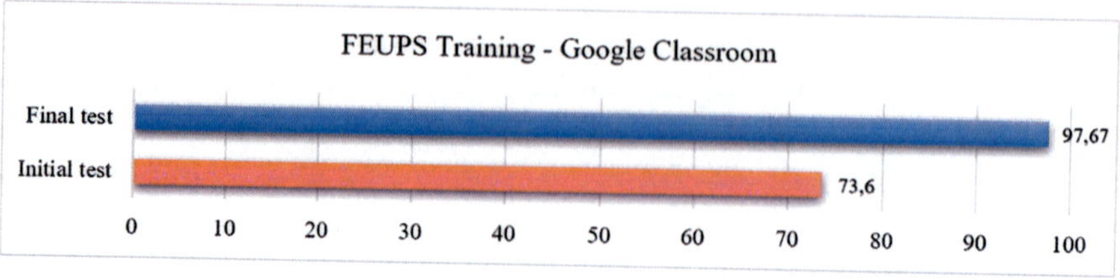

Fig. 4. Average grade for initial assessment and final evaluation.

Regarding Phase 3, the results of the course satisfaction survey completed by the target group are the following: 73.3% (11) of candidates are older than 21 yr old. 20% (3) are between 20 and 21 yr old and 6.7% (1) of candidates are between 17 and 19 yr old (see Fig. 5). 60% of participants "only study", and 40% "work and study". Based on this data it can be affirmed that most students interested in and able to participate in the FEUPS are currently in higher levels of their studies and often both work and study.

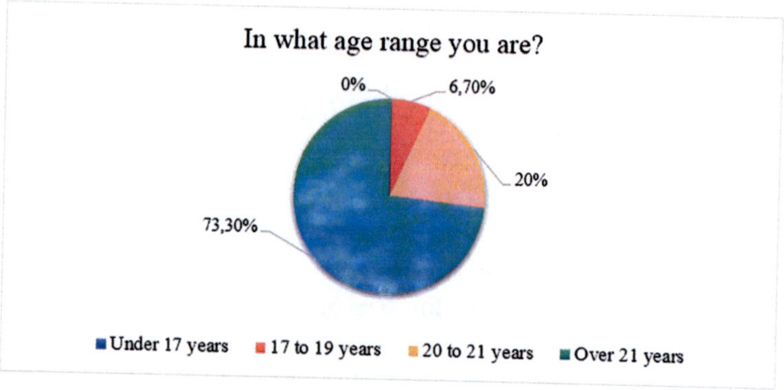

Fig. 5. Target group age distribution

As indicated by FEUPS bylaws, all participants have been continually enrolled in the university since their first semester, have maintained a sufficiently high grade point average (80/100) and have not failed any classes. They have complied with their Major's curriculum, and they will soon be looking for work. 100% of participants have a smart device.

Participants are asked if they are familiar with Blended Learning and M-Learning methodologies: 53% are familiar with blended learning methodologies, whereas 47% are familiar with M-learning (Fig. 6).

Participants were asked which characteristics students need to be able to learn through blended learning and M-learning methodologies, specifically considering their time management, space, access to internet and transportation needs. 40% indicated that responsibility is a fundamental characteristic to be able to learn with these methodologies, whereas 26.7% indicated order and 26.7% indicated discipline (see Fig. 7).

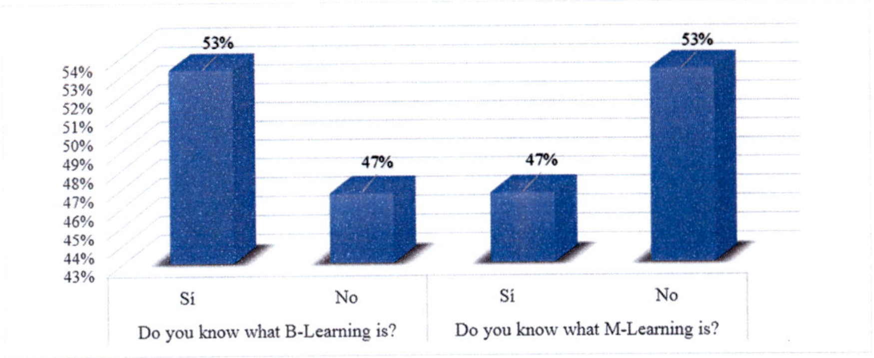

Fig. 6. Percentage of participants that are familiar with Blended Learning and Mobile Learning.

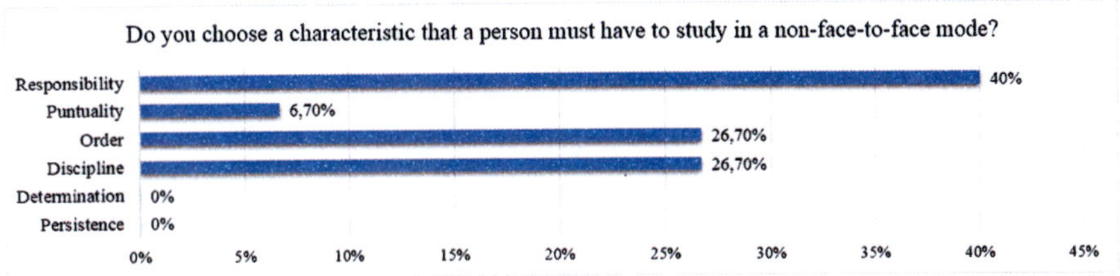

Fig. 7. Distribution of the characteristics students should exhibit to learn successfully with Blended Learning and M-learning methodologies.

The results of the survey allow us to obtain important information about the use of collaborative technology to improve the learning process of FEUPS board members. Each term the new board generates information, tools, activities, and training material that are not documented, digitalized, and archived in an organized and shared fashion and hence the next board that enters starts working from zero. Similarly, participants manifest that training is needed to able to be effective in their governing positions in the FEUPS. The results of the survey indicate that without exception all members of the FEUPS should receive training (see Fig. 8).

By choosing Google Classroom as a virtual platform for the course, the study ensures access to an adequate infrastructure as the totality of participants have access to the internet via their smartphones (most of which use Android operating systems). Results show that the experience in the platform was decisive: 46.7% of participants had never received training through this platform, whereas 53.3% are familiar with the platform and had received some previous training. 100% of participants finish the FEUPS preparation and training course (see Fig. 9).

Students give the course a satisfaction rating of 93.3%. The high rating is due to the fast and comfortable training modality which combines both classroom and e-learning; for the knowledge acquired through the chosen blended methodology; and for the comfort, availability and efficiency of the course in their mobile devices, chosen infrastructure and appropriate content.

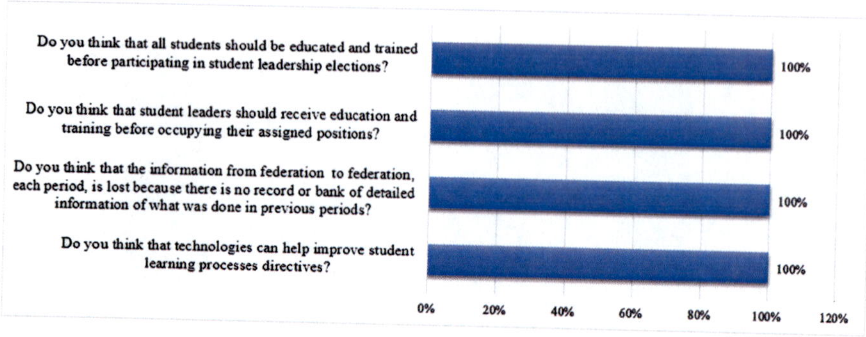

Fig. 8. Desired situation of youth participating in the FEUPS, (percentage).

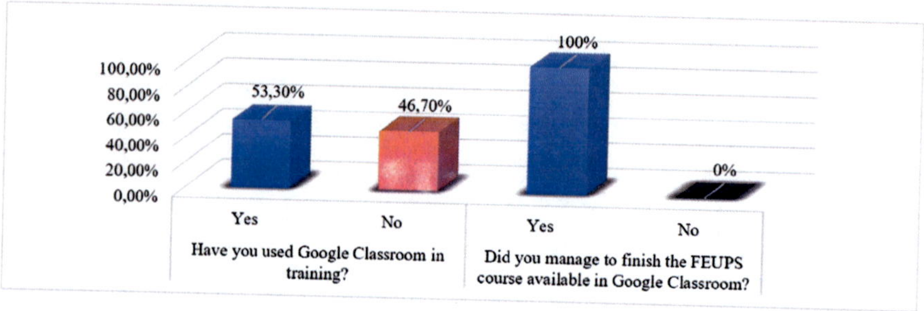

Fig. 9. Percent of study participants that are familiar and have previously used Google Classroom.

Course satisfaction was measured using the Likert scale (see Fig. 10). 80% of participants indicated that they were very satisfied, and 13% indicated that they were somewhat satisfied. Together 93% of course participants were satisfied with the preparation and content of the course.

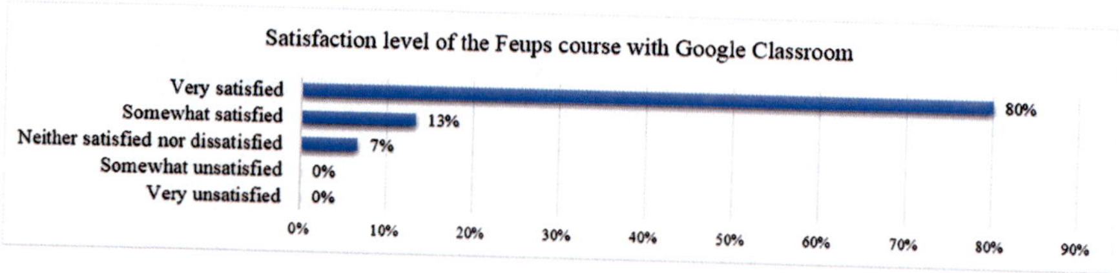

Fig. 10. Course satisfaction rating (percentage).

The course was very well received and highly valued with 53% of participants indicating that the course was extremely useful, 40% very useful and 7% somewhat useful: 93% (see Fig. 11) confirm that it is useful for future members of the FEUPS to participate in the training course with the technological support of Google Classroom.

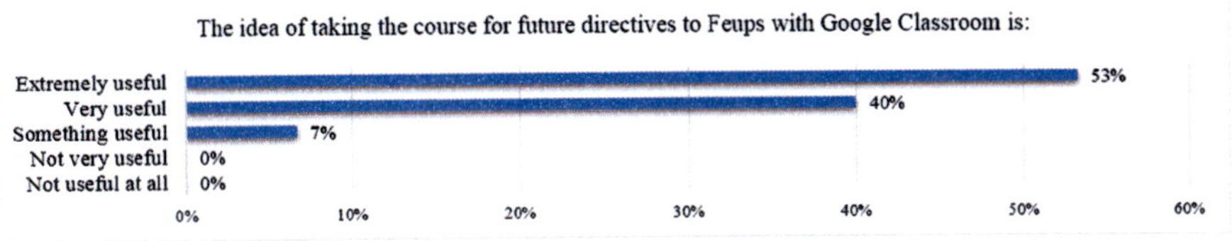

Fig. 11. Participants' evaluation of course usefulness (percentage).

It is indisputable that FEUPS members currently require more training and preparation; this study shows that 67% totally agree and 27% agree that students who aspire to participate in the FEUPS should receive more training, a 94% consensus (see Fig. 12).

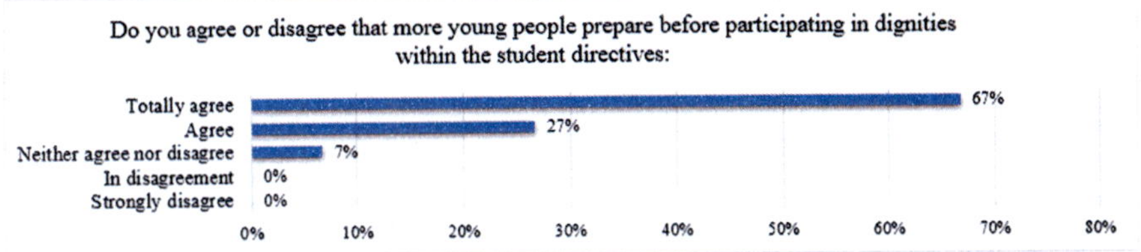

Fig. 12. Consensus that students who aspire to participate in the FEUPS should receive more training.

4 Conclusions

This year is the 25th anniversary of the founding of the Salesian Polytechnic University (Guayaquil, Ecuador); the students of the Student Federation of the University see this study as an efficient blended learning and m-learning proposal to improve training and preparedness of participants in the student government, and recognize the Google Classroom platform as an adequate tool for this type of training.

The proposal achieved a high level of satisfaction among participants, and the training proposal was adequate for a diverse group of students with different ages and time constraints. This process has proven effective as students voluntarily and enthusiastically chose to participate in the training process and were able to participate through their internet enabled smartphones.

The study proposes an adequate training program for future FEUPS Executive Board members, a training program which today is considered innovative in the UPS for its teaching methodology and accessibility, giving new generations of leaders the opportunity to acquire the skills and abilities needed to represent their peers in the Student Federation.

Acknowledgements. We thank the Salesian Polytechnic University for allowing us to conduct this study in its Guayaquil branch of the university; the team of youth aspiring to participate in the FEUPS; the members of the innovation in education group GIETICEA and of the research group GIEACI (https://gieaci.blog.ups.edu.ec/) for their unconditional support while conducting this study.

References

1. Andrew, M.: Collaborating online with four different Google apps: benefits to learning and usefulness for future work. J. Asia TEFL. **16**, 1268–1288 (2019). https://doi.org/10.18823/asiatefl.2019.16.4.13.1268
2. Cheung, S.K.S., Wang, F.L.: Blended learning in practice: guest editorial. J. Comput. High. Educ. **31**, 229–232 (2019). https://doi.org/10.1007/s12528-019-09229-8
3. Hrastinski, S.: What do we mean by blended learning? TechTrends. **63**, 564–569 (2019). https://doi.org/10.1007/s11528-019-00375-5
4. Bringle, R.G., Clayton, P.H.: Integrating service learning and digital technologies: examining the challenge and the promise. RIED. Rev. Iberoam. Educ. a Distancia. **23** (2020). https://doi.org/10.5944/ried.23.1.25386
5. Llerena, J., Mendez, A., Sanchez, F.: Analysis of the factors that condition the implementation of a backhaul transport network in a wireless ISP in an unlicensed 5 GHz band, in the Los Tubos sector of the Durán Canton. In: Proceedings - 2019 International Conference on Information Systems and Computer Science (INCISCOS 2019), pp. 15–22. Institute of Electrical and Electronics Engineers Inc. (2019). https://doi.org/10.1109/INCISCOS49368.2019.00012
6. Soeiro, A., Poças Martins, J., Zavrski, I., Theodossiou, N., Meszaros, P., Sidani, A.: CSETIR—construction safety with education and training using immersive reality. In: Presented at the (2020). https://doi.org/10.1007/978-3-030-41486-3_6
7. Llerena, J., Andina, M., Grijalva, J.: Mobile application to promote the Malecón 2000 tourism using augmented reality and geolocation. In: Proceedings - 3rd International Conference on Information Systems and Computer Science (INCISCOS 2018) (2018). https://doi.org/10.1109/INCISCOS.2018.00038
8. Llerena-Izquierdo, J., Cedeño-Gonzabay, L.: Photogrammetry and augmented reality to promote the religious cultural heritage of San Pedro Cathedral in Guayaquil, Ecuador. In: Presented at the, 3 December 2020. https://doi.org/10.1007/978-3-030-42520-3_47
9. Olelewe, C.J., Agomuo, E.E., Obichukwu, P.U.: Effects of B-learning and F2F on college students' engagement and retention in QBASIC programming. Educ. Inf. Technol. **24**, 2701–2726 (2019). https://doi.org/10.1007/s10639-019-09882-7
10. Malik, S.I., Mathew, R., Al-Nuaimi, R., Al-Sideiri, A., Coldwell-Neilson, J.: Learsning problem solving skills: comparison of E-learning and M-learning in an introductory programming course Educ. Inf. Technol. **24**, 2779–2796 (2019). https://doi.org/10.1007/s10639-019-09896-1
11. Izquierdo, J.L., Alfonso, M.R., Zambrano, M.A., Segovia, J.G.: Aplicación móvil para fortalecer el aprendizaje de ajedrez en estudiantes de escuela utilizando realidad aumentada y m-learning. Rev. Ibérica Sist. e Tecnol. Informação. **E22**, 120–133 (2019)
12. Thomas, C.E.: Teacher and student experiences in learning. Pac. J. Technol. Enhanc. Learn. **2**, 3 (2019). https://doi.org/10.24135/pjtel.v2i1.21
13. Rosita, N., Saun, S., Mairi, S.: Google classroom for hybrid learning in senior high school. J. Learn. Teach. Digit. Age. **5**, 35–41 (2020)
14. Singh, H., Miah, S.J.: Design of a mobile-based learning management system for incorporating employment demands: case context of an Australian University. Educ. Inf. Technol. **24**, 995–1014 (2019). https://doi.org/10.1007/s10639-018-9816-1

15. García, A., Vidal, E.: A B-learning/M-learning experience as support for improving the capabilities of the English area for engineering students. In: Proceedings of the LACCEI international Multi-conference for Engineering, Education and Technology. Latin American and Caribbean Consortium of Engineering Institutions (2019). https://doi.org/10.18687/LACCEI2019.1.1.30
16. Ata, R., Cevik, M.: Exploring relationships between Kolb's learning styles and mobile learning readiness of pre-service teachers: a mixed study. Educ. Inf. Technol. **24**, 1351–1377 (2019). https://doi.org/10.1007/s10639-018-9835-y

Introducing Gamification to Improve the Evaluation Process of Programing Courses at the Salesian Polytechnic University (Guayaquil, Ecuador)

Joe Llerena-Izquierdo(✉) and Jamileth Idrovo-Llaguno

Universidad Politécnica Salesiana, Guayaquil, Ecuador
jllerena@ups.edu.ec, jidrovol@est.ups.edu.ec

Abstract. Promoting student learning, especially the learning of first year students, is a continuous challenge for university authorities due to the large gap between secondary school preparedness and university requirements. The established programmatic content and evaluation processes of each subject are part of the analytical planning that teachers must conduct each semester to restructure and reformulate courses based on course pass rates. The academic evaluation identifies which activities have significantly enhanced or improved student learning and specifically have helped them achieve the expected group and individual learning objectives. In the university's programming courses, the incorporation of games through technological platforms using mobile devices strengthens the teaching-learning process by reinforcing knowledge. In an enthusiastic and fun way participation in the platform helps strengthen previous knowledge, improves student communication, encourages socialization among peers and allows for the collective identification of solutions to problems related to class material. Kahoot is a digital, web-based education and gamification tool that can be used to implement this new complementary tool for learning and assessment in the classroom. This study uses gamification as an educational resource and incorporates it into the curriculum of the first year Programming course (2019 curriculum adjustment), improving the learning and assessment activates as well as the teaching – learning process as measured by specific goals defined in the course's analytical plan.

Keywords: Gamification · Technology enhanced learning · Participatory attitude

1 Introduction

Increasingly, research shows how information and communication technologies (ICT) in combination with learning management systems (LMS) offer students virtual learning environments (VLE) to access meaningful materials and activities that facilitate the acquisition of new knowledge [1]. If these systems are integrated with social network applications (SNA) to support learning and incorporate gamification tools or web services, they open the way for the development of an innovative classroom learning

M. Botto-Tobar et al. (Eds.): ICCIS 2020, AISC 1273, pp. 402–412, 2021.
https://doi.org/10.1007/978-3-030-59194-6_33

process that favors retention and appropriation of knowledge by students [2]. Clear initiatives that use technologies as didactic tools to support the educational pedagogical processes [3–5], contribute to the development of game designs with diverse elements and characteristics that awaken students' motivation to learn [6, 7].

Technologies that support enhanced teaching and learning ("technology enhanced learning", TEL), facilitate the diversification of validation or evaluation strategies in educational institutions and even in companies incorporating new personnel [8, 9]. There are several key factors when using gamification as an evaluation tool that need to be studied more in-depth: the correct elaboration of gamification environments; teacher motivation; user experience; effective technological structure for its adequate incorporation in the planned comprehensive process of evaluation [10, 11]. These challenges are at the forefront for researches studying effective learning and evaluation methods [12–15].

At the Salesian Polytechnic University (Guayaquil, Ecuador), first semester Engineering students (from Computer, Electronic, and Electrical engineering majors for this study) take a Programming in Python course. The course is made up of three units: Unit 1 - Mathematical Logic, Unit 2 - Algorithms and Unit 3 - Programming Language. Two achievement indicators are established for Unit 3: the student identifies syntax and semantics in the Python language and can design and implement programs. In previous semesters, class abandonment levels reached 5% and class repetition up to 25%, high values for students in their first year of studies.

Gamification begins with the adequate design of a specific game to achieve specific learning goals. The game design will be different for students who participate in classroom learning or E-learning and may or may not incorporate technology (see Fig. 1). Games that incorporate technology may be connected to the internet and are generally known as "Plugged Gamification" as opposed to older, traditional games known as "Unplugged Gamification" [16].

Fig. 1. Plugged gamification and unplugged gamification

This study explores a learning process using the educational gamification web platform Kahoot[1] to bolster the teaching-learning process by reinforcing previous knowledge, strengthening students' participatory attitude, improving students' communication

[1] Available at https://kahoot.com/.

skills, encouraging socialization among students and finding solutions to overcome the difficulties of the Programming in Python course, specifically the acquisition of knowledge required in the course's Unit 3, validating learning in an enthusiastic and fun way.

2 Materials and Methods

The research presented in this paper is conducted at the Salesian Polytechnic University (Guayaquil, Ecuador), with students who attend the School of Engineering in the traditional classroom modality. Teachers' work in the classroom is accompanied by a virtual learning environment (VLE) that allows the incorporation of technologies to improve learning inside and outside the classroom.

Activities are planned by units according to the Kolb experimental learning cycle. The process is divided into four parts (see Fig. 2): the first part is Abstract Conceptualization [motivation - presentation of the topic - development], where the teacher captivates the student's attention, provides a general explanation of the topic and explains learning goals. The second part is Active Experimentation [learning fixation - integration]. In this part of the process the students' knowledge is expanded by integrating theoretical knowledge into concrete knowledge through experiences.

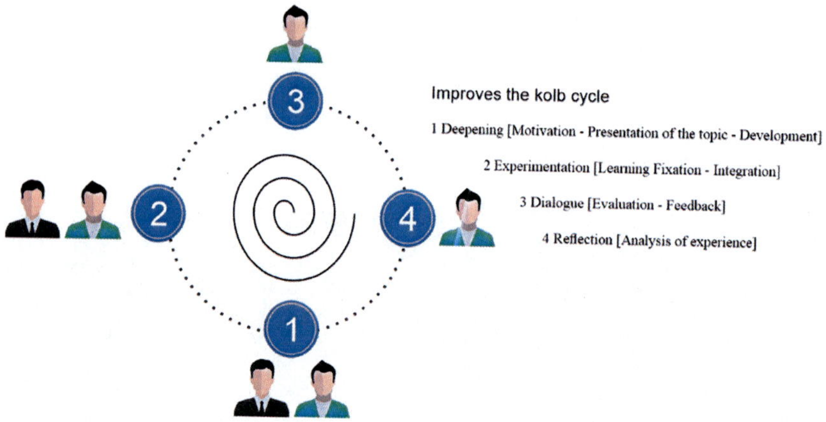

Fig. 2. Kolb experimental learning cycle used in the learning process proposed by the Salesian Polytechnic University

The process is strengthened as it is completed in a community setting and students work together with their peers. The third part of the cycle is Dialogue [evaluation - feedback]. In this part of the cycle learning is appraised through didactic evaluations and feedback and knowledge is integrated and consolidated through concrete experiences. The research presented in this paper is focused on this part of the cycle. The fourth part of the cycle is Reflection [Analysis of experience]. In this part of the cycle new knowledge is generated and interiorized through a deep analysis, reflection, and comparison of experience. Compliance with the guidelines provided by the university is ensured and space is left for teacher initiative through the implementation of a set of significant planned activities for each part of the teaching-learning process.

The Salesian Polytechnic University (Guayaquil, Ecuador) has adopted the Kolb experimental learning cycle as its teaching-learning methodology. The research presented in this paper uses gamification, through the Kahoot platform (see Fig. 3) as a technique to strengthen the Dialogue component (part 3) of the teaching-learning cycle, specifically strengthening evaluation activities. The student thinks, does, acts, feels and finally observes learning achievements. In this study gamification is added as a tool in the third part of the process. Knowledge is validated by using a variety of appropriate technologies that facilitate the validation of previously planned learning objectives.

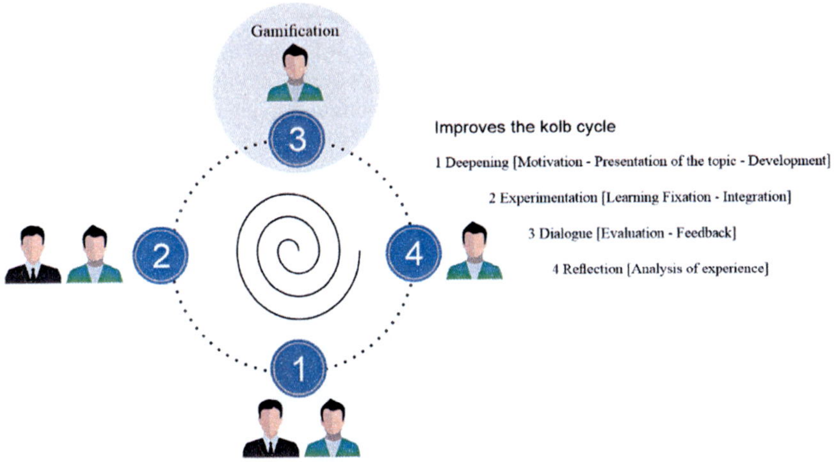

Fig. 3. Application of gamification in the Kolb experimental learning cycle

Considering the Learning line defined for this study (see Fig. 4), to successfully use gamification as an evaluation tool to measure learning goals it is of utmost importance that teachers are extremely motivated in their participation considering that they will have to learn new gamification tools, and prepare new material, and prepare themselves for the challenge of introducing new learning methodologies in their classrooms.

This research uses an empirical-analytical methodology with a quantitative approach. It is quasi-experimental and uses the survey technique to sample a group of students enrolled in the Programming in Python course as well as a group of teachers who teach the course, aiming to understand their satisfaction with the method and willingness to adapt it to other courses. The study is divided into three phases: in Phase 1 the Kahoot platform is used to create a gamification environment for the Programming in Python course. In Phase 2 specific learning achievement indicators and questions based on the content of the analytical program are defined together with teachers who teach the course. In Phase 3 learning is evaluated.

154 students and 5 teachers from 6 courses of Programing in Python participate in this study, two courses from each of the participating Computer, Electronic, and Electrical

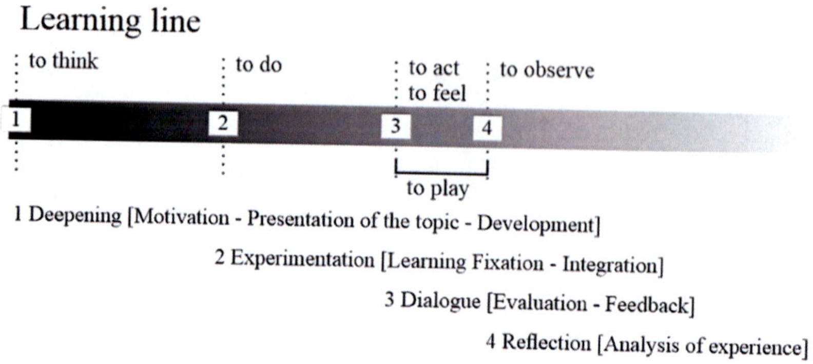

Fig. 4. Learning line used to plan each class

engineering majors. The teachers (see Fig. 5c) who participate in the study receive prior training to learn about the benefits of the platform and the process of creating the corresponding game, as well as specific instructions to execute their role in two moments as participants (see Fig. 5a) and as moderators of the designed game. Upon completing their training teachers are invited to create their own virtual evaluation space to test their mastery of the Kahoot platform (see Fig. 5b).

Fig. 5. a) Teachers participate with students b) Making their own game c) Team of participating teachers

Within the teacher training process on the Kahoot platform, teachers can access a gaming environment (see Fig. 6a). They define a set of objective questions and answers (see Fig. 6b) according to the course's achievement indicators. Trial runs are carried out in the participants' environment[2] (see Fig. 6c).

Trial runs are conducted with students to familiarize them with the gaming environment. The trails can be viewed on YouTube[3]. In Phase One questions are selected from a question bank provided by teachers (see Fig. 7a). In Phase Two the game environments are created on the Kahoot platform (see Fig. 7b). In Phase Three students are trained through the platform activities and then evaluated (see Fig. 7c).

[2] Available at https://kahoot.it/.

[3] Available at http://bit.ly/evaluation_process.

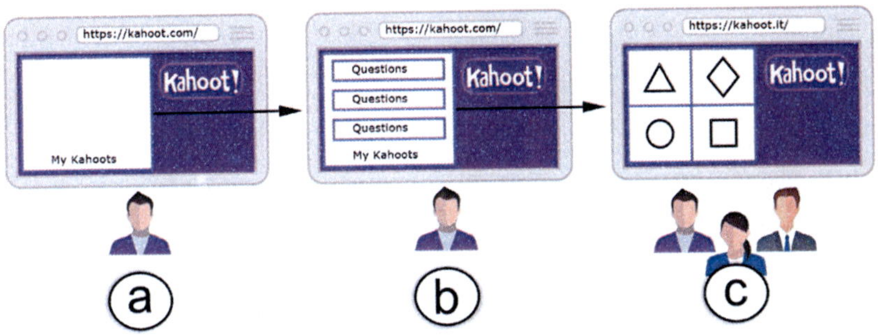

Fig. 6. a) Gamification web platform; b) Question elaboration; c) Real time game application

Fig. 7. a) Selected evaluation questions; b) Platform tests; c) Induction of students to the platform and final exam

To use the gamification platform as an evaluation tool one must establish course specific basic criteria such as the objectives, rules, norms, target group, activity space and the appropriate stimuli according to the pedagogical criteria for the course [17].

3 Results

Two evaluation strategies were implemented with participating groups: for the first group the initial and final evaluations were conducted without the platform. For the second group the initial evaluation was conducted without the platform followed by an final evaluation through the gamification platform. Then a smaller group was randomly selected from the second group to participate in a second evaluation through the gamification platform. Results show that after using the gamification platform information retention increases and stimulus improves: the improvement in these indicators leads to an average decrease of 25.15% incorrect responses in the final evaluation (see Fig. 8).

The participating group of professors maintain periodic staff meetings to supervise the correct accompaniment of each course. Through these meetings the professors can integrate the platform into the evaluation process, defining both the physical space for the evaluation and the evaluation schedule.

The evaluation results (see Table 1) reveal a notable improvement in students' test scores, up to a 25% improvement in participating groups.

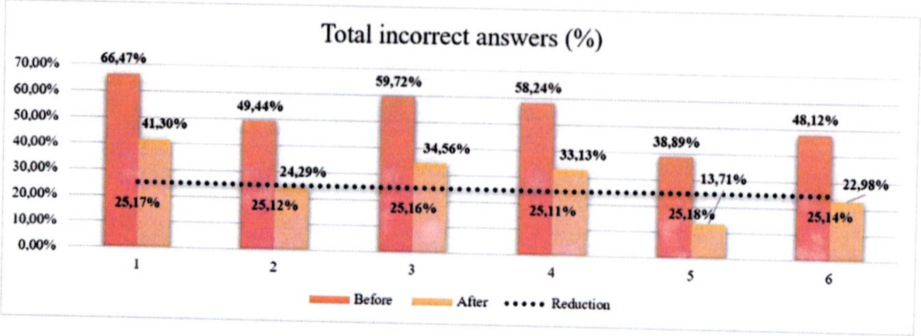

Fig. 8. Percentage of participating group's incorrect responses

Table 1. Percentage improvement per final exam per major while using the platform

Group	Career	Improvement	Percentage
Group 1	Electronics engineering	33,53%–58,70%	25,17%
Group 2	Electronics engineering	50,56%–75,71%	25,12%
Group 3	Electrical engineering	40,28%–65,44%	25,16%
Group 4	Electrical engineering	41,76%–66,87%	25,11%
Group 5	Computer engineering	61,11%–86,29%	25,18%
Group 6	Computer engineering	51,88%–77,02%	25,14%

Similarly, 81% of students who use the gamification method during the course passed the course and 19% failed the course, increasing the pass rate by 6% compared to previous semesters. Regarding the survey applied to the group of teachers, they identified that the most important characteristics of the method are its functionality and ease of configuration of the games offered by the gamification platform (Kahoots), thus easily favoring the incorporation of gamification into planned activities as a learning validation tool (see Fig. 9).

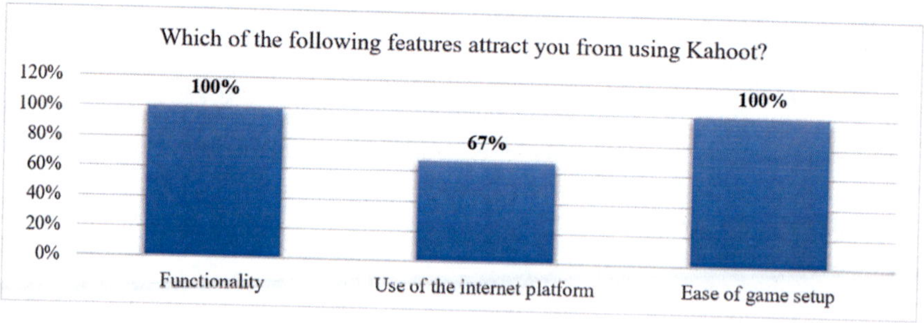

Fig. 9. Characteristics that attract professors (percentage)

Finally, factors are identified that are important for both teacher and student participants (see Fig. 10), differentiating traditional evaluation activities from gamification evaluation activities, all of which are significant during the evaluation process. These factors are decisive to assess learning. For teachers, important criteria include: the ability to communicate usage procedures, present the rules for the evaluations, have concrete criteria to measure levels of difficulty. The preparation of the environment and ability to establish an organizational mode for the items is important to them as is the ability to create strategies for game usage, establish progress parameters and finally incorporate surprise factors during game execution.

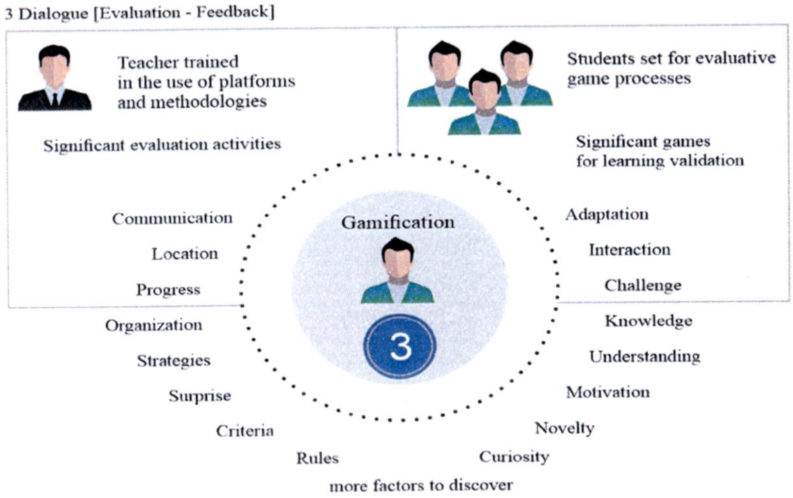

Fig. 10. Factors discovered during the gamification process

Likewise, attitudinal, cognitive and procedural factors were detected in the students, Among the most notable, spontaneously visible and observable factors were: the capacity to adapt while playing, the increase in personal interaction with peers while working individually and collectively, the proposal of competitions, observable knowledge reinforcement, a positive attitude towards understanding, an increased motivation to participate, enthusiasm for novelty and contagious curiosity among colleagues.

The application of a gamification technique, in this case through the Kahoot platform, improves academic performance within the teaching-learning process as a motivational complement of the evaluation phase of the Kolb cycle. It strengthens the acquisition of knowledge proposed by the instructor, even when this knowledge is proposed with traditional classroom teaching techniques (see Table 2). The use of the gamification platform increases students' interest and participation, the specifically appreciate the use of multiple environments, environment design, ambient music, interactive design, synchronization of sound and visual effects, which make the evaluation different and interesting [18].

Table 2. Aspects of the Kahoots platform valued by teachers

Area	Aspect valued by teachers
Cognitive	Creates interactive moments with the competition partners, motivating the generation of different solutions
Procedural	Facilitates the interoperability of technological devices with internet access for proper interactions
Attitudinal	Permits continuous practice even after the game ends, strengthening understanding through repetition

The gamification tool used meets certain criteria [19] defined specifically for virtual environments for students who participate in a traditional classroom education. Students' perception of the platform (see Fig. 11) is contrasted with evaluation satisfaction, motivation to participate, and finally learning achievements.

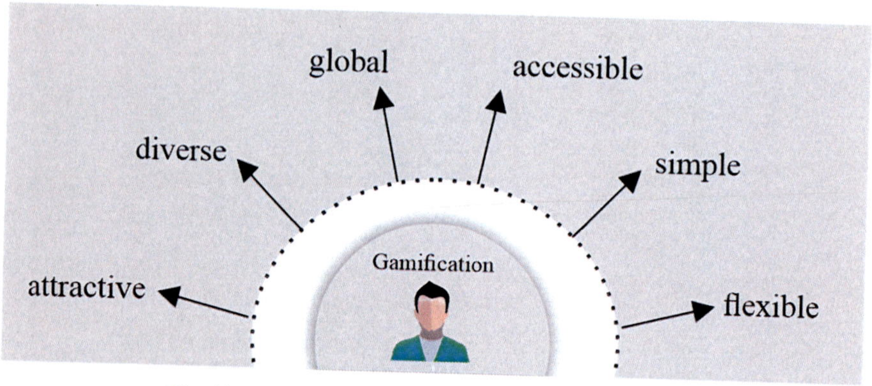

Fig. 11. Platform characteristics as perceived by students

4 Discussion

The results of this study suggest that the application of gamification in education can generate favorable effects, though there are some considerations to be considered when incorporating this technique into daily classes. One risk when introducing gamification is that students and teachers are not familiar with virtual spaces, generating fear and uncertainty and hindering the use of new and innovative technology that can help improve evaluation and academic achievement.

This study also suggests that professors generate greater student acceptance, interest, motivation, and class pass rates when they use innovative methods to evaluate the knowledge they impart in the classroom, specifically algorithmic logic in this study. The study also shows that positive effects were generated in student performance as a result of the enthusiasm generated when students use games that invite them to think, reason, learn, enjoy themselves and interact with their peers. Learning is more effective when electronic devices, which are an important part of students lives today, are incorporated into the learning process.

The Salesian Polytechnic University (Guayaquil, Ecuador), is looking to continually improve its teaching-learning process through the incorporation of new activities and evaluation methodologies to favor the integral learning of its students, particularly its first-year students. When introducing innovative evaluation methods planning and preparation is required, by both teachers and students, to achieve expected results. The study is open to the possibility of using different gamification platforms within the evaluation process.

5 Conclusions

Each semester more than 20% of first year students are required to repeat programming courses. The joint work of teachers in this study has allowed them to create a staff group, favoring the planning and implementation of coordinated activities for the implementation of the gamification platform.

The use of gamification tools to conduct participative individual and group activities increases class comradery, generates positive competition between participants and pushes students to correct learning difficulties that they themselves can detect. Previous knowledge is strengthened and built upon and students' knowledge retention is improved. The gamification platform in no way replaces classroom learning, rather it functions as a classroom accompaniment and demonstrates that learning in a specific way can be carried out through a participatory environment that excites and elevates aspects of learning.

Acknowledgments. We thank the Salesian Polytechnic University (Guayaquil, Ecuador) for allowing us to carry out this research in the first-year programing courses of different Majors, the team of teachers in the Programming faculty for participating in the study, the first year students of the 55 university cycle for their enthusiastic participation, and finally the innovation in education group GIETICEA-GIEACI (https://gieaci.blog.ups.edu.ec/) for their support in the methodological process of this research.

References

1. Wijesooriya, C., Heales, J., Rohde, F.: Evaluation of virtual learning environment characteristics in learning performance. In: ACM International Conference Proceeding Series, pp. 259–263. ACM Press, New York (2019). https://doi.org/10.1145/3306500.3306535
2. Zhao, F., Guo, D.: Rewards in gamification. Lecture Notes in Computer Science (including subseries Lecture Notes in Artificial Intelligence and Lecture Notes in Bioinformatics), pp. 453–462. Springer, Cham (2019). https://doi.org/10.1007/978-3-030-22602-2_34
3. Llerena Izquierdo, J.F.: El Uso De Grabaciones Por Video Como Recurso De Evaluación De Conocimientos De Aprendizajes (2018). https://ups.elsevierpure.com/es/publications/the-use-of-video-recordings-as-a-resource-for-assessing-learning

4. Izquierdo, J.L., Alfonso, M.R., Zambrano, M.A., Segovia, J.G.: Aplicación móvil para fortalecer el aprendizaje de ajedrez en estudiantes de escuela utilizando realidad aumentada y m-learning. Rev. Ibérica Sist. e Tecnol. Informação. **2019**(E22), 120–133 (2019). http://www.risti.xyz/issues/ristie22.pdf
5. Daniela, L.: Smart pedagogy for technology-enhanced learning. In: Daniela, L. (ed.) Didactics of Smart Pedagogy, pp. 3–21. Springer, Cham (2019). https://doi.org/10.1007/978-3-030-01551-0_1
6. Yordanova, Z.: Gamification for handing educational innovation challenges. Advances in Intelligent Systems and Computing, pp. 529–541. Springer (2020). https://doi.org/10.1007/978-3-030-11367-4_53
7. Deterding, S., Dixon, D., Khaled, R., Nacke, L.: From game design elements to gamefulness. In: Proceedings of the 15th International Academic MindTrek Conference on Envisioning Future Media Environments - MindTrek 2011, p. 9. ACM Press, New York (2011). https://doi.org/10.1145/2181037.2181040
8. Alberto, G., Samperio, T., Árcega, A.F., De Jesús Gutiérrez Sánchez, M., Navarrete, A.S.: La gamificación en los ambientes de realidad virtual móvil. The gamification in the environments of mobile virtual reality. Pistas Educativas (133) (2019). http://www.itcelaya.edu.mx/ojs/index.php/pistas/article/view/2054
9. Simons, A., Wohlgenannt, I., Weinmann, M., Fleischer, S.: Good gamers, good managers? A proof-of-concept study with Sid Meier's Civilization. Rev. Manag. Sci. 1–34 (2020). https://doi.org/10.1007/s11846-020-00378-0
10. Guerrero-Roldán, A.E., Noguera, I.: A model for aligning assessment with competences and learning activities in online courses. Internet High. Educ. **38**, 36–46 (2018). https://doi.org/10.1016/j.iheduc.2018.04.005
11. Noguera, I., Guerrero-Roldán, A.E., Rodríguez, M.E., Baneres, D.: Students' and instructors' perspectives regarding e-assessment: a case study in introductory digital systems. Int. J. Eng. Educ. **35**, 473–490 (2019)
12. Zainuddin, Z., Chu, S.K.W., Shujahat, M., Perera, C.J.: The impact of gamification on learning and instruction: a systematic review of empirical evidence. Educ. Res. Rev. **30**, 100326 (2020). https://doi.org/10.1016/j.edurev.2020.100326
13. Holguín, F., Holguín, E., García, N.: Gamificación en la enseñanza de las matemáticas: Una revisión sistemática. Gamification in mathematics education: a systematic review. Telos: Revista de Estudios Interdisciplinarios en Ciencias Sociales **22**, 62–75 (2020). https://dialnet.unirioja.es/servlet/articulo?codigo=7200001
14. Thomas, C.E.: Teacher and student experiences in learning. Pac. J. Technol. Enhanc. Learn. **2**, 3 (2019). https://doi.org/10.24135/pjtel.v2i1.21
15. Liu, Q., Geertshuis, S., Grainger, R.: Understanding academics' adoption of learning technologies: a systematic review. Comput. Educ. **151**, 103857 (2020). https://doi.org/10.1016/j.compedu.2020.103857
16. Lo, C.K., Hew, K.F.: A comparison of flipped learning with gamification, traditional learning, and online independent study: the effects on students' mathematics achievement and cognitive engagement. Interact. Learn. Environ. (2018). https://doi.org/10.1080/10494820.2018.1541910
17. Martins, E.R., Geraldes, W.B., Afonseca, U.R., Gouveia, L.M.B.: Using Kahoot as a learning tool. Lecture Notes in Information Systems and Organisation, pp. 161–169. Springer, Heidelberg (2019). https://doi.org/10.1007/978-3-030-14850-8_11
18. Wang, A.I., Tahir, R.: The effect of using Kahoot! for learning – a literature review. Comput. Educ. **149** (2020). https://doi.org/10.1016/j.compedu.2020.103818
19. Orhan Göksün, D., Gürsoy, G.: Comparing success and engagement in gamified learning experiences via Kahoot and Quizizz. Comput. Educ. **135**, 15–29 (2019). https://doi.org/10.1016/j.compedu.2019.02.015

Security

Technique for Information Security Based on Controls Established by the SysAdmin Audit, Networking and Security Institute

Flavio Morales(✉), Yanara Simbaña, Rosario Coral, and Renato M. Toasa

Universidad Tecnológica Israel, Quito, Ecuador
{fmorales,rcoral,rtoasa}@usirael.edu.ec, yanyjss@gmail.com

Abstract. Information security has become a worldwide trend; however, there is little or no mention of how the incorporation of security-related concepts should be initiated from the network design phase. The lack of a secure network means that losses amount to millions of dollars from cybercrime, much more if this same scenario were taken into account in the globalized environment. The need to evaluate the method by which the designs are executed becomes crucial, a before and after of the networks designed based on security. The present work proposes an alternative in information security with the methodology of design of communication and telecommunication networks, implementing 2 security controls, the 8 and 12, malware defense and perimeter respectively. These controls are listed by the SysAdmin Audit, Networking and Security Institute (SANS), evaluating the compliance with the recommendations that this institute delivers and demonstrating at the end that a functional and secure network can be obtained. The applicability of the described is measured in the implementation of the mentioned controls, since the antimalware provided adequate protection, administration and configuration in all the endpoints, thus avoiding zero-day threats. While the configuration at the perimeter level has allowed blocking of attacks, web filtering with new generation security parameters, maintenance of operational business roles and protected information.

Keywords: Security controls · Secure network · SANS

1 Introduction

With technological innovation, companies tend to change where they are most competitive, agile and profitable, but at the same time there can be the risk of allowing dangerous cyber threats into organizations [1], thus raises the need to ensure adequate protection of computer systems, in turn, ensure the continuity of the work that allows the fulfillment of the functions of all users.

SysAdmin Audit, Networking and Security Institute (SANS) [2] aims to gather information on network security, communication equipment such as routers, switches, firewalls, applications, databases, etc, has developed recommendations on 20 minimum

M. Botto-Tobar et al. (Eds.): ICCIS 2020, AISC 1273, pp. 415–426, 2021.
https://doi.org/10.1007/978-3-030-59194-6_34

controls required to manage a network that provides availability, integrity and reliability in the data circulating, stored and processed. Thus, since there is no direct correlation in the network design methodology and these 20 minimum controls, it is considered necessary to research and develop the design methodology, implementing a network whose bases, estimates and calculations are made with these controls.

There are works that address related topics such as this one, for example "Implementation of a perimeter security system as an information security strategy [4], which proposed to make a technological renovation with virtualized equipment that allows to extend the capacity when it is necessary and to maintain its effectiveness in the control of threats, working in HA (High Availability), presenting a topology with four equipment, applying policies in real time, working with modules of filtering URL, IPS, antimalware, antivirus, sandboxing and control of applications, obtaining like result an effective protection to all the technological silver-form of the institution, on the other hand several proposals of security of information exist in the work [5] mentions the importance of making anonymous communications and through a case study demonstrates the integrity of information in anonymous communications.

In Ecuador, there are several cases in which certain systems were violated, one of which is mentioned in [6], where a virus attempted to penetrate personal and corporate accounts to take over computers and gain full control of them. Experts mention that Ecuador is the third most vulnerable country in the region with WannaCry during the last decade [7].

For all the mentioned, it is necessary to carry out the implementation of mechanisms of security of information, therefore this work proposes to develop a process of protection of information against malicious attacks, using two essential controls proposed by the SANS, like the control of defense of malware and the control of perimetral defense. The defense of malware involves the defense of various types of viruses, protects the endpoint at the level of layer 2, for your protection is considered to have an effective antimalware on the computers avoiding damage to them, information and data theft, depends on the characteristics that each has for the scope of security controls and control [8]. On the other hand, the main objective of the perimeter defense is to protect the network from attackers, spoofing, mail spam and all events that may occur at the network level; in other words, a security perimeter is established that protects and isolates the internal local network and the local network from uncontrolled intrusions by third parties and prevents them from fraudulently accessing the assets that reside within the identified perimeters [9].

This document is organized as follows Sect. 1 presents the introduction to this work, Sect. 2 addresses the silver proposal, then Sect. 3 details the implementation of the safety mechanisms, the results obtained are mentioned in Sect. 4 and finally the conclusions and future work are described.

2 Proposal

It is proposed to implement controls 8 and 12, for internal endpoint protection there are different hardware and software tools that allow you to protect against malware, exploits, zero-day threats, worms, trojans and other attacks. After performing an analysis and taking into account the number of users in the network and the requirements for control compliance for malware protection and defense, centralized software of last generation is used. The responsibility is to provide technological services to all users, to provide technical advice, and to strengthen the quality and efficiency of its development. The software system must comply with minimum requirements, be multiplatform; that is, support Operating Systems such as: Mac, OSx, Windows 7 SP1, Windows 8–8.1 and Windows 10, Vista, as for server operating systems must be compatible with Windows Server 2012 or higher.

To implement the control 8 the minimum requirements to be met are: centralized cloud protection, centralized cloud protection, host-based IPS, peripheral control, real-time protection, navigation protection, email malware protection, file control by extension.

To comply with security control 12, perimeter defense recommended by the SANS standard, a security platform for perimeter protection is available to meet security demands with state-of-the-art protection against evolving threats. Control modules and policies applicable to the network, the minimum requirements are: allows to control access to the network for authorized and unauthorized devices, allows to control the use of network applications, has a network antivirus engine, email protection is available, has protection and control for safe navigation, port and protocol control in layer 4, allows networks to be sectioned and provide access by specific users or equipment, network perimeter security, integration with endpoint security.

Together with the two platforms it ensures that the attackers do not achieve their goal and cover the security gap. Figure 1 shows the design to make use of the Control 8, in this antimalware is implemented where the administration is centralized for workstations inside and outside the network, and compatible with several platforms according to the minimum requirements.

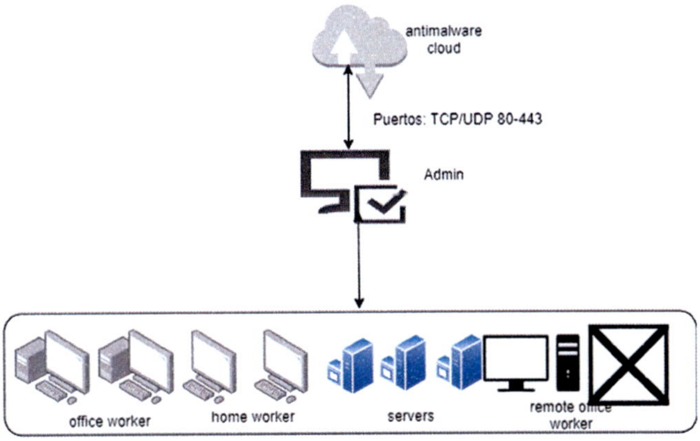

Fig. 1. Anti-malware implementation architecture. Source: Own elaboration

The anti-malware protects the endpoint and is managed by a person with knowledge of networking and computer security. To follow Control 8 recommendations, it is necessary to implement centralized security as anti-malware.

For the fulfillment of Control 12 focused on perimeter security, a firewall is used which will protect the network, Fig. 2. Indicates the architecture for the implementation of a firewall. It is used in gateway mode where it needs to support routing between multiple networks and zones, and is the point of entry and exit to the network.

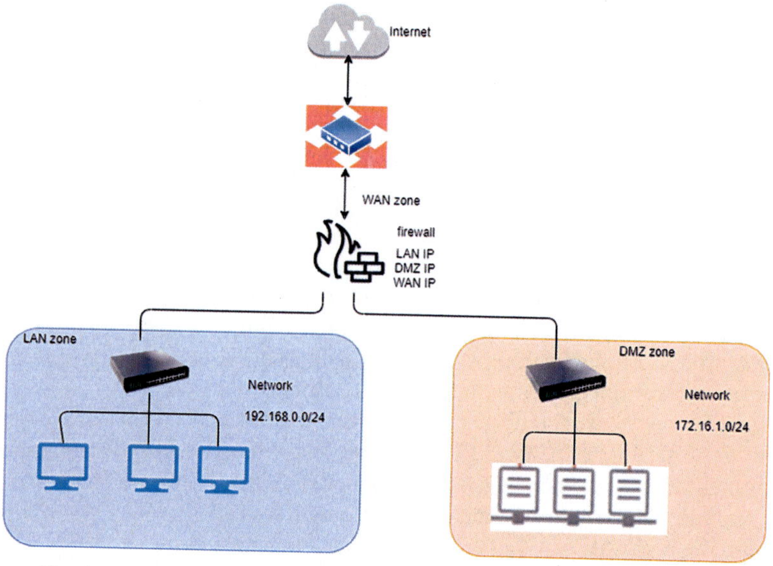

Fig. 2. Implementation architecture with firewall. Source: Own elaboration

3 Implementation

SANS shows 20 controls [10]; however, this paper refers to the implementation of two controls, 8 and 12, malware defense and perimeter respectively. As shown in Tables 1 and 2.

Table 1. Control 8 - malware defense

Family	Control	Description
System	8.1	Use of automated tools to continuously monitor workstations, servers, and mobile devices with anti-virus, anti-spyware, personal firewalls, and host-based IPS functionality. All malware detection events should be sent to the company's anti-malware management tools and event log servers
System	8.2	Employment of anti-malware software that provides a centralized infrastructure that collects file reputation information or that administrators manually push updates to all machines. After applying an update, automated systems must verify that each system has received its signature update
System	8.3	Limit the use of external devices to those with an approved and documented business need Monitor the use and attempted use of external devices. Configuration of laptops, workstations, and servers to not to automatically run removable media content and network shares
System	8.4	Enable anti-exploitation features such as Data Execution Prevention (DEP), Address Space Mapping (ASLR), virtualization/containerization, etc
System	8.5	Use of network-based anti-malware tools to identify executables in all network traffic and use techniques other than signature-based detection to identify and filter malicious content before it reaches the endpoint
System	8.6	Enable logging of do-minio name system (DNS) queries to detect host name lookups for known malicious C2 domains

3.1 Implementation: Controls 8 and 12

Control 8, Malware Defense, should control the installation and execution of malicious codes at multiple points. This control recommends the use of automated tools to continuously monitor workstations, servers and mobile devices with antivirus, antispyware, personal firewalls and host-based intrusion prevention system (IPS) functionality, as shown in Fig. 3.

Table 2 Control 12 – perimetral defense

Family	Control	Description
Network	12.1	Deny communications with known malicious IP addresses (blacklists) or limit access to only trusted sites (whitelists). Testing can be done periodically by sending packets from bogon source IP addresses (non-routable) to the network to verify that they are not transmitted across network perimeters. The bogon address lists publicly available on the internet
Network	12.2	In DMZ networks, configuration of monitoring systems to record at least packet header information, and preferably full packet header and payloads of traffic destined for or passing through the network edge. This traffic should be sent to a log analysis system so that events can be correlated with all devices on the network
Network	12.3	Implementation of network-based IDS sensors in Internet and extranet DMZ systems that look for unusual attack mechanisms and detect the compromise of these systems. These sensors can detect attacks through the use of signatures, network behavior analysis, or other mechanisms to analyze traffic
Network	12.4	Network-based IPS devices should be deployed to complement IDS by blocking poorly known signatures or the behavior of potential attacks. As attacks become automated, methods such as IDS often delay the amount of time it takes someone to react to an attack
Network	12.5	Design and implement network perimeters so that all outgoing network traffic to the Internet passes through at least one application layer that filters the proxy server. The proxy must support decryption of network traffic, logging of individual TCP sessions, blocking of specific URLs, domain names, and IP addresses to implement blacklisting, and whitelisting of permitted sites that can be accessed through the proxy
Network	12.6	Require all remote login access (including VPN, dial-up, and other forms of access that allow logging into internal systems) to use two-factor authentication
Network	12.7	All business devices that register remotely on the internal network must be managed by the company, with remote control of their configuration, installed software and patch levels. For third party devices, it publishes minimum security standards for access to the enterprise network and performs a security exploit before allowing access
Network	12.8	Periodically analyze post-internet channel connections that bypass the DMZ, including unauthorized VPN connections and dual-site hosts connected to the corporate network and other networks via wireless modems, dial-up or other mechanisms
Network	12.9	Implement NetFlow collection and analysis of DMZ network flows to detect anomalous activity

(*continued*)

Table 2 (*continued*)

Family	Control	Description
Network	12.10	To help identify covert channels that exfiltrate data through a firewall, configure the built-in firewall session tracking mechanisms included in many commercial firewalls to identify TCP sessions that last an unusually long time for the organization and the firewall device, alerting staff to the source and destination addresses associated with these long sessions

Fig. 3. Workstation implementation. Source: Own elaboration

For the installation of the anti-malware it is considered a console in the cloud activated under an email account and for the client machines the installation of an agent that runs in each machine (see Fig. 4).

Control 12, Perimeter Defense, needs to detect, manage and control the flow of information between networks of different levels of confidence, with an analytical approach. It has reviewed data that could damage the availability, reliability or integrity of information. The best defense is technologies that provide deep visibility and control over data flow throughout the environment, such as intrusion detection or prevention, as well as deep packet analysis at network boundaries. The number of users, peak traffic, controls and sub controls to be met is evaluated for equipment, model and brand selection, see Fig. 5 and Fig. 6.

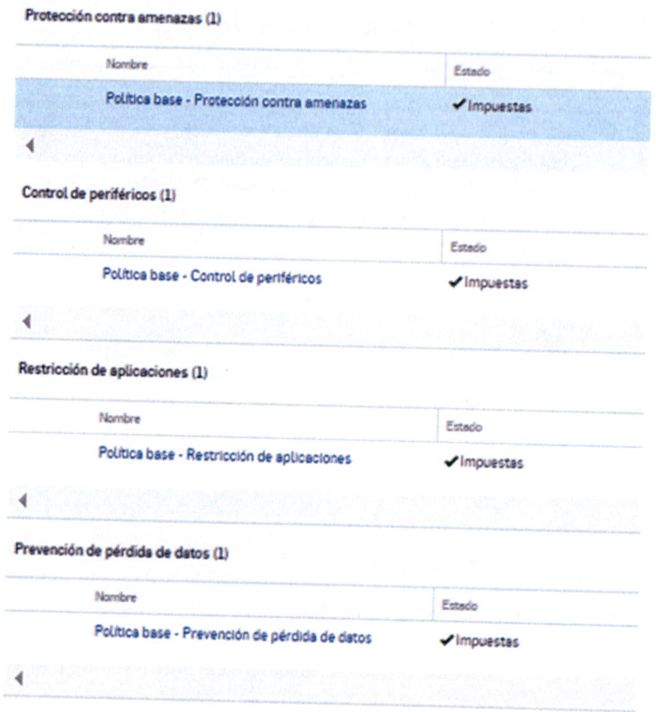

Fig. 4. Configured rules and controls Control 8. Source: Own elaboration

Fig. 5. Rules and policies Control 12. Source: Own elaboration

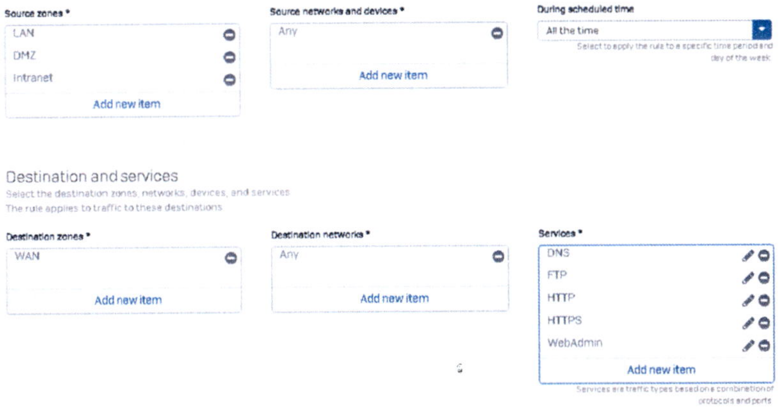

Fig. 6. Zone configuration. Source: Own elaboration

4 Results

After the implementation of an antimalware, the compliance of the controls and the identification of malware, viruses, PUAS can be evaluated by means of the delivered events, which are controlled, detected and eliminated.

According to control 12, a firewall was implemented in which access zones, network security policies, and traffic filtering were configured. The configurations of both the antimalware and the firewall can be made by scanning whether it is an end point or a network, since traffic can be avoided depending on the assigned configurations. Figure 7 shows the main dashboard of a perimeter tool with the shortcuts of the applied rules and controls assigned to each zone, Fig. 8 show the categories.

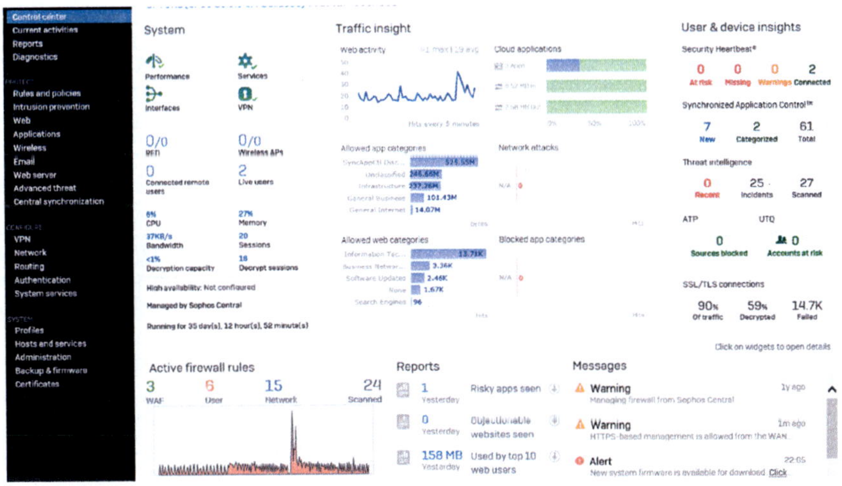

Fig. 7. Configuration and tool status dashboard. Source: Own elaboration

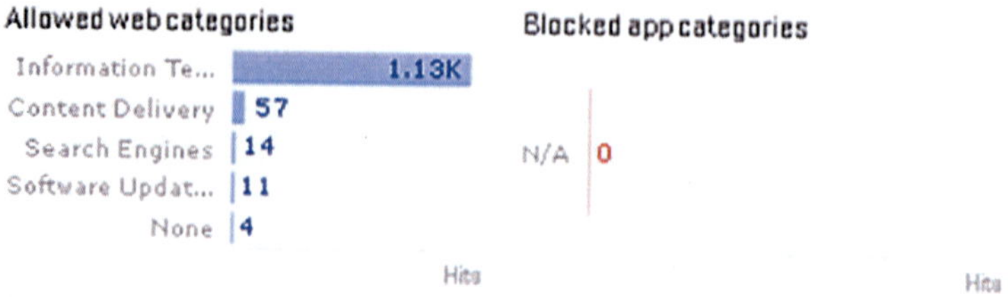

Fig. 8. Allowed categories according to recommendations. Source: Own elaboration

Figure 9 shows the number of synchronized users and devices covered, additional, application control synchronization, threat intelligence, ATP, SSL/TLS connections, which comply with the control recommendations.

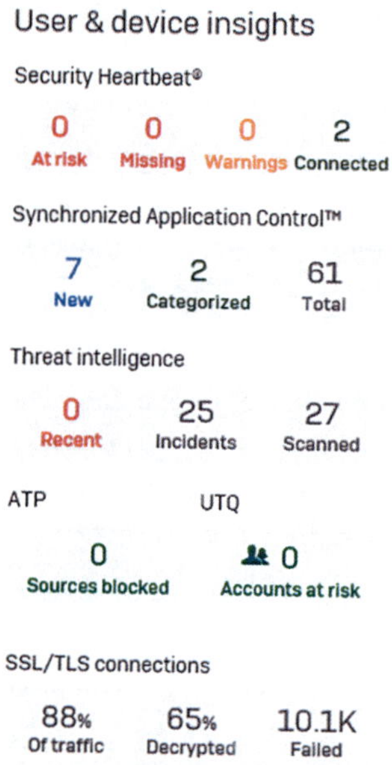

Fig. 9. User and device information. Source: Own elaboration

For the analysis of results, the compliance with the technical parameters of the solutions that were implemented is evaluated. For the malware defense according to control 8, anti-malware was installed with the following compliance:

- Centralized cloud protection
- Host-based IPS
- Peripheral control
- Real-time protection
- Navigation protection
- Email malware protection

– File control by extension

It has been possible to identify attacks on the network, malware among other parameters that affect the day-to-day work and cause losses to organizations.

For the fulfillment of the control 12 a firewall was installed, this fulfills the proposed in a successful way and the due justification. The implementation of the firewall allows:

– Control the use of network applications
– It has an antivirus engine for network
– It has email protection
– It has protection and control for safe navigation
– Performs port and protocol control at Layer 4
– Allows to section the networks and provide access by specific users or equipment
– Provides network perimeter security
– Integration with endpoint security

The advantages of each of the products for internal and perimeter security that are used are based on compliance with control 8 and 12 respectively, which have allowed to protect a network that is not very secure or not at all, reaching the end user considered vulnerable to new trends in malware and protecting the perimeter from network attacks and traffic, however, the current trend of hacking, malware and other infections with an exponential growth forces large companies to implement and comply with security standards to cover security gaps, in addition to having integration tools for a correlation of events, very important for the visibility of logs, according to the controls mentioned the network is being protected but in turn be applied to different companies for proper implementation is considered user number and the different applications, segmentations and architecture to be covered. In this way it is considered with the fulfillment that a network must be secure, scalable and reliable.

5 Conclusions

The solution applied within the project through the two controls 8 and 12, complies with a design based on the standard controls proposed by the Institute-SANS, these controls enforce the principles of confidentiality, integrity and availability, as well as considers a scalable implementation, modular and easy to manage.

It defines the different methodologies to establish critical security controls established by the SANS Institute for the administration, management of a network considered as safe, so each evaluation is concluded based on the needs of the employee and the delivery of services available within and outside the company, in this way we have the efficiency, control and proper administration.

The implementation of an anti-malware is considered reliable with adequate protection, administration and configuration at all endpoints, according to the established standard, the teams must have the signature updates up to date to avoid zero-day threats.

The perimeter level configuration complies with the protection of the network allowing blocking of attacks, web filtering with new generation security parameters, maintenance of operational business roles and protected information.

The established controls have been previously evaluated according to requirements and information gathering by the network administrator and these can be evaluated according to the company's role.

References

1. Manual de supervisión de riesgos cibernéticos para juntas corporativas (2017). https://www.oas.org/es/sms/cicte/docs/ESP-Manual-de-Supervision-de-riesgos-ciberneticos-para-juntas-coporativas.pdf. Accessed 15 Apr 2020
2. Instituto SANS: Acerca de. https://www.sans.org/about/. Accessed 15 Apr 2020
3. Groš, S.: A critical view on CIS controls, October 2019
4. Morales, F., Toapanta, S., Toasa, R.M.: Implementación de un sistema de seguridad perimetral como estrategia de seguridad de la información. Rev. Ibérica Sist. e Tecnol. Informação **E**(27), 553–565 (2020)
5. Leon, D., Mayorga, F., Vargas, J., Toasa, R., Guevara, D.: Using of an anonymous communication in e-government services: in the prevention of passive attacks on a network. In: Iberian Conference on Information Systems and Technologies, CISTI, June 2018, pp. 1–4 (2018)
6. Chuquilla, A., Guarda, T., Ninahualpa Quina, G.: Ransomware - WannaCry security is everyone's, pp. 1–4 (2019)
7. Alexandrovna, G.M., Christian, F.: Kaspersky security bulletin 2013. Boletín Segur. Kaspersky 2013, vol. 3, p. 4 (2014)
8. Qué es el malware y cómo prevenirlo. https://www.elespanol.com/ciencia/tecnologia/20180912/malware-prevenirlo/325467916_0.html. Accessed 23 Apr 2020
9. Como Logitek implementa la defensa perimetral - Ciberseguridad Logitek. https://www.ciberseguridadlogitek.com/como-logitek-implementa-la-defensa-perimetral/. Accessed 23 Apr 2020
10. Guía de Controles Críticos de Ciberseguridad. https://www.cert.gov.py/index.php/download_file/view_inline/1375?cv=1. Accessed 15 Apr 2020

Software Engineering

Maintainability and Portability Evaluation of the React Native Framework Applying the ISO/IEC 25010

Darwin Mena[1](✉) and Marco Santorum[2]

[1] Escuela Politécnica Nacional, Observatorio Astronómico, Quito, Ecuador
darwin.mena@epn.edu.ec

[2] Facultad de Ingeniería en Sistemas, Escuela Politécnica Nacional, Quito, Ecuador
marco.santorum@epn.edu.ec

Abstract. Creating native mobile applications generates twice as much work for developers, as a solution appears Cross-Platform development. React Native (RN) is a framework developed by Facebook and is proposed as an alternative to develop native applications for iOS and Android. However, there are still unknowns such as how easy it is to modify and transfer a software product from one environment to another. The present project proposes the evaluation of the maintainability and portability of RN through a case study using an existing standardized quality model; we specifically focus on that proposed in the ISO/IEC 25010 standard. All Sub-characteristics were evaluated using metrics generating quantitative results which show how maintainable and portable is the RN framework. The study also yields data that enables good development practices that significantly increase the quality of the software product.

Keywords: Maintainability · Portability · ISO25010 · React native · Software

1 Introduction

Currently, the development of mobile applications is in constantly growing. Mobile devices are always increasing in efficiency and performance, with 3.8 billion smartphone users estimated by 2021 [1].

Developers have to satisfy the needs of different brands and models of mobile devices, which is feasible with Cross-Platform development [2]. In this context, the React Native framework (RN) is an alternative that allows the development of native applications for iOS and Android [3].

A study published by "BIT Computer Training" [4] analyzes the performance of RN reaching quotas close to a native application developed for iOS. However, there are not enough studies that analyze the quality of the software product for the framework.

The mobile application development industry, when opting for a multiplatform framework, asks the following questions: how feasible is it to modify a software product

M. Botto-Tobar et al. (Eds.): ICCIS 2020, AISC 1273, pp. 429–439, 2021.
https://doi.org/10.1007/978-3-030-59194-6_35

so that it works efficiently in different Operating Systems, and what capacity it has to be transferred from an environment to another. [5].

The solution to the questions can be posed through the ISO/IEC 25010 standard [6], that raises the software product quality model. It contains the characteristics of maintainability and portability, which can both be precisely measured, meaning we can determine how transferable the product is from one environment to another [7].

This project proposes the evaluation of maintainability and portability of the React Native framework using the ISO 25010 standard through a case study. As a result, an evaluation matrix is generated for easy use in future research.

2 Methodology

To evaluate the maintainability and portability of the React Native framework, we proceeded as follows. First, the ISO/IEC 25010 standard was revised, identifying the maintainability and portability Sub-characteristics, metrics, and equations. Second, the case study is described with the necessary parameters for the equations. Finally, the results are presented with their corresponding analysis. Next, the case study and the evaluation matrix are presented (see Tables 3 and 4).

2.1 Description of the Study Case

The case study application is developed in React Native and aims to manage events. It gives the user the possibility to access the information of the events, asking questions to the speaker and scheduling the conferences of interest.

The use case diagram presented in Fig. 1 illustrate the case study. Once a user registers he becomes an assistant and can see next events, manage his schedule, ask questions to the exhibitor and receive notifications of next events.

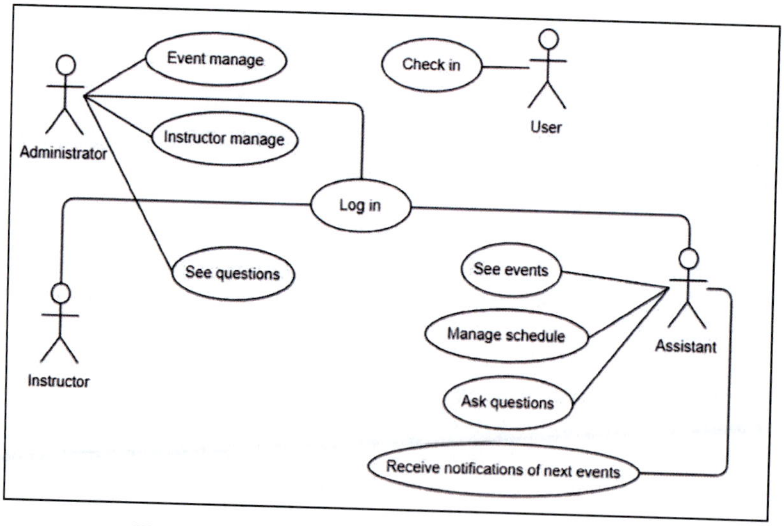

Fig. 1. Use case diagram of the case study application.

2.2 Evaluation Matrix Based on ISO/IEC 25010

Internal and External Quality Model of Software Product [8]

The model ISO/IEC 25010 defines 8 characteristics for the internal and external quality of a software product: Adequacy, Functionality, Performance Efficiency, Compatibility, Usability, Reliability, Security, Maintainability and Portability. Which are subdivided into Sub-characteristics described (see Fig. 2).

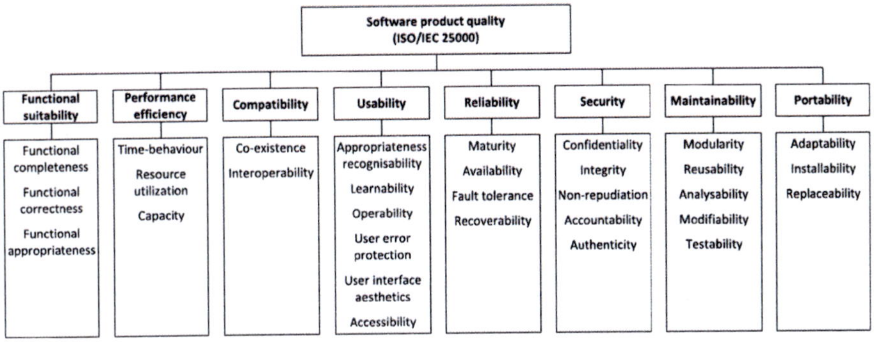

Fig. 2. ISO/IEC quality model for software products [9].

The objective of this project is the evaluation of maintainability and portability, and the ISO/IEC 25010 defines them as follows:

Maintainability [6]

Maintainability represents the ability of the software product to be effectively and efficiently modified. This characteristic is subdivided into the following Sub-characteristics (see Table 1).

Table 1. Maintainability Sub-characteristics [10].

Code	Sub-characteristics	Definition
MS1	Modularity	The capacity of a computer system or program (made up of discrete components) that allows a change in one component to have a minimal impact on the others
MS2	Reusability	Capacity of an asset that allows it to be used in more than one system or in the construction of other assets
MS3	Ability to be analyzed	Facility with which you can evaluate the impact of a certain change on the rest of software, diagnose deficiencies or causes of software failures, or identify the parts to be modified
MS4	Ability to be modified	Product capacity that allows it to be modified effectively and efficiently without introducing defects or degrading performance
MS5	Ability to be tested	Facility with which test criteria can be established for a system or component and with which tests can be carried out to determine if those criteria are met

Portability [6]

Portability analyzes the ability of the product or component to be transferred effectively and efficiently from a hardware, software or user environment to another. This characteristic is subdivided into the following Sub-characteristics (see Table 2).

Table 2. Portability Sub-characteristics [10].

Code	Sub-characteristics	Definition
PS1	Adaptability	Product capacity that allows it to be adapted effectively and efficiently to different hardware, software, operational or usage environments
PS2	Ability to be installed	Ease with which the product can be installed and/or uninstalled successfully in a certain environment
PS3	Ability to be replaced	Ability of the product to be used instead of other software product for the same purpose and in the same environment

To evaluate the maintainability and portability Sub-characteristics, the ISO/IEC 25010 model uses metrics that are described in the ISO/IEC 25022 standard [10]. Each metric defines an equation with parameters A and B [11]. It also details the range and the best value to obtain in response (see Tables 3 and 4).

The first step to obtain the value corresponding to each metric, is to identify each value of A and B, for those in Tables 3 and 4 detailed how to obtain each parameter, in some cases only the number of functions or components that each metric detail in other cases for example the complexity of modification intervenes time (T) as a variable to be measured [12]. The resources needed for the evaluation are the source code and information provided by the developer and the tester, so the use of a case study is necessary.

Once we have the values of A and B, we proceed to calculate the value of X according to each metric (see Tables 3 and 4). Here it is important to consider that since the range and the best value differ in each metric, for this reason is necessary to weight the result on the same scale. For ease of calculating general percentages, in this project the values will be weighted on the scale of 0 to 1 (see Tables 5 and 6).

Furthermore, the methodology to be applied will allow correcting errors and generating good development practices that would improve the quality of the software product.

Table 3. Equations of maintainability metrics [10].

Metric & equation	Description
MS1M1- Condensation capacity X = A/B	A = Number of components that are not affected by changes of other components B = Total number of specific components Where: B > 0. 0 <= X <= 1; The closest to 0, is the best
MS1M2- Coupling classes X = A	A = Number of relations that a function has respect to other classes. 1 <= X <= 4; The closest to 1, is the best
MS2M1- Execution of reusability X = A/B	A = Number of reused elements; B = Total number of items in the reusable library; Where: B > 0 0 <= X <= 1; The closest to 1, is the best
MS3M1- Audit trail capacity X = A/B	A = Number of data actually recorded during operation; B = Number of expected data to be recorded to control the state of the system during the operation. 0 <= X <= 1; The closest to 1, is the best
MS3M2- Diagnosis of sufficient functions X = A/B	A = Number of diagnostic implemented functions; B = Number of diagnosis functions required in the specification of requirements (B > 0). 0 <= X <= 1; The closest to 1, is the best
MS4M1- Cyclomatic complexity X = A + 1	A = Number of conditional instructions that a function has [13] 1 <= X < 15; The closest to 1, is the best
MS4M2- Inheritance depht X = A	A = Number of hierarchies used for a given function 0 <= X <= 4; The closest to 0, is the best
MS4M3- Impact correction localization degree X = A/B	A = Number of failures appearing after a fault has been solved; B = Number of solved faults. (B > 0) 0 <= X <= 1; The closest to 0, is the best
MS4M4- Modification complexity X = A/T	A = Number of modifications; B = Work time that takes the developer to modify. (T > 0) X = A/T The furthest to 0/t is the best
MS4M5- Modification success rate X = A/B	A = Number of problems within a certain period before maintenance B = Number of problems in the same period after maintenance (B > 0). 0 <= X <= 1; The closest to 0, is the best
MS5M1- Functional complexity of test functions X = A/B	A = Number of implemented test functions; B = Number of required test functions. (B > 0) 0 <= X <= 1; The closest to 1, is the best
MS5M2- Autonomous testing ability X = A/B	A = Number of tests that are depending on of other systems; B = Total number of tests depending of other systems. (B > 0) 0 <= X <= 1; The closest to 0, is the best
MS5M3- Test restart ability X = A/B	A = Number of cases in which the maintainer can pause and restore the test; B = Number of pause cases in the execution tests. (B > 0). 0 <= X <= 1; The closest to 1, is the best

Table 4. Equations of portability metrics [10].

Metric & equation	Description
PS1M1- Adaptability in hardware environment X = A/B	A = Number of operational functions that have not been completed during operational tests with the hardware B = Total number of functions that have been tested. (B > 0) 0 <= X <= 1 The closest to 0, is the best
PS1M2- Adaptability in software environment X = A/B	A = Number of operational functions of tasks that have not been completed during operational tests with the system B = Total number of functions that have been tested. (B > 0) 0 <= X <= 1 the closest to 0, is the best
PS1M3- Adaptability in an organizational environment X = A/B	A = Number of operational functions of tasks that have not been completed during operational tests with users of the business environment. B = Total number of functions that have been tested. (B > 0) 0 <= X <= 1 The closest to 0, is the best
PS2M1- Installation time efficiency X = A/T	A = Number of retries when installing the system. T = Total time elapsed when installing the system; Where: T > 0 X = A/T; The farthest to 0/t, is the best
PSSM2- Ease of installation X = A/B	A = Number of cases in which users succeeded in installing the system changing the installation process for their convenience. B = Total number of cases in which users have attempted to change the installation process for their convenience. (B > 0). 0 <= X <= 1; The closest to 1, is the best
PS3M1- Consistency in user support function X = A/B	A = Number of new functions that are considered as inconsistent by the user. B = Number of new functions. (B > 0) 0 <= X <= 1; The closest to 0, is the best
PS3M2- Functional inclusivity X = A/B	A = Number of functions that previously produced similar results and no changes have been required. B = Number of tested functions that are similar to functions provided by other software to be replaced 0 <= X <= 1; The closest to 1, is the best
PS3M3- Continuous use of data X = A/B	A = Number of data that is continuously only usable by the software to be replaced. B = Number of data that is reusable by the software to be replaced. (B > 0) 0 <= X <= 1; The closest to 1, is the best

3 Results

For the maintainability and portability evaluation of React Native, the source code of the case study described above was used, and for certain metrics we used data provided by the developer and tester.

3.1 Weighted Results on a Scale of 0 to 1

Because the maintainability and portability metrics have different ranges and the desired value in some cases is closest to 0 or 1 (as indicated in Table 5 for maintainability and Table 6 for portability), the results are scaled in the range of 0 to 1 to facilitate the analysis.

For example, for the "MS4M1-Cyclomatic Complexity" metric, the desired value as a response is 1 and the value obtained as a response after the evaluation of the case study was 13.

Table 5. Maintainability results in scale 0 to 1.

Metric	Desired value	Range	X	Result 0 to 1
MS1M1- Condensation capacity	0	0 to 1	0.5	0.5
MS1M2- Coupling classes	1	1 to 4	1	1
MS2M1- Execution of reusability	1	0 to 1	0.2	0.2
MS3M1- Audit trail capacity	1	0 to 1	0.5	0.5
MS3M2- Diagnosis of sufficient functions	1	0 to 1	0.9	0.9
MS4M1- Cyclomatic complexity	1	1 to 15	13	0.12
MS4M2- Inheritance depth	0	0 to 4	1	0.75
MS4M3- Impact correction localization degree	0	0 to 1	0.4	0.6
MS4M4- Modification complexity	None	None	0.5	0.5
MS4M5- Modification success rate	0	0 to 1	0.7	0.3
MS5M1- Functional complexity of test functions	1	0 to 1	0.4	0.4
MS5M2- Autonomous testing ability	0	0 to 1	0.8	0.2
MS5M3- Test restart ability	1	0 to 1	0.6	0.6

Table 6. Portability results in scale 0 to 1.

Metric	Desired value	Range	X	Result from 0 to 1
PS1M1- Adaptability in hardware environment	0	0 to 1	0.1	0.9
PS1M2- Adaptability in software environment	0	0 to 1	0.2	0.8
PS1M3- Adaptability in an organizational environment	0	0 a 1	0.1	0.9
PS2M1- Installation time efficiency	None	None	0.07	0.93
PSSM2- Ease of installation	1	0 to 1	0.67	0.67
PS3M1- Consistency in user support function	0	0 to 1	0.3	0.7
PS3M2- Functional inclusivity	1	0 to 1	0.4	0.4
PS3M3- Continuous use of data	1	0 to 1	0.5	0.5

The range defined for the result ranges from 1 to 15; 13 represents 0,12 on the scale from 0 to 1. In this way, we proceeded with the rest of the results (see Table 5 and 6).

Finally, with the average for maintainability and portability is calculated with weighted results on a scale of 0 to 1 (see Table 7).

Table 7. Maintainability and Portability averages.

Characteristic	Summary	Average
Maintainability	6.57	50.53%
Portability	5.85	73.13%

4 Discussion of Results

Measuring the degree of quality for maintainability and portability of the software product for the RN framework is complex, although there are studies that analyze performance at the hardware level such as the one proposed by Lelli and Bostrand [14] where speed, memory and energy efficiency are analyzed. There are still few studies of maintainability and portability in software product quality.

However, there is a study by Heden and Backlund [15] where the product quality of RN and Progressive Web App software is evaluated using the ISO 25010 standard, but it is done in a general way, that is no metrics are applied. The present project presents a matrix to evaluate maintainability, (see Tables 1 and 3) and can serve as a reference for future studies.

Evaluating portability is a complement to this project so it is important to analyze the ability of a product or framework to be transferred from one software or hardware environment to another. In this context, there is a study by Yuhana and Saptarini [16] where the portability of an academic application is evaluated using metrics, but it only focuses on adaptability.

The most relevant maintainability metrics of this study are presented by the next.

MS1M1-Condensation Capacity. The result obtained for the case study is 0.5, on a scale of 0 to 1 is a good result, as it confirms that RN is a modular framework, that is the change in most components is not affected by other components.

MS2M1-Reusability Execution. The evaluation of this metric is important because it answers the question that many companies and developers in general ask themselves. In other words, how reusable are the elements when we talk about two operating systems such as Android and iOS.

The value obtained is 0.2, which means that 20% of the elements are reusable. This implies that you can easily have an average application with similar characteristics and functionality in the two operating systems.

MS3M2-Diagnosis of Sufficient Functions. This metric analyzes the number of diagnostic functions to be implemented to detect any error, either by the developer or the test team. The value obtained was 0.9 out of a maximum of 1, which is a high value that implies that RN is versatile when creating test functions.

MS4M2-Depth to Inheritance. This metric analyzes the hierarchical level of the functions and the less hierarchy there is, the better. The value obtained is 0.75 on a scale from 0 to 1, which confirms the high efficiency of RN because everything is inherited from the Component class which facilitates the handling of events, data and states.

MS4M4-Modification Complexity. The metric analyzes the time that it takes the developer to modify the application for the case study. A value of 0.5 was obtained which is moderate because it is neither fast nor slow for the developer to modify the RN code. Taking into account that the framework is new the result is acceptable.

The overall result for maintainability of the evaluation is 50.53%, that is good considering that in the case study many metrics depend on the developer and the test team.

Similarly, the most important portability metrics are presented by the next.

PS1M1-Adaptability in Hardware Environment. The metric measures how adaptable the functions are to a hardware environment. The value obtained was 0.9 that is high and confirms that React Native is adaptable to any hardware environment.

PS1M2-Adaptability in Software Environment. This metric measures how adaptable a function is to a software environment. The present project analyzed in two environments, Android and IOS, resulting in a value equal to 0.8, therefore RN fits perfectly.

PS1M3-Adaptability in an Organizational Environment. This metric analyzes whether the operational functions were completed with users of the system. The value obtained was 0.9 that is excellent, since most of the user requirements were completed.

PS2M2-Ease of Installation. This metric analyzes users who succeeded in installing application. The value obtained was 0.67 which guarantees that users would have no problem installing the mobile application.

PS3M1-Consistency of the User Support Function. This metric analyzes the number of functions considered to be inconsistent by the user, the value obtained was 0.75, as a function that users consider to be inconsistent was identified.

In general, after evaluating portability the result obtained was 73.13% for the case study, which is excellent and guarantees that RN can be easily adapted to any current mobile device model regardless of the software or operating system.

5 Conclusions

This project evaluates the maintainability and portability characteristics of the React Native framework using the ISO 25010 standard, it is carried out quantitatively evaluating each metric of the characteristics through a case study, thus generating indirect results and benefits of the RN framework using it as a multiplatform language instead of conventional development methods.

The methodology used in this project can be used as a tool during development to identify how maintainable a product is after putting it into a production environment, even more so if it is a new framework such as React Native. And in turn become a measure of how well you are coding to overcome future mistakes.

Maintainability makes it possible to measure how modifiable a software product is from one environment to another. According to the results of this study, RN is flexible and modifiable, it has several elements that work in multiplatform environments, but rather the recommendations and good practices are for the development team and tests to cover aspects that go beyond the requirements, such as the correct use of inheritance and functions that can be updated without affecting other existing ones.

Portability measures the ability of a software product to be transferred from one hardware and software environment to another, according to the study conducted RN can adapt to any operating environment. The applied methodology also allows identifying installation problems for example and solving them effectively.

The applied methodology undoubtedly not only produces maintainability and portability measures, but also generates good practices that significantly improve the quality of the software product.

Finally, indirectly through a case study, it is shown that RN is an excellent alternative for multiplatform development according to current mobile operating systems, being able to modify and maintain itself from one environment to another throughout the life cycle of the software product.

References

1. Number of smartphone users worldwide. https://www.statista.com/statistics/330695/number-of-smartphone-users-worldwide/. Accessed 23 Mar 2020
2. Heitkötter, H., Hanschke, S., Majchrzak, T.: Evaluating cross-platform development approaches for mobile applications. In: International Conference on Web Information Systems and Technologies, pp. 120–138 (2012)
3. Johansson, E., Söderberg, J.: Evaluating performance of a React Native feature set (2018)
4. React Native: Una nueva arquitectura para la generación de Apps – Bit. https://www.bit.es/knowledge-center/react-native-una-nueva-arquitectura-para-la-generacion-de-apps-moviles/. Accessed 24 Apr 2019
5. Axelsson, O., Carlström, F.: Evaluation targeting react native in comparison to native mobile development (2016)
6. ISO_25010: ISO/IEC 25010:2011: Systems and Software Engineering - Systems and Software Product Quality Requirements and Evaluation - System and Software Quality Models, 4.1.3.2, ISO, p. 4 (2011)

7. Paz, J.A.M., Gómez, M.Y.M.: Análisis sistemático de información de la Norma ISO 25010 como base para la implementación en un laboratorio de Testing de software en la Universidad Cooperativa de Colombia Sede Popayán, Memorias de Congresos UTP, pp. 149–154 (2017)
8. Gordieiev, O., Kharchenko, V., Fominykh, N.: Evolution of software quality models in context of the standard ISO 25010. In: Ninth International Conference on Dependability and Complex Systems, pp. 223–232 (2014)
9. Oriol, M., Marco, J., Franch, X.: Quality models for web services: a systematic mapping. Inf. Softw. Technol. **56**(10), 1167–1182 (2014)
10. ISO_25022, ISO/IEC 25022:2016 [ISO/IEC 25022:2016] Systems and software engineering — Systems and software quality requirements and evaluation (SQuaRE)—Measurement of quality in use. p. 41 (2016)
11. Lenhard, J.: Improving process portability through metrics and continuous inspection. In: Advances in Intelligent Process-Aware Information Systems, pp. 193–223 (2017)
12. Banker, R.D., Davis, G.B.: Software development practices, software complexity, and software maintenance performance: a field study. Manage. Sci. **44**(4), 433–450 (1998)
13. Tiwari, U., Kumar, S.: Cyclomatic complexity metric for component based software. ACM SIGSOFT Softw. Eng. Notes **39**(1), 1–6 (2014)
14. Lelli, A., Bostrand, V.: Evaluating application scenarios with react native: a comparative study between native and ReactNative development (2016)
15. Bäcklund, L., Hedén, O.: Evaluating react native and progressive web app development using ISO 25010 (2018)
16. Yuhana, U., Saptarini, I., Rochimah, S.: Portability characteristic evaluation academic information System assessment module using AIS quality instrument. In: 2nd International Conference on Information Technology, Computer, and Electrical Engineering (ICITACEE), pp. 133–137 (2016)

Web Information Systems and Applications

A Comprehensive Solution for Electrical Energy Demand Prediction Based on Auto-Regressive Models

Juan-José Sáenz-Peñafiel[1(✉)], Jorge E. Luzuriaga[2], Lenin-Guillermo Lemus-Zuñiga[2], and Vanessa Solis-Cabrera[1]

[1] Universidad de Cuenca Dirección de Investigación (DIUC), Av. 12 de Abril y Av. Loja s/n, 010107 Cuenca, Ecuador
juan.saenz@ucuenca.edu.ec,
vsolis@pros.upv.es

[2] ITACA Institute - Universitat Politècnica de València ICT against Climate Change Research Group, Camí de Vera s/n, 46022 València, Spain
{jorluqui,lemus}@upvnet.upv.es
https://ictvscc.webs.upv.es/en

Abstract. Energy consumption and demand are two widely used terms necessary to understand the functioning of the different mechanisms used in electrical energy transactions. In this article, the design and construction of a comprehensive solution to forecast future trends in electricity transactions using the historical data and two auto-regressive models were considered. Simple linear regression and a complete model such as ARIMA. We compared these models to find which one best suits the type of data considering their strengths and weaknesses for this specific case. Finally, to complete the comprehensive solution, the results are presented to the final user. This solution is mainly aimed at professionals who carry out activities related to contracting and managing electricity supply in public institutions. This solution pretends to collaborate to reduce energy demand and therefore, consumption.

Keywords: Prediction · ARIMA · Energy · Energy demand · Data capture · Auto-regressive models

1 Introduction

Administrators of municipalities and municipalities have as one of their tasks the management of resources, such as the supply of energy, water, among others. However, either for lack of resources or for a number of reasons, it may be the case that those responsible for the aforementioned entities do not know the information (quantity or distribution) on the consumption made in their locations that is available to them as happened with the covid-19 emergency that generated a lack of accurate information. In this work we consider that this knowledge is a fundamental basis for developing initiatives related to proper

M. Botto-Tobar et al. (Eds.): ICCIS 2020, AISC 1273, pp. 443–454, 2021.
https://doi.org/10.1007/978-3-030-59194-6_36

management, specifically energy management. For example, the cost generated by the consumption of energy in public buildings involves a fairly high expense in the budgets of the municipalities [2]. To try to reduce this drawback, alternatives are sought to reduce consumption or the cost at which energy is acquired, so that the distribution of consumption within a set time, called the load curve [2], is less than the of previous consumption.

With all the information on the behavior of energy consumption in relation to the public buildings considered for this work, the aim is to present the user with an interactive interface that helps the user, as a public decision maker, to make decisions for the purchase of energy necessary electrical, obtained based on the prediction. Likewise, observe the consumption history in such a way that it can carry out measures in the face of the possible decrease in consumption within the buildings it is in charge of, or failing that, analyze alternatives for the use of renewable energies that cover the total demand in its buildings. This article defines the methodology to be used in Sect. 2, then the stages of the prediction model design are presented in Sect. 3. Then in Sect. 4 the development of the prediction model is presented. Finally results and conclusions are shown in Sect. 5 and 6.

2 Methodology

The methodology used for the design of electricity demand starts from a clear definition of the objectives towards the energy transition. To address this problem, the methods used are described below.

- Problem statement and objectives: The reasons for carrying out this work are determined, as well as the scope of this work, delimited by the objectives.
- State of the art about work: Search and compilation of related works and useful tools that can be used for the specific case.
- Data capture: Use of the infrastructure for data collection that is available to date.
- Presentation of historical and predictions: Graphically show the user the captured data, as well as those of the predictions in the form of a scorecard.
- Data transformation: From the formats obtained in the capture, the necessary data is extracted in text files or databases according to the case, so that they can be used in the following steps with ease.
- Data prediction: Development of the energy demand prediction tool with the data obtained in the transformation.
- Results: Presentation of results obtained in the prediction with the changes of parameters and types of predictors used.
- Conclusions: The conclusions of the work are obtained and the options found to extend the work carried out are presented.

3 State of the Art

3.1 Demand Prediction Models

Several studies have sought to provide demand prediction alternatives, in such a way that those in charge of using the electricity market can only purchase the necessary energy based on the prediction obtained. For this, many techniques and models have been used, from mathematical models based on probabilities such as the one proposed by Hong in [3] that uses the probabilistic model for the prediction of demand and also of price; or in other cases comparing statistical models, both linear and stochastic for the prediction of demand [4]. Neuronal network-based models for demand prediction have also been implemented, as is the case of Kankal in [5], who seeks to train various neuronal network models for long-term prediction. Or based on fuzzy logic relating the demand values by grouping these values by dates similar to those to be predicted and giving a value similar to the group [6].

When it comes to energy consumption, there have been many works related to its prediction, having notable results with the use of neuronal networks, as mentioned by Hu in [7], who predicts consumption with various techniques, giving positive results and prediction percentages with error less than 4%. Or in other cases with comparative results with regressions, decision trees and neuronal networks, as proposed by Tso and Yau [8], which as a result obtain very similar values in the techniques mentioned above. In the same way as in demand, the ARIMA models have also given good results (mean absolute percentage error less than 3%)t in the prediction of energy consumption, as are the cases of Pao and Tsai [9] and, Yuan, Liu and Fang [10] that obtain prediction results with said model and compare it with their own models, giving close results between them.

3.2 Data Clustering

Clustering techniques vary according to the type of data being handled; In the specific case of time series clustering, there are techniques based on the hierarchy of the data, which group the data based on their importance [11], or techniques based on the relocation of the prediction center, as in the case of [12] that perform energy data clustering by moving the data centroid to get groups in relation to the distance to said centroid.

Furthermore, for time series works such as Niennattrakul et al. [13] or Astakhova et al. [14] that use the k-means technique that separates the data into k groups in such a way that the data is classified in the group with the closest mean data. In [13], clustering with k-means in multimedia time series data is successfully used to group the data by time, but generating their own time intervals. While in [14] they use k-means grouping to make price data prediction.

3.3 Data Analysis

For data analysis, various statistical, computational or visual methodologies have been used, which vary their use according to the type of data and results that are desired in the study. In the specific case of data with statistical models, techniques are used with autoregressive models to the time series, correlation, or graphic models as detailed by George et al. in [15], specifying the correct type of analysis for each type of series of time.

Regarding computational models for time series, techniques such as data mining have been widely used as indicated [16] in his review of the use of the techniques, highlighting examples of clusterization and data grouping, in addition to tools that have been used data mining using visualizers.

In specific cases of electrical energy data analysis, [17] they perform the price prediction analysis using the ARIMA autoregressive model, obtaining error values of up to 10% with energy price data.

3.4 Software Development Methodologies

As for the software development that will perform the prediction and presentation of the data, it must be separated in relation to the type of application to be developed, being complex software methodologies oriented to business software such as Rational Unified Process (RUP) which is a guide to software construction based on a set of adaptive processes for software [18]; and Service Oriented Architecture (SOA), which guides the construction of software based on reusable services, as exemplified by the authors González et al. [19].

In turn, agile methodologies for experimental software development such as the one based on iterative and incremental processes called Dynamic Systems Development Method (DSDM); or methodologies also based on iterative but change-tolerant processes such as Adaptive Software Development, which have been part of comparative analysis in [20].

As a methodology adaptable to experimental software, SCRUM is also found, which has been applied to various prediction projects, such as the work by Batarseh and González [21] that uses SCRUM to predict failures in data analysis and presents statistics on failures in the use of the methodology.

4 Design and Development Methodology

For the development of the application, a methodology was required that repeats at each stage and that in each iteration the working group can make the relevant changes. For this reason, the SCRUM methodology was selected for its ability to adapt to changes in requirements and iterations within the process.

4.1 Data Collection

In the capture stage, the Web Scraping technique was used, which consists of automating the extraction of data from a certain web page and storing it in our repository, be it a database or files separated by commas. The scraping technique was used to simulate the behavior of a user in downloading the PDF and XML files of data consumption from the electricity company website, performing the automation process of web browsing shown In the diagram in Fig. 1. The PDFs and XMLs obtained in the capture process correspond to the period between January 2015 and December 2018.

To process these files, they were first separated by years and then grouped based on the type of invoice of each file as It is explained in Fig. 2 because there are certain important differences that characterize each type of invoice. As a result of this process, a set of PDF files corresponding to invoices for each contract is obtained. The process must iterate one by one across all con-tracts required for the application.

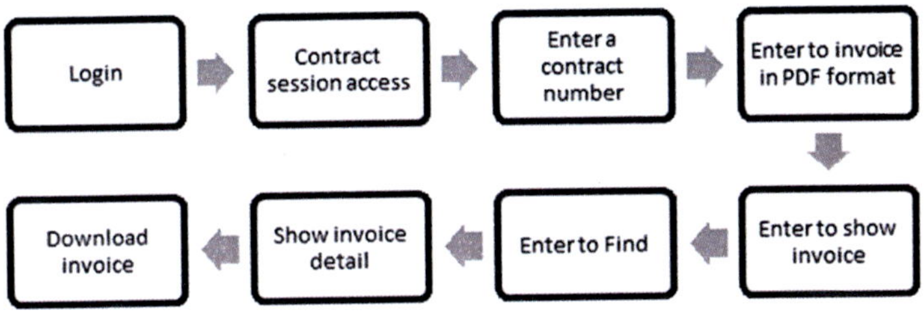

Fig. 1. Process diagram for downloading invoices.

4.2 Transformation

In order to obtain the data from the invoices issued and downloaded from the electricity company website, a module was carried out in charge from the organization of raw data to the loading in the database; This database has been prepared in relation to the available data.

To process the files, they were first separated by years and then grouped based on the type of invoice of each file as explained in Fig. 2 due to that there are certain important differences that characterize each type of invoice.

Within the data transformation process, different problems were evidenced, among which the diversity of formats among the invoices, special characters in the content of the invoice, different sizes in the invoices, finally the variation of information within the content of the invoice. which caused the download in XML format to be activated within the electricity company platform. Subsequently, the data was extracted from the XML files based on the diagram in Fig. 3.

4.3 Data Analysis

After transforming and organizing the data we rely on common methodologies used in publications and research regarding data analysis such us such as the

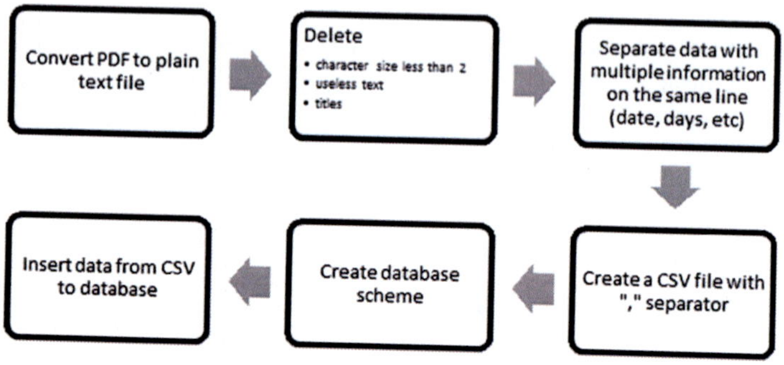

Fig. 2. Process diagram for extracting data from a PDF file.

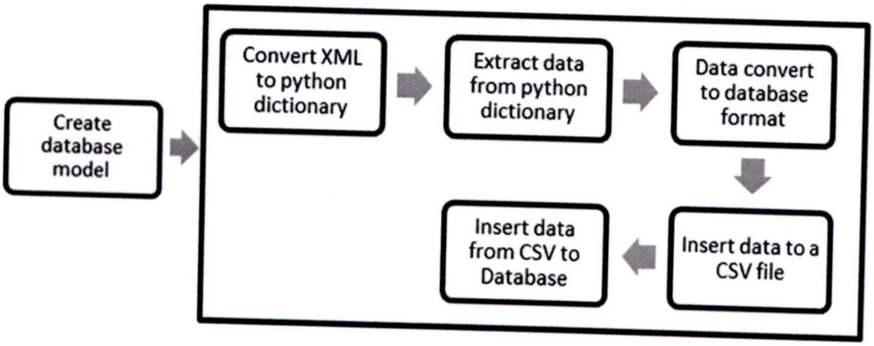

Fig. 3. Process diagram for extracting data from a XML file

time series analysis described in [14] or the grouping of data as indicated in [13]. Basically we follow the scheme shown in Fig. 4 described below.

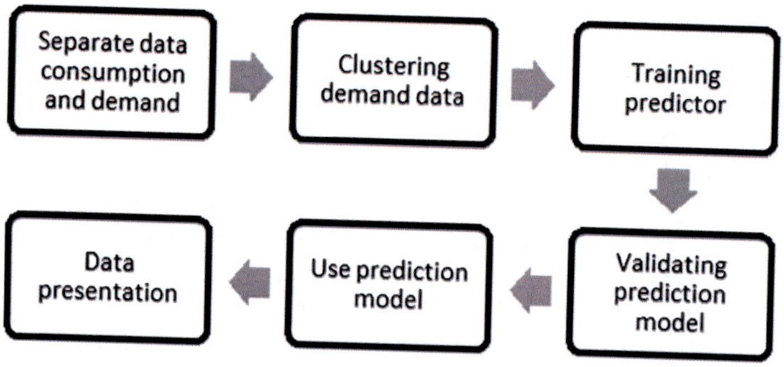

Fig. 4. Data analysis process diagram.

On the one hand, we have historical consumption data that helps us to observe and analyze the behavior in years, months, days and hours of each electrical service contract. On the other hand, the energy demand prediction data with which the prediction model will be trained and the corresponding tests.

4.4 Data Prediction

For the prediction of energy demand, a data set consisting of one year and one month has been prepared, from February 1, 2017 to February 28, 2018, because it is the complete data set. The units of the data are watts and the periodicity of each reading this data is schedules (every hour).

For the prediction of the energy demand, two models were selected from the aforementioned in the state of the art, a simple linear model [8] and an ARIMA autoregressive model [10,17] that will be described during the development of the prediction process presented in the process diagram of Fig. 5.

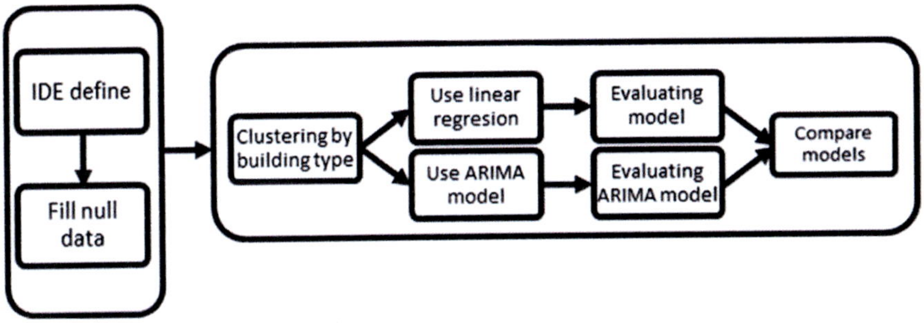

Fig. 5. Demand prediction process diagram.

As a next step, to classify/cluster data according to their behavior, the data was separated into two groups based on the closest mean value to the observation (K-means) [22]. In the specific case, each electrical service contract is separated by taking its aver-age value and comparing each other to finally generate two consumption profiles that will be used to create the prediction model. The first one represents the public lightning, while the second one represents the consumption of the offices themselves. In Fig. 6 the dispersion of consumption means can be observed from the demand data, thus distinguishing two groups marked for classification.

From the classification into groups that can be seen in Fig. 6 the consumption profiles on a specific day as can be seen in Fig. 7 for instance.

As can be seen in Fig. 7, the profile of Cluster 1, which corresponds to public lighting, has an on-time behavior at night and off during the day; while Cluster 2 has an office profile in which its energy demand increases during working hours, but at night it maintains a low consumption. Once the data is classified, we continue with the prediction process and for this, a first prediction approach is sought based on a simple linear regression model of progressive forecasting, that is, based on current data, predict the next step, add it to the set data and

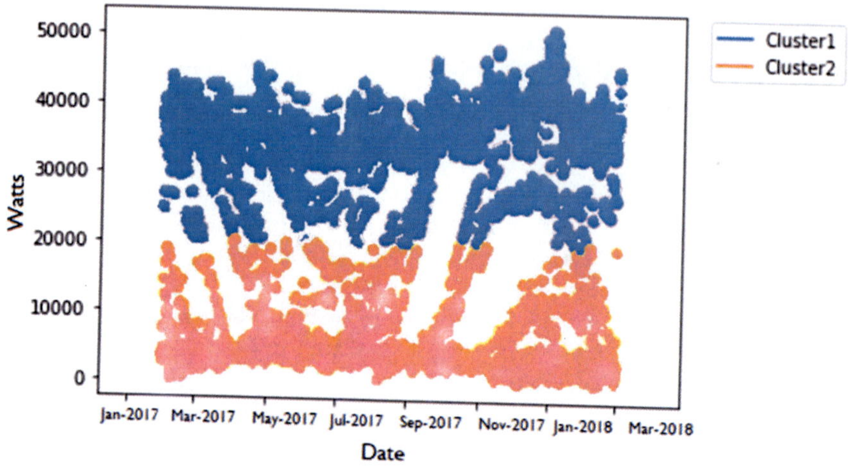

Fig. 6. Data classification from clustering.

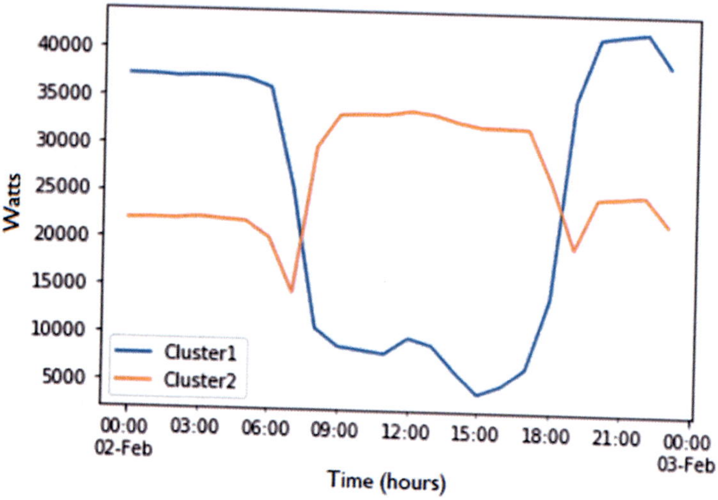

Fig. 7. Consumption profiles on a given day based on ranking.

redo the next prediction procedure and so on until the total data needed in the prediction is obtained [22]. For this first approach, 50% of the total set is used to train and then evaluate it with the next half of the data.

Next, for the evaluation of the simple linear regression model, the root mean square error (RMSE) was used in such a way that the extreme error values have a greater weight within the evaluation of the model. In addition, the RMSE gives us a value in watts, which makes it easier for the user to read. It is also evaluated with the mean absolute error (MAE) so as not to give excessive values to extreme values and the mean absolute percentage error (Mean Absolute Percentage Error (MAPE)), to handle an unrelated percentage of error to the units of measurement used (watts). The simple linear regression model gives us a starting point for prediction, but it can be improved by means of parameters or configurations that give us better results, as is the case with autoregressive linear models such as ARIMA [23]. ARIMA has the strength in prediction using time series [24]. For the ARIMA model, parameterize the model according to the number of jumps we want to make in the model regression or the number of means we want to have within the total data set. To obtain the best result, the combination of ARIMA parameters should be performed sequentially one by one until the prediction error is lowered. To achieve this, the model was evaluated for 15 times, combining each of the parameters with values from 0 to 4 as recommended by Grillenzoni [25]. Based on the parameters obtained, the model should be evaluated using RMSE, MAE and MAPE and in case of having very erroneous predictions, reconfigure the aforementioned parameters until having low error and the model can be used as definitive.

Once the model is configured, the results are evaluated graphically, using part of the data and graphing real values against prediction values.

4.5 Data Presentation

The data are presented in an orderly, interactive and easy to interpret manner in an interface that helps the user to make decisions. For this, the work process described in the diagram in Fig. 8 was followed.

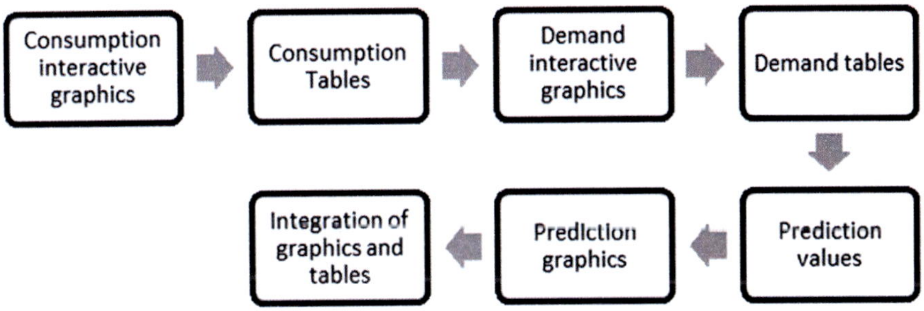

Fig. 8. Data presentation process diagram.

5 Results

To obtain the predictions, two models were used, simple linear regression and an ARIMA model. RMSE, was used for evaluation thereof MAPE and MAE, which are presented in Table 1.

Table 1. Comparison of statistical results of the prediction models.

Model	RMSE (KW)	MAE	MAPE
Single linear regression	5.888	4.187	14.24%
ARIMA	6.275	3.457	10.74%

Taking these results into account, it can be seen that the linear model works better for predicting extreme values because RMSE values this type of values more and gives us a value less than 6 kilowatts (kW), but both MAE and MAPE give us a value lower in ARIMA so that in general said model works better for the largest number of values except for the aforementioned extremes. This can be seen in Fig. 9 with 48 48 h prediction with the data used.

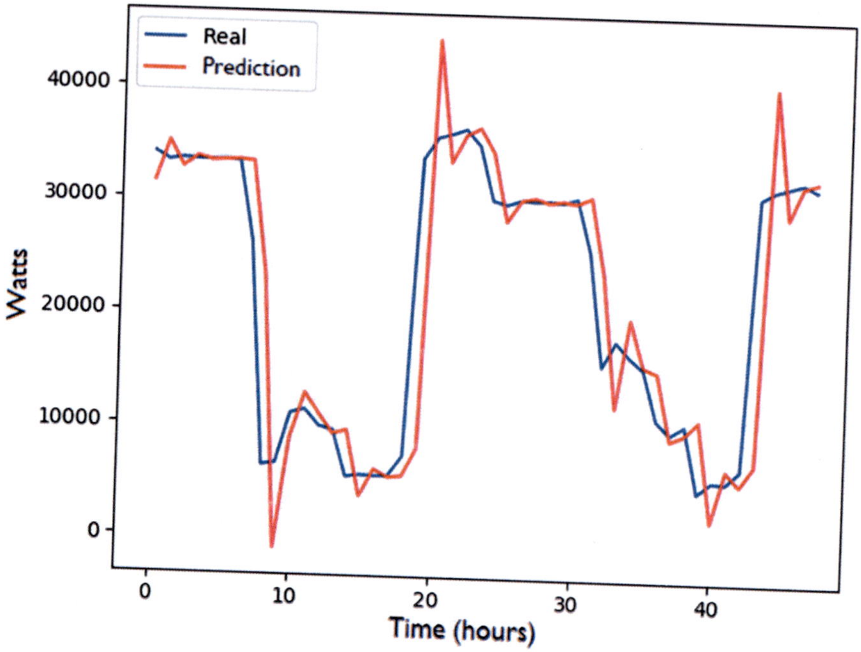

Fig. 9. Real data and prediction data.

6 Conclusions

In this article, we have presented an integral solution for electrical energy demand prediction aimed mainly at professionals responsible for contracting and managing electricity supply in public institutions. The treatment of large amounts of energy demand and consumption data was presented thought a solution. A solution that uses this data in a transparent way to users using a dashboard visual tool so that end-users can know the consumption presented by the different electrical contracts enabling decision-making processes according to this information.

Historical energy demand data were used as an input to our forecasting system. The prediction, whether of demand or energy consumption, has many alternatives to be worked on and with the passage of time it generates more significant variations that are influenced by current regulations or by the capacities of the equipment that generate or manage energy; this is why the updating of data collectors and predictors must be constant to improve the quality of data and predictions.

Two auto-regressive models were used to predict demand. The ARIMA model is a reasonably good alternative for predicting demand with time series. The ARIMA model meets the prediction needs and approximates its prediction to the real data, including abnormal values.

References

1. Alonso, G.: Análisis del mercado eléctrico español **41** (2014)
2. Diez, E.R., Pérez, M.G.: Políticas y medidas contra la pobreza energética a quién le corresponde? Areas. Rev. Internacional de Ciencias Sociales **36**, 153–165 (2017)
3. Hong, T., Pinson, P., Fan, S., Zareipour, H., Troccoli, A., Hyndman, R.J.: Probabilistic energy forecasting: global energy forecasting competition 2014 and beyond. Int. J. Forecast. **32**(3), 896–913 (2016). https://doi.org/10.1016/j.ijforecast.2016.02.001
4. Verdejo, H., Awerkin, A., Becker, C., Olguin, G.: Statistic linear parametric techniques for residential electric energy demand forecasting. A review and an implementation to Chile. Renew. Sustain. Energy Rev. **74**, 512–521 (2017). https://doi.org/10.1016/j.rser.2017.01.110
5. Kankal, M., Uzlu, E.: Neural network approach with teaching-learning based optimization for modeling and forecasting long-term electric energy demand in Turkey. Neural Comput. Appl. **28**(1), 737–747 (2017)
6. Kucukali, S., Baris, K.: Turkey's short-term gross annual electricity demand forecast by fuzzy logic approach. Energy Policy **38**(5), 2438–2445 (2010). https://doi.org/10.1016/j.enpol.2009.12.037
7. Hu, Y.C.: Electricity consumption prediction using a neural-network-based grey forecasting approach. J. Oper. Res. Soc. **68**(10), 1259–1264 (2017)
8. Tso, G.K., Yau, K.K.: Predicting electricity energy consumption: a comparison of regression analysis, decision tree and neural networks. Energy **32**(9), 1761–1768 (2007)
9. Pao, H.T., Tsai, C.M.: Modeling and forecasting the CO2 emissions, energy consumption, and economic growth in Brazil. Energy **36**(5), 2450–2458 (2011). https://doi.org/10.1016/j.energy.2011.01.032
10. Yuan, C., Liu, S., Fang, Z.: Comparison of China's primary energy consumption forecasting by using ARIMA (the autoregressive integrated moving average) model and GM(1, 1) model. Energy **100**, 384–390 (2016). https://doi.org/10.1016/j.energy.2016.02.001

11. Rodrigues, P.P., Gama, J., Pedroso, J.P.: Hierarchical clustering of time series data streams. IEEE Trans. Knowl. Data Eng. **20**(5), 615–627 (2008)
12. Lavin, A., Klabjan, D.: Clustering time-series energy data from smart meters. Energy Effi. **8**(4), 681–689 (2015)
13. Niennattrakul, V., Ratanamahatana, C.A.: On clustering multimedia time series data using k-means and dynamic time warping. In: 2007 International Conference on Multimedia and Ubiquitous Engineering (MUE 2007), pp. 733–738 (2007)
14. Astakhova, N.N., Demidova, L.A., Nikulchev, E.V.: Forecasting method for grouped time series with the use of k-means algorithm. Appl. Math. Sci. **9**(97), 4813–4830 (2015). http://www.m-hikari.com/ams/ams-2015/ams-97-100-2015/55391.html
15. George, G.M.L., Box, E.P., Jenkins, G.M., Reinsel, G.C.: Time Series Analysis: Forecasting and Control, vol. 4. Wiley, Hoboken (2015)
16. Fu, T.C.: A review on time series data mining. Eng. Appl. Artifi. Intell. **24**(1), 164–181 (2011). https://doi.org/10.1016/j.engappai.2010.09.007
17. Contreras, J., Espínola, R., Nogales, F.J., Conejo, A.J.: ARIMA models to predict next-day electricity prices. IEEE Trans. Power Syst. **18**(3), 1014–1020 (2003)
18. Paper Rational Software White: Rational unified process best practices for software. In: Development, pp. 1–21 (2004)
19. González, L., Piedrabuena, F.: Reporte Técnico RT 06-16 Desarrollo de aplicaciones con enfoque SOA Andrea Delgado (2006)
20. Letelier, P., Canós, M., Sánchez, E., Penadés, M.: Métodologías ágiles en el Desarrollo de Software. España, Valencia, pp. 1–8 (2003). http://www.cyta.com.ar/ta0502/v5n2a1.htm
21. Batarseh, F.A., Gonzalez, A.J.: Predicting failures in agile software development through data analytics. Softw. Qual. J. **26**(1), 49–66 (2018)
22. Wagstaff, K., Cardie, C., Rogers, S., Schroedl, S.: Constrained k-means clustering with background knowledge. In: International Conference on Machine Learning, pp. 577–584 (2001)
23. Zou, K.H., Tuncali, K., Silverman, S.G.: Correlation and simple linear regression. Radiology **227**(3), 617–628 (2003). https://doi.org/10.1148/radiol.2273011499
24. Jakasa, T., Androcec, I., Sprcic, P.: Electricity price forecasting - ARIMA model approach. In: 2011 8th International Conference on the European Energy Market (EEM), pp. 222–225, May 2011
25. Grillenzoni, C.: ARIMA processes with ARIMA parameters. J. Bus. Econ. Stat. **11**(2), 235–250 (1993)

Construction and Leverage Scientific Knowledge Graphs by Means of Semantic Technologies

Teresa Santamaria[1], Mariela Tapia-Leon[1], and Janneth Chicaiza[2(✉)]

[1] Univesidad de Guayaquil, Guayaquil, Ecuador
{teresa.santamarial,mariela.tapial}@ug.edu.ec
[2] Universidad Técnica Particular de Loja, Loja, Ecuador
jachicaiza@utpl.edu.ec

Abstract. Scientific production is a key issue in the assessment of universities' research. Due to the information overload in the publishing domain, it is increasingly difficult for experts, and decision-makers to find relevant information to carry out their research or management tasks. To address this problem, Knowledge Graphs are used to explore and connect publishing sources by increasing their value. In this paper, the authors propose a process to construct a scientific knowledge graph based-on publications' metadata. A case study implemented in a given university is explained. Finally, some visual representations are created from results returned by SPARQL queries. Thus, people will find relationships and patterns underlying scholarly production.

Keywords: Higher education institution · Ontology · Scientific knowledge graph · Semantic web · University of Guayaquil · SPARQL

1 Introduction

Scientific production is a key issue in the assessment of universities' research [13]. But, the explosion of publications difficulties to [4] decision-makers to find relevant information to carry out their management tasks. To address this issue, in recent years, Knowledge Graphs (KGs) are being used as the core of information systems because they facilitate the discovery of information [1].

To build a KG, the Resource Description Framework (RDF) is used. In RDF graphs, nodes represent domain's entities and they are connected by edges that represent relationships. The set of entities and relationships are organized in an ontology which defines their semantic and their usage restrictions [15].

To publish scientific production as knowledge graph has the potential to increase their value and help to discover interesting relationships amongst scholarly documents [7]. Furthermore, a KG published as RDF format supports complex, multidimensional and path-based queries. Therefore, research activities and their results, such as papers and data, are ideal candidates to publish as Scientific Knowledge Graphs (SKGs).

M. Botto-Tobar et al. (Eds.): ICCIS 2020, AISC 1273, pp. 455–466, 2021.
https://doi.org/10.1007/978-3-030-59194-6_37

An ecosystem of Linking SKGs, built from metadata of scholarly documents, provides interesting answers about the production of institutions, researchers, knowledge areas, sources, topics among other entities. For example, (1) what papers or authors do address a given subject or research problem? (2) who are authors that require more support, empowerment, or acknowledgment? (3) what are the production and the publication patterns in a specific area? The answers to these and other questions are relevant to (1) better conduct the efforts of researchers and universities, and (2) bring out transparency, re-usability, and knowledge sharing, by encouraging the flow of information through the scientific community [5].

In this paper, authors present a process to create SKGs from scientific publications' metadata, and then they discuss how knowledge graphs can help institutions and researchers to have an overview and, at the same time, a deeper understanding of the quantity and quality of their research's production. The authors explain the process by applying it to a well-known university in Ecuador.

2 Publishing Scholarly Production as Knowledge Graphs

Scientific production is the materialized part of the knowledge generated from research activities [11]. Today, it is possible to find descriptive data (metadata) of scholarly documents published using formal languages like RDF. In this way, semantic of data is explicitly codified and it can be used to facilitate access, integration, sharing and automatic processing of data by machines. Under a Semantic Web approach, we design and reuse ontologies as a framework for the representation of knowledge, and we use languages such as RDF, to construct graphs of scientific knowledge readable by computers.

2.1 Technologies for Semantic Description

In this section, the authors describe two technologies for building a knowledge graph. First, the well-known ontologies suite called SPAR is described, and second, some features of RDF as a standard for describing specific pieces of information are identified.

Semantic Publishing and Referencing (SPAR[1]) Ontology Network enables the semantic description of different entities of the scholarly publishing domain [10]. There are other ontological models to describe scholarly material such as Bibliographic Ontology Specification (BIBO[2]). But SPAR provides a more comprehensive collection of classes and properties than BIBO, thus SPAR enables the creation of more precise descriptions of bibliographic entities [9]. SPAR, by means of FaBiO (ontology which is part of SPAR) defines the structure to describe resources, according to the FRBR (Functional Requirements for Bibliographic Records), the recommendation of the International Federation of Library

---

1 www.sparontologies.net.
2 bibliontology.com.

Associations and Institutions. Therefore, SPAR provides greater expressivity, that BIDO, for unambiguously describing a bibliographic resource.

The current SPAR Ontology suite is composed of some ontologies. Each group of ontologies addresses a specific function such as descriptions of textual publications (e.g., books, conference proceedings, journals); description of document components both structural (e.g., paragraph, section, chapter) and rhetorical (e.g., introduction, discussion, reference list, figure, appendix); characterization of several agent roles (authors, editors, reviewers, publishers) in the publication process, bibliometric data and workflow processes, among others.

Once the ontological model has been chosen, it is time to use RDF to create semantic descriptions, of each entity of the domain, in a machine-readable format. RDF organizes information in sentences subject-predicate-object called triples. This is a way of representing and linking knowledge according to a small set of design criteria. RDF is simpler for machines because of triples can be queried and joined from different sources [1]. As a result, an RDF knowledge graph can be traversed and linked with others, regardless of its provenance and used ontologies.

2.2 Scientific Knowledge Graphs

A Knowledge Graph represents structural relationships between entities of a domain, i.e. it provides semantically structured information that can be interpreted by computers. Many tasks related to question answering, recommendation, and information retrieval systems can be addressed by using this model. Therefore, KG is the core component to build more intelligent services or applications. Although many approaches about knowledge graphs are focused on construction techniques, there are many applications in other areas [15]. Today, several KGs have been built by using semantic technologies to describe resources of the scientific domain.

Scientific Knowledge Graphs represent facts of the scholarly domain. Typically, they contain metadata describing research publications such as title, abstract, authors, venues, organizations, research topics, citations, among others [2], i.e., some initiatives of SKG construction are focused on the representation of papers' metadata. Between these initiatives are Open Academic Graph[3], Scholarlydata.org[4], OpenCitations[5], and Microsoft Academic Knowledge Graph (MAKG[6]).

2.3 Related Works

According to [15], researches on KGs are classified into two categories, researches on graph construction techniques and based-on KGs applications. In this paper,

[3] www.openacademic.ai/oag/.
[4] http://www.scholarlydata.org.
[5] https://opencitations.net.
[6] http://ma-graph.org.

both topics are covered. First, the authors present a process to construct a KG from scholarly production metadata. Second, authors try to leverage this data, from an analytical point of view, by providing more suitable information for summarizing and assessment this production.

In relation to KGs construction, one of the most interesting proposals is developed by Färber et al. in [6]. Here, authors present the Microsoft Academic Knowledge Graph (MAKG) which is a large RDF dataset that describes more than 210 million publications. Using this graph, researchers could (1) make entity-centric exploration of papers, researchers, or affiliations, (2) integrate data using RDF as a common data model and link to other data sources; and (3) analyze and discover scholarly knowledge. Although MAKG offers valuable use cases, in this paper, authors are interested in analyzing certain characteristics of the research production doing in a given institution. Therefore, attention is focused on a more limited domain.

Finally, in relation to analytics over SKG, Papadaki et al. in [8] present an approach for applying analytics to RDF data, based on a high-level functional query language called HIFUN. This implies to make transformations on queries. Unlike that proposal, here, authors directly use the W3C recommendation, SPARQL, to extract data from a KG, and use data analytics tools well-known as R to create visual results.

3 Construction of SKGs Based on Metadata

Creating a knowledge graph based-on metadata is the first task to publish the scientific production of an institution. In this section, authors describe the data flow to achieve this goal.

3.1 Process of Construction

Figure 1 shows the UML diagram sequence that identifies the main six stages of the graph construction.

A Data collection. Nowadays, this task is simple to automate because some academic databases provide access to data through APIs. Before, it is important to define the scope, goals, potential users, and applications that will walk around the graph.

B Data pre-processing. Collected data can be stored in a temporary database in order to identify their quality. Depending on the results, the most appropriate cleaning techniques should be applied. One of the main issues to deal with is to disambiguate text due to lexical differences in the names of authors and institutions.

C Data modeling. SPAR can be the base to define a new model or an application profile that describes the cleaned data and thus add value to the project.

D RDF conversion. At this stage, the goal is to transform data into an RDF format. There are some tools to convert data from an ontological model. The final task at this stage is to link and store the generated data in an RDF triplet repository.

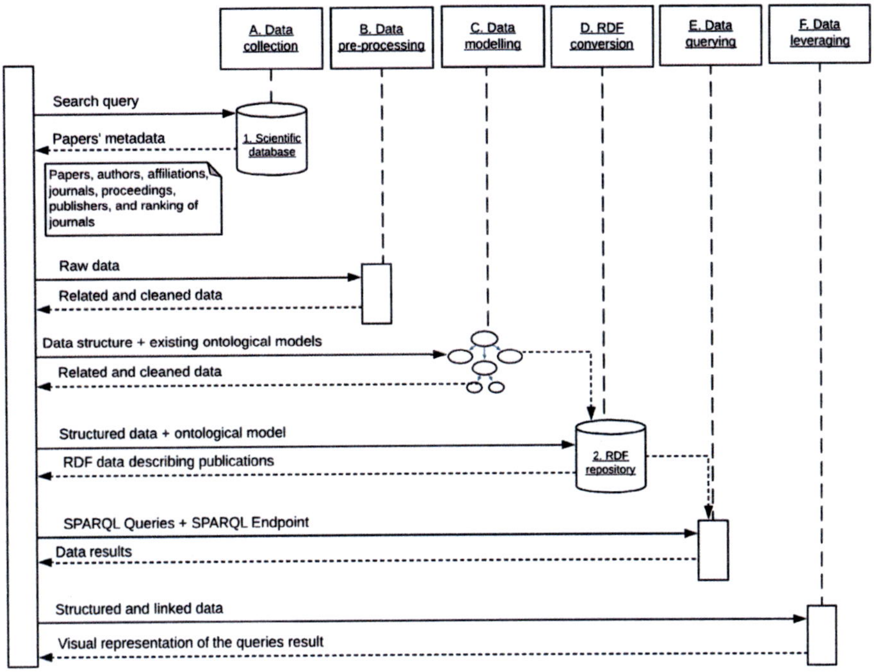

Fig. 1. SKG construction based-on publishing' metadata

E Data querying. SPARQL query templates provide answers to requirements over the graph for all potential consumers of the graph (users and applications). Data about authors, publications, affiliations, topics, and their respective metrics will be commonly requested.

F Data leveraging. This stage includes the construction of all kinds of services or applications that will allow end-users to understand or access knowledge in the graph.

3.2 UG-SKG Construction

According to the defined process, authors built a scientific knowledge graph from publications of the University of Guayaquil (UG), Ecuador, that is called UG-SKG. Of the 6 stages explained in Fig. 1, the first 4 were implemented in two previous works:

- In [14], authors focused on data modeling (stage C of the process). In this work, the authors analyzed the SPAR's capacity to describe academic publications based on requirements of the "Scientific Production" criteria of the "Model Generic of Evaluation of the Environment of Learning of Careers in Ecuador" by CACES - Ecuador [3]. From a set of functional requirements, authors applied the NeOn Methodology [12] for finding requirements not covered by SPAR. As a result, a set of the domain's terms not cover by SPAR were discovered. Therefore, an extension for the BiDO ontology was proposed by authors. The extension comprised the creation of new classes, properties, and individuals.

- In [13], authors performed the stages, collection (A), pre-processing (B), conversion (C), and linking of (D) data. In the first stage (A), data about documents of UG indexed by Scopus were gathered. Elsevier APIs were consumed to get metadata of papers, authors from UG and external co-authors, affiliations, sources, among others. From the Scimago Journal and Country Rank (SJR), authors collected data about journals' ranking according to knowledge areas, and disciplines. Next, the collected data were pre-processed before to transform in RDF format. For this stage, the OpenRefine tool, and FaBiO and BiDO ontologies, that are part of SPAR, were used to generate RDF files. Finally, data were upload in the project's repository. Table 1 summarizes RDF data that is available online through a SPARQL Endpoint service[7].

In this study, authors try to complete the data processing cycle presented in Fig. 1, by addressing the stages of data querying (E) and data leveraging (F). Taking advantage of the data from the two sources (Scopus and SJR) published

Table 1. Summary of elements of the UG-SKG

Source	RDF element	Number
Scopus	time:Instant	93
	fabio:Article, fabio:ConferencePaper, fabio:BookChapter, fabio:Review, fabio:Letter	430
	foaf:Organization	473
	bido:BibliometricDataInTime, fabio:BibliographicMetaData, bido:NumericBibliometricData, time:Interval	8981
	foaf: Person	1446
SJR 2017	fabio:SubjectDiscipline, fabio:SubjectTerm	337
	fabio:Journal	34157
	bido:BibliometricDataInTime	34157
	bido:QuartileRank	79043
RDF data	Classes	22
	Properties	35
	Individuals	159206
	RDF triplets	625897

[7] http://spar.linkeddata.es/sparql.

in RDF format, in the following section, some potential uses of the knowledge graph are explained.

4 Leveraging Knowledge from the UG-SKG

To discover relevant information from SKGs, attributes from documents, authors, and sources should be combined. In this section, the authors present four queries executed in the UG-SKG. These cases try to demonstrate the potential of KG to find patterns between quantitative and qualitative indicators related to scientific production. From this information, UG leaders can make decisions about investments, support for researchers.

Queries are executed by accessing the SPARQL Endpoint and their results are plotted using the R Studio tool. For the first case, the SPARQL statement is presented[8], and the visual results for the four cases analyzed are shown below.

4.1 What Type of Document Is Published by UG Researchers and What Is Its Impact According to the Source Where They Publish?

To answer this question, the authors identify the type of publication of each indexed document. Then this data are merged with the rank of the source where it has been published. Since a source (journal, book series, etc.) may has different ranks in different areas of knowledge, authors filter the best rank for each publication (see the SPARQL query).

```
SELECT DISTINCT ?type, SUM(IF(?rank <= 10, 1, 0)) AS ?rank1,
  SUM(IF(?rank > 10 && ?rank <= 50, 1, 0)) AS ?rank2,
  SUM(IF(?rank > 50 && ?rank <= 100, 1, 0)) AS ?rank3,
  SUM(IF(?rank > 100 && ?rank <= 500, 1, 0)) AS ?rank4,
  SUM(IF(?rank > 500, 1, 0)) AS ?rankMore500
WHERE{?paper a ?t ; frbr:partOf ?source.
  BIND(REPLACE(STR(?t),'http://purl.org/spar/fabio/','') AS ?type)
  FILTER (?type != "Expression")
  {SELECT ?paper, MIN(?rankA) AS ?rank
    WHERE { ?paper a ?type ; frbr:partOf ?source.
      ?source bido:holdsBibliometricDataInTime ?SMY.
      ?SMY bido:withBibliometricData/bido:rankInQuartile ?r ; bido:accordingTo [].
      BIND(xsd:integer(?r) AS ?rankA)
      BIND(REPLACE(STR(?type),'http://purl.org/spar/fabio/','') AS ?type)
      FILTER (?type != "Expression")}
    GROUP BY ?paper }
}
```

[8] The technical report is available in https://archive.org/services/purl/scipub/queryReport.

Figure 2 shows results for this question. Here appears some interesting data (1) three out of four documents, published in ranked sources, are journal's articles, on the other hand, the least amount of publications corresponds to book chapters (0.2%); (2) the ranking distribution by document type shows that a minimum number of the sources where the documents are published, corresponds to journals ranked in the top 10 according to a certain area of knowledge, on the other hand, most of the sources chosen for publication are located in ranks between 101 and 500. Crossing these data with authors' information such as h-index or career length, some guides to improve the productivity of UG researchers can be established.

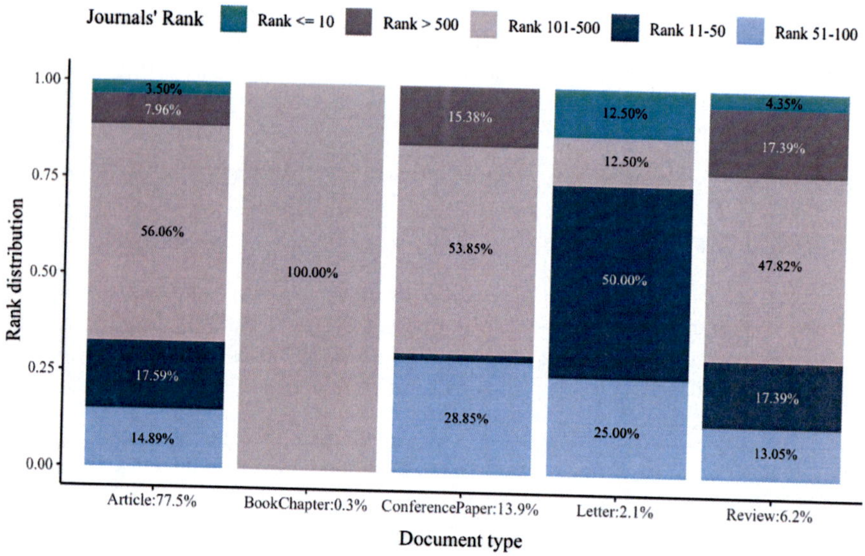

Fig. 2. Distribution of publications by document type and rank of sources

4.2 What Are the Most Popular Publication Sources and What Is Their Impact Factor Quartile?

To answer this question, the most popular sources where UG researchers publish were identified, and then the best quartile achieved by each journal was established.

Figure 3 shows that the most popular journal is ESPACIOS with 54 papers, that is, this number represents 8% of all the publications that the UG. This best quartile found this journal is 3 (Q3). On the other hand, the only journal Q1 is PLOS One, where only 4 papers have been published. These results make us suspect that the impact factor quartile is not the main criterion that researchers values when choosing the source to publish their research.

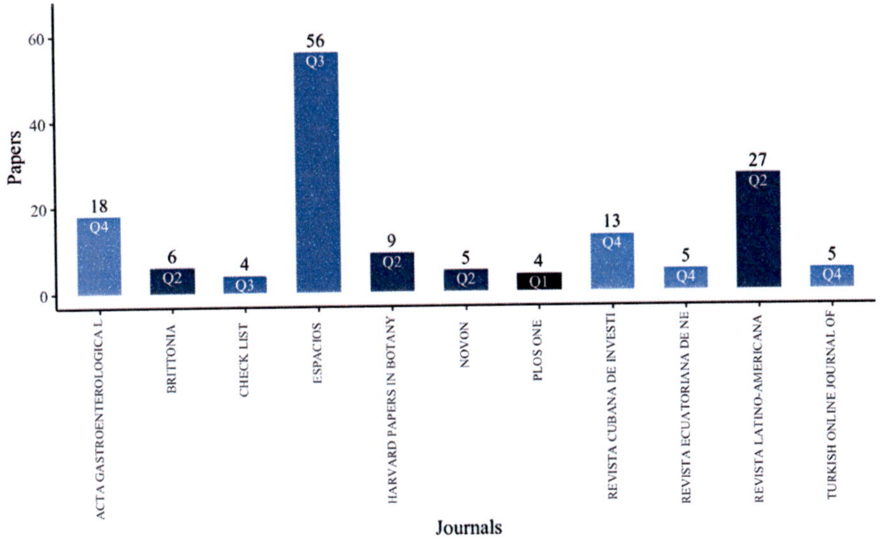

Fig. 3. Common sources and their impact

4.3 Who Are the Researchers Working in the Disciplines of a Specific Area and What Is Their Production?

Knowing the answer to this question can be useful to (1) to find pairs of researchers in the same discipline, (2) to identify the most prominent researchers in a particular topic, (3) to promote research in emerging disciplines that are not being investigated. Figure 4 shows the authors' network for the Arts and Humanities area. A sub-tree of the dendrogram shows people who publish in the same discipline. The size of the circle associated with each researcher indicates his/her productivity based on the published papers.

Results show that two authors are the only researchers in a discipline (History and Philosophy). Facts like this could have two implications (1) to redirect researchers' efforts, or (2) to motivate more colleagues to research these disciplines little explored.

4.4 What Are the Main Interests of the UG Researchers?

In this question, the goal is to identify the main topics in which UG authors research (see Fig. 5). To achieve this goal, the papers' titles were processed in order to create a word cloud. Some text processing techniques were applied on titles, such as, filtering stop words and numbers, and stemming.

According to the results, the most used term is "Ecuador" (filtered from the word cloud). In fact, approximately 2 out of 10 papers have this word in their title. The rest of terms that are shown in Fig. 5 give us an idea of the research carried out at UG: (1) studies have a local and/or national focus because the

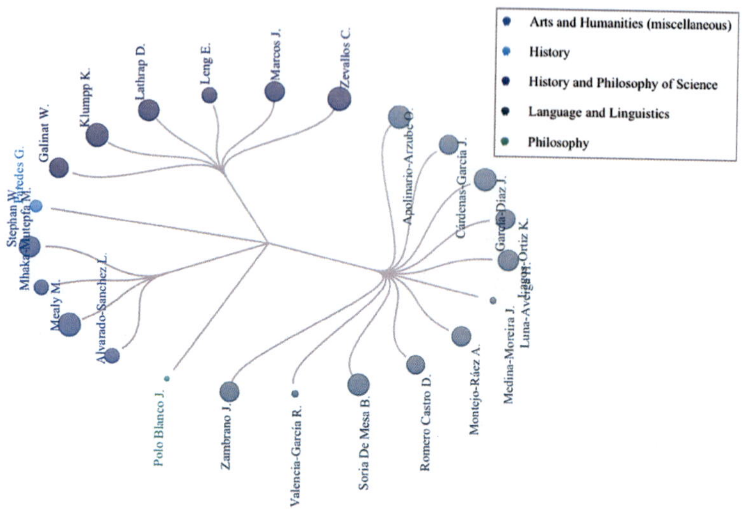

Fig. 4. Authors by knowledge area

Fig. 5. Word cloud based-on papers' titles

terms Guayaquil, and Ecuador or Ecuadorian are highlighted; (2) research is starting in certain disciplines because the terms, analysis, and study are the most popular.

Conclusions

Scientific Knowledge Graphs in RDF format are suitable for the consumption of machines but not for humans. In this paper, authors try to leverage the UG-SKG, a graph built with documents of researchers of the University of Guayaquil and indexed by Scopus. Through a graph it is possible to create tools based-on analysis that leverage the special features of RDF graphs, i.e. their semantics and to support flexible, complex, and multidimensional queries.

In this paper, SPARQL was used to retrieve data from an RDF SKG. Although this technology is not suitable to be directly handled by decision-makers, this reason doesn't limit its use as a language to query KG. In this project, authors created query templates, and, in a transparent way, they were executed from an analytical tool. Then, results were to be filtered, analyzed, and visualized in a traditional way. Visual representations can facilitate people to find relationships and patterns underlying scholarly production. The code source and the collected data are available on the Web. Thus, the reproduction or reuse of each part of this project is enabled.

Finally, as stated in [4], understanding, and encoding knowledge underlying in scientific production is a complex task posing several challenges. Therefore, the construction of a graph is an iterative task, in this project, the authors proposed a process to create an SKG based on metadata. As future work, authors will focus on creating links with external data sets and provide interactive graphics that users can easily interact with. Thus, the SKG will support better decision-making by researchers and managers of academic and research activities of higher education institutions.

Acknowledgement. This work was partially supported by the scholarship provided by the "Secretaría Nacional de Educación Superior, Ciencia y Tecnología" of Ecuador (SENESCyT).

References

1. Auer, S., Mann, S.: Towards an open research knowledge graph. Ser. Libr. **76**(1–4), 35–41 (2019). https://doi.org/10.1080/0361526X.2019.1540272
2. Buscaldi, D., Dessì, D., Motta, E., Osborne, F., Reforgiato Recupero, D.: Mining scholarly publications for scientific knowledge graph construction. Lecture Notes in Computer Science. vol. 11762, pp. 8–12. Springer, June 2019. https://doi.org/10.1007/978-3-030-32327-1_2
3. CACES: Modelo genérico de evaluación del entorno de aprendizaje de carreras en Ecuador (2017)
4. Constantopoulos, P., Pertsas, V.: From publications to knowledge graphs. In: Communications in Computer and Information Science. CCIS, vol. 1197, pp. 18–33. Springer (2020). https://doi.org/10.1007/978-3-030-44900-1_2
5. Edelstein, E., Pan, J.Z., Soares, R., Wyner, A.: Knowledge-driven intelligent survey systems towards open science. New Gener. Comput. (2020). https://doi.org/10.1007/s00354-020-00087-y
6. Färber, M.: The Microsoft academic knowledge graph: a linked data source with 8 billion triples of scholarly data. Lecture Notes in Computer Science (including subseries Lecture Notes in Artificial Intelligence and Lecture Notes in Bioinformatics). LNCS, vol. 11779, pp. 113–129. Springer (2019). https://doi.org/10.1007/978-3-030-30796-7_8

7. Omitola, T., Koumenides, C.L., Popov, I.O., Yang, Y., Salvadores, M., Correndo, G., Hall, W., Shadbolt, N.: Integrating public datasets using linked data: challenges and design principles. In: Future Internet Assembly Future Internet Assembly, pp. 1–10 (2010)
8. Papadaki, M.E., Tzitzikas, Y., Spyratos, N.: Analytics over RDF graphs. In: Communications in Computer and Information Science. CCIS, vol. 1197, pp. 37–52. Springer (2020). https://doi.org/10.1007/978-3-030-44900-1_3
9. Peroni, S., Shotton, D.: FaBiO and CiTO: ontologies for describing bibliographic resources and citations. J. Web Semant.: Sci. Serv. Agents World Wide Web **17**, 1–15 (2012)
10. Peroni, S., Shotton, D.: The SPAR ontologies. In: 17th International Semantic Web Conference, pp. 119–136 (2018)
11. Piedra Salomón, Y., Martínez Rodríguez, A.: Scientific production. Ciencias de la Información **38**(3), 33–38 (2007)
12. Suarez-Figueroa, M.D.C.: NeOn methodology for building ontology networks: specification, scheduling and reuse. Ph.D. thesis (2010)
13. Tapia-Leon, M., Chicaiza Espinosa, J., Espinoza Arias, P., Santana-Perez, I., Corcho, O.: Using the SPAR ontology network to represent the scientific production of a university: a case study. In: 7th World Conference on Information Systems and Technologies (2019)
14. Tapia-Leon, M., Santana-Perez, I., Poveda-Villalón, M., Espinoza Arias, P., Chicaiza, J., Corcho, O.: Extension of the BiDO ontology to represent scientific production. In: 8th International Conference on Educational and Information Technology (2019)
15. Zou, X.: A survey on application of knowledge graph. In: Journal of Physics: Conference Series, vol. 1487 (2020)

Author Index

M. Botto-Tobar et al. (Eds.): ICCIS 2020, AISC 1273, pp. 467–468, 2021.
https://doi.org/10.1007/978-3-030-59194-6